Latitude					income per capita (US$)	
Spitsbergen	77-				...rmuda	30 000
Wrangel					...waii	26 030
Iceland	63-				...eat Britain	21 200
Aleutian	52-				...wfoundland	19 360
Hokkaido	42-				...iwan	14 000
Azores	37-				...prus	13 500
Canaries	27-				...alta	12 900
Hispaniola	18-				...auritius	10 300
Socotra					...uru	10 000
Sri Lanka	6-				...ychelles	7 000
Galápagos	1°...				...union	4 300
Ascension	8°S	Marquesas	AD 300	Madagascar	19	Sri Lanka	3 800
Tahiti	18°S	Maldives	400 BC	Easter	16.4	Jamaica	3 660
Rodrigues	20°S	Canaries	c. 500 BC	Socotra	12	Solomon	3 000
Pitcairn	25°S	Fiji–Samoa	1 500 BC	Tasmania	7.3	Tonga	2 250
Lord Howe	32°S	Majorca	c. 2 040 BC	Iceland	2.7	Maldives	1 800
Amsterdam	38°S	Crete	c. 6 250 BC	Galápagos	1.9	Cape Verde	1 360
Tasmania	42°S	Corsica	c. 6 700–8 600 BC	Spitsbergen	0.1	Sao Tomé	1 000
Kerguélen	50–48°S	Hispaniola	7 000 BC	Kerguélen	0.01	Madagascar	730
Macquarie	54°S	Cyprus	c. 8 320 BC			Comoros	685
		Tasmania					
			20 000–30 000 BC				
		New Guinea					
			40 000–50 000 BC				

Island Biogeography

Front cover image: Rapa (Austral Islands, French Polynesia) from Mt. Perahu. Photo by R. G. Gillespie, taken on an expedition funded by La Délégation à la Recherche and l'Institut Louis Malardé, Tahiti.

Back cover image: Kosrae (Micronesia) from the summit of Mt. Finkol. Photo by G. K. Roderick and R. G. Gillespie, taken on an expedition funded by the National Geographic Society.

Island Biogeography

Ecology, evolution, and conservation

Second Edition

Robert J. Whittaker
School of Geography, University of Oxford, UK

AND

José María Fernández-Palacios
Grupo de Investigación de Ecología Insular, Universidad de La Laguna, Tenerife

OXFORD
UNIVERSITY PRESS

OXFORD
UNIVERSITY PRESS

Great Clarendon Street, Oxford OX2 6DP

Oxford University Press is a department of the University of Oxford.
It furthers the University's objective of excellence in research, scholarship,
and education by publishing worldwide in

Oxford New York

Auckland Cape Town Dar es Salaam Hong Kong Karachi
Kuala Lumpur Madrid Melbourne Mexico City Nairobi
New Delhi Shanghai Taipei Toronto

With offices in

Argentina Austria Brazil Chile Czech Republic France Greece
Guatemala Hungary Italy Japan Poland Portugal Singapore
South Korea Switzerland Thailand Turkey Ukraine Vietnam

Oxford is a registered trade mark of Oxford University Press
in the UK and in certain other countries

Published in the United States
by Oxford University Press Inc., New York

British Library Cataloguing in Publication Data
Data available

Library of Congress Cataloging in Publication Data
Applied for

Typeset by Newgen Imaging Systems (P) Ltd., Chennai, India
Printed in Great Britain
on acid-free paper by
Antony Rowe Ltd., Chippenham, Wiltshire

ISBN 0–19–856611–5   978–0–19–856611–3
ISBN 0–19–856612–3   978–0–19–856612–0

10 9 8 7 6 5 4 3 2 1

Preface and acknowledgements

Island biogeography is an important subject for several reasons. First, it has been and remains a field which feeds ideas, theories, models, and tests of same into ecology, evolutionary biology, and biogeography. This is because islands provide natural scientists with model systems—replicated and simplified contexts—allowing us to isolate particular factors and processes and to explore their effects. Secondly, some of these theories have had great weight placed upon them in applications in nature conservation, as scientists and conservationists attempt to understand, predict and manage the biodiversity impacts of habitat loss and fragmentation. Thirdly, in our modern age of anthropogenic extinctions, islands qualify as 'hotspots': combining the attributes of high levels of unique biodiversity, of recent species extinctions, and of likely future species losses. The protection of the unique biological features of island ecosystems presents us with a considerable challenge, not only ecologically, but also because of the fragmented nature of the resource, scattered across all parts of the globe and all political systems, and generally below the horizons of even global media networks. It is our hope that this book will foster an increased interest in island ecology, evolution, and conservation and that it will be of value for students and researchers working in the fields of the life and environmental sciences.

This second edition is built upon the foundations of the first edition but has been substantially reorganized and updated to reflect what we consider the most important developments in island biogeography over the last decade. As will become evident to those who dip into this volume, we cover a great deal more than the biology of the systems. Indeed, we have expanded our coverage of the developmental history and environmental dynamics of islands in this second edition. Much fascinating new work has been published in this arena, and it is proving to be fundamental to improving our understanding of island evolution and ecology.

Another feature of this revision is the inclusion of a great deal of material on the island region of Macaronesia (*the Happy Islands*), and particularly of the Canaries. These islands are the Atlantic equivalent of Hawaii and the Galápagos, providing a rich mix of geological and evolutionary–ecological insights on the one hand and biodiversity conservation problems on the other. Much new and exciting work has been published on these islands since the first edition of this book was written, and we were keen to bring some of this work to the attention of a wider audience of students and scholars.

Island biogeography is a dynamic field. Whilst many ideas and themes have long pedigrees, new ideas, and insights continue to be generated, often building on long-running debates. We have attempted to reflect the diversity of viewpoints and interpretations within the field, although inevitably the selection of material reflects our own biases and interests.

There are many people we would like to thank, not least our students and the members of our research groups, with whom we have enjoyed illuminating discussions on many island themes. Ian Sherman, our editor at OUP, provided encouragement, help, and good advice at all stages of the project, and we thank him, Stefanie Gehrig, and their colleagues at the Press, for all their efforts. We thank the following colleagues for variously commenting on draft material, supplying answers to queries, and discussion of ideas: Gregory H. Adler, Rubén Barone, Paulo Borges, Pepe Carrillo, James H. Brown, Juan Domingo Delgado, Lawrence Heaney, Scott Henderson, Paco Hernán, Joaquín

Hortal, Hugh Jenkyns, Richard Ladle, Mark Lomolino, Águedo Marrero, Aurelio Martín, Bob McDowall, Leopoldo Moro, Manuel Nogales, Pedro Oromí, Jonathan Price, Mike Rosenzweig, Dov Sax, Ángel Vera, and James Watson. All errors and omissions are of course our own to claim. Most of the figures were drawn by Ailsa Allen. Sue Stokes helped in compilation of material. We are grateful to those individuals and organizations who granted permission for the reproduction of copyright material: the derivation of which is indicated in the relevant figure and table legends, supported by the bibliography. Finally, we thank our families, Angela, Mark, and Claire and Neli, José Mari, and Quique, for all their support and tolerance during the preparation of this book.

R.J.W.
J.M.F.P.
Oxford and La Laguna, 28 April 2006

Contents

PART I

Islands as Natural Laboratories

The basaltic massif of Teno is one of the oldest parts of the
island of Tenerife (Canary archipelago), dating back about
8 million years. It is only within the last 1.5 Ma or so that the
great acidic volcanic cycle of Las Cañadas unified the old
basaltics massifs of Teno, Adeje and Anaga to form what we
know today as the island of Tenerife.

CHAPTER 1

The natural laboratory paradigm

... it is not too much to say that when we have mastered the difficulties presented by the peculiarities of island life we shall find it comparatively easy to deal with the more complex and less clearly defined problems of continental distribution ...

(Wallace 1902, p. 242)

These words taken from Alfred Russel Wallace's *Island life* encapsulate an over-arching idea that could be termed the central paradigm of island biogeography. It is that islands, being discrete, internally quantifiable, numerous, and varied entities, provide us with a suite of natural laboratories, from which the discerning natural scientist can make a selection that simplifies the complexity of the natural world, enabling theories of general importance to be developed and tested. Under this umbrella, a number of distinctive traditions have developed, each of which is a form of island biogeography, but only some of which are actually about the biogeography of islands. They span a broad continuum within ecology and biogeography, the end points of which have little apparently in common. This book explores these differing traditions and the links between them within the covers of a single volume.

This book is divided into four parts, each of three chapters. The first part, *Islands as Natural Laboratories*, sets out to detail the properties of these natural laboratories, without which we can make little sense of the biogeographical data derived from them. The second part, *Island Ecology*, is concerned with pattern and process on ecological timescales, and is focused on properties such as the number and composition of species on islands and how they vary between islands and through time. The chapters making up the third part, on *Island Evolution*, focus on evolutionary

pattern and process, at all levels from the instantaneous loss of heterozygosity associated with the colonizing event on an island, through to the much deeper temporal framework associated with the great radiations of island lineages on remote oceanic archipelagos like Hawaii. The final section of the book, *Islands and Conservation*, incorporates together two contrasting literatures, concerned respectively with the threats to biodiversity derived from the increased insularization of continental ecosystems, and the threats arising from the loss of insularization of remote islands.

We cannot begin our investigation of island ecology, evolution, and conservation problems, until we have explored something of the origins, environments, and geological histories of the platforms on which the action takes place. This forms the subject matter of the second chapter, *Island environments*. There are many forms of 'islands' to be found in the literature, from individual thistle plants (islands of sorts for the arthropods that visit them) in an abandoned field, through to remote volcanic archipelagos like the Galápagos and Hawaiian islands (Fig. 1.1). The former are very temporary islands in an ecological context, and in practice, individual volcanic islands are also quite short-lived platforms when judged in an evolutionary time frame. The extraordinary environmental dynamism of these remote platforms requires much greater attention than it has hitherto received in the island evolutionary literature. Consider the following example. The Juan Fernández archipelago consists of two main islands, Masatierra, an island of 48 km^2 and 950 m height, lying some 670 km from mainland Chile, and Masafuera, an island of 50 km^2 and 1300 m

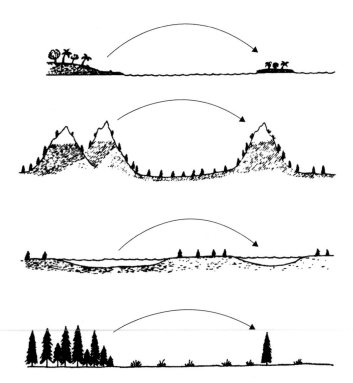

Figure 1.1 There are many different types of islands in addition to those found in the world's oceans. This figure illustrates just a few of these (based on an original in Wilson and Bossert 1971).

height, lying a further 180 km further out in the Pacific. We might be interested in analysing how biological properties of these islands, such as species richness and endemism, relate to properties such as island area, altitude, and isolation. However, when Masatierra first arose from a hotspot on the Nazca plate some 4 Ma (millions of years ago), it probably formed an island of some 1000 km² and perhaps 3000 m in height (Stuessy *et al*. 1998). Since then, it has been worn down and diminished in area by sub-aerial erosion, wave action, and subsidence, losing both habitats and species in the process. Modelling efforts that ignore this environmental history would be liable to produce quite misleading accounts of factors controlling biotic diversification on such islands (Stuessy *et al*. 1998).

The third chapter, on *The biogeography of island life*, concentrates on the biogeographical affinities and peculiarities of island biotas—a necessary step before the processes of evolution on islands are tackled. The chapter thus begins in the territory of historical biogeographers concerned with tracing the largest scales from the space–time plot (Fig. 1.2) and with working out how particular groups and lineages came to be distributed as they are. This territory has been fought over by the opposing schools of dispersalist and vicariance biogeography. Island studies have been caught up in this debate in part because they seemingly provide such remarkable evidence for the powers of long-distance dispersal, whereas rejection of this interpretation requires alternative (vicariance) hypotheses for the affinities of island species, invoking plate movements and/or lost land bridges, to account for the breaking up of formerly contiguous ranges. Some of the postulated land-bridge connections now appear highly improbable; nevertheless the changing degrees of isolation of islands over time remains of central importance to understanding the biogeography of particular islands. As will be seen, the vicariance and dispersalist hypotheses have been put into too stark an opposition; both processes have patently had their part to play (Stace 1989; Keast and Miller 1996).

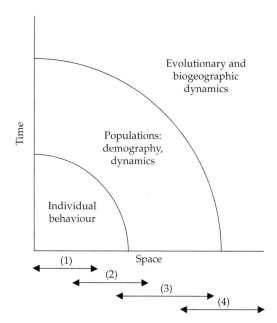

Figure 1.2 A scheme of different time–space scales of ecological processes and criteria that define corresponding scales of insularity. (1) individual scale; (2) population scale 1: dynamics; (3) population scale 2: differentiation; (4) evolutionary scale. (Redrawn from Haila 1990, Fig. 1.)

Table 1.1 Some prominent island biogeographical theories and the geographical configurations of islands for which they hold greatest relevance

Type of archipelago	Prominent theories
Large, very distant	Adaptive Radiation
Large, distant	Taxon Cycle
Medium, mid-distance	Assembly Rules
Small, near	Equilibrium Model of Island Biogeography
Small, very near	Metapopulation dynamics

Remote island biotas differ from those of continents in a number of ways, being generally species-poor and disharmonic (peculiar in taxonomic composition), yet rich in species found nowhere else, i.e. endemic to those islands. Although many of these features are well known through the more popular illustrations from island groups such as the Galápagos, the global significance of island biodiversity is not so well appreciated. Data for a number of taxa are available that unequivocally demonstrate that islands, particularly large and remote islands, contribute disproportionately to global biodiversity, i.e. they are biodiversity 'hotspots'. Island biogeographers have, in recent decades, continued to discover 'new' species on islands, some from extant (i.e. living) populations, others from fossils or subfossils. These new discoveries require a reappraisal of island biogeographical ideas and models. They also underline the accelerated rate of attrition of island biotas through human action, which qualifies many islands today as 'threatspots' as well as centres of endemism.

Having explored the properties of the laboratories, the second and third parts of the book examine the ecological and evolutionary insights provided by islands, moving from the finer spatial and temporal frames of reference, to the coarser scales over which evolutionary change operate (Fig. 1.2). Table 1.1 picks out some of the dominant island theories or themes and the typical island configurations that, as will become clear later, appear to match these themes (cf. Haila 1990). The first two, adaptive radiation and the taxon cycle, are each forms of evolutionary change: typically, the best examples of island evolution are from very isolated, high islands. The last three themes fall within what may be termed island ecology. Island ecological theories have applications to 'habitat islands' within continental land masses as well as to real islands and, indeed, form an important part of the contribution of biogeography to the problems of conservation science (conservation biogeography *sensu* Whittaker *et al.* 2001).

The separation of ecology from evolution is a rather arbitrary one: evolution cannot work without ecology to drive it; ecological communities operate within the constraints of evolution. Nonetheless, we find it helpful to distinguish theory that treats species as essentially fixed units of analysis as 'island ecology', and this we consider within the three chapters of Part II. The first is an analysis of the macroecology of islands (Chapter 4), focused on what we light-heartedly term *Species numbers games*. Undoubtedly, the most influential contribution to this literature has been Robert H. MacArthur and Edward O. Wilson's (1967) *The theory of island biogeography*. MacArthur and Wilson were not the first to recognize or theorize about the species–area effect—that plots of species number versus area have a characteristic form, differing between mainland and island biotas—but it was they who developed a fully fledged theory to explain it: the equilibrium theory of island biogeography, based on a dynamic relationship between forces accounting for immigration to islands (afforced on remote islands by speciation) and losses of species from them.

Whether on islands or continents, the form of the relationship between area and species number is of fundamental biogeographical importance. Over the past half century, there have been numerous island biogeographical studies that have set out to describe and understand the form of the species–area curve, and the influence of numerous other variables on species richness. What have these studies shown? Are we any closer to understanding the factors that control species richness in patches and islands, and how they vary in significance geographically? We will show that MacArthur and Wilson's equilibrium theory has actually proved remarkably difficult to test and that while its heuristic influence remains strong, its predictive value appears limited. In large measure this may be because of insufficient attention to the scale of the study system.

The study of *Community assembly and dynamics on islands* (Chapter 5) has interested natural scientists since at least the mid-nineteenth century, which explains why the botanist Melchior Treub took the opportunity to begin monitoring the recolonization of the Krakatau islands just 3 years after they were sterilized in the eruptions of 1883 (Whittaker *et al.* 1989). It emerges that island biotas typically are not merely a 'random' sample of mainland pools. There is generally some degree of structure in the data. Important insights into the processes producing such structure have come from Jared Diamond's work on birds on islands off the coast of New Guinea. Diamond (1975*a*) formulated a set of 'assembly rules' based largely on distributional data. In this work he invoked a strong role for interspecific competition in structuring ecological assemblages on islands, but he also recognized a role for long-term ecological (and

evolutionary–ecological) processes. As we will see later, this approach became embroiled in heated controversy, focused both on our ability to measure pattern and on the causal interpretation of the patterns detected. Other approaches have since been developed that aim to detect and interpret structure in island assemblages. One important form of structure is nestedness, the term given to the situation where upon ranking a set of island species lists by species richness, each species set is found to represent a proper subset of the next larger species set. Again, the measurement of nestedness and its causal interpretation turn out to be less than straightforward, but it appears clear that a significant tendency to nestedness is common within island (and habitat island) archipelagos.

Much of the research on island assembly has been based around 'snapshot' data, capturing distributions at a particular point in time. However, the dynamics of island biotas have also been the focus of study and debate. For instance, Diamond developed the idea from his New Guinea studies that some species are 'supertramps', effective at colonizing but not at competing in communities at equilibrium, whereas others are poor colonists but effective competitors. This is a form of successional structuring of colonization and extinction. Successional structure can be recognized through other approaches, for instance by analysis of the dispersal characteristics of the flora of the Krakatau islands through time, or by analysis of the habitat requirements of birds and butterflies appearing on and disappearing from the same islands. Krakatau is unusual in having species–time data for a variety of taxa, and to some extent they tell different stories, yet stories linked together by hierarchical links between taxa and by the emergent successional properties of the system.

Within this book, a recurring theme is the importance of scale, and we open the final chapter of this section, on *Scale and island ecological theory*, with a consideration of the scale of relevance of theoretical constructs such as the MacArthur–Wilson model. There has been an increasing realization that ecological phenomena have characteristic spatial and temporal signatures, which tend to be linked (Delcourt and Delcourt 1991; Willis and

Whittaker 2002). Placing island ecological studies into a scale framework, helps us reconcile apparently contradictory hypotheses as actually being relevant to different spatio-temporal domains (Fig. 1.2, Haila 1990). We suggest that alongside islands conforming to the dynamic, equilibrium MacArthur–Wilson model, some islands may show dynamic but non-equilibrial behaviour, while other island biotas persist unchanged for substantial period of time, constituting ecologically 'static' systems that can again be regarded as either equilibrial or non-equilibrial. Island ecological theories should also be capable of accommodating the hierarchical links that form within ecological systems (e.g. with respect to feeding relationships, pollination, dispersal, etc): this represents a considerable challenge given a literature hitherto focused mostly on pattern within a single trophic level, or major taxon (mammals, birds, ponerine ants, etc.).

The three chapters in Part III, *Island Evolution*, set out to develop the theme of island evolution from the micro-evolutionary changes following from initial colonization through to the full array of emergent macro-evolutionary outcomes. In Chapter 7, *Arrival and change*, we start with the founding event on an island and work in turn through the ecological and evolutionary responses that follow from the new colonist encountering the novel biotic and abiotic conditions of the island. A series of traits, syndromes, and properties emerge as characteristic of islands, including loss of dispersal powers, loss of flower attractiveness, the development of woodiness, characteristic shifts in body sizes of vertebrates, and rather generalist pollination mutualisms. These shifts are detectable to varying degrees of confidence, and in a wide variety of taxa, sometimes but not always involving attainment of endemic species status by the island form.

The attainment of speciation, the emergence of separate species from one ancestor, is of central interest within biology, and so we devote Chapter 8, *Speciation and the island condition*, to a brief account of the nature of the species unit, and of the varying frameworks for understanding the process of speciation. Thus, we examine first the geographical context of speciation, in which islands

conveniently provide us with a strong degree of geographical separation between the original source population and the island theatre, but in which the extent of intraarchipelago and intraisland isolation is often harder to discern. Second, we examine the various mechanistic frameworks for understanding speciation events, and, finally, we consider phylogenetic frameworks for describing the outcome of evolutionary change. Having developed these frameworks, we go on to look at the *Emergent models of island evolution* (Chapter 9), which provide descriptions and interpretations of some of the most spectacular outcomes of island evolutionary change, including the classic examples of the taxon cycle and of adaptive radiation. However, as more studies become available of phylogenetic relationships within lineages distributed across islands, archipelagos, and even whole ocean basins, new ways of looking at island evolution are emerging. From this work it is becoming increasingly evident that the ever-changing geographical configuration and environmental dynamism of oceanic archipelagos are crucial to understanding variation in patterns and rates of evolutionary change on islands.

Anthropogenic changes to biodiversity are evident across the planet, and we may turn to islands for several important lessons if we care for the future of our own human societies (Diamond 2005). In the final section of the book, *Islands and Conservation*, we start with the theme of *Island theory and conservation* (Chapter 10), which debates the contribution of island ecological thinking within conservation science. As we turn continents into a patchwork quilt of habitats, we create systems of newly insularized populations. What are the effects of fragmentation and area reduction within continents? Not just the short-term changes, but the long-term changes?

As concern over this issue grew in the 1970s and 1980s, the obvious place to derive such theory was from the island ecological literature, and in particular the dynamic equilibrium model put forward by MacArthur and Wilson (1967). We review this literature, showing that although we can often develop good statistical models for the behaviour of particular species populations, and for how species numbers may change in fragments following isolation, we still have a poor grasp of the overall implications for diversity on a regional scale of the ongoing processes of ecosystem fragmentation. Nonetheless, it is clear that habitat loss and fragmentation threatens many species with local, regional and ultimately global extinction. It is also evident that the processes of ecosystem response can take considerable periods (often decades) to play out, meaning that today's fragmentation is storing up numerous extinctions for the future. However, taking a more positive approach to the problem, the fact that responses are often lagged means that there is an opportunity for mitigation measures to be put in play, to reduce the so-called 'extinction debt', providing society cares enough to act. Insofar as island effects are driving species losses, the key is to prevent previously contiguous and extensive ecosystems from becoming too isolated: reserve systems need to be embedded in wildlife friendly landscapes wherever possible.

If the catalogue of known extinctions is drawn up for the period since AD 1600, it can be seen that for animal taxa that are relatively well known (e.g. mammals, birds, and land snails), the majority of losses have been of island species (Fig. 1.3). Today, some of the greatest showcases for evolution are in peril, with many species on the verge of extinction. Why is this? These issues and some conservation responses are explored in the final two chapters of the book, *Anthropogenic losses and threats*, and *Island remedies*. There is mounting evidence that humans are repeat offenders when it comes to extinguishing island endemics. Wherever we have colonized islands, whether in the Pacific, the Caribbean, the Atlantic, the Indian Ocean, or the Mediterranean, we have impacted adversely on the native biota, and often on the ecosystem services on which we ourselves rely (Diamond 2005). Here we ask the question: are islands inherently fragile, or are island peoples peculiarly good at extinguishing species? In some senses, island biotas are indeed fragile, but very often the demise of endemic taxa can be traced to a series of 'hard knocks' and the so-called synergistic interactions between a number of alien forces, such that the

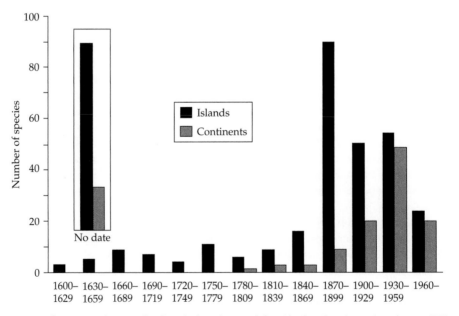

Figure 1.3 Time series of extinctions of species of molluscs, birds, and mammals from islands and continents since about AD 1600. (Redrawn from Groombridge 1992, Fig. 16.5.)

nature of the final demise of the endemic may be unrelated to the driver(s) of the initial range collapse. Typically, there have been at least two major waves of extinctions, one associated with the aboriginal or prehistoric human colonizations, and the second following contact with modern European societies.

Whereas, on the continents, the increasing insularization of habitats is a prime cause for concern, the problem for oceanic island biotas today is the increasing breakdown of the insularization of theirs. The prime agents of destruction are introduced exotic species (especially mammalian browsers and predators), habitat loss, predation by humans, and the spread of disease. The forces involved are now well established, and many of the biological management options are firmly founded in experience, but attaining the twin goals of sustainable development and management for conservation requires attention to fields largely outside the scope of the present volume, concerning the politics, economics, sociology, and culture of islands and island states: continental solutions may be poorly transferable to islands.

What is an island, anyhow?

CHAPTER 2

Island environments

... from cowpats to South America, it is difficult to see
what is not or at some time has not been an island.

(Mabberley 1979, p. 261)

The islands move horizontally and vertically and thereby
grossly modify the environment on and around them. Life
forms too must evolve, migrate, or become extinct as the
land changes under them.

(Menard 1986, p. 195)

2.1 Types of islands

Islands come in many shapes and sizes, and their
arrangement in space, their geology, environments,
and biotic characteristics are each extremely variable, simply because there are vast numbers of
them in the world. If this makes them marvellous
experimental laboratories for field ecologists and
biogeographers, it also means that attempting to
make generalizations about islands holds the danger—almost the certainty—that one will be wrong!

Defining the term 'island' for the purposes of
this book is not as straightforward as it seems.
Many biogeographical studies treat isolated
patches of habitat as *de facto* islands, but if we take
the simplest dictionary definition then an island
is a 'piece of land surrounded by water'. This
seems simple enough, yet some authors regard
land areas that are too small to sustain a supply of
fresh water as merely beaches or sand bars rather
than proper islands: the critical size for a supply
to be maintained being about 10 ha (Huggett
1995). At the other end of the scale, the distinction
between a continent and an island is also fuzzy, in
that larger islands assume many of the characteristics of continents. Indeed, Australia as an entity

is considered an island continent and is rarely
treated in biogeographical analysis as being an
island (for an exception, see Wright 1983).

Where ought the line to be drawn? If New
Guinea is taken as the largest island—somewhat
arbitrarily, given that many would consider
Greenland an island (in fact it is a set of three
islands unified by an icecap)—then islands in the
sea constitute some 3% of the Earth's land area
(Mielke 1989). Much of island biogeography—and
of this book—is concerned with islands significantly smaller than New Guinea, and commonly
treats such substantial islands as New Guinea and
Australia merely as the 'mainland' source pools. To
some extent this reflects a distinction between
island evolutionary biogeography, mostly concerned with larger (and typically oceanic) islands,
and island ecological biogeography, mostly concerned with other types of island.

For present purposes we may divide islands
into two broad types: **true islands**, being land
wholly surrounded by water; and **habitat islands**,
being other forms of insular habitat, i.e. discrete
patches of habitat surrounded by strongly contrasting habitats (Table 2.1). True islands can in
turn be subdivided into island continents
(Australia), oceanic islands, continental fragments, continental shelf islands, and islands in
lakes or rivers.

- **Oceanic islands** are those that have formed over
oceanic plates and have never been connected to
continental landmasses.

- **Continental fragments** are those islands that by
their location would pass for oceanic islands but in
terms of their origin are actually ancient fragments

Table 2.1 A simple classification of island types distinguishing: (1) classic types of 'real' island, being land surrounded by open water, from (2) habitat islands, for which the contrast between the 'island' and the surrounding matrix is less stark but still sufficient to represent a barrier or filter to population movements. Australia, given its huge size, is essentially continental in character and in practice is not treated as an island in the present work

Type of island	Examples
Land surrounded by water	
Island continent	Australia
Oceanic islands	Hawaii, Canaries
Continental fragments	Madagascar, New Caledonia
Continental shelf islands	British Isles, Newfoundland
Islands in lakes or rivers	Isle Royale (Lake Superior), Barro Colorado island (Lake Gatún),
	Gurupá island (River Amazon)
Habitat islands	
Patches of a distinct terrestrial habitat	
isolated by a hostile matrix	Great-Basin (USA) mountain tops surrounded by desert
	Woodland fragments surrounded by agricultural land
	Thistle heads in a field
	Continental lake (Baikal, Titicaca)
Marine habitat islands	The fringing reef around an isolated oceanic island
	Coral reefs separated from other reefs by stretches of seawater
	Seamounts (submerged or not yet emerged mountains below sea level)
	Guyots (submerged flat-topped former islands, i.e. a type of seamount)

of continental rock stranded out in the oceans by plate tectonic processes.

• **Continental shelf islands** are those islands located on the continental shelf. Many of these islands have been connected to mainland during the Quaternary ice ages (formally, the last 1.8 million years), although cooling actually began earlier), as these were periods of significantly lower sea levels. The most recent period of connection for these so called **'land-bridge'** islands ended following the transition from the Pleistocene into the Holocene. The Holocene began about 11 500 years ago, but seas took several thousand years to rise to their present levels.

• Finally, **islands occurring within freshwater bodies**, both lakes and large rivers, are likely to be more comparable to islands in the sea than to habitat islands, and hence can also be considered 'true' islands.

Habitat islands are essentially all forms of insular system that do not qualify as being 'real islands'. This means, for terrestrial systems, discrete patches of a particular habitat type surrounded by a matrix of strongly contrasting terrestrial habitats. For aquatic systems we can consider similarly discrete habitat types separated by strongly contrasting aquatic environments (e.g. shallow benthic environments isolated by deep water) also to constitute habitat islands. Both habitat and lake islands can, of course, occur within the true islands. Moreover, a more refined classification of marine islands has to recognize that there are, for example, many forms and ages of continental shelf islands, and that there are anomalous islands midway in character between oceanic and continental.

The literature and theory of island biogeography has been built up from a consideration of all forms of island, from thistle-head habitat islands (Brown and Kodric-Brown 1977) to Hawaii (e.g. Wagner and Funk 1995). True islands have the virtue of having clearly defined limits and properties, thus providing discrete objects for study, in which such variables as area, perimeter, altitude, isolation, age, and species number can be quantified with some degree of objectivity, even if these properties can vary hugely over the lifespan of an island (Stuessy *et al.* 1998). In contrast, habitat islands exist typically within complex landscape matrices, which are often rapidly and dramatically changing over just a few years. Matrix landscapes may be hostile to some but not all species of the habitat islands, and isolation is thus of a different nature from islands in the sea. It cannot necessarily be assumed that what goes for real islands also works for habitat islands and vice versa.

As much of this book is concerned almost exclusively with islands in the sea, their biogeographical peculiarities, and their problems, it is important to consider their modes of origin, environmental characteristics, and histories. These topics form the bulk of this chapter.

2.2 Modes of origin

Are islands geologically distinctive compared with most of the land surface of the Earth? If we take this question in two parts, continental shelf islands are a pretty mixed bag, a reflection of their varied modes of origin, whereas oceanic islands are fairly distinctive geologically, being generally composed of volcanic rocks, reef limestone, or both (Darwin 1842; Williamson 1981). The analysis that follows is intentionally a simplified account of an extremely complex reality.

The development of the theories of continental drift and, more recently, plate tectonics, has revolutionized our understanding of the Earth's surface and, along with it, our understanding of the distribution and origins of islands. According to the latter theory, the Earth's surface is subdivided into some seven major plates, each larger than a continent, and a number of smaller fragments (Fig. 2.1). With the exception of the Pacific plate and its subordinates (Nazca, Cocos, Juan de Fuca, and Philippines), the plates themselves are typically made up of two parts, an oceanic part and a continental part. The silicon/aluminium-rich granitic parts of the plates are of relatively low density and these form the continental parts, supporting the continents themselves (consisting of a highly varied surface geology), extending to about 200 m below sea level (Fig. 2.2). This zone, from 0 to −200 m, forms the continental shelf and supports islands such as the

British Isles and the Frisian Islands (off Germany, the Netherlands, and Denmark), typically involving a mix of rock types and modes of formation, such that any combination of sedimentary, metamorphic, or igneous rocks may be found. As Williamson (1981) noted, about the only generalization that can be made is that the geological structures of continental shelf islands tend to be similar to parts of the nearby continent.

In places, continental plate can be found at much greater depths than 200 m below sea level, and can then be termed **sunken continental shelf**. Islands on these sections of shelf (ancient continental islands *sensu* Wallace: Box 2.1) are thus formed of continental rocks; examples include Fiji and New Zealand (Fig. 2.3; Williamson 1981). Typically, however, there is a steeply sloping transition zone from shallow continental shelf down to *c*.2000 m or more, where the **basaltic** part takes over and the true oceanic islands occur. These are all volcanic in origin, although in certain cases they may be composed of sedimentary material, principally limestones, formed as the volcanic core has sunk below sea level. True **oceanic plate islands** have never been attached to a greater land mass. They may grow through further volcanism, or subside, erode, and disappear. In a geological sense, they tend to be transitory: some may last only a few days, others some millennia; relatively few last tens of millions of years. Thus, in addition to the islands that sustain terrestrial biota today, there are also a large number of past and future islands, or **seamounts**, found at varying depths below sea level. Flat-topped seamounts, formed through the submergence of limestone-topped volcanoes, are called **guyots**, after the Swiss geologist Arnold Guyot (Jenkyns and Wilson 1999).

Islands may originate by existing areas of land becoming separated from other land masses to which they were formerly connected, by erosion, or by changes in relative sea level from a variety of causes. Many islands have originated from volcanism associated with plate movements. The form this volcanism takes and the pattern of island genesis involved, depend crucially on the nature of the contact zone between plates, that is whether the plates are moving apart, moving towards each

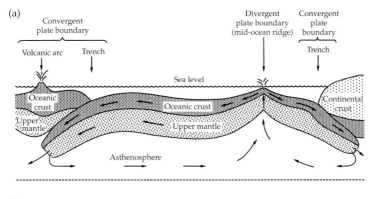

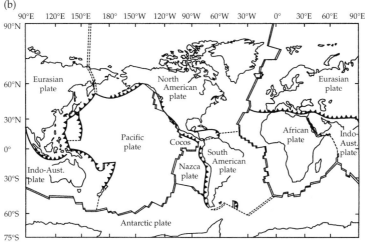

Figure 2.1 Model and map of crustal plates. (a) The basic plate tectonics model, exemplified by the South Pacific. New oceanic crust and upper mantle are created along the divergent plate boundary and spread east and west from here. The eastward moving plate (the Nazca plate) is subducted beneath the continental lithosphere of South America. The westward movement of the much larger Pacific plate ends with subduction beneath the oceanic lithosphere of the Indo-Australian plate. At both trenches, the subducted plate begins to melt when it reaches the asthenosphere. (b) The major plates of the world. Divergent plate boundaries (mid-ocean ridges), at which plates move apart, are represented by parallel lines. Convergent plate boundaries (mostly marked by trenches) are represented by lines with teeth along one side: the teeth point from the downward moving plate towards the overriding or buoyant plate. Transverse plate boundaries are shown by solid lines. Broken lines indicate boundaries of an uncertain nature. (Redrawn from Nunn 1994, Fig. 2.4.)

other, or moving past each other laterally. Relatively few islands are ancient biogeographically; some have experienced a complex history of both lateral movements and alternating emergence and submergence, with profound importance to the resulting biogeography, not just of the particular island in question, but also of the wider region (Keast and Miller 1996).

Plate tectonic processes give rise to islands by three main means. First, the breaking away of pieces of continent by sea-floor spreading has carried New Zealand, Madagascar, and a few other ancient islands off into isolation (Box 2.2). Second, in connection with plate boundaries, volcanic islands may arise to form an archipelago of islands, as in the case of the Greater and Lesser Sundas—the many islands of the Indonesian region. Third, volcanic islands may arise from hotspots (e.g. the Hawaiian islands) and certain parts of mid-ocean ridges. Hawaii is of hotspot origin, and Iceland is

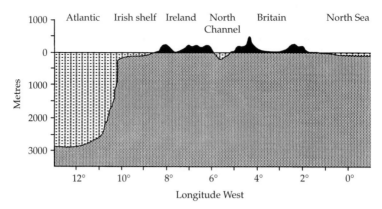

Figure 2.2 A section across the British Isles at 55°N. The section passes through Londonderry in Northern Ireland, Cairnsmore of Fleet in Scotland, and Newcastle upon Tyne in England. (Redrawn from Williamson 1981, Fig. 1.2.)

Box 2.1 Islands in the ocean: an appraisal of Alfred Russel Wallace's classification

In his seminal book entitled *Island life* (3rd edition 1902, first published 1880) Alfred Russel Wallace offered a first classification of 'true' islands, according to their geological origin and biological properties. Although derived long before the development of plate tectonic theory and at a time where scientists were only just beginning to understand the importance of the glaciation events, his classification, illustrated in the table below, remains fundamental.

• **Continental islands** (or recent continental islands *sensu* Wallace) constitute emergent fragments of the continental shelf, separated from the continents by narrow, shallow waters. This separation is often recent, as a consequence of postglacial sea-level rise (c.130 m), which has isolated the species that were already on these islands from their mainland conspecifics. Pronounced sea-level change connected with glaciation has occurred repeatedly during the Pleistocene, resulting in reduced isolation of all continental shelf islands and, for many, joining them to the adjacent continental land mass. In such cases, they are termed **land-bridge islands**, and their effective age as an island is some 10 000 years or less. They will remain islands only until the next glaciation event, perhaps in another 10 000 years or so. Because of their origin, continental islands are very similar in geological and biological features to the continents.

• **Continental fragments** or micro-continents (ancient continental islands *sensu* Wallace): once part of the continents, tens of millions of years ago, tectonic drift started to separate these fragments from the mainland, with the species they carried. Now the waters between them and the continents are wide and deep and the long isolation has allowed both the persistence of some ancient lineages and the development of new species *in situ*. These continental fragments lose their status only when, after tens of millions of years of drift, they finally collide with a different continent, forming a new peninsula. This was the case for the Indian subcontinent, which split from the Gondwanaland supercontinent during the Early Cretaceous (c.130 Ma) and began a lonely trip northwards, finally colliding with Eurasia c.50 Ma, in the process forming the Himalayan range (Stanley 1999; Tarbuck and Lutgens 2000).

• **Oceanic islands** originate from submarine volcanic activity, mostly with basaltic foundations. Oceanic islands have never been connected to the continents and so are populated initially by species that have dispersed to the islands from elsewhere—subsequently enriched by speciation. They form in

Examples of islands corresponding to the three types identified by Wallace, from different oceans of the world

Ocean	Continental islands	Continental fragments	Oceanic islands
Arctic	Svalbard Novaya Zemlya Baffin Ellesmere		Iceland Jan Mayen
North Atlantic	Britain Ireland Newfoundland		Azores Madeira Canaries Cape Verde
Mediterranean	Elba Rhodos Djerba	Balearic Archipelago Corsica–Sardinia Sicily Crete Cyprus	Santorini Eolie
Caribbean	Trinidad Tobago	Cuba Jamaica Hispaniola Puerto Rico	Martinique Guadeloupe Montserrat Antigua
South Atlantic	Falkland Tierra del Fuego	South Georgia	Ascension St Helena Tristan da Cunha South Sandwich
Indian	Zanzibar Sri Lanka Sumatra Java	Madagascar Seychelles Kerguelen Socotra	Réunion Mauritius Saint Paul Diego García
North Pacific	Vancouver Queen Charlotte St Lawrence Sakhalin		Aleutian Kuril Hawaii Marianas
Central and South Pacific	Borneo New Guinea Tasmania Chiloé	New Zealand New Caledonia	Galápagos Society Islands Marquesas Pitcairn

NB: Japan and the Philippines are good examples of islands with a mixed (continental–oceanic) origin

connection with plate boundaries, and in certain circumstances, within a tectonic plate, leading to different volcanic island types. There may today be some 1 million submarine volcanoes, from which only several thousands have been able to reach the sea level to form volcanic islands (Carracedo 2003). Volcanic islands typically are short-lived and even substantial ones may only exist for a few million years before subsiding and eroding back into the ocean. Where sea temperature is adequate oceanic islands may persist despite subsidence and erosion, through the formation of coralline rings, or atolls.

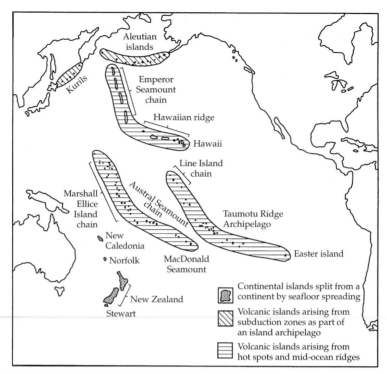

Figure 2.3 Map of the Pacific Ocean, depicting the major modes of origin of islands. NB: New Guinea and Tasmania are continental islands of the Pleistocene Sahul Mainland, which included these islands plus Australia. The New Zealand land mass has been physically isolated from Australia and Antarctica by up to 2000 km for 60 million years, as a result of Late Cretaceous to Late Palaeocene sea-floor spreading (Pole 1994). (Redrawn from Mielke 1989, Fig. 9.1., slightly modified)

thought to have been formed by both mid-ocean ridge and hotspot activity.

Setting aside islands of continental origin for the moment, a two-tier classification relating specifically to oceanic islands has been proposed by Nunn (1994). At the first level are the plate boundary and intraplate island types, each of which may also be subdivided into a number of major types, based on their geographical configuration in relation to the plate boundaries (Table 2.2). The recognition of these distinctive configurations of islands is a helpful starting point. However, Nunn (1994) warns against an over-simplistic use of the classification, pointing out that, whereas many islands with a common origin may be found in proximity to each other, islands of a particular geographical group do not necessarily share a common origin (this is well illustrated by the Mediterranean islands; Schüle 1993). A brief account of this classification (drawn largely from Nunn 1994) now follows.

Plate boundary islands

Islands at divergent plate boundaries

Divergent plate boundaries produce islands in two rather different circumstances, along mid-ocean ridges and along the axes of back-arc or marginal basins behind island arcs that are themselves associated with convergent plate boundaries. Although divergent plate boundaries are constructive areas involving more magma output than occurs in any other situation, seamounts, rather than islands, tend to form in connection with these boundaries. This is believed to be because of the relative youthfulness of the seamounts, the likelihood that the supply of magma decreases as the seamounts drift away from the plate boundary, and the increasing depth of the ocean floor away from these boundaries. In some cases, mid-ocean ridge islands may also be associated with hotspots in the mantle, providing the seemingly exceptional

Box 2.2 The splitting of Gondwanaland and the debate between vicariant and dispersalist explanations in island biogeography

The origin of many contemporary continental-fragment islands is directly linked with the breakup of Gondwanaland, the southern supercontinent that split apart from Pangaea, the last unique landmass, some 160 Ma. At the start of the Cretaceous period (c.140 Ma) Gondwanaland was still intact. By the Late Cretaceous (c.80 Ma), however, South America, Africa, and peninsular India were already discrete entities (Stanley 1999). It was during the Cretaceous that first Madagascar, and later the Seychelles and Kerguelen micro-fragments, began their trip to their present isolated geographic locations in the south-west, north-west, and south of the Indian Ocean, respectively. Australia and New Zealand rifted from Antarctica about 100 Ma, and subsequently, New Zealand broke first from Australia and then from Antarctica about 80 Ma (Lomolino *et al.* 2005). While other elements of Gondwanaland drifted toward the lower latitudes, the Antarctic slowly moved poleward, so that by about 24 Ma it was located over the South Pole, triggering the development of a great icecap.

Although Kerguelen is too far south (50° latitude) to possess a diverse biota, other ancient

(a)

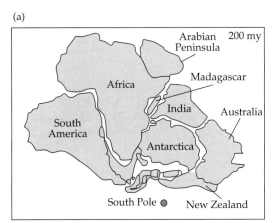

(b)

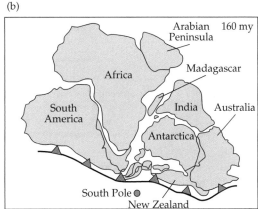

(c)

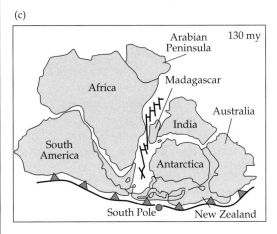

(d)

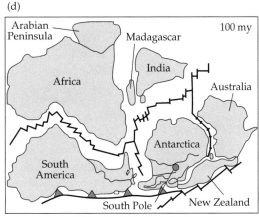

The stages in the breaking up of Gondwanaland. Bold lines show zones of sea-floor spreading. (From Storey 1995.)

continental fragment islands like Madagascar, New Zealand, and New Caledonia, are characterized by high endemism and some interesting ancient lineages, which biogeographers attribute to their ancient origins. Yet, the biotas of these ancient fragments of continent may not necessarily be as ancient as their rocks suggest. For instance, several Madagascan mammal taxa such as the Lemurs, the Tenrecidae, and the Viverridae, although relatively primitive, are nonetheless estimated to have diverged from their ancestral lineages during the Tertiary period, roughly between 45 and 26 Ma: much later than the separation from Africa. McCall (1997) points to geological evidence as suggesting that areas of the Mozambique channel were dry land during this period of the Tertiary, and that subsequent subsidence has opened up what is currently a wide clear water channel. His interpretation has been disputed by e.g. Rogers *et al.* (2000), who contend that at best there were merely a few isolated dots of land in the channel during this period.

The idea of vanished intercontinental land-bridges and island arcs across ocean basins to explain disjunct distributions has in general been discarded in favour of plate-tectonic mechanisms, afforced by the processes of long-distance dispersal known to populate remote oceanic island groups. However, the evidence of subaerial exposure on a few seamounts within the Mozambique channel suggests the possibility that Madagascar's mammals may well have used, if not a solid land bridge, at least a set of now-vanished stepping stones, in colonizing the island

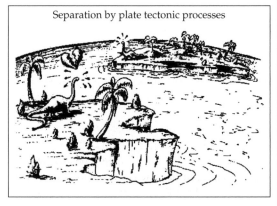

Four hypothetical means of species reaching islands: rafting, land-bridges, jump-dispersal, and the separation of a fragment of continent by plate tectonic processes (sketch by John Holden from Tarbuck, E. J. and Lutgens, K. (1999) *Earth: introduction to physical geology*, 6th edn, © 1999, p. 472, Fig. 19.4. Reprinted by permission of Pearson Education, Inc., Upper Saddle River, NJ)

long after this continental fragment began its ocean voyaging. Similarly, changing sea level, especially the reductions associated with major periods of glaciation, can turn islands into extensions of continents (such that for lengthy periods during the Quaternary, Tasmania, Australia, and New Guinea have formed one land mass, called Sahul), while providing additional 'stepping stone' islands that will have assisted the colonization of otherwise remote islands.

Debates between explanations for island biotas that postulate vicariance (the breaking of a past land connection) and those based on long-distance dispersal continue to generate considerable controversy and are sometimes not easily resolved. Perhaps the most celebrated example is that of

New Zealand. The case that New Zealand's biota constitutes a Gondwanan 'time-capsule' entirely explicable through long isolation of a full assemblage of species (a vicariance model) has been contested on the basis of evidence indicating more recently colonization of many lineages. As McGlone (2005) has put it ' . . . for New Zealanders in particular, abandoning "Time Capsule of the South Seas" for "Fly-paper of the Pacific" will be a wrench. But it has to happen . . . ' And indeed many features of New Zealand's biogeography are comparable with those of true oceanic island archipelagos such as Hawaii, strongly suggesting that much of its biota has colonized after the break-up of Gondwanaland, by long-distance dispersal (Pole 1994; Cook and Crisp 2005).

Table 2.2 P. D. Nunn's (1994) genetic classification of oceanic islands, with examples

Level 1	Level 2	Examples
Plate boundary islands	Islands at divergent plate boundaries	Iceland, St Paul (Indian Ocean)
	Islands at convergent plate boundaries	Antilles, South Sandwich (Atlantic)
	Islands along transverse plate boundaries	Cikobia and Clipperton (Pacific)
Intra-plate islands	Linear groups of islands	Hawaii, Marquesas, Tuamotu
	Clustered groups of islands	Canaries, Galápagos, Cape Verde
	Isolated islands	St Helena, Christmas Island (Indian Ocean), Easter Island

conditions necessary for their conversion from seamounts to large islands. Iceland is the largest such example (103 106 km^2), being a product of the mid-Atlantic ridge and a hotspot that may have been active for some 55 million years. Another context in which mid-ocean ridge islands can be found is in association with triple junctions in the plate system, a notable example being the Azores, where the North American, Eurasian, and African plates meet. The second form of divergent plate boundary island, exemplified by the Tongan island of Niuafo'ou, is that sometimes formed in back-arc basins, which develop as a result of plate convergence, but which produce areas of sea-floor spreading (Nunn 1994).

Islands at convergent plate boundaries
Where two plates converge, one is subducted below the other. This results in a trench in the ocean floor

at the point of subduction. Beyond the trench, a row of volcanic islands develops parallel with the trench axis on the surface of the upper of the two plates (Fig. 2.4). This mechanism accounts for some of the classic island arcs in the Pacific and Caribbean. Most commonly, where islands are formed in connection with subduction zones, the edges of the two converging plates are formed exclusively of oceanic crust. The magma involved in island arc volcanism derives from the melting of the subducted ocean crust, complete with its sedimentary load. Its composition thus owes much to the nature of the subducted crust. It is believed that the subduction of basaltic crust and water-bearing sediment leads to explosive andesitic volcanism. The Sunda island arc, including the Indonesian islands, is predominantly of this form. Basaltic volcanism in island arcs is less common, possibly because of a relative

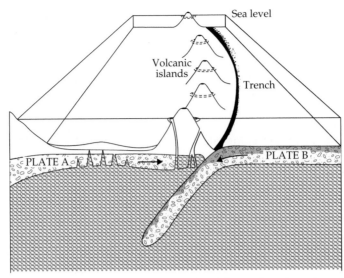

Figure 2.4 Simplified representation of the formation of an island arc, by the subduction of one plate under another (from an original in Williamson 1981).

paucity of sedimentary load in the subducted crust. The Sandwich arc in the south Atlantic involves both basaltic and andesitic volcanism.

Islands along transverse plate boundaries
This is a fairly rare context for island formation as, by definition, less divergence or convergence of plates is involved. However, strike–slip movement, compression, or both, can occur between adjacent parts of plates. An example of an island believed to have been produced by these forces is the Fijian island of Cikobia, in the south-west Pacific.

Islands in intraplate locations

This class of islands contains some of the most biogeographically fascinating islands of all, notably the Hawaiian group. They are equally fascinating geologically.

Linear island groups
The Hawaiian chain is the classic example of a series of islands with a significant age sequence arranged linearly across a plate (Table 2.3, Fig. 2.3). Among the larger islands, Kauai is the oldest at more than 5 million years, whereas Hawaii itself is less than 1 million years old. Islands older than

Table 2.3 Approximate ages of selected islands and of the Meiji seamount. This table illustrates the wide variation in the ages of oceanic islands, the simple age sequence of members moving north-west along the Hawaii–Emperor chain, and the lack of a similarly simple sequence moving south in the Austral–Cook cluster. For fuller listings, error margins, and original sources see Nunn (1994); for Hawaii, in addition see Wagner and Funk (1995)

Island group	Age (Ma)
Hawaii–Emperor chain	
Mauna Kea, Hawaii (the Big Island)	0.38
Kauai	5.1
Laysan	19.9
Midway	27.7
Meiji Seamount	74.0
Austral–Cook cluster	
Aitutaki	0.7
Rarotonga	1.1–2.3
Mitiaro	12.3
Rimatara	4.8–28.6
Rurutu	0.6–12.3
Isolated intra-plate islands	
Ascension	1.5
St Helena	14.6
Christmas Island (Indian Ocean)	37.5

Kauai stretch away to the north-west, but through combinations of erosion, subsidence, and coral formation, they are now reduced to forming coral atolls, or guyots. The chain extends from the Loihi Seamount, which is believed to be closest to the hotspot centre, through the Hawaiian islands themselves and along the Emperor Seamount Chain, a distance of some 6130 km (Keast and Miller 1996). The oldest of the submerged seamounts is more than 70 million years old, so that there must have been a group of islands in this part of the Pacific for far longer than is indicated by the age of the present Hawaiian islands. This may have allowed for colonization of the present archipelago from populations established considerably earlier on islands that have since sunk. However, recent geological findings indicate that there was a pause in island building in the Hawaiian chain, and molecular analyses support the notion that few lineages exceed 10 million years—that is, the pre-Kauai signal in the present biota is actually rather limited (Wagner and Funk 1995; Keast and Miller 1996).

The building of linear island groups such as the Hawaiian chain is explained by J. T. Wilson's (1963) **hotspot hypothesis**, which postulates that 'stationary' thermal plumes in the Earth's upper mantle lie beneath the active volcanoes of the island chain (Wagner and Funk 1995). A volcano builds over the hotspot, then drifts away from it as a result of plate movement and eventually becomes separated from the magma source, and is subject to erosion by waves and subaerial processes, and to subsidence under its own weight. It has been calculated that the Hawaiian islands have been sinking at a rate of approximately 2.6–2.7 mm a year over the last 475 000 years (cited in Whelan and Kelletat 2003). As each island moves away, a new island begins to form more directly over the hotspot. This process is thought to have been operating over some 75–80 million years in the case of the Hawaiian hotspot.

The change in orientation of the Hawaiian chain (Fig. 2.3) has been attributed to past changes in the direction of plate movement about 43 Ma, although recent work has questioned whether in fact hotspots are really such fixed reference points, raising the possibility of a combination of plate movement and hotspot migration to explain the distribution of hotspot island chains (Christensen 1999). The Society and Marquesas island groups, also in the Pacific, provide further examples of

hotspot chains, and Nunn (1994) provides a critique of several other postulated cases.

Clustered groups of islands

Many island clusters, once regarded as hotspot island chains that had become slightly less regular than the classic examples, have since been realized to differ significantly from the hotspot model, such that different groups and different islands within the groups may require distinctive models. The **tectonic-control model**, attributed to Jackson *et al.* (1972), postulates that, instead of lying along a single lineation, the islands lie along shorter lines of crustal weakness (termed *en echelon* lines), which are sub-parallel to each other. This line of reasoning may explain clusters of islands and chains in which there is no neat age–distance relation along the length of the chain, an example being the Line Islands in the central Pacific (Nunn 1994).

Two of the largest intraplate island clusters are the Canary and Cape Verde island groups, both in the central Atlantic and including a number of active volcanoes. Until recently, it was held possible that the Canaries were of mixed origins, with the easternmost—Lanzarote and Fuerteventura—being landbridge islands, once connected to Africa (Sunding 1979). However, it is now established beyond doubt that the entire archipelago is oceanic in origin, and that the gap of 100 km and more than 1500 m depth between the eastern islands and the African continent has never been bridged. Although thus clearly oceanic in the strict sense (Box 2.1), the Canaries do not appear to conform with the classical hotspot model as (1) they lack a clear lineal geographic distribution, (2) they do not present a simple age sequence, and (3) all the islands, with the exception of La Gomera, have been volcanically active in the last few thousand of years, with Lanzarote, Tenerife, and La Palma active within the last two centuries.

One explanation for their origin is the propagating fracture model, which relates the origin of the Canaries to fractures generated in the oceanic crust of this region when the Atlas chain began to form some 70–80 Ma as a result of the collision of the African and Eurasian plates (Anguita and Hernán 1975). However, some authors continue to favour the hotspot concept, suggesting a locus somewhere between the two youngest islands (La Palma and

El Hierro) (Carracedo *et al.* 1998). It seems that neither idea on its own can explain the origins of this particular archipelago, and that the roles of a residual hotspot and the Atlas tectonics must be considered simultaneously (Anguita and Hernán 2000).

The Cape Verde islands are also of uncertain origin, but Nunn (1994) favours the tectonic-control model. The Galápagos deserve a mention in this section as being a group of great significance to the development of island evolutionary models. They lie in the western Pacific just south of the divergent plate boundary separating the Nazca and Cocos plates. Again, Nunn regards them as an intraplate cluster rather than mid-ocean ridge islands.

Isolated islands

As knowledge of the bathymetry of the sea floor has improved, a number of islands believed to be isolated have been shown rather to be part of a seamount-island chain or cluster. In general, it can be taken that truly isolated intraplate islands form only at or close to mid-ocean ridges. They result from a single volcano breaking off the ridge with part of the sub-ridge magma chamber beneath it. Nunn (1994) identifies this condition as essential if the island is to continue to grow once it is no longer associated with the ridge crest. Examples include Ascension, Gough, and St Helena in the Atlantic, Christmas Island in the Indian Ocean, and Guadalupe in the Pacific. Isolated intraplate islands or island clusters may also, in certain cases, be the product of small continental fragments being separated from the main continental mass. The granitic Seychelles provide the best example, where there is good evidence of a continental basement affiliated to the Madagascar–India part of Gondwanaland (Box 2.2).

2.3   Environmental changes over long timescales

In the case of these islands we see the importance of taking account of past conditions of sea and land and past changes of climate, in order to explain the relations of the peculiar or endemic species of their fauna and flora.

(Wallace 1902, p. 291)

The separation of environmental changes from island origin is artificial. Islands may be built up

by volcanism over millions of years; in such cases, active volcanism is thus a major part of the environment within which island biotas develop and evolve. The preceding section may give the impression that all islands have been formed by volcanism, or at least fairly directly by the action of plate-tectonic processes. However, the formation of an island may come about either by connecting tracts of land disappearing under rising water, or by land appearing above the water's surface, either by depositional action or by some other process, of which tectonic forms are a large but not exclusive class. Indeed, Nunn (1994) offers the observation that changes of sea level are perhaps the most important reason why islands appear and disappear. This section therefore deals with long-term changes in the environments of islands, focusing particularly on changes in sea level, but also on locational shifts and changes in climate.

Changes in relative sea level—reefs, atolls, and guyots

As already noted, islands may come and go as a consequence of sea-level changes. Some of these sea-level changes are eustatic (i.e. they are due to the changing volume of water in the sea), and others are due to relative adjustment of the elevation of the land surface (isostacy). This can be brought about by the removal of mass from the land causing uplift, as when an icecap melts, or by tectonic uplift. Subsidence of the lithosphere can be due to increased mass (e.g. increased ice, water, or rock loading) or may be due to the movement of the island away from mid-ocean ridges and other areas that can support anomalous mass. In the right environment, coral reefs build around subsiding volcanoes, eventually forming atolls—an important category of tropical island.

Darwin (1842) distinguished three main reef types: fringing reefs, barrier reefs, and atolls. He explained atolls by invoking a developmental series from one type into the next as a result of subsidence of volcanic islands: fringing reefs are coral reefs around the shore of an island, barrier reefs feature an expanse of water between reef and island,

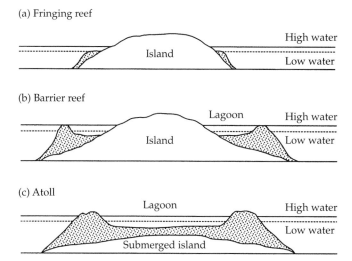

(a) Fringing reef

(b) Barrier reef

(c) Atoll

Figure 2.5 The developmental sequence of coral reefs as a result of subsidence, from (a) to (b) to (c), as hypothesized by Darwin (1842). (Redrawn from Mielke 1989, Fig. 7.10.)

and the final stage is the formation of an atoll, where the original island has disappeared, leaving only the coral ring surrounding a shallow water lagoon (Fig. 2.5). As Ridley (1994) noted, Darwin thought out the outline of his theory on the west coast of South America before he had seen a true coral reef! Although the theory may not be globally applicable and requires modification, for instance in the light of contemporary understanding of sea-level change, his basic model remains viable for most oceanic atolls and can account for most of the massive coral reefs (Steers and Stoddart 1977).

Coral reefs are built by small coelenterate animals (corals) that secrete a calcareous skeleton. Within the tissues and calcareous skeleton, numerous algae and small plants are lodged. The algae are symbionts critical to reef formation, providing the food and oxygen supplement necessary to account for the energetics of coral colonies, while obtaining both growth sites and nutrients from the coral (Mielke 1989). Reef-building corals generally grow in waters less than 100 m deep (exceptionally they can be found as deep as 300 m); they require water temperatures between 23 °C and 29 °C, and are thus found principally in tropical and subtropical areas, notably in the Indo-Pacific Ocean and in the Caribbean Sea (Fig. 2.6).

Coral growth has been found to vary between 0.5 and 2.8 cm/year, with the greatest growth rates occurring in water of less than 45 m depth (Mielke 1989). Historically, these growth rates have been great enough to maintain reefs in shallow water as either the underlying sea floor subsides or the sea level has risen (or both). Drilling evidence from islands with barrier reefs demonstrates this ability, just as predicted by Darwin's (1842) theory. For example, the thicknesses of coral that have accumulated on the relatively young islands of Moorea (1.5–1.6 Ma), Raiatea (2.4–2.6 Ma), and Kosrae (4.0 Ma) vary between 160 and 340 m, whereas the rather older islands of Mangareva (5.2–7.2 Ma), Ponape (8.0 Ma), and Truk (12.0–14.0 Ma) have accumulated depths of 600–1100 m of coral (Menard 1986). However, it should be noted that massive reef accretion also occurs in the western Atlantic on continental shelves that have not experienced long-term subsidence (Mielke 1989). This lends support to another theory: that reefs established on wave-cut platforms during periods of lowered sea level associated with glaciation, and that reef development then kept pace with rising sea levels in the interglacial period. This illustrates that several processes—subsidence, eustatic sea-level changes, temperature changes, and wave action—must each be considered with respect to the influence that they have had on the growth of coral reefs (cf. Nunn 1997, 2000).

Given the vagaries of tectonic processes and sea-level changes, coral-capped islands can become

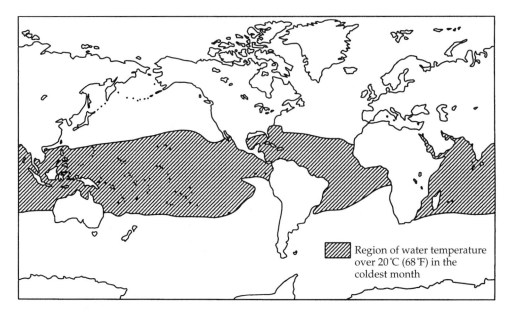

Figure 2.6 Region of water temperatures exceeding 20 °C in the coldest month, which corresponds roughly to the major reef regions of the world's oceans. (Redrawn from Mielke 1989, Fig. 7.9.)

elevated. Christmas Island in the Indian Ocean exemplifies this. It has been elevated to form a plateau of 150–250 m above sea level, but in parts reaches 360 m. The general aspect of the island is of coastal limestone cliffs 5–50 m high, which carries with it the implication that the area currently available for seaborne propagules to come ashore is considerably less than the island perimeter, a factor in common with many islands of volcanic origin.

It is becoming apparent that the relative elevation of particular islands may be subject to influence by the behaviour of others in the vicinity. The loading of the ocean floor by a volcano appears to cause flexure of the lithosphere, producing both near-volcano moating and compensatory uparching some distance from the load; thus one repeated pattern in the South Pacific is the association within island groups of young volcanoes (< 2 million years) with raised reef, or makatea islands, as exemplified respectively by Pitcairn and Henderson Islands (Benton and Spencer 1995). There are many other examples of islands that contain substantial amounts of limestone as a product of uplift: one that will be considered below is Jamaica.

Over time, sea levels have fluctuated markedly, particularly during the Quaternary glacial–interglacial cycles. These fluctuations, in combination with island subsidence, wave action, and subaerial erosion, have resulted in many islands declining below the current sea level. In nautical parlance, they are known as **banks** if they are less than 200 m from the surface (Menard 1986). Once they have sunk below about 200 m, they are effectively below the range of eustatic sea level and in this situation will rarely re-emerge as islands. Flat-topped sunken islands are termed **guyots**, and although they were once thought to be predominantly erosional features, most are in fact flat-topped through accretion of carbonate sediments (Jenkyns and Wilson 1999). The distribution of guyots, and particularly of banks, may be crucial to an understanding of the historical biogeography of contemporary islands (Diamond and Gilpin 1983; and compare McCall 1997 with Rogers *et al.* 2000).

Eustatic changes in sea level

From a biogeographical viewpoint, it does not matter greatly whether the changing configuration of an

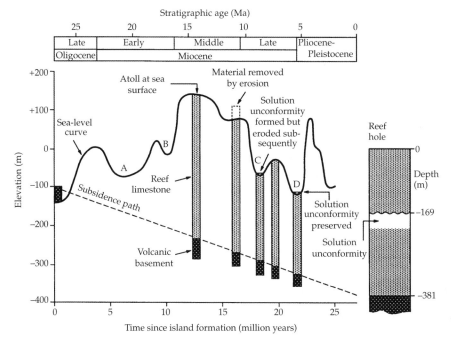

Figure 2.7 The complex sequence of sea-level changes proposed for Midway Island for the last 25 million years. Solution unconformities formed during low stages A–C were all removed during lowering of the atoll surface by subaerial processes during stage D (low sea level). The solution unconformity formed at this stage has remained preserved because subsidence carried it below the reach of subsequent periods of atoll surface lowering. (Redrawn from Nunn 1994: original sources given therein.)

archipelago is primarily a result of isostatic or of eustatic changes. However, it is important to recognize these complexities because of the need to understand past land–sea configurations (Keast and Miller 1996). That isostatic effects associated with Quaternary glaciations have not been confined to high-latitude continents and their margins, but also extended to low latitudes and ocean basins, was something that became clear only in the 1970s, necessitating a cautious approach to the construction of regional eustatic curves (Nunn 1994). Stratigraphic data from isolated oceanic islands have been of particular value in such analyses (Fig. 2.7).

Given the general emphasis on Quaternary events, it is noteworthy that certain of the sea-level oscillations in the Tertiary appear to have been of greater amplitude, principally as a consequence of high stands at 29, 15, and 4.2 Ma (Nunn 1994)—the wider biogeographical consequence of which are only beginning to be explored (Nores 2004). Within the Quaternary the pattern has been one of glacial

episodes in the northern latitudes corresponding with lowered sea levels, and interglacials associated with levels not dissimilar to those of the present day, yet superimposed on a falling trend. At a more detailed and precise level of analysis it has, however, proved difficult to construct global or even regional models of sea-level change for the Quaternary (Stoddart and Walsh 1992). Data for the previous and current interglacial indicate that sea-level maxima have varied in magnitude and timing across the Earth's surface (Nunn 2000). One intriguing, if controversial, explanation for some of the irregularities in sea-level changes relates to the configuration of the oceanic geoid surface (i.e. the sea surface itself), which, rather than being perfectly ellipsoid, is actually rather irregular, with a vertical amplitude of about 180 m relative to the Earth's centre. Relatively minor shifts in the configuration of the geoid surface, which might be produced by underlying tectonic movements, would be sufficient to cause large amounts of noise in the glacio-eustatic picture (Nunn 1994; Benton and

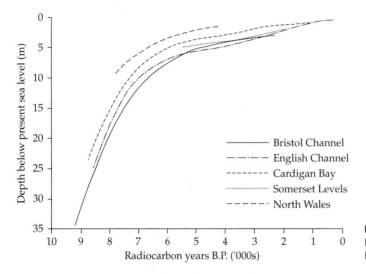

Figure 2.8 Holocene sea-level rise in southern Britain. (Redrawn from Bell and Walker 1992, Fig. 4.10, after Shennan 1983).

Spencer 1995). However, there is some measure of agreement that stand levels have not exceeded present levels by more than a few metres within the past 340 000 years. The most widely accepted figure for Pleistocene minima is of the order of −130 m, although in places it may have been greater than this (Bell and Walker 1992; Nunn 1994). This order of sea-level depression, given present lithospheric configuration, is sufficient to connect many present-day islands—such as mainland Britain—to continents, thus allowing biotic exchange between land areas that are now disconnected. Equally important, many islands existed which are now below sea level. However, as will be clear, simply drawing lines on maps in accord with present-day −130 m contour lines is not a sufficient basis for reconstructing past island–mainland configurations (Kayanne *et al.* 1993).

The rise from the late glacial minima to present levels was not achieved overnight, nor was it a steady or uniform pattern. As a broad generalization, at around 14 500 BP (Before Present, where Present is by convention designated as AD 1950), sea levels stood at about −100 m, rising by some 40 m over the next millennium (Bell and Walker 1992). A second major phase of glacial melting around 11 000 BP caused an eustatic rise to about −40 m by the beginning of the Holocene, when ice volumes had been reduced by more than 50%. The pattern for the British Isles for the past 9000 years can be seen in Fig. 2.8. Both eustatic and isostatic elements were

involved, leaving a legacy in raised shorelines and drowned valleys. In the North Sea, the Dogger Bank was breached by 8700 BP and the Straits of Dover by 8000 BP. The present configuration of the southern North Sea coastlines was more or less established by 7800–7500 BP and that of the British coastline by *c.*6000 BP, although slow adjustments continue today. The severing of Britain from Ireland took place some 2000–3000 years before that of Britain and the rest of Europe. In short, the switch from glacial to interglacial conditions took place some 2000–3000 years before 'mainland' Britain became an island once again, with different parts of the British Isles being separated from one another at quite different times. These events left a legacy in the biotic composition of the islands that has long been noted by biogeographers (Williamson 1981). Oscillations in effective sea level over the last few thousand years, although comparatively small, have nonetheless been sufficient to have significant impacts on human societies in island regions such as the Pacific (Nunn 2000), and there is currently considerable concern that future sea-level rises in association with predicted global warming will have serious implications for island peoples (see Chapter 12).

Climate change on islands

An important point made by Williamson (1981) is that short-term variations in climate have lower amplitude than long-term variation, i.e. the variance within decades is less than the variance within centuries, which is again less than the variance within millennia. This can be summed up in the phrase: climatic variation has a reddened spectrum. Within large landmasses one important way in which species respond to high-amplitude climate change is by range displacement. The possibilities for this mode of response within isolated oceanic islands are of course limited, with the corollary that persistent island endemic forms must find environments they can cope with within the limited confines of their island(s). If remote islands have indeed experienced pronounced climate change within the last two or three million years, has this driven enhanced extinctions on islands, or has it selected for broad-niched, adaptable genotypes? At present, we don't have the data to answer these questions.

Over the past several million years, the global climate system has undergone continual change, most dramatically illustrated by the glacial–interglacial cycles of the Pleistocene (e.g. Bell and Walker 1992). For much of the last glacial period, for instance, large areas of the British Isles resembled arctic tundra, while the northern and western regions supported extensive glacial ice. Subpolar islands such as the Aleutians (north Pacific) and Marion Islands (south-west Indian Ocean) also supported extensive icecaps at the last glacial maximum, and there is evidence that the Pleistocene cold phases caused extinction of plant species on remote high-latitude islands, such as the sub-Antarctic Kerguelen (Moore 1979). The classical model of four major Pleistocene glaciations has long been replaced by an appreciation that there have been multiple changes between glacial and interglacial conditions over the past 2 million years (Bell and Walker 1992). Although high-latitude islands have been the most dramatically affected by these climatic oscillations, it would be unwarranted to assume that low-latitude oceanic islands have been so effectively buffered that their climates have been essentially stable.

The Galápagos Islands are desert-like in their lower regions, with moist forests in the highlands; however, palaeoenvironmental data from lacustrine sediments demonstrate that the highlands were dry during the last glacial period. The moist conditions returned to the highlands about 10 000 BP, but the pollen data for El Junco lake on Isla de San Cristóbal indicate a lag of some 500–1000 years before vegetation similar to that of the present day occupied the moist high ground (Colinvaux 1972). This delay may reflect the slow progress of primary succession after expansion from relict populations in limited refugia in more moist valleys, or the necessity of many plants having to disperse over great expanses of ocean (the group is approximately 1000 km west of mainland Ecuador) to reach the archipelago.

Analyses of pollen cores from subtropical Easter Island also show the local effects of global climate change. The data indicate fluctuating climate between 38 000 and 26 000 BP, cooler and drier conditions than those of the present day between 26 000 and 12 000 BP, and the Holocene being generally warm and moist, but with some drier phases (Flenley *et al.* 1991). Contrary to the evidence from Easter Island and the Galápagos, it appears from studies of snowline changes on tropical Hawaii that conditions were both cooler and, quite possibly, wetter during the last glaciation (Vitousek *et al.* 1995). Given the dominant influence of ocean and atmospheric current systems on the climates of oceanic islands (below), it is unwise to assume a straightforward relationship between continental and island climate change over the Quaternary.

The developmental history of the Canaries, Hawaii, and Jamaica

The Canaries
The Canaries constitute a comparatively ancient system for an oceanic archipelago. The increasingly well-specified geological history provides a crucial background to understanding the biogeography of these islands, and the distributions of the many regional, archipelagic, and single-island endemic species. The oldest basaltic shields (Northern, Betancuria, and Jandía) of the island of Fuerteventura are estimated to have emerged

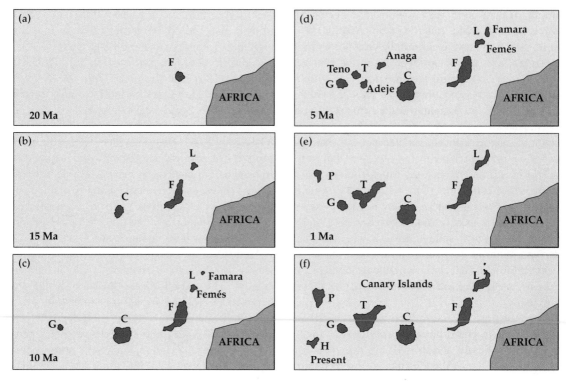

Figure 2.9 Sequence of emergence of the islands of the Canarian archipelago (Modified from Marrero and Francisco-Ortega 2001, Fig. 14.1). L = Lanzarote, F = Fuerteventura, C = Gran Canaria, T = Tenerife, P = La Palma, G = La Gomera, H = El Hierro.

above sea level some 20 Ma (Fig. 2.9a), although their construction process began almost 70–80 Ma in the Late Cretaceous Era (Anguita *et al.* 2002). This first island was likely colonized and populated predominantly from the nearby African mainland, some 100 km away. Some 4 million years later (*c.*16 Ma), the basaltic massives of Femés (Los Ajaches) emerged, forming the first shield of what became the island of Lanzarote, which was thus available for colonization simultaneously from the mainland and the older island.

The expansion of the archipelago towards the Atlantic Ocean began 2 million years later, with the emergence of Gran Canaria (Fig. 2.9b), and was consolidated with the appearance of La Gomera *c.*12 Ma and Famara *c.* 10 Ma (Fig. 2.9c). These islands, much larger and higher than today, have suffered many catastrophic landslides during their geological history (below). The next major additions to the archipelago were the basaltic massifs of Teno,

Adeje, and Anaga, which emerged about 8 Ma between La Gomera and Gran Canaria, forming the corners of what was to become the island of Tenerife (Fig. 2.9d). It appears likely that Teno and Adeje were mainly populated from La Gomera, whereas Anaga was mainly colonized from Gran Canaria. Interestingly, some 10% of the endemic plants of the Canaries are restricted to one or more of the palaeoislands of Tenerife, likely including a mix of early diverging and recently diverging lineages (Trusty *et al.* 2005).

Some 3.5 Ma, the catastrophic Roque Nublo ash flow is thought to have almost completely sterilized Gran Canaria, perhaps with the exception of two localized refugia (Marrero and Francisco-Ortega, 2001*a, b*). If so, the recolonization process will have started both from those refugia, and, especially for the mountain and summit ecosystems, from the nearby island massifs of Anaga and Jandía, respectively 60 and 90 km away.

The last 2 million years has been a period of great environmental dynamism in the Canaries. Volcanic mountain building activity has constructed the two westernmost islands, La Palma (1.5 Ma) (Fig. 2.9e) and El Hierro (1.1 Ma). Both have been prone to catastrophic landslides, related principally to the instability of their high volcanic summits. The last of them (El Golfo in El Hierro) occurred as recently as 15 000 years ago, cutting away the northwest half of the island, and distributing it across some 1500 km² of the ocean floor (Canals *et al.* 2000).

It is only within the last 1.5 Ma that the great acidic volcanic cycle of Las Cañadas unified the old basaltic massifs of Teno, Adeje, and Anaga into today's Tenerife. The new land surface enabled enhanced biotic exchange between the old massifs. Several landslides (Icod, Las Cañadas, La Orotava, Güímar) followed the construction of this edifice, having catastrophic effects on both Tenerife and the nearby islands. The building of the Teide stratovolcano is very recent, with its highest tip (3718 m) being built in the prehistoric period, approximately 800 years ago. Such events leave their imprint on the genetic architecture of lineages, through repeated subdivision of formally contiguous populations (e.g. Moya *et al.* 2004).

Finally, some 50 000 BP, the small islets north of Lanzarote (La Graciosa, Montaña Clara, and Alegranza) and Fuerteventura (Lobos) were built, giving the Canary archipelago its present shape (Fig. 2.9f). With the exception of La Gomera, which has been dormant for some 2–3 Ma, the islands have remained active, with some 15 volcanic eruptions in Lanzarote, Tenerife, and La Palma in the last 500 years (Anguita *et al.* 2002).

In addition to the building, collapse, and erosion of the islands, eustatic sea-level changes during the Pleistocene have alternately doubled and halved the emerged area of the archipelago, from approximately 7500 km² during interglacials to some 14 000 km² during stadials (García-Talavera 1999). During low sea-level stands, such as coincided with the peak of the last Ice Age *c.*13 000 BP, the islands were about 130 m higher in altitude. Lanzarote and Fuerteventura were joined together, along with nearby islets, into a single large island

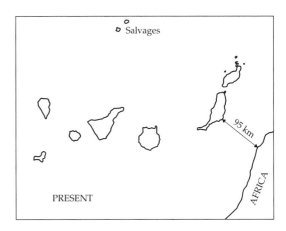

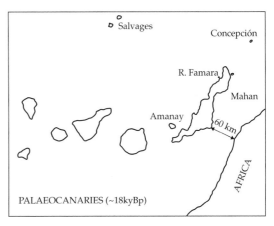

Figure 2.10 The Canaries at the time of the sea-level minimum of the Last Glacial period and in the present day (Modified from García-Talavera 1999).

(Mahan, *c.*5000 km² in area). The shortest distance from the archipelago to the African mainland was reduced from about 100 km to 60 km, the present-day submarine bank of Amanay, north of Jandía, formed an island of some 100 km² (Fig. 2.10), and a 'stepping-stone' corridor of islands provided enhanced connectivity between the Canaries, Madeira and the Iberian Peninsula (Fig. 2.11).

Hawaii

The Hawaiian chain is characterized by the growth of shield volcanoes that go through a known life cycle, with four well-defined stages (Wilson 1963; Price and Clague 2002):

Figure 2.11 A reconstruction of PalaeoMacaronesia at the time of the sea-level minimum of the Last Glacial period (Redrawn from García-Talavera 1999).

1 Volcanoes form over the hotspot and are subsequently removed from it by the movement of the Pacific plate, producing a linear array of volcanic summits increasing in age to the north-west.

2 After achieving their maximum heights perhaps 0.5 Ma after their emergence, volcanoes rapidly subside as they move away from the hotspot (Moore and Clague 1992).

3 Erosive processes reduce volcanic peaks to sea level over several million years.

4 Small atolls may remain while conditions are suitable for coral growth, or they are reduced to guyots if they drown and sink below the sea surface, carried down by the spreading and subsiding plate (Jenkyns and Wilson 1999).

Recently, sonar surveys of the sea floor have enabled the reconstruction of past island configurations (Moore *et al.* 1994, Clague 1996). As Hawaiian volcanoes grow, the lava deposited below the sea level forms a steeper slope than that deposited subaerially. The location of the now submerged breaks-in-slope enables the estimation of the maximum shorelines and thus areas of islands throughout the archipelago (Fig. 2.12). Furthermore, the maximum altitude achieved by an island can also be reconstructed assuming a 7°angle for subaerial lava deposits. Knowing the age, the original altitude of rocky outcrops and the subsidence rate, it is possible to estimate the rate of erosion of each particular island. This is, of course, a fairly approximate science, as erosion can vary through time as a result of climatic fluctuations, the occurrence of mega-landslides, and enhanced precipitation around higher peaks associated with orographic cloud formation.

Using all this information, Price and Clague (2002) have reconstructed the history of the Hawaiian Chain, from 32 Ma, which marks the appearance of the first Hawaiian island (the earlier islands of the Emperor seamount chain were either submerged or just atolls by this time). The island configurations at 5 Ma intervals for the Hawaiian chain are given in Fig. 2.13a, and Fig. 2.13b summarizes the reconstructed distributions of island altitudinal range. During the first half of Hawaiian chain history (*c.*32–18 Ma) a few volcanoes (Kure, Midway, Lisianski, Laysan), briefly exceeded 1000 m above sea level (ASL) although at any given moment most islands have been small. The most important island that pre-dated the present high islands was Gardner, which formed around 16 Ma and was likely comparable in size (10 000 km²) and height (> 4000 m) to today's Big Island (also known simply as Hawaii). After its - formation, a series of mid-sized volcanoes formed (French Frigate Shoals, La Perouse Pinnacle), culminating with Necker, *c.*11 Ma. The period between 18 and 8 Ma, is described by Price and Clague (2002) as the 'first peak period', because of the existence of multiple peaks over 1000 and some over 2000 m, contributing to a substantial archipelago, with several zonal (coast to summit) ecosystems.

After the formation of Necker, there was a period when only smaller islands formed, leading to an archipelago diminished in height and area and thus impoverished in habitats. The emergence of Kauai (5 Ma) constituted the beginning of the 'second peak period', which continues today, with

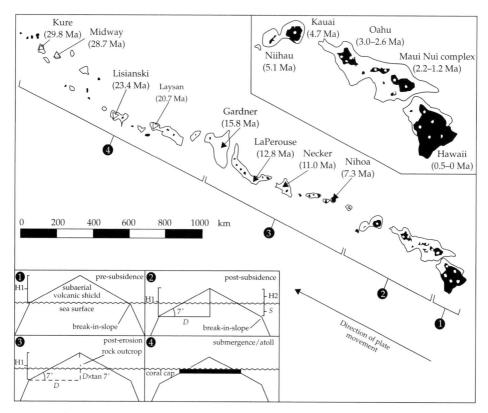

Figure 2.12 Features of the Hawaiian ridge, life stages of volcanoes, and estimates of island altitude over time. The main islands are detailed in the inset. Ages are given in parentheses for selected volcanic peaks. The numbered sequence of boxes depicts life stages of islands and associated features used to estimate life histories. (1) The break-in-slope, marking the maximum shoreline of a volcano as it formed, is depicted by a black outline on the map. (2) The depth of break-in-slope (S) indicates the amount of subsidence that has occurred since shield formation; the present elevation of an uneroded volcano (H2) is then added to determine its original height (H1). The original height can also be calculated by using the distance from summit to shore (D) and assuming a 7°angle. Summits of the main islands are shown as white dots on the map over the island areas shown in black. (3) Slower subsidence and erosion reduce volcanic peaks to sea level. Four rocky islets (filled triangles on the map) are near the end of this stage, with the amount of erosion (E) estimated. (4) Finally, with no more rock above sea level, only seamounts (black dots) and small coralline islands (open triangles) remain (Redrawn from Price and Clague 2002, Fig. 1).

multiple volcanic island summits over 1000 and 2000 m (Kauai, Oahu, Molokai, Lanai, Maui, Kahoolawe, and Hawaii) and with more emerged land than ever. Only some 1.2 million years ago, the now separated islands of Maui, Lanai, Molokai, and Kahoolawe, as well as the Penguin Bank seamount, constituted the Maui-Nui complex ('Big Maui' in Polynesian), an island larger than the current island of Hawaii, which was divided by subsidence some 0.6 Ma (Price and Elliott-Fisk 2004). The Maui-Nui complex has been reunited several times by sea-level falls associated with Pleistocene

glaciation. This increasingly well-specified environmental history has considerable relevance to understanding the biogeography of Hawaii, as will be discussed in later chapters.

Jamaica

Several of the themes of this section are well illustrated by Ruth Buskirk's (1985) analysis of the history of Jamaica in relation to the rest of the Caribbean. The changing configurations of Antillean land masses during the Cenozoic, as suggested by a variety of authors, are presented in

(a)

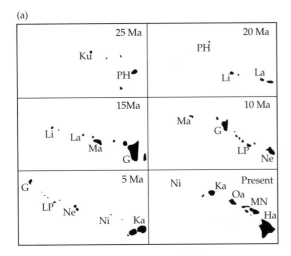

(b)

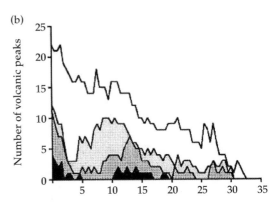

Figure 2.13 (a) Hawaiian island configurations at 5 Ma intervals. Ku: Kure, PH: Pearl and Hermes, Li: Lisianski, La: Laysan, Ma: Maro, G: Gardner, LP: La Perouse, Ne: Necker, Ni: Nihoa, Ka: Kauai, Oa: Oahu, MN: Maui Nui, Ha: Hawaii. (Redrawn from Price and Clague, 2002, Fig 3). (b) Distribution of estimated island maximum altitudes over time (black areas >2000 m ASL, dark grey: 1000–2000 m, light grey: 500–1000 m, white: <500 m) (Redrawn from Price and Clague 2002, Fig. 2a).

Fig. 2.14. Beginning in the Eocene, left-lateral faulting along the northern Caribbean plate margin slowly moved Jamaica eastward relative to the North American plate, the island varying in size and becoming increasingly distant from the North American continent in the process. More crucially still, from a biogeographical perspective, in the Late Eocene, it was entirely submerged, with extensive marine deposits being laid down and total subsidence in Jamaica amounting to some 2800 m, before its eventual re-emergence as an island in the early Miocene. The uplift began in the northern and northeastern coastal region, with general uplift beginning in the middle Miocene. Maximum uplift and faulting occurred in the Pliocene (the final stage of the Tertiary), with as much as 1000 m of uplift in the Blue Mountains area since the middle Pliocene. Even after the major period of emergence began, the island continued to move away from Central America, possibly by as much as 200 km in the past 5 million years.

The message that must be drawn from this and the foregoing sections is that such basic environmental features as island elevation, area, geology, location, isolation, and climate are each subject to significant change in the long term. On evolutionary time scales, volcanic islands in particular are remarkably dynamic platforms. If we are to understand the biogeography of such island systems, it is clearly necessary to integrate both historical and contemporary ecological factors (e.g. Stoddart and Walsh 1992; Wagner and Funk 1995; Price and Elliott-Fisk 2004; Price and Wagner 2004).

2.4 The physical environment of islands

Topographic characteristics

Topographic characteristics for a selection of islands in the New Zealand area are provided by Mielke (1989), who notes that, although the largest islands have the highest peaks, smaller islands display no consistent pattern (Table 2.4). In fact, topographic characteristics depend on the type of island. Volcanic islands tend to be steep and relatively high for their area and, through time, become highly dissected through erosion (Fig. 2.15). As long ago as 1927, Chester K. Wentworth calculated the age of the Hawaiian volcanoes as a function of the degree

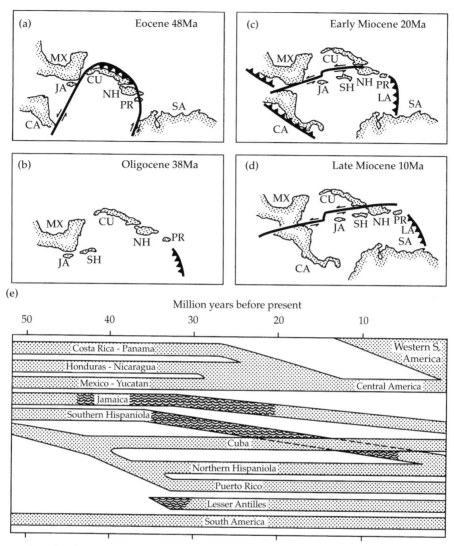

Figure 2.14 Hypothesized configurations of the Caribbean area during the past 50 million years. (a–d) Maps of changing positions. Present-day land outlines are used for ease of recognition and do not indicate shorelines of the time. CA = Central America, MX = Mexico, JA = Jamaica, CU = Cuba, NH = northern Hispaniola, SH = southern Hispaniola, PR = Puerto Rico, LA = Lesser Antilles, SA = South America. (e) Simplified scheme showing the relative positions of the land areas, increased distance apart indicating increased barriers to dispersal. Areas that were largely inundated are designated with water wave symbols. (Redrawn from Buskirk 1985, Figs 2 and 4.)

of dissection. More recently, Menard (1986) compared the values resulting with the ages determined by potassium–argon dating (up to an age of 1.6 million years), thus demonstrating that dissection is indeed a function of age of the volcano. Height of island can be important in respect to changes in sea level, receipt of rainfall, and to other climatic characteristics. In addition, large, high volcanic islands

are subject to periodic catastrophic landslips, which can radically reshape substantial areas of the island (Hürlimann *et al.* 2004; below).

Coral or limestone islands and atolls tend to be very low-lying and flat. There are clear implications here with respect to future sea-level changes (Chapter 12). Those which have been uplifted to more than a few metres above sea level are termed

Table 2.4 Area and relief of New Zealand and neighbouring islands (from Mielke 1989, Table 9.1)

Island	Area (km²)	General relief
South Island	148 700	50 peaks over 2750 m
North Island	113 400	Three peaks over 2000 m
Stewart	1720	Granite; three peaks, up to 987 m
Chatham	950	Cliffs to 286 m
Auckland	600	Volcanic; one peak 615 m
Macquarie	118	Volcanic; 436 m, glacial lakes
Campbell	113	574 m, glaciated
Antipodes	60	Volcanic; peak of 406 m
Kermadec	30	Volcanic; peak of 542 m
Three Kings	8	Volcanic
Snares	2.6	Granite; cliffs to 197 m

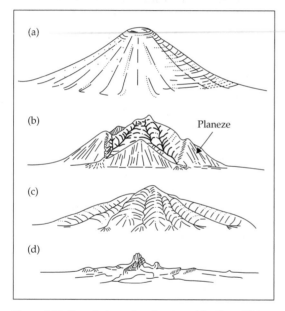

Figure 2.15 Stages in the erosion of a volcano (after Nunn 1994, based on an original in Ollier 1988). (a) An intact volcano, with radial grooves (e.g. Tristan da Cunha); (b) Planeze stage, in which some of the radial drainage channels have undertaken headwall capture, leaving triangular-shaped remnants relatively untouched by fluvial erosion (examples can be seen on St Helena); (c) Residual volcano, where planezes have been removed by erosion, but the original form of the volcano is still evident (examples can be seen in the Canaries and Cape Verde islands); (d) Volcanic skeleton, where only the necks and dykes remain as prominent features (e.g. Bora-Bora, French Polynesia).

makatea islands. Examples include Makatea itself (Tuamotu archipelago), Atiu (Cook Islands), and most of the inhabited islands of Tonga. This is an important class of islands. They are characterized by rocky coralline substrates, but some are partly volcanic. Many have had commercially exploitable deposits of phosphate (guano) formed through the ages from droppings of seabirds (e.g. Nauru; see Chapter 12). For a discussion of the complexities of volcanic, limestone, and makatea (composite) island landscapes, see Nunn (1994).

Climatic characteristics

Island climates have, self-evidently, a strong oceanic influence, and quite often are considered anomalous for their latitude, as a consequence of their location in the path of major ocean- or atmospheric current systems (e.g. the Galápagos; Darwin 1845). Low islands tend to have relatively dry climates. High islands tend to generate heavy rainfall, although they may also have extensive dry areas in the rain shadows, providing a considerable environmental range in a relatively small space. Even an island of only moderate height, such as Christmas Island (Indian Ocean), with a peak of 360 m and general plateau elevation of only 150–250 m, benefits from orographic rainfall, allowing rain forest to be sustained through a pronounced dry season (Renvoize 1979). Islands may also be expected to receive rainfall that has a generally different chemical content than that experienced over continental interiors (cf. Waterloo *et al.* 1997).

As noted by J. D. Hooker in his lecture to the British Association in 1866, the climate and biota of islands tend to be more polar than those of continents in the same latitude, while intraannual temperature fluctuation is reduced (Williamson 1984). Islands near the equator frequently exhibit annual average temperature ranges of less than 1 °C, and even in temperate latitudes, the annual average ranges are less than 10 °C: for example, 8 °C in the case both of Valentia (Ireland) and the Scilly Isles. Some islands do, however, experience quite large interannual variations in other features of their weather. Variability in rainfall, for example related

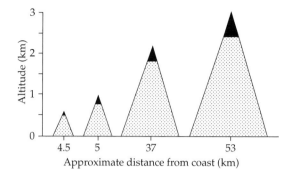

Figure 2.16 The transition to mossy forest on mountains in Indonesia occurs at lower elevations on small mountains near the sea than on larger mountains inland. From left to right: Mt Tinggi (Bawean), Mt Ranai (Natuna Island), Mt Salak (West Java), and Mt Pangerango (West Java). (After van Steenis 1972.)

to El Niño–Southern Oscillation (ENSO) phenomena, can be of considerable ecological importance. Islands may also experience periodic extremes such as hurricanes (below) (Stoddart and Walsh 1992).

Contraction, or 'telescoping', of altitudinal zones is another marked feature of smaller islands. For example, on Krakatau, Indonesia, the plentiful atmospheric moisture and the cooling influence of the sea result in the near permanent presence of cloud in the upper parts, further lowering temperatures and allowing the development of a montane mossy forest at around 600 m above sea level, a much lower altitude than the continental norm (Whittaker *et al.* 1989). Taylor (1957) suggested 2000 m as the height at which the vegetational transition occurs in interior Papua New Guinea, and Whitmore (1984) observed that clouds normally form around tropical peaks higher than 1200 m, but that this altitudinal limit is reduced on offshore islands. In his review of upper limits of forests on tropical and warm-temperate oceanic islands, Leuschner (1996) cited a range of 1000–2000 m for the lowering of the forest line compared with that for continental areas. Factors that have been invoked for these reductions include steepened lapse rates (and associated trends in atmospheric humidity levels), droughts on trade-wind-exposed island peaks with temperature inversions, the absence of well-adapted high-altitude tree species on some islands, and even the immaturity of volcanic soils (Bruijnzeel *et al.* 1993; Leuschner 1996). Whatever the precise cause(s) of the telescoping, it is a common feature of tropical islands, and it has the effect of (potentially) increasing the number of species

that an island can support, by compressing habitats and allowing a relatively low island to 'sample' additional upland species pools (Fig. 2.16).

The range of climatic conditions of an island is determined in large measure by the elevation of the highest mountain peaks. Thus regression studies often find altitude to be an important variable in explaining species numbers on islands, in some cases ranking only second to, or even ahead of, island area (Fernández-Palacios and Andersson 2000). Tenerife, in the Canaries, is an excellent example of a volcanic island with a steep central ridge system and a pronounced rain shadow. The Canaries lie on the subsiding eastern side of the semi-permanent Azores anticyclone at about 28°N, 100 km off Africa. The subsidence produces a warm, dry atmosphere aloft that is separated at 1500–1800 m from a lower layer of moist, southward-streaming air, producing what is known as a **trade-wind inversion** (Fernandopullé 1976). Tenerife's climate is one of warm, dry summers and mild, wet winters: in essence, the island experiences a Mediterranean-type climate, despite its latitudinal proximity to the Sahara. The north-east trades bring in moisture from the sea and the mountain backbone forces the air to rise. It cools, and this produces a zone of orographic clouds, locally known as *mar de nubes*, between about 600 and 1500 m, typically 300–500 m thick, on the windward slope of the island—a feature that builds up more or less daily (Fig. 2.17), providing the laurel forest that grow at that altitude with the moisture surplus (through fog drip and reduced evapotranspiration) needed to overcome the dry Canarian summers (Fernández-Palacios 1992, Fig. 1). The southern sector, in the rain shadow, receives far less precipitation. As a consequence of the differences in precipitation and temperature, the lower and upper limits of forest growth are higher on the southern side. Indeed, the southern sector lacks a dense forest zone at mid-altitude and is much more xerophytic. A further consequence of the trade-wind inversion is the existence of well-defined vegetation belts on the windward slope, whereas on the leeward slopes, beyond the influence of the orographic cloud layer, the vegetational landscape seems to be closer to a coenocline, i.e. a vegetation continuum (Fernández-Palacios and Nicolás 1995).

Water resources

Availability of water shapes the ecology and human use of islands (Whitehead and Jones 1969; Ecker 1976; Menard 1986). Most oceanic islands, whether they are high volcanoes or atolls, contain large reservoirs of fresh water. Fresh lava flows are highly permeable, but over time the permeability and porosity of the rock decreases as a result of weathering and subsurface depositional processes. The residence time of groundwater in the fractured aquifers (Fig. 2.18) of large volcanic islands may be from decades to centuries. We can divide these

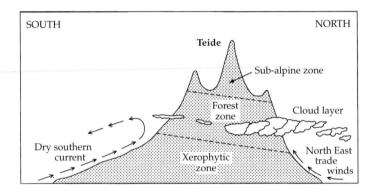

Figure 2.17 Climatic/vegetation zones of Tenerife. (Redrawn from Bramwell and Bramwell 1974.)

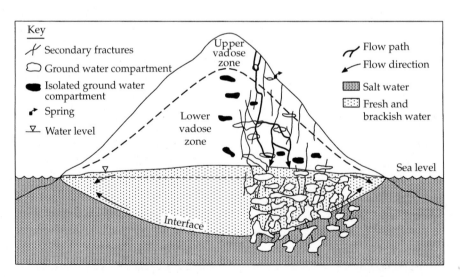

Figure 2.18 Scheme of the vertical distribution and flow of groundwater through a volcanic island. The vadose zones contain groundwater compartments and chains, interspersed with dry zones. The Ghyben–Herzberg lens of fresh water under an oceanic island rests on the denser salt/brackish water that permeates the base of the island. (Modified from Ecker 1976, Fig. 4, with permission from Elsevier.)

systems into two zones: the **vadose zone** and the **saturated or basal water zone**. The vadose zone contains groundwater compartments, often linked into chains, interspersed with dry zones. The basal zone is characterized by many closely placed groundwater compartments and by a high percentage of saturated secondary fractures. Both fresh water and seawater occur in this zone, with tides influencing the water level up to 4–5 km from the coast (Ecker 1976). In the absence of rain, the seawater within islands would be at sea level; however, rainwater percolating through an island floats on the denser salt or brackish water that permeates the base of the island, in what is termed a **Ghyben–Herzberg lens** (Menard 1986).

As indicated above, the average annual rainfall varies greatly within the island of Tenerife, from over 800 mm in the highest parts of the Anaga peninsular (a sum that might be doubled if fog drip were considered), to less than 100 mm in the extreme south of the island. The rain shadow thus has profound effects both on the ecology and on the potential human use of the island. The steep north side, which benefits from the majority of the precipitation, is intensively cultivated, whereas much of the flatter east and west coastal zones is essentially useless for cultivation without irrigation. Thus, even on this large, high island, groundwater (which increasingly has been tapped from aquifers deep in the volcano) is the most important source of water for the human inhabitants, and water constitutes a key limiting resource for development (Fernández-Palacios *et al.* 2004*a*). Even low-lying atolls maintain a freshwater lens but, as noted in the introduction to this chapter, very small islets (below the order of about 10 ha) can lack a permanent lens. Such habitats are liable to be hostile to plants other than those of strandline habitats, thus limiting the variety of plant species that can survive on them (Whitehead and Jones 1969).

Tracks in the ocean

One of the intriguing features of the island biogeographical literature is that isolation is, in general, rendered merely as distance to mainland. This ignores the geography of the oceans. Figure 2.19 displays the pattern of surface drifts and ocean currents in January. This pattern is, of course, variable on an

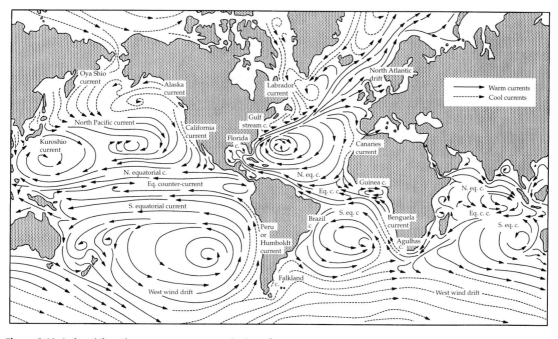

Figure 2.19 Surface drifts and ocean currents in January. (Redrawn from Nunn 1994, Fig. 4.)

interannual and an intraannual basis. However, in some areas, the ocean currents and wind currents are strongly directional and persistent. In some cases, knowledge of current systems provides a parsimonious explanation of biogeographical patterning (Cook and Crisp 2005), as in the case of the butterflies of the island of Mona in the Antilles. The island is equidistant between Hispaniola and Puerto Rico and is 62 km^2 in area. Its 46 species of butterflies feature 9 subspecies in common with Puerto Rico and none with the larger Hispaniola, whereas the ratio of source island areas is 9:1 in favour of Hispaniola. The explanation appears to lie in a remarkably constant bias in wind direction, which is from Puerto Rico towards Hispaniola (Spencer-Smith *et al.* 1988). However, it should be recalled that palaeoenvironmental data from islands such as Hawaii and the Galápagos indicate that significant changes in ocean and atmospheric circulation patterns occur over time (Vitousek *et al.* 1995).

2.5 Natural disturbance on islands

One of the themes that we develop in this book is that the significance of naturally occurring disturbances has not been given due recognition in the development of island biogeographical theory. We have already seen that any ancient island will, over the course of tens of thousands of years, have experienced substantial environmental changes. This section is concerned with shorter-term changes in environment, such as the individual volcanic eruptions that may intermittently add to the bulk of an island while, at least temporarily, reducing its plant and animal populations (e.g. Partomihardjo *et al.* 1992; Hilton *et al.* 2003).

Ecologists have defined disturbance as 'any relatively discrete event in time that removes organisms and opens up space which can be colonized by individuals of the same or different species' (Begon *et al.* 1990, p. 754). Such a definition is rather inclusive and, in the context of island biogeography, it will be recognized that different scales of perturbation are relevant in differing contexts (Fig. 2.20). Increasingly, ecologists are recognizing that natural systems are largely structured by disturbance (Pickett and White 1985; Huggett

1995). Often, a single event such as a hurricane (tropical cyclone), will impact on both mainland and island systems; however, because of the geographical and geological idiosyncrasies of small islands and their location in oceans, their disturbance regimes when scaled to island size are atypical of continental land masses.

In order to incorporate the disturbance regime into island biogeography, some form of classificatory framework is required (Whittaker 1995). Figure 2.20 represents an example of one attempt to develop such a framework. Another scheme, devised for the Caribbean islands and reproduced in Table 2.5, indicates the main types of disturbance phenomena, the area liable to be affected, the primary nature of the impact, the duration, and the recurrence interval. In an approach owing much to H. T. Odum, Lugo (1988) has classified the disturbance phenomena into five types, each characterized by their impact on energy transfers.

- **Type 1 events** change the nature or magnitude of the island's energy signature before the energy can be 'used' by the island; for example, shifts of weather systems through ENSO events, leading to droughts or heavy rainfall (Stoddart and Walsh 1992; Benton and Spencer 1995).

- **Type 2 events** are those acting on the major biogeochemical pathways of an island, for example as a result of changes caused by an earthquake.

- **Type 3 events** are those that remove structure from an island ecosystem, but without altering the basic energy signature, so that recovery can proceed rapidly after the event. Hurricanes are Type 3 events. A well-researched example is Hurricane Hugo, which caused complete defoliation of a large part of the Luquillo forest on Puerto Rico in 1989, although one might note that, despite a rapid 'greening-up' of the forest, the signature of such an event will be evident in the unfolding vegetation mosaic for many decades (Walker *et al.* 1996).

- **Type 4 events** alter the 'normal' rate of material exchange between the island and either the ocean or atmosphere. For instance, if the trade winds are inhibited by changes in atmospheric pressure, exchanges may be reduced.

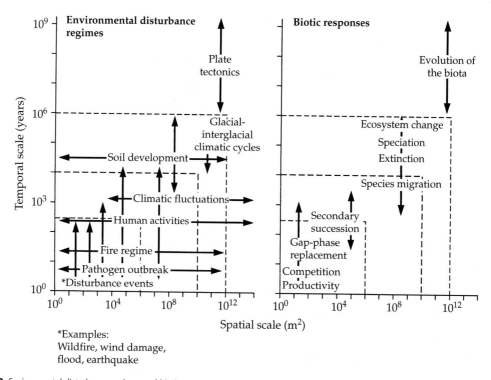

Figure 2.20 Environmental disturbance regimes and biotic responses, viewed in the context of four space–time domains (shown here bounded by dashed lines), named micro-, meso-, macro-, and mega-scales by the scheme's authors. (Redrawn from Delcourt and Delcourt 1991, Fig. 1.6; from an original in Delcourt and Delcourt 1988, with kind permission from Kluwer Academic Publishers.)

Table 2.5 Disturbance phenomena affecting Caribbean islands (after Lugo 1988). See text for explanation of the five types

Disturbance phenomena	Type	Area affected	Primary impact	Duration	Recurrence
Hurricanes	3, 5	Large	Mechanical	Hours–days	20–30 years
High winds	3, 4, 5	Large	Mechanical	Hours	Annual
High rainfall	4	Large	Physiological	Hours	Decade
High-pressure systems	1	Large	Physiological	Days–weeks	Decades
Earthquakes	2, 5	Small	Mechanical	Minutes	10^2 years
Volcanism	All	Small	Mechanical	Months–years	10^3 years
Tsunamis	3, 4, 5	Small	Mechanical	Days	10^2 years
Extreme low tides	1	Small	Physiological	Hours–days	Decades
Extreme high tides	3, 4, 5	Small	Mechanical	Days–weeks	1–10 years
Exotic genetic material	2, 3	Large	Biotic	10^2 years	Decades
Human, e.g. energy	1	Small	Biotic	Years	1–10 years
Human, e.g. war	5	Small	Mechanical	Months–years	?

• **Type 5 events** are those that destroy consumer systems, by which is meant human systems, possibly with subsequent repercussions for the rest of the island's ecology; examples again include hurricanes and earthquakes. The eruptions of Montserrat in 1995–1998 provide a classic illustration of the extensive cross-cutting impact that within-island volcanic action can have in transforming the physiography, ecology and human use of islands (Hilton *et al.* 2003; Le Friant *et al.* 2004).

Magnitude and frequency

Between 1871 and 1964, an average of 4.6 hurricanes per year was recorded in the Caribbean, resulting in, for example, a return time of 21 years for the island of Puerto Rico (Walker *et al.* 1991*a*; Fig. 2.21). With minimum wind speeds of 120 km/h and paths tens of kilometres wide, hurricanes can have a profound impact, fundamental to an understanding of the structure of natural ecosystems in the region. Moreover, the Caribbean by no means corners the market in tropical storms. Hurricanes develop in all tropical oceanic areas where sea surface temperatures exceed 27–28 °C, although they are generally absent within 5°north and south of the equator, where persistent high pressure tends to prevent their development (Nunn 1994). Wind defoliation and large

blow-downs are important and frequent disturbances for forested tropical islands throughout the hurricane belts, centred 10–20°north and south of the equator. They may have particularly destructive impacts on high islands, an example being the devastation of large areas of forest on the Samoan island of Upolu by tropical cyclones Ofa, Val, and Lyn in 1990, 1991, and 1993, respectively (Elmqvist *et al.* 1994). Although storm damage to lower islands can also be severe, storms can be important to island growth by throwing up rubble ramparts (Stoddart and Walsh 1992). Thus some very small islands, such as the so-called *motu* (sand islands), may be in a kind of dynamic equilibrium with both extreme and normal climatic phenomena (Nunn 1994). A comprehensive analysis of disturbance regimes requires quantification of both magnitude and frequency of events (Stoddart and Walsh 1992). Lugo (1988) attempted this for a subset of the Caribbean disturbance phenomena given in Table 2.5. He concluded that, whereas hurricanes have the highest frequency of recurrence among his major 'stressors', the susceptibility of Caribbean islands to hurricanes is intermediate, partly because they do not change the base energy signature over extensive periods. Also, with the larger Caribbean islands, hurricane damage is unlikely to impact catastrophically on the entire island area. Given their recurrence interval, these island ecosystems should be

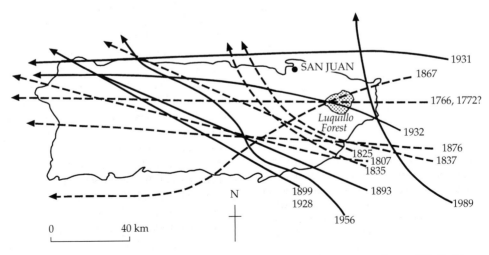

Figure 2.21 Paths of hurricanes that have crossed Puerto Rico since AD 1700. (Redrawn from Scatena and Larsen 1991, Fig. 1.)

evolutionarily conditioned by such storm events (Whittaker 1995; Walker *et al*. 1996).

Not all extreme weather phenomena take the form of storms. ENSO phenomena are large-scale interannual events that are the product of variations in air pressure and associated rainfall patterns between the Indonesian and South Pacific regions, coupled with ocean-current and temperature variation. In the Asia–Pacific region, ENSO events are associated with heightened amplitude of interannual climatic variation—that is, more intense droughts and wet periods. As an illustration, the island of Aututaki (Cook Islands) received 3258 mm rainfall in December 1937, which is in excess of the average annual total (Nunn 1994). That events of the magnitude of ENSO phenomena have a detectable ecological impact, even on populations conditioned evolutionarily to the island way of life, was shown by the significant increase in numbers of four endemic Galápagos bird species (three finches and a mockingbird) as a consequence of the increased food supply resulting from the exceptionally high rainfall during 1982–1983 (Gibbs and Grant 1987). Analyses of historical data for hurricanes show that the frequency of climatic phenomena such as hurricanes and ENSO events varies over decadal time scales, further complicating analysis. This goes to illustrate that, despite the reddened spectrum of climatic variation, short-term fluctuations deserve the attention of ecologists and biogeographers (Stoddart and Walsh 1992; Benton and Spencer 1995; Schmitt and Whittaker 1998).

There is a premise that, because of the comparative simplicity of geology and climate of many oceanic islands, their landscapes should attain a condition of dynamic equilibrium. Nunn (1994) argues that this notion cannot be supported as a generality:

. . . as is evident from the often catastrophic impacts of certain irregular climatic phenomena on oceanic island landscapes, the degree to which particular environments can accommodate the impacts of particular uncommon or extreme events without fundamental alteration, while variable, is generally low. For those oceanic island landscapes in a state of dynamic equilibrium, the effects of such events may be to cause a threshold of landscape development to be crossed . . . . The effects of irregular

climatic phenomena are so variable, so site-specific, that it is pointless to attempt a broad generalization. (pp. 159–160)

These observations might prompt us to ask ourselves, first, with what portion of the magnitude–frequency spectrum of climatic variation might oceanic island biotas best be considered to be in dynamic equilibrium and, second, might not extreme climatic events have caused ecological or evolutionary thresholds to be crossed in the same manner as described above for their landscapes?

Disturbance from volcanism and mega-landslides

In contrast to the geological stability of continental fragments or continental islands, oceanic islands are characterized by the repetitive incidence of geological catastrophes, amongst which volcanic eruptions and gravitational flank collapses or landslides (with the tsunamis they generate affecting nearby islands) are the most frequent.

Volcanic eruptions
Volcanic eruptions are the main constructive forces of oceanic islands: they bring them into existence, and subsequently make them larger and higher, whereas the processes of erosion (by rain, wind, and sea) and subsidence provide the opposing forces that reduce them to sea-level or below (e.g. Figures 2.11, 2.15). However, the same volcanic processes that enlarge islands simultaneously damage or destroy island ecosystems. The impacts can be diffuse, whereby volcanic ash is deposited in thin layers over a wide area, or can be intense and entirely destructive. In some cases, destructive volcanic ash flows can be deposited over the entire island, more or less completely eliminating the existing biota. This is known as **island sterilization**, and implies that colonization has to begin from nothing. It is, of course, very difficult to demonstrate complete sterilization, even for historical events such as the 1883 Krakatau event (Backer 1929; Docters van Leeuwen 1936; Whittaker *et al*. 1989; Thornton 1996). The challenge is even greater for events as distant as Gran Canaria's *c*.3.5 Ma event (Marrero and Francisco-Ortega 2001*a,b*), although in this case it can be inferred from phylogenetic evidence that some

resetting of the biological clock has occurred, with renewed colonization from other islands. Santorini in the Aegean Sea, whose collapse in 1628 BC may have destroyed the Minoan culture on Crete, or, more recently, Martinique and Montserrat in the Caribbean, provide further examples of the destructiveness of island volcanoes.

It has been established that volcanic islands are typically active over an extensive period, often spanning many millions of years, and that there is a variety of distinctive tectonic situations in which volcanic islands are built (above). A corollary of this is that a number of major types of volcanic eruption can be identified. In increasing degrees of explosiveness (Decker and Decker 1991) they are as follows:

- **Icelandic eruptions** are fluid outpourings from lengthy fissures, and they build flat plateaux of lava, such as typify much of Iceland itself.
- **Hawaiian eruptions** are similar, but occur more as summit eruptions than as rift eruptions, thereby building **shield volcanoes**.
- **Strombolian eruptions** take their name from a small island off Sicily that produces small explosions of bursting gas that throw clots of incandescent lava into the air.
- **Vulcanian eruptions**, named after the nearby island of Vulcano, involve the output of dark ash clouds preceding the extrusion of viscous lava flows, thus building a **stratovolcano** or **composite cone**.
- **Peléan eruptions** produce pyroclastic flows termed **nuées ardentes**, high-speed avalanches of hot ash mobilized by expanding gases and travelling at speeds in excess of 100 km/h.
- **Plinian eruptions** are extremely explosive, involving the sustained projection of volcanic ash into a high cloud. They can be so violent, and involve so much movement of magma from beneath the volcano, that the summit area collapses, forming a great circular basin, termed a **caldera**.

Like all generalizations, this classification is oversimplistic; for example, calderas have formed by collapse in Iceland and Hawaii as well as through major explosive volcanoes. However, this classification serves to illustrate that there are great differences in the nature of volcanism, both between islands and within a single island over time.

The eruptive action within Hawaii over the past few centuries has been both more consistent and less ecologically destructive than that of Krakatau, which in 1883 not only sterilized itself but, through caldera collapse, created a tsunami killing an estimated 36 000 inhabitants of the coastlines of Java and Sumatra. This death toll was tragically surpassed by an order of magnitude on 26 December 2004, by a tsunami that swept across coastal settlements in an arc from north Sumatra, in this case generated not by caldera collapse but by submarine faulting. Even fairly small volcanic eruptions can have very important consequences for island ecosystems. For instance, eruptions on Tristan de Cunha in 1962 covered only a few hectares in ejecta, but toxic gases affected one-quarter of the island area. The evacuation of the human population left behind uncontrolled domestic and feral animals, which transformed the effects of the eruption and produced a lasting ecological impact (Stoddart and Walsh 1992).

Mega-landslides

The progressive accumulation of volcanic material upon young oceanic islands tends to build up relatively steep slopes within a short geologic time span, and at a faster rate than 'normal' subaerial erosion processes can bring into dynamic equilibrium (cf. Le Friant et al. 2004). Exceeding critical slope values produces gravitational instabilities that lead to the collapse of the slopes through debris avalanches that transfer hundreds of cubic kilometres into the sea (Carracedo and Tilling 2003, Whelan and Kelletat 2003). Such collapses may occur unexpectedly and suddenly and may lead to the disappearance of a significant part of an island, as much as a quarter of it, in just minutes (Carracedo and Tilling 2003), achieving mass movements with velocities that have been estimated as over 100 km/h (Lipman et al. 1988).

Although they were introduced into modern vulcanology by Telesforo Bravo (1962) almost half a century ago to explain the origin of the Las Cañadas caldera in Tenerife, our knowledge of catastrophic landslides has developed relatively recently, with improvements in the three-dimensional bathymetric analysis of the ocean floor. Such analyses have led to the discovery of the blocks and debris avalanches resulting from the landslides collapses, often extending over areas of hundreds of square kilometres around the islands. Lipman *et al.* (1988) and Moore *et al.* (1989) were amongst the first authors contributing indisputable evidence of the existence of many giant landslides in the Hawaiian archipelago. Later, such giant landslides have been identified in several volcanic islands around the world (Table 2.6). Extensive deposits in the sea around oceanic islands can also be produced by pyroclastic flows (Hart *et al.* 2004), or indeed by a combination of landslips and volcanic ejecta.

Major volcanic slope failures are fairly common globally, occurring at least four times a century and they are a particular feature of volcanic islands in excess of 1000 or 1500 m height (Hürlimann *et al.* 2004). As many as 20 mega-landslides, with volumes varying between 30 and 5000 km^3, have been recognized both for the Canaries (Canals *et al.* 2000) and for Hawaii (Fig. 2.22; Moore *et al.* 1994). These events are reiterative in actively forming islands. For example, El Hierro, the youngest of the Canaries, with an age slightly greater than 1 million years, has already suffered four such catastrophic events, the last of them (El Golfo landslide *c.*15 000 BP), carried more than half the island into the sea (Carracedo *et al.* 1999). Similarly, the La Orotava valley slide on Tenerife is

estimated to have been up to 1000 m in thickness, and created an amphitheatre up to 10 km wide and 14 km long and an offshore prolongation several kilometres in length (Hürlimann *et al.* 2004). The frequency of recurrence of large landslides on the Canaries has been about one per 100 000 years across the whole archipelago, or once every 300 000 years for each single island (Masson *et al.* 2002).

Hürlimann *et al.* (2004) suggest that there are four key factors controlling where mega-landslides occur on volcanic islands:

• the development of deep erosive canyons on the flanks of the volcano
• the presence of high coastal cliffs eroded at the base by wave action
• the presence of widespread residual buried soils from an earlier phase of the development of the volcano, providing a line of weakness
• the orientation of structural axes.

Such landslides can modify the subsidence regime of the affected islands, as well as promoting new volcanic activity, mainly centred in the summit scars (Ancochea *et al.* 1994). When not eroded or buried by the emission of new volcanic material, the scars

Table 2.6 Examples of large debris-avalanche landslides occurring around young volcanic islands (slightly modified from Canals *et al.* 2000 and Whelan and Kelletat 2003)

Island	Landslide name	Vertical distance moved (m)	Sea floor area covered (km²)	Volume (km³)	Event age (ky)
Réunion	Ralé-Poussée	1700	200	30	4.2
Réunion	Eastern Plateau	3000	–	500	15–60
					100–130
Fogo (C. Verde)		1200	–	100	>10
El Hierro	El Golfo	5000	1500	400	15–20
					and 100–130
El Hierro	El Julan	4600	1800	130	>160
El Hierro	Las Playas II	4500	950	<50	145–176
La Palma	Cumbre Nueva	6000	780	200	125–536
Tenerife	Güímar	4000	1600	120	780–840
Tenerife	La Orotava (Icod)		2100	<500	540–690
Tenerife	Las Cañadas	>4000	1700	150	150–170
Kauai	South Kauai	>4000	6800	–	>13
Kauai	North Kauai	>4000	14000	–	–
Oahu	Nuuanu	5000	23000	5000	–
Hawaii	Pololu	>4000	3500	–	370
Hawaii	Alika 1–3	4800	4000	2000	247

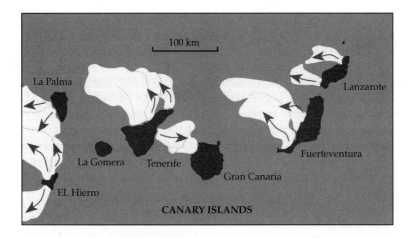

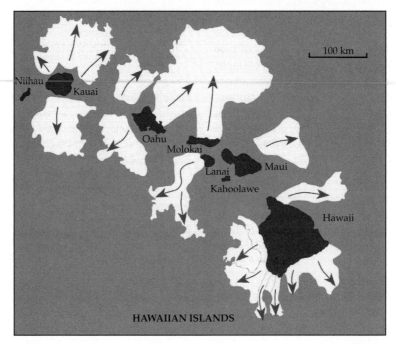

Figure 2.22 Mega-landslides in two oceanic island archipelagoes: (a) Landslides recognized in the Canaries (b) Landslides recognized in Hawaii (Redrawn from Carracedo and Tilling 2003, Fig. 2.18).

created by such collapses are easily recognized. They include calderas (Taburiente in La Palma or Las Cañadas in Tenerife), valleys with knife-cut cliffs—called *laderas* in Spanish—delimiting the collapse-formed valleys (Laderas de Güímar or Ladera de Tigaiga, in Tenerife) or concave cliff-archs such as those in the Anaga (Tenerife), Tamadaba (Gran Canaria), or Jandía (Fuerteventura) massifs.

The sudden collapse of volcanic island slopes and consequent debris avalanches has the potential to generate tsunamis of extreme magnitude (mega-tsunamis), which may not only have highly destructive impacts on the coasts and low areas of nearby islands but also impact far-distant land-masses. For example, it is estimated that the El Golfo collapse (El Hierro) produced a mass movement with a volume of $c.400$ km^3, generating a mega-tsunami that transported boulders weighing more than 1000 t on to elevations of 11 m on the Bahamas Archipelago, located more than 3000 km westwards (Hearty 1997, in Whelan and Kelletat 2003). Ward and Day (2001) have highlighted the great instability of the Cumbre Vieja massif, the active southern summit of La Palma, where seven volcanic eruptions have been concentrated in the last five centuries. It has been predicted that Cumbre Vieja will collapse to the west in the coming millennia, generating a mega-tsunami likely to devastate the east coast of North America on the far side of the Atlantic.

2.6 Summary

Two broad classes of island are identified, true islands surrounded by water, and habitat islands surrounded by a contrasting matrix. True islands can be subdivided into: island continent, oceanic islands, continental fragments, continental shelf islands, and islands within lakes. This chapter is concerned almost exclusively with islands located in the oceans of the world. These islands are rarely ancient in geological terms, and in many cases are significantly less ancient biologically than geologically. Oceanic islands have volcanic foundations, are concentrated in a number of distinctive inter- and intraplate settings and have commonly experienced a dynamic history involving varying degrees of lateral and vertical displacement, the latter confounded by, but often in excess of, eustatic sea-level changes. In the tropics and subtropics, upward growth of reef-forming corals at times of relative or actual subsidence has led to the formation of numerous islands of only a few metres elevation, contrasting with the generally steep topography of the high volcanic islands, to

a large degree according with Darwin's 1842 theories on coral reef formation.

Some of the more obviously important environmental features of islands are directly related to these geological factors and include characteristic topographic, climatic, and hydrological phenomena. Subsurface water storage is a particularly important feature of both volcanic and low, sedimentary islands: the possession of a subsurface freshwater lens is a characteristic of all but the smallest islets. Island environments might be thought to have been shielded from the full amplitude of continental climatic fluctuations, but the palaeoecological record indicates significant climatic changes within the late Pleistocene and Holocene on a range of islands, including examples from the tropics and subtropics. With limited scope for range adjustments within the island setting, the biogeographical significance of such environmental changes should not be ignored.

Shorter term environmental variation, perturbation, or disturbances (broadly overlapping categories) characterize many island environments, not least for those islands which lie in the tropical cyclone belt and for those which remain volcanically active. Recent bathymetric, geomorphological, and volcanological advances have revealed just how dynamic oceanic island environments are. Island surfaces are constantly in flux, not only built by volcanism, and reduced by subaerial and coastal erosion, but also subject to transforming catastrophic slope failures. Some appreciation of the distinctive nature of the origins, environmental characteristics, and history of islands is almost self-evidently important to an understanding of the biogeography of islands, although, in this scene-setting chapter, the biogeographical content has intentionally been kept within limits.

CHAPTER 3

The biogeography of island life: biodiversity hotspots in context

... the scarcity of kinds—the richness in endemic forms in particular classes or sections of classes,—the absence of whole groups, as of batrachians, and of terrestrial mammals notwithstanding the presence of aerial bats,—the singular proportions of certain orders of plants,—herbaceous forms having developed into trees, &c.,—seem to me to accord better with the view of occasional means of transport having been largely efficient in the long course of time, than with the view of all our oceanic islands having been formerly connected by contiguous land with the nearest continent ...

(Darwin 1859, p. 384)

3.1 Introduction: the global significance of island biodiversity

The term **biodiversity** is a contraction of biological diversity, and broadly defined refers to the variability of life from all sources, including within species, between species, and of ecosystems (Matthews *et al.* 2001). We may at times be concerned with higher or lower points in the taxonomic hierarchy, such that familial, generic, subspecific, or even gene frequency data may be analysed under the header 'biodiversity', but the most commonly used diversity unit is undoubtedly the species. As a first-order generalization, islands are species poor for their size but rich in forms found nowhere else, i.e. **endemic** to that island or archipelago. They are thus 'hotspots' of biodiversity (see Box 3.1). Given that the total number of living species on the planet is not yet known to within an order of magnitude (Groombridge 1992), it is difficult to provide a firm measure of the relative contribution of islands to

global biodiversity. Yet, for particular taxa there are sufficient data to demonstrate that, *taken collectively*, islands contribute disproportionately for their area to global species totals.

In order to appreciate the special significance of island biotas it is important to consider some of their general properties, and in what ways they are peculiar. This chapter sets out to do that, distinguishing the notion of their compositional distinctiveness or 'disharmony' from simple species poverty. We also explore aspects of the historical biogeographical context and the importance of dispersal limitations in determining the balance of island assemblages.

The affinities of island biotas to continental source areas have fascinated biogeographers for more than a century, providing some key insights for evolutionary theory (e.g. Darwin 1859; Wallace 1902). First steps are to establish how island biotas are related to continental biotas, where they ultimately derive from, and how far the unique (i.e. endemic) forms found on islands have developed from the colonizing forms. Once this context has been developed, we can move on to look at the mechanisms and processes involved in the evolution of island biotas, which we do in later chapters. In addition, as discussed later in this volume, islands are repositories for many of the world's threatened species and are worthy of special attention in conservation prioritization decisions. Scientific guidance as to where particular islands fit in global and regional planning frameworks is to be derived from the application of biogeographical techniques and theories, under the recently recognized (but fairly long-developing)

Box 3.1 Islands as hotspots and their place in Conservation International's hotspots scheme

There are at least three separate hotspot concepts employed within this book (and there are others around in the literature). First, is a geological usage, whereby hotspots are localized but long-lived areas of volcanism overlying plumes in the mantle (Chapter 2). Second, in the introduction to this chapter we have applied the term **biodiversity hotspot** to signify the concentration of endemic species on isolated islands. This usage is exemplified in BirdLife International's Endemic Bird Area (EBA) scheme, which delimits areas supporting two or more species of birds with breeding ranges of <50 000 km². Of the 218 EBAs designated in the 1996 version of the scheme, 113 are island EBAs (21 continental islands, 21 large oceanic islands, 55 small oceanic islands) (Long et al. 1996). It is beyond doubt that the world's islands, and especially the tropical islands, hold a disproportionately high degree of global diversity.

The third usage stems from an approach to conservation prioritization put forward by the British environmentalist Norman Myers in the late 1980s and recently promoted by the international environmental NGO Conservation International (Myers et al. 2000). Myers' approach was to combine a measure of the concentration of biodiversity, with a simple index of threat, i.e. it is really a hotspot–threatspot scheme. The version of the CI hotspot scheme published in 2000 identified as 'hotspots' 25 areas of the world that met two criteria: (1) the area should possess at least 0.5% (1500) of the world's plant species as endemics, and (2) the area should have lost 70% or more of its primary vegetation. Islands featured prominently among the 25 CI 2000 hotspots, which is appropriate given the high proportion of global species losses over recent centuries that have been island species. The CI hotspot approach attracted unprecedented levels of funding following the 2000 paper, and has become extremely influential in strategic conservation planning.

Partly as a result of the phenomenal fund-raising success of the organization (aided by the publication of the 2000 scheme in *Nature* magazine), partly in response to new data on diversity and habitat loss, and partly in an effort to improve the fit of their scheme with another major conservation planning effort (the WWF ecoregions approach), CI revisited the boundaries of their hotspots in a review carried out in 2004, which resulted in a new map published on their web site (CI 2005). We term this version the CI 2005 hotspots. It depicts 34 hotspots, and increases the coverage of the world's islands, particularly in the Pacific.

Although the stated reason for the addition of the East Melanesian islands as one of the new hotspots is recent habitat loss, to a large degree the increased attention to islands represents merely a widening of their cast to include islands of high biodiversity value (e.g. the Galápagos and Juan Fernández islands) that, as they put it, might otherwise have slipped through the net of conservation priorities. The approach is acknowledged on their web site (CI 2005) to be pragmatic, 'with full recognition that the floristic affiliations of these islands with their associated land masses are often tenuous at best.'

Within the 34 CI 2005 hotspots, 9 areas (see map) are composed exclusively of islands. Organized by geographical region these are: (1) the Caribbean islands (American 'region'); (2) Madagascar and adjacent islands (the Comoros, Mascarenes, and Seychelles) (African region); (3) East Melanesia, (4) Japan, (5) New Caledonia, (6) New Zealand, (7) Philippines, (8) Polynesia-Micronesia, and (9) Wallacea (all Asia-Pacific region). Three more have a substantial proportion of their diversity within islands: (10) the Mediterranean basin (including the Atlantic islands of Macaronesia) (Europe and Central Asia region); (11) the Western Ghats and Sri Lanka, and (12) Sundaland (both Asia-Pacific region). In this expanded version of the scheme, almost all tropical and subtropical islands of high biodiversity value have been included, although it is evident that were many of the individual archipelagos considered as separate biogeographical entities they would not have reached the qualifying diversity-plus-threat criteria. Examples of newly included islands in the 2005 scheme include the Galápagos and Malpelo islands, lumped into a Tumbes–Chocó–Magdalena hotspot; the Juan

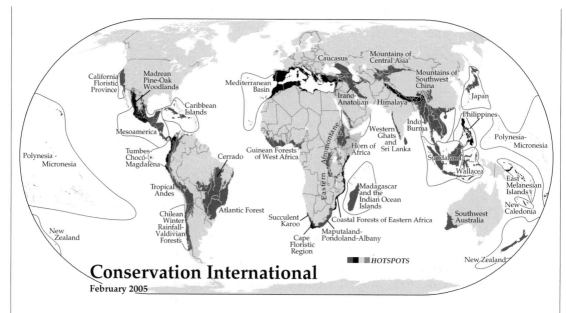

Map of the CI-2005 hotspots.

Fernández and Las Desventuradas islands, grouped with the Chilean Winter Rainfall–Valdivian Forests; Revillagigedo, Cocos and Las Tres Marías islands, included in the Mesoamerica hotspot; and the Macaronesian islands, which form an intriguingly shaped intrusion of the Mediterranean region into the Atlantic ocean. Biogeographically, this final example is not as peculiar as it might first seem (e.g. Blondel and Aronson 1999), although the islands—especially the Cape Verde islands—also have African affinities (see text).

CI claim that the CI 2005 hotspots scheme covers just 2.3% of the Earth's surface, but incorporates as endemics over 50% of the world's plant species, and 42% of terrestrial vertebrates. The fashion in which the hotspots have been delimited in both the 2000 and 2005 versions of the scheme may be rather arbitrary, but long-recognized biogeographical patterns are nonetheless detectable in their map products, which now serve to guide their conservation investment strategy. The world's islands not only represent disproportionate amounts of diversity, but they also account for a high proportion of recorded global extinctions over the last few hundred years, and a high proportion of globally threatened species. A stress on islands is thus an appropriate part of any global conservation assessment based on the currency of species and indeed is common to schemes promoted by all the major international conservation NGOs at the present time (e.g. the WWF, Conservation International, BirdLife International).

subfield of **conservation biogeography** (Whittaker *et al.* 2005). We illustrate some of the issues for conservation biogeographers involved in assigning island systems within faunal and floral regions by considering the affinities of the Macaronesian islands. Of course, in practice, political and economic frameworks and affinities may be as or more important in developing and realizing strategic conservation planning, but here we lay these pragmatic concerns to one side.

Not only do islands have many threatened species today, but they have also contributed disproportionately to species extinction in historic and prehistoric times, such that in places much of the underlying biogeographical structure may already have been lost or distorted. This is problematic for biogeographers, in that we may misinterpret fragmentary data and reach erroneous conclusions. Before we embark on a consideration of speciation patterns and processes, we provide at the end of this chapter some brief illustrations of island losses as a cautionary note.

3.2 Species poverty

Islands typically have fewer species per unit area than the mainland, and this distinction is more marked the smaller the area of the island, i.e. interarchipelago species–area curves are steeper than curves constructed by subdividing a large mainland area (Rosenzweig 1995). This will be examined more fully in Chapter 4, and the following examples should suffice to introduce the theme here. First, Figure 3.1 illustrates that no matter whether the Californian mainland plant data are built up in spatially nested sets or non-nested data are considered, the points lie above the regression line for the Californian island data. A second example (after Williamson 1981) is provided by Jarak island in the Straits of Malacca (96 km from Sumatra, 64 km from Malaya, and 51 km from the nearest island). This 40 ha island is forested but lacks dipterocarps, the family that provides the most common dominant species in the forests of Malaya, but whose members have poor long-distance dispersal ability. Within a 0.4 ha plot on the island,

Wyatt-Smith (1953) recorded 34 species of trees having a trunk diameter greater than 10 cm, compared with figures of 94 and 102 species, respectively, for two Malayan mainland plots. These mainland figures actually exceeded the total spermatophyte (not just tree!) flora noted by Wyatt-Smith in his survey of Jarak, and although it would be a surprise if this was found to be a complete inventory of all the species on the island (a notoriously difficult feat to achieve), it nonetheless gives an indication of the extent of impoverishment of a not particularly remote island. A third example is St Helena (15°56′S, 5°42′W), a remote island of about 122 km², and 14.5 million years old, which has a known indigenous flora of just 60 species, of which 50 are endemic, with 7 of them extinct, and others endangered (Davis *et al.* 1994).

The successful spread of species transported to islands such as St Helena in recent centuries would appear to indicate that such islands are undersaturated with species and could support more in total (D'Antonio and Dudley 1995; Sax *et al.* 2002). However, some turnover is often involved, with endemic plant and especially animal species becoming extinct as part of this process. As discussed later in this volume, the assessment of such changes in diversity is problematic, as, for instance, the losses and gains are also tied up with other human influences, such as forest clearance and horticultural and agricultural

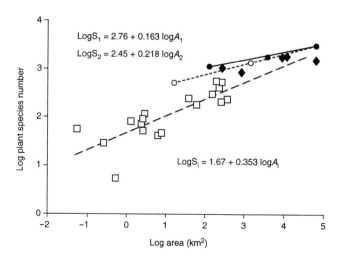

$\log S_1 = 2.76 + 0.163 \log A_1$

$\log S_2 = 2.45 + 0.218 \log A_2$

$\log S_i = 1.67 + 0.353 \log A_i$

Figure 3.1 Species–area curves for Californian plants, showing the steeper slope for islands (log S_i open squares), compared with two alternative nested sets from within the mainland (log S_1 dots, San Francisco area; log S_2, open circles, Marin County). The diamonds represent other areas from the mainland, which lie in neither nested set. (Redrawn from Rosenzweig 1995, Fig. 2.9.)

activities. Moreover, it may well be the case for St Helena, as for some other island systems (e.g. Easter Island—Flenley 1993), that some native and endemic species became extinct before botanical investigations were undertaken (Cronk 1989), so that the known native biota is an incomplete record of what once was present.

3.3 Disharmony, filters, and regional biogeography

This section deals with the regional biogeographical setting of islands, beginning with some fairly simple patterns, and progressing through increasingly complex scenarios.

Filtering effects, dispersal limits, and disharmony

Islands tend to have a different balance of species compared to equivalent patches of mainland: they are thus said to be **disharmonic**. There are two aspects to this disharmony (Williamson 1981). First, as Hooker noted, both the climate *and the biota* of islands tend to be more polar than those of nearby continents (Williamson 1981). Thus, the Canaries, located west of the Sahara, have a generally Mediterranean flora, Kerguelen in the Indian Ocean is bleak and Antarctic-like for its latitude, and the Galápagos, although equatorial, are closer to desert (i.e. subtropical) islands, being influenced by upwellings of cool subsurface waters. Hence, there is a climatic 'filter' involved in the assemblage of species on islands. Secondly, islands are disharmonic in that effectively they sample only from the dispersive portion of the mainland pool: i.e. there is an isolation or dispersal filter. This effect has of course, to be distinguished from simple impoverishment, i.e. it should not be merely a random subset of a potential mainland pool that is missing.

It is hard to quantify the maximum dispersal capability of particular taxa across oceans, although we attempt just that in Table 3.1. At the core of the problem are the following complications:

- Present-day distances may not reflect distances at the time of colonization (Chapter 2, Myers 1991; Galley and Linder 2006).

- There is much variation in dispersal powers between related species.
- Dispersal powers change through time following island colonization.

Readers should be aware that some historical biogeographers regard extreme long-distance dispersal with enormous scepticism, instead arguing for 'vicariant' events, i.e. the past formation of barriers (by a variety of geotectonic and eustatic processes) separating islands from previously connected landmasses with which they share biogeographical affinities (e.g. contrast the analyses of the biogeography of the Falklands offered by Morrone 2005 and McDowall 2005). However, the evidence for the occurrence of long-distance dispersal is clearly established. What remains at issue is not whether it occurs at all, but exactly how far can it reach for particular taxa, and what is the relative importance of vicariance and dispersal for particular biogeographically challenging island groups like New Zealand (see Winkworth *et al.* 2002)?

Table 3.1 provides an analysis at a coarse taxonomic level, and it is important to realize that within taxa such as birds, or higher plants, there is considerable structure to the dispersal reach of subtaxa (Fig 3.2). Thus, for instance, Krakatau, sterilized in 1883 and isolated by only *c.*40 km of ocean, is already home to one quarter of the ferns of West Java, but lacks altogether certain tree families and genera that are common on the mainland, but have no obvious ocean-going dispersal mechanisms (Whittaker *et al.* 1997). The partial sampling of mainland pools is also exemplified by the flora of Hawaii, which, relative to other tropical floras, has few orchids, only a single genus of palms (*Pritchardia*), and is altogether lacking in gymnosperms and primitive flowering plant families. The largest eight families of flowering plants on Hawaii—Campanulaceae, Asteraceae, Rutaceae, Rubiaceae, Lamiaceae, Gesneriaceae, Poaceae, and Cyperaceae—together constitute over half of the native species (Davis *et al.* 1995).

Many other examples could be given. For instance, the Azores and the islands of Tristan da Cuhna possess no land mammals or amphibians. In this case, the disharmony is clearly related to dispersal ability. Tristan da Cuhna also lacks birds of

Table 3.1 Apparent limits to long-distance dispersal of selected taxonomic groups based on water gap distances. Considerable caution should be observed when inferring dispersal capability from water gap distances for the following reasons: (a) For continental fragments there is the possibility of a vicariance explanation due to geotectonic processes rather than a dispersal event; (b) eustatic sea level changes may in the past have diminished the distances between islands and the mainland, in the case of many continental shelf islands removing the barrier altogether. (Data compiled from various sources, including: Carlquist 1965, 1974; Gorman 1979; Menard 1986.)

Taxonomic group	Water gap (km)	Example
Freshwater fishes*	5	British Columbia–Vancouver Island
Large mammals	70–150	Pliocene Mediterranean megafauna
Small mammals (excepting rodents)	410	Africa–Madagascar
Rodents, land tortoises, and snakes	920	South America–Galápagos
Amphibians	1265	Africa–Seychelles
	(1641)	(Australia–New Zealand (Palaeoendemisms?))
Gymnosperms	1356	Europe–Azores
Lizards (geckos)	1641	Australia–New Zealand (Palaeoendemisms?)
Bats, land birds, snails, arthropods, fungi, mosses, ferns, angiosperms	3650	North America–Hawaii

*Excluding diadromous fishes, for which almost anything appears possible (R. McDowall, personal communication).

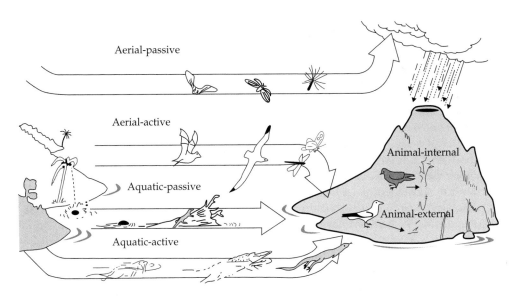

Figure 3.2 The various modes of dispersal by which species may naturally disperse to and colonize a remote island through or across a sea-water gap (redrawn from Aguilera *et al.* 1994, based on an original in Rodríguez de la Fuente 1980). Alternative nomenclature for the top three categories as they apply to plants: seed dispersal by wind = anemochory; in the gut of an animal = endozoochory; attached to the fur or feathers = exozoochory; flotation in the sea = thalassochory.

prey: with very few species of land birds, no land mammals, and a small land area, it is doubtful whether a predator population could maintain itself, were one to reach the island (Williamson 1981). This interpretation suggests that impoverishment among one group may lead through food-chain links to disharmony in another.

The notion of disharmony has been criticized by Berry (e.g. 1992) who has written:

It is a somewhat imprecise concept indicating that the representation of different taxonomic groups on islands tends to be different to that on the nearest continent, and carrying the implication that there is a proper—or harmonious—composition of any biological community. This is an idea closer to the 'Principle of Plenitude' of mediaeval theology than to modern ecology. (Berry 1992, pp. 4–5)

Berry has a point: that island environments have 'sampled' distinctive subsets of the mainland pool does not necessarily imply that there is something wrong with island assemblages and that they are further from balance with their physical environment than those of continents (this is a question worth posing, but it can be distinguished as a separate question). Nonetheless, consideration of remote oceanic islands shows that they are distinctive, and there seems little point in developing a longer phrase to capture the same idea as the term disharmony.

Another classic example of the filtering out of particular groups of animals with increased isolation is provided by Fig. 3.3, which shows the distributional limits of different families and subfamilies of birds with distance into the Pacific from New Guinea. Such patterns have sometimes been termed **sweepstake dispersal**. The data in Fig. 3.3 seem to be interpretable broadly in terms of the differing dispersal powers of the groups of birds concerned, illustrating that insular disharmony can be identified not only at the level of orders but also below this at the family level (cf. Whittaker *et al.* 1997). However, as pointed out by Williamson (1981), within Fig. 3.3 there is a general thinning out of islands and decline in size to the east. This raises the question of whether dispersal difficulties alone have caused the filter effects—a problem that, using presently available data, appears insoluble (but see Keast and Miller 1996).

The New Guinea–Pacific study provides a basically unidirectional pattern, but in other cases a two-way filter appears to operate (Carlquist 1974), best exemplified for certain linearly configured archipelagos. A classic example is shown in Fig. 3.4, which quantifies the decline in reptiles and birds of oriental affinity and the increase in Australian species going from west to east along the Lesser Sunda Islands. A similar sort of two-way pattern can also be detected in certain circumstances on mainlands, particularly on peninsulas (e.g. Florida; Brown and Opler 1990), but island archipelagos generally provide the clearest filter effects.

Biogeographical regionalism and the vicariance/dispersalism debate

These two-way filter effects evidence a merging of faunas or floras between two ancestral regions. They indicate the biogeographical imprint of extremely distant events in the Earth's history, intimately connected with the plate-tectonic processes that have seen the break-up of supercontinents, parts of which have come back into contact with one another many millions of years later. Early students of biogeography, notably Sclater and Wallace, recognized the discontinuities and, on the basis of the distribution patterns of particular taxa, were able to divide the world into a number of major world zoogeographic or phytogeographic regions. One such scheme is shown in Fig. 3.5.

The boundary between the Oriental and Australian zoogeographic regions is marked by what has long been known as **Wallace's line** (Fig. 3.6), which marks a discontinuity in the distribution of mammals, and divides the Sunda islands between Bali and Lombok—not an altogether obvious place to draw the line from the point of view of the present-day configuration of the islands. Modifications, and additions in the form of **Weber's line**, distinguishing a barrier for Australian mammals like Wallace's line for Oriental mammals, have also been proposed (Fig. 3.6). However, the precise placing of lines is problematic, in part because the area between the Asian and Australian continental shelves ('Wallacea') actually contains

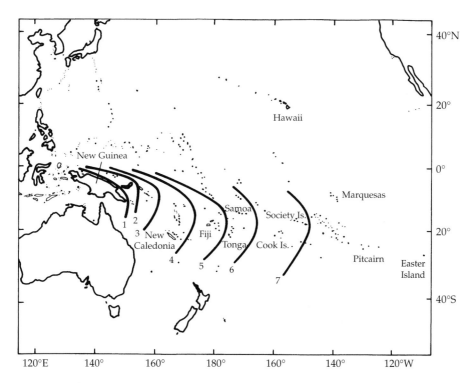

Figure 3.3 Eastern limits of families and subfamilies of land and freshwater breeding birds found in New Guinea. The decline in taxa is fairly smooth, and shows both differences in dispersal ability and that there is a general decline in island size to the east. Not beyond: 1, New Guinea (14 taxa), pelicans, snakebirds, storks, larks, pipits, logrunners, shrikes, orioles, mudnesters, butcherbirds, birds of paradise, bowerbirds, Australian nuthatches, Australian tree-creepers; 2, New Britain and Bismarck islands (2 taxa), cassowaries, quails, and pheasants; 3, Solomon Islands (10 taxa), owls, frogmouths, crested swifts, bee-eaters, rollers, hornbills, pittas, drongos, sunbirds, flower-peckers; 4, Vanuatu and New Caledonia (7 taxa), grebes, cormorants, ospreys, button-quails, nightjars, wren warblers, crows; 5, Fiji and Niuafo'ou (4 taxa), hawks, falcons, brush turkeys, wood swallows; 6, Tonga and Samoa (7 taxa), ducks, cuckoo-shrikes, thrushes, whistlers, honeyeaters, white-eyes, and waxbills; 7, Cook and Society islands (3 taxa), barn owls, swallows, starlings; 8 (beyond 7), Marquesas and Pitcairn group, herons, rails, pigeons, parrots, cuckoos, swifts, kingfishers, warblers, and flycatchers. Others: owlet-nightjars, one species from New Caledonia, otherwise limit 1; ibises, one species from the Solomon Islands, otherwise limit 1; kagu, endemic family of one species from New Caledonia. (From Williamson 1981, Fig. 2.5. We have not attempted to update this figure in the light of more recent fossil finds.)

relatively few Asian or Australian mammals (Cox and Moore 1993), and in part because other groups of more dispersive animals, such as birds, butterflies, and reptiles, show the filter effect (see Fig. 3.4) rather than an abrupt line. Another problem for the drawers of lines is that the insects of New Guinea are mainly of Asian origin. There are several such enigmas of distribution in relation to these lines: for a fuller discussion see Whitmore (1987) and Keast and Miller (1996). Perhaps the best option is to restrict the Oriental and Australia regions to the limits of the respective continental shelves, and to treat the intervening area of Wallacea as essentially outside the framework of continental biogeography (Cox 2001).

The placing of distant oceanic islands into this sort of traditional biogeographical framework has drawn criticism from Carlquist (1974, p. 61) who has argued that because of the isolation of islands, and their differing histories from continents, their biotas provide very poor fits with regional biogeographical divisions. Commonly, oceanic islands

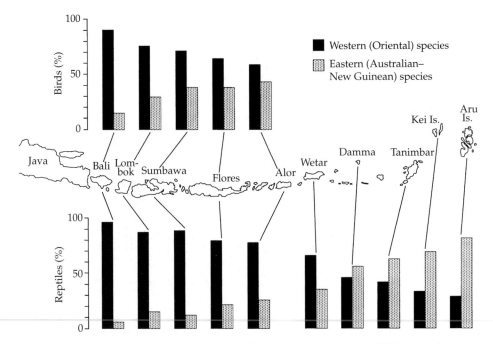

Figure 3.4 The two-way filter effect, showing trends in the proportion of Oriental and Australian species of birds and reptiles across the Sunda islands, South-East Asia. In contrast to the gradual changes shown here, mammals 'obey' Wallace's line, a sharp line of demarcation between Bali and Lombok (see Fig. 3.5). (Modified with permission, from Carlquist, S. (1965). *Island life: a natural history of the islands of the world*. Natural History Press, New York, Fig. 3.7.)

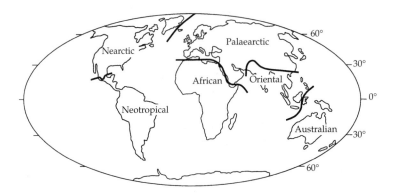

Figure 3.5 Based on the distribution of extant mammals, A. R. Wallace in 1876 divided the world into zoogeographical regions. (The map shown here has been redrawn from Cox and Moore 1993, Fig. 8.1.)

will have gained significant proportions of their species from more than one land mass and more than one biogeographical province. For instance, Fosberg (1948) calculated that the flowering plants of Hawaii could have descended from some 272 original colonists, of which 40% were believed to be of Indo-Pacific origin; 16–17% were from the south (Austral affinity); 18% were from the American con-tinent; less than 3% were from the north; 12–13% were cosmopolitan or pantropical; and the remain-der could not be assigned with confidence. Can the Hawaiian islands logically be put into an Oriental province when appreciable portions of their flora and fauna are American in origin? Most phytogeo-graphers, in fact, assign the flora to its own floristic region (Davis *et al.* 1995).

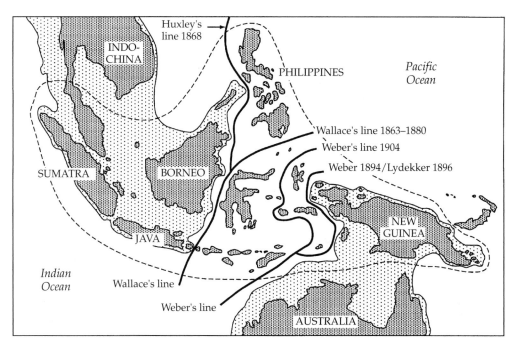

Figure 3.6 Map of South-East Asia and Australasia showing some of the boundaries proposed by A. R. Wallace and others, dividing the two faunal regions. The continental shelves are shown lightly shaded. In contrast to the sharp discontinuity first noted by Wallace in the distribution of mammals between Bali and Lombok—marked by the original version of Wallace's line—birds and reptiles show a more gradual change between the two regions (see Fig. 3.3). (From various sources.)

Renvoize's (1979) analysis of the dominant regional phytogeographical influences within the Indian Ocean (Fig. 3.7) tells a similar story. His paper illustrates that many island biotas are not the product of simple one- or two-way filter effects, nor do they enjoy full affinity with a particular continental region, but instead they consist of a mix of species of varying sources and origins. For instance, Christmas Island, close to Indonesia, is dominated by colonists from South-East Asia, whereas the Farquhar Group has an African component, a reflection of the location of these low-lying islands north-east of the northern tip of Madagascar.

Renvoize's (1979) analysis of Christmas Island was based on a list of 145 indigenous vascular plants from the end of the nineteenth century. The breakdown of this flora was: 31 pantropical, 21 palaeotropical, 76 tropical Asian (Madagascar to Polynesia and Australia), and 17 endemic species. This contrasts with the indigenous flora of the Farquhar Group, which consisted of 23 pantropical, 18 Indo-Pacific, and 5 African species—which seems

to be a fairly typical sort of mix for the low islands of the Indian Ocean. Leaving aside introduced species, the pantropical species constitute a far greater proportion of the flora of the Farquhar Group than of Christmas Island (respectively 50% and 21%). Although several factors may be involved in this comparison, the most obvious differences are location and in the range of habitats available on the high as opposed to low-lying islands (Renvoize 1979).

Christmas Island is an elevated coral-capped island, having a plateau at 150–250 m ASL, a peak of 361m, and a land area of 135 km^2. On low islands few areas are beyond the direct influence of the sea. Strandline species thus dominate the floras and, since they are nearly all sea-dispersed and since the most constant agent of plant dispersal is the sea, the coastal habitats receive a steady flow of possible colonizers and a steady gene flow, not conducive to the evolution of new forms. High islands do have coastal habitats, but these habitats and their characteristically widespread sea-dispersed species do not dominate the floras as they do on the low islands.

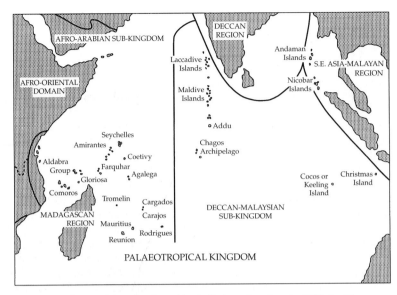

Figure 3.7 Phytogeographical classification of the Indian Ocean islands. (Redrawn from Renvoize 1979, Fig. 1.)

Those oceanic islands with large proportions of endemic plants are typically high islands with varied habitats (Box 3.2; Humphries 1979). Recent work, based on molecular phylogenies, supports the more traditional biogeographical analyses undertaken by Renvoize (1979) in indicating multiple influences on particular islands, and a variety of stepping-stone routes back and forth across the Indian Ocean (e.g. Galley and Linder 2006; Rocha et al. 2006).

Renvoize (1979) declined to attribute the low islands of the Indian Ocean below the level of Palaeotropical Kingdom because they contain too high a proportion of very widespread species. This restrained approach to demarcation is realistic, and although the categorization of islands into particular biogeographical provinces or regions is clearly problematic (Cox 2001), such phytogeographical and zoogeographical analyses have some value. They establish the broad biogeographical context within which the compositional structure of a particular island or archipelago needs to be placed (cf. van Balgooy et al. 1996).

Distance from a larger source pool can be a poor indicator of biogeographical affinity. First, there may be historical factors, as in the cases of Madagascar and the granitic islands of the Seychelles, which are ancient continental fragments from Gondwanaland and so began their existence as islands millions of years ago, complete, we assume, with a full complement of species. In the case of Madagascar the split from Africa has been estimated to have occurred as long ago as 165 million years (Davis et al. 1994), but the effective isolation date may be considerably more recent (see Box 2.2). Secondly, ocean or wind currents (or even migration routes) may have determined colonization biases (Cook and Crisp 2005), such as apparently in the case of the butterflies of the Antillean island of Mona (Chapter 2). Another classic example is that of the flora of Gough Island (Moore 1979). The predominant wind circulation system around the South Pole provides a plausible explanation for the affinities with New Zealand and South America (Fig. 3.8). Changes in relative sea level, landmass positions, and oceanic and atmospheric circulation over geological time (Chapter 2) can mean that such analyses are difficult to bring to a definite conclusion. We illustrate the problems involved in more detail below, with reference to Macaronesia.

As already mentioned, within historical biogeography two opposing camps have formed up over recent decades, the **dispersalist** and **vicariance biogeographers**, each concerned with how disjunct

Box 3.2 Islands in the ocean: general features

In Box 2.1 we outlined the differing origins of three classes of island. Here we summarize aspects of their geology, geography, and biology.

- **Oceanic islands** are built over the oceanic plate, are of volcanic or coralline formation, are remote and have never been connected to mainland areas, from which they are separated by deep sea. They generally lack indigenous land mammals and amphibians but typically have a fair number of birds and insects and usually some reptiles.
- **Continental islands** are more varied geologically, containing both ancient and recent stratified rocks. They are rarely remote from a continent and always contain some land mammals and amphibians as well as representatives of the other classes and orders in considerable variety (Wallace 1902), typically having been connected

to other land masses before the Holocene transgression.

- **Continental fragment islands** are in some respects intermediate between the other two classes, but being typically ancient and long-isolated they are in other respects biologically peculiar.

The biogeographical distinction between islands of differing geological origins of course depends greatly on the ability of potential inhabitants of an island to disperse to it, whether over land or sea. As a simplification, we may conceptualize an oceanic island as one for which evolution is faster than immigration, and a continental island as one where immigration is faster than evolution (Williamson 1981, p. 167). Thus, a particular island may be thought of as essentially continental for some highly dispersive groups of organisms (e.g. ferns (Bramwell 1979) and some types of birds),

General features of islands of different origin of relevance to their biogeography (modified from Fernández-Palacios 2004)

Features	Continental islands	Continental fragments	Oceanic islands
Origin	Interglacial sea-level rise	New mid-ocean rift creation	Between or within plate submarine volcanic activity
End	Glacial sea-level drop	Collision with a continent	Erosive break up and subsidence
Degree of isolation	Small	Variable; small for Cuba, large for New Zealand	Large (except Canaries)
Size	From very small to very large	Some small, mostly large	Small (except Iceland)
Longevity	Short (20–30 Ka) as separate entities	Long (50–150 Ma)	Variable (hours–20 Ma)
Water gap depth	Small ($\leqslant$130 m)	Large ($>$1000 m)	Large ($>$1000 m)
Parent rocks	Granites	Granites	Basalts
Erodibility	Low	Low	High
Fragmentation	Low	Low (except Balearics and Seychelles)	High (except lonely islands such as Ascension or St. Helena)
Original biota	Present	Present	Absent
Relictualism	Lacking	High	Moderate
Speciation process	Limited	Vicariance	Founder event, etc.
Endemism	Variable according to latitude (low in high latitudes, often high in low latitudes)	Very high	High

yet essentially oceanic in character for organisms with poor powers of dispersal through or across seas (e.g. conifers, terrestrial mammals, and freshwater fish). Moreover, there are some groups of islands of mixed origins and types, e.g. the Seychelles, or which appear to have been partially connected at times of lowered sea levels, such that they have biological characteristics part way between oceanic and continental, land-bridge islands (as discussed for the avifauna of the Philippines by Diamond and Gilpin, 1983).

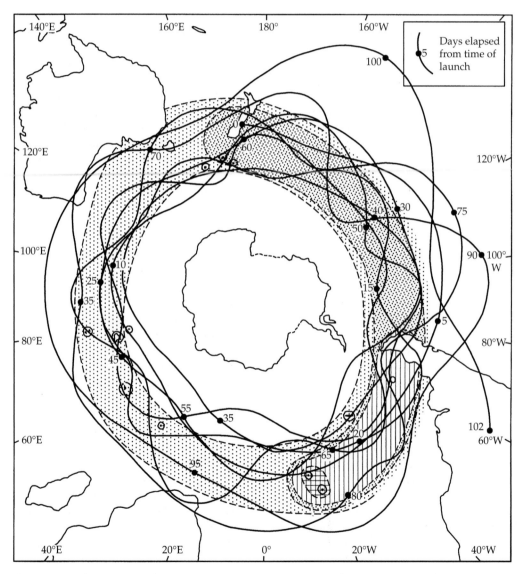

Figure 3.8 Floristic affinities of Gough Island (Tristan da Cunha) in relation to predominant wind direction as revealed by the flight of a balloon: areas sharing 20–25 species, cross hatch; 15–20 species, single hatching; 10–15 species, heavy stipple; 5–10 species, light stipple. (After Moore 1979, Figs 1 and 2.)

distributions arise (see Box 2.2). In essence, dispersal hypotheses have it that descendent forms are the product of chance, long-distance dispersal across a pre-existing barrier, whereas vicariance biogeographers envisage species ranges being split up by physical barriers, often followed by speciation in the now separate populations (Myers and Giller 1988). Some vicariance biogeographers seem remarkably dismissive of long-distance dispersal (e.g. Heads 2004), given the evident importance of the process in island colonization (Wagner and Funk 1995, McGlone *et al.* 2001; Winkworth *et al.* 2002; Cook and Crisp 2005, McGlone 2005; Rocha *et al.* 2006) and indeed in post-separation exchanges between continental land-masses (e.g. Galley and Linder 2006). Although there are contexts in which the relative contribution of both processes can be exceedingly difficult to discern, providing plenty of room for genuine debate, the polarization of debate (and especially the dismissal of long-distance dispersal) is unnecessary and has become a real hindrance to progress. Available data concerning island origins and environmental histories illustrate that both types of process operate, and both contribute to the complexities of biogeographical patterning (Stace 1989; Keast and Miller 1996; Winkworth *et al.* 2002; Galley and Linder 2006).

This point can be exemplified with reference to the geological history of the Caribbean, introduced in Chapter 2. Buskirk (1985) set out to employ this framework in interpreting the phylogenetic relationships of modern beetles, reptiles, amphibians, and other terrestrial animals in the Greater Antilles. The evidence indicates that the Antilles have been islands throughout most of the Cenozoic and that therefore most of their faunas were derived from over-water colonists and not from the fragmentation of original populations on a large land mass. However, Buskirk suggests that tectonic events have affected dispersal and colonization distances. The shear along the northern Caribbean plate boundary has moved Jamaica further away from Central America, and thus modern Jamaican species that are relicts or radiations from relatively early colonists have more Central American affinities than do the endemic species of neighbouring Antillean islands. Jamaica was largely submerged in the early to mid-Tertiary, and its modern endemic fauna lacks the groups that invaded the other Antilles at about that time. As well as the change in sea level, the accompanying climatic changes have been important, and in Jamaica have apparently selected against members of some animal groups that required more mesic habitats (Buskirk 1985). Thus, Buskirk shows that to explain the uniqueness of Jamaica's fauna, it is necessary to refer to tectonic events (horizontal movement and extensive late Tertiary uplift), combined with Pleistocene cycles of climate and sea-level change. Both dispersal and vicariance processes are involved in this history.

Over the past few million years, vertical changes of land and sea have been of more general significance to Caribbean biogeography than horizontal land movements. An example is provided by Williams' (1972) studies of anoline lizards. Anoles are small green or brown lizards that are more or less the only diurnal arboreal lizards found in most of the region. Much of the distributional pattern within this group is related to the submarine banks on which today's islands stand. Each bank has many islands on it, but as recently as about 7000 BP as a generalization, many of these islands were joined together, as they would have been for lengthy periods during the Pleistocene. The present-day distributions of anoline lizards thus owe much to events during the low sea-level stands (Williams 1972).

The unravelling of past species movements in order to determine how and when a species reached its present insular distribution presents considerable challenges. Much can depend on assumptions concerning the powers of long-distance dispersal of particular species. Many questions in island biogeography come back to the issue of dispersal, as it is inherently extremely difficult to determine the effective long-distance range of propagules ('potentially viable units') (but see Hughes *et al.* 1994; Whittaker *et al.* 1997). We will return to this theme again; suffice to note for the moment that although dispersal limitations may provide parsimonious interpretations of data such as the anoline lizard distributions, it is important to remain open to new data and insights into dispersal powers of particular species. This point was driven home by Johnson's (1980) documentation, complete with photographs, of the ability of elephants to swim across ocean gaps, using the trunk as a sort of snorkel. It had been generally assumed that this behaviour was beyond the ability or, if not the ability, then the inclinations of elephants. Thus, the existence of elephants on an island was formerly taken to be proof of a former land connection. As we discuss in the next section, the increasing application of modern phylogeographical methods is gradually bringing improved resolution to such matters.

Macaronesia—the biogeographical affinities of the Happy Islands

From the beginning of the nineteenth century it was already evident to the European naturalists who called in at what we now term the Macaronesian islands in the course of their transatlantic voyaging, that these islands possessed an interrelated biota. The term Macaronesia (derived from the classic Greek macarion = islands and nesoi = happy) was first used by the British botanist Philip Barker Webb in 1845, with reference to the Canaries (in turn following their Roman name Insulae Fortunatae; Stearn 1973). Today, however, the term is more broadly utilized for the designation of a biogeographical entity including all the north-west Atlantic archipelagoes off the European and North African coasts, thus comprising the Azores, Madeira with the Salvage Islands, the Canaries, and the Cape Verde Islands (Fig. 3.9). To the initial core formed by these archipelagos, some authors have suggested adding a narrow coastal strip of the African continent, extending approximately from Agadir (Morocco) to Nouadhibou (Mauritania), including the valleys and wadis (dry riverbeds) of the Anti-Atlas mountain chain (Sunding 1979).

With the exception of the Azores, where the volcanic activity relates to the Mid-Atlantic ocean ridge, the origins of the Macaronesian islands lie within the African plate. Their formation began perhaps 80 Ma, gradually building the submarine platforms for the East Canarian and Salvage Islands. Their emergence above sea level began some 27 Ma in the Salvage Islands, which today are almost completely eroded back to sea level. Later, the Canaries (from 20 Ma) and Madeira (15 Ma)

Figure 3.9 The biogeographical area known as Macaronesia consists of the island groups as shown and a narrow coastal strip of north-west Africa (after García–Talavera 1999). The affinities with the Iberian peninsula are also recognized by some as warranting a link to the peninsula.

emerged as the nucleus of the region (Central Macaronesia), and finally the southern and northern archipelagos, the Cape Verde and Azorean archipelagoes, arose some 10 and 8 Ma, respectively. Eruptions in recent decades attest to the active nature of these volcanic systems (e.g. Capelinhos, Faial in the Azores in 1957; Teneguía, La Palma in the Canaries in 1971; and Caldeira do Pico, Fogo, in the Cape Verde archipelago in 1995).

The Macaronesian archipelagoes range across a considerable latitudinal (40 °N (Corvo, Azores) to 15° N (Brava, Cape Verde), and climatic gradient, from the cool-oceanic climate of the Azores to the oceanic tropical monsoon-drift climate of the Cape Verde islands, with the Mediterranean climates of Madeira and the Canaries in between (Lüpnitz 1995, Fernández-Palacios and Dias 2001). Moreover, the distances from the African and European mainlands vary hugely. For example, only 96 km separates Fuerteventura (Canaries) and Stafford Point

(Western Sahara), but Boa Vista (Cape Verde) is 570 km from the nearest mainland at Dakar, Senegal, and Porto Santo in Madeira is 630 km from the nearest mainland at Cape Sim (Morocco). Sao Miguel in the Azores is some 1370 km from mainland landfall at Lisbon, in south-west Europe. Furthermore, the islands of Corvo and Flores (in the Azores), although occurring within the North American plate, are virtually equidistant from Cape Race (Newfoundland) and Lisbon.

Macaronesia has been considered a distinct phytogeographical region within the Holarctic Kingdom for more than a century. Floristically, although there are shared elements, it is now recognized that there is a transition from strong Eurosiberian–Atlantic affinities for the Azores, to a Mediterranean flavour for Madeira and the Canaries and, finally, to a Saharan–Sudanian character for the Cape Verde islands (Sunding 1979; Nicolás et al. 1989; Kunkel 1993), leading some to question the validity of the Macaronesian region (Lobin 1982; Lüpnitz 1995).

Kunkel's (1993) solution is a hierarchical framework recognizing Central Macaronesia (including the Canaries and Madeira), Lauri Macaronesia (including Central Macaronesia, Azores, and the south-western part of the Iberian peninsula) and finally Great Macaronesia (including Lauri Macaronesia, the Cape Verde islands, and a strip of the African coast). Lobin (1982) prefers dropping the term Macaronesia altogether, placing the Azores in the Atlantic region of the Holarctic Kingdom, Madeira and the Canaries in the Mediterranean Region (cf. Blondel and Aronson 1999), and finally the Cape Verde islands in the Saharan region of the Palaeotropic Kingdom. Lüpnitz (1995) goes still further, advocating the inclusion of the Canaries in the Palaeotropics. In addition, although the nucleus of the Canaries and Madeira are recognized to have zoogeographical affinities, on the whole Macaronesia is a phytogeographical rather than a zoogeographical concept (Pedro Oromí, personal communication).

What becomes clear from the above is that these widely spaced archipelagos have connections of varying strength to different mainland regions. Recent developments in our understanding of the environmental history of the islands (Chapter 2)

and the phylogeography of Macaronesian lineages shed considerable light on these phytogeographical arguments and provide fascinating scenarios of the colonization history of the region. The general picture is as follows. The islands emerged more recently than traditionally thought, at a point when the continents (Africa, Europe, and North America) were clearly well separated. So, the islands have been colonized by long-distance dispersal across open ocean, in a structured fashion, resulting in varying island–mainland affinities across Macaronesia: a scenario some historical biogeographers dismiss as improbable.

Successful, structured long-distance dispersal becomes easier to understand and accept as reasonable when we consider the following factors (Nicolás et al. 1989):

- First, as a result of continuing volcanic activity, new territory has been provided for colonization throughout the history of the various archipelagos.
- Secondly, because of the existence in this part of the North Atlantic Ocean of well-consolidated wind and marine current systems since the closure of the Panama strait (c.3.5–5 Ma), the north-easterly trade winds and the cool Canarian marine current, respectively, have favoured the constant arrival of propagules from the adjacent continents.
- Thirdly, during the repeated and lengthy low sea-level stands in the Pleistocene, a number of submarine banks have emerged as islands, serving as stepping stones at intervals of 200 km or less, facilitating the flow of colonists to and among these archipelagos (Fig. 2.10, 2.11; Carine et al. 2004).

These stepping stones may have been particularly important for birds, which in turn have been responsible for introducing many plant species to the islands. The banks existing to the north of Madeira (Seine, Ampere, Gettysburg, Ormonde) provided stepping-stone connections between Madeira and the Iberian peninsula, and the Dacia and Concepción banks similarly linked Madeira with Africa (Fig. 2.11). Dispersal opportunities between the Canaries and Madeira were also enhanced by these banks (Dacia and Concepción) and by the Salvage Islands. These connections may play some role in explaining the greater biotic richness of Madeira and the Canaries and how it is that these islands come to be regarded as the authentic core of Macaronesia (e.g. Kunkel 1993), sharing some 50 Canarian–Madeiran endemic plant species (Sziemer 2000), as well as 8 endemic genera (*Bystropogon, Heberdenia, Isoplexis, Marcetella, Normania, Phyllis, Semele and Visnea*) (Santos-Guerra 2001). In contrast, the Azores and Cape Verde islands were less well served by former stepping stones, again consistent with the pattern of affinities (Dias 1996; Brochmann *et al.* 1997).

Recent molecular phylogenetic analyses have greatly advanced our understanding of these relationships, indicating the most likely colonization pathways for particular taxa both between the mainlands and Macaronesia and within the region. For example, it appears that some plants of Canarian origin (certainly species of *Aeonium, Echium, Sonchus,* and likely *Lavandula* and *Limonium* as well) dispersed across the *c.*1300 km from the Canaries to the Cape Verde islands, where endemic forms developed (Santos-Guerra 1999). Furthermore, at least six plant taxa appear to have colonized Madeira from the Canaries: *Bystropogon, Aeonium, Convolvulus, Crambe, Pericallis,* and the woody *Sonchus* alliance (Trusty *et al.* 2005). Some lineages (e.g. *Pericallis*) subsequently colonized the Azores, most likely via the southernmost Azorean island, Santa Maria, some 850 km from Madeira. In contrast, *Argyranthemum* and *Crambe,* two Asteraceae genera that have radiated profusely in Macaronesia, are believed to have colonized the Canaries from Madeira (Santos-Guerra 1999). Amongst birds, it has been suggested that the chaffinches colonized the Azores from the Iberian peninsula, and then colonized the western Canaries via Madeira, whereas the Canarian blue tit appears to have colonized Tenerife, in the centre of the Canaries, and then to have spread to the other islands of the Canaries (Kvist *et al.* 2005).

In terms of African links, the recent discovery in the Anti-Atlas valleys of typical Macaronesian floristic elements, including the Dragon tree (*Dracaena draco*) and the laurel (*Laurus* cf. *novocanariensis*), highlight the role of northwest Africa in the colonization of Macaronesia. Another very interesting piece of the jigsaw is the discovery of North African endemic species of genera considered to be Canarian in origin. Specific examples include the following species of *Aeonium*: *A. arboreum, A. kornelius-lemsii* (both in Morocco), *A. leucoblepharum,* and *A. stuessyi* (both in Ethiopia). A further example is a species of *Sonchus* (*S. pinnatifidus*), still extant in the easternmost Canaries, from where it has colonized the African coast, indicating a return trip for lineages that developed on the Canaries having originally colonized from Africa some 4 Ma (Santos-Guerra 2001).

Summing up current understanding based on recent molecular studies, Carine *et al.* (2004) report that in the majority of genera investigated so far, Macaronesian endemic taxa form a single monophyletic group, in most cases with western Mediterranean sister groups, but also with affinities revealed with North Africa and the Iberian peninsula, and indeed with more widely separated regions, such as the New World (Madeiran *Sedum*), East Africa (*Solanum*) and southern Africa (*Phyllis*). They also note cases of multiple congeneric colonizations, and plausible cases of back-colonization from Macaronesia to continental areas (e.g. in *Tolpis* and *Aeonium*).

How robust these scenarios will prove to be in the light of future collecting and phylogeography is unknowable at the present time. However, we can draw several conclusions from these studies:

- There have been multiple and indeed conflicting colonization pathways followed by different taxa.
- The Macaronesian flora comprises lineages diverging from their closest continental relatives at different times, some ancient and some relatively recent.
- There may well have been distinct waves of colonization related to the greatly changing geological and climatic history of the region.
- Following from this, although the islands are truly oceanic, such that long-distance dispersal is necessary to the exchange of lineages with the mainlands, this is not to rule out a role for past variation in the strength of barriers between these

islands and mainland area, and in this sense both dispersal and vicariance reasoning has relevance to the region's biogeography.

• Because of the complexity of the colonization scenarios within Macaronesia, assigning these islands to a particular phyto- or zoo-geographic region, or mainland source area, is inevitably going to be problematic; nevertheless, there is clearly a strong Mediterranean flavour to the biota.

3.4 Endemism

Neo- and palaeoendemism

A fair number of island endemics belong to groups that formerly had a more extensive, continental distribution. Since colonizing their island(s) in the remote past (over either land or sea), they have failed to persist in their original form in their mainland range (Box 3.3). The implication is that not all

---

Box 3.3 Palaeoendemism of the Macaronesian laurel forest

We illustrate the idea of palaeoendemism with an example drawn, once again, from the islands of Macaronesia. The laurel forests of the Azores, Madeira and the Canaries are examples of the evergreen broadleaved subtropical biome that can be found between 25 and 35° latitude in both hemispheres (i.e. Texas, Florida, Atlantic islands, Chile, Argentina, South Africa, South East Asia, Japan, South Australia, and New Zealand) growing, at least in Macaronesia, under the influence of orographic cloud layers (Santos-Guerra 1990). The Macaronesian laurel forests are considered to be the only surviving remnants of a once rather richer forest flora, which was distributed in central Europe (fossils found in Austria and Hungary) and southern Europe (fossils found in France, Italy, and Spain) during the Oligocene–Miocene (Axelrod 1975; Sunding 1979). Today we know that the laurel forest has been present on the Atlantic islands at least from the late Pliocene (a minimum of 2 Ma) (Sziemer 2000) and, very likely, much longer.

The failure of these forests to persist in Europe has been attributed to a range of dramatic geological and climatic events that took place in Europe following the transition of the Tethys ocean seaway into the largely enclosed Mediterranean basin during the Tertiary. In particular, around the end of the Tertiary, the connection between the Mediterranean and Atlantic closed between c.5.6 and 5.3 Ma, leading to the Mediterranean drying out almost completely, with associated drought and salinization. This spectacular event was followed by the glacial cycles of the last 2–3 million years and the desertification of the Saharan region,

which began 5–6 Ma (Blondel and Aronson 1999). With the closure of the Panama strait, and the associated shifts in global ocean currents, the present-day Mediterranean climate developed some 3 Ma and with it emerged the Mediterranean sclerophyllous forest, which was better adapted to the new prevailing climatic conditions of hot, dry summers and cool, wet winters.

The explanation for the survival of this 'relict' forest flora on the Atlantic islands, albeit in rather species-poor form, has been attributed to three factors.

• The islands may have been buffered from the climatic extremes of mainland southern and central Europe and the Mediterranean by their mid-Atlantic location.

• The topographic amplitude of the islands has enabled vertical migrations of several hundred metres in response to such changes in climate as the islands have experienced.

• The existence in the higher islands of a stable orographic cloud layer has counteracted the aridity of the generally Mediterranean climate, especially so in the northern hemisphere summer.

As a result, the laurel forests of the Atlantic islands possess a number of plant species that are considered to be palaeoendemics. These include the following Madeiran/Canarian species known from Tertiary deposits in Europe (5.3 Ma in South France): *Persea indica*, *Laurus azorica*, *Ilex canariensis*, *Picconia excelsa*. However, this is not to say that they are absolutely identical to the ancestral forms (see table), and as Carlquist (1995)

Selected Central and South Europe fossil taxa and corresponding palaeoendemic Macaronesian laurel forest taxa (source: Sunding 1979, slightly modified). Sunding's interpretation of the Canary Islands biogeography is now regarded as flawed in a number of respects: e.g. the supposition of past land connections between the eastern islands and the mainland, and the notion of woodiness in normally herbaceous genera as a relictual trait (Carlquist 1995)

Central–South European fossil taxon	Corresponding Macaronesian taxon (if different from the fossil)
Adiantum reniforme	
Apollonias aquensis	*Apollonias barbujana, A. ceballosi*
Asplenium hemionitis	
Clethra berendtii	*Clethra arborea*
Ilex canariensis	
Laurus azorica	*Laurus azorica, L. novocanariensis*
Maytenus canariensis	
Myrsine spp.	*Myrsine canariensis*
Ocotea heerii	*Ocotea foetens*
Persea indica	
Picconia excelsa	
Smilax targionii	*Smilax canariensis*
Viburnum pseudotinus	*Vinurnum tinus* ssp. *rigidum*
Woodwardia radicans	

Other laurel forest species still extant in Iberia and present in Macaronesia:
Culcita macrocarpa, Davallia canariensis, Erica arborea, Myrica faya, Prunus lusitanica, Trichomanes speciosum, Umbilicus heylandianus

cautions, some authors have attributed relictualism to features (e.g. woodiness in herbaceous taxa) and lineages that are actually recent products of *in situ* evolutionary change on islands. Indeed, the Canaries in particular are characterized by plenty of examples of neoendemic speciation and radiation (Santos-Guerra 1999).

In the last few hundred years, human settlement and deforestation within Macaronesia has led to the loss of most of the laurel forest, especially in the Azores, and today large remnants are only to be found in Madeira and the western Canaries, particularly on La Gomera.

species now endemic to an island or island group have actually evolved *in situ*. Cronk (1992) refers to such relict or stranded forms as **palaeoendemics**, whereas species that have evolved *in situ* are **neoendemics**. As with all such distinctions, this is an oversimplification, as it implies a lack of change in either island or mainland forms, respectively, which may not strictly be the case.

It should be noted that the significance of relict forms or relictual ground plans as against *in situ* evolution is a topic that generates considerable heat, and indeed some of the more keenly contested issues in biogeography (cf. Heads 1990; Pole 1994; Carlquist 1995; Winkworth *et al*. 2002). Cronk (1992)

argues that ancient, taxonomically isolated, relict species should hold a special importance in conservation terms, in that their extinction would cause a greater loss of unique gene sequences and morphological diversity than the extinction of a species with close relations.

There is no doubt as to the importance of endemic forms on particular islands, although as taxonomic work continues, estimates of the degree of endemism for an island group vary. Figure 3.10 demonstrates this for the flora of the Galápagos. Since fairly complete inventories became available, the number of known endemics has changed relatively little: the refinements come from

(a)

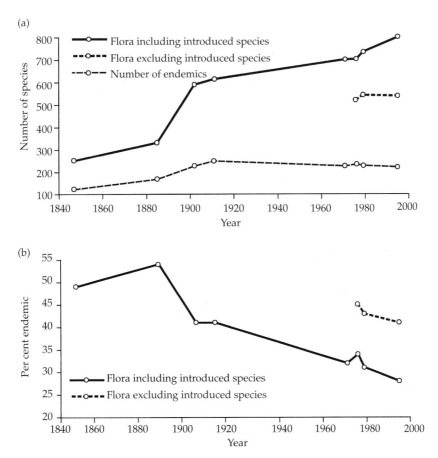

(b)

Figure 3.10 Changing estimates of the Galápagos flora through time: (a) the size of the flora; and (b) the proportion of endemics. (Data from Porter 1979 and Davis *et al.* 1995.)

improved knowledge of mainland source areas and as a result of taxonomic revisions. Although the proportion of endemics appears to have declined (Fig. 3.10b), this is mostly the result of increases in numbers of introduced species. If these species are excluded, the proportion of endemics in the flora has remained reasonably constant. The basic pattern established by Hooker and Darwin (Ridley 1994) has thus proved to be robust. As a proportion of the indigenous flora of the Galápagos today, the endemics represent 43% of the indigenous taxa (species, subspecies, or varieties); allowing for recent extinctions and uncertainties in taxonomy, the figure could fall to 37% (Porter 1979). Recently published estimates for Hawaii are that the following proportions of

native taxa are endemic: birds, 81%; angiosperms, 91%; molluscs, 99%; and insects, 99% (Sohmer and Gustafson 1993).

The following sections provide a few estimates that set such figures in the broader context of regional or global diversity.

Endemic plants

Groombridge (1992) provides the following 'best estimates' for the number of higher plant species in the world: 12 000 pteridophytes (ferns and their allies); 766 gymnosperms (conifers, cycads, and their allies); and 250 000 angiosperms (flowering plants). This makes for a grand total of about 263 000 species. It will be recalled that if

Table 3.2 Native higher plant species richness and endemism of selected islands or archipelagos. (From Groombridge 1992; Davis *et al.* 1995, Sziemer 2000, Izquierdo *et al.* 2004, Traveset and Santamaría 2004, Arechavaleta *et al.* 2005.) These figures should be treated cautiously as different values for some islands can be found, in cases even within a single publication (see Groombridge 1992, Tables 8.3, and 14.1)

Island or archipelago	Number of species	Number of endemic species	% Endemic
Borneo	20 000–25 000	6000–7500	30
New Guinea	15 000–20 000	10 500–16 000	70–80
Madagascar	8000–10 000	5000–8000	68.4
Cuba	6514	3229	49.6
Japan	5372	2000	37.2
Jamaica	3308	906	27.4
New Caledonia	3094	2480	80.2
New Zealand	2371	1942	81.9
Seychelles	1640	250	15.2
Fiji	1628	812	49.9
Balearic	1359	89	6.5
Canaries	1300	570	44.3
Hawaii	1180	906	89.8
Mascarenes	878	329	37.5
Madeira	793	118	14.9
Galápagos	529	211	39.9
Cook Islands	284	3	1.1
Cape Verde	224	66	29.4
Azores	197	67	34.0
St Helena	74	59	79.7

New Guinea is taken as the world's largest, islands constitute about 3% of the land surface area of the world. The small selection of 20 islands and archipelagos in Table 3.2 provides a minimum estimate of about 35 500 species endemic to those same islands, amounting to some 13.5% of the world's higher plant species, indeed, just 10 of these islands/archipelagos account for 12.8%. It has been calculated that the Pacific islands contain about 22 000 vascular plant species, nearly half of which may have been introduced by humans. Of the 11 000–12 000 native species, about 7000 are considered to be endemic, most being found within a single island or island archipelago (Davis *et al.* 1995).

All such figures come with large error margins. Indeed, estimates of the number of flowering plant species in the world have varied between 240 000 and 750 000 (Groombridge 1992, p. 65). However, island species are as much a part of these uncertainties as are continental species (see data for New Guinea and Borneo, Table 3.2), so the relative importance of island species is unlikely to be seriously diminished by further refinements of the estimates. It is a reasonable estimate that about one in six of the Earth's plant species grows on islands, and one in three of all known threatened plants are island forms (Groombridge 1992, p. 244). Clearly, in terms of plant biodiversity, the islands of the world make a disproportionate contribution for their land area and are also suffering further disproportionate pressure in terms of the maintenance of that biodiversity.

The percentage of endemics varies greatly among islands. The highest proportions are often associated with ancient continental islands, such as Madagascar and New Zealand. Large numbers of endemics are also associated with the larger, higher oceanic islands in tropical and warm-temperate latitudes. Hawaii, to provide a classic example, has about 1180 native vascular plants, and of the 1000 angiosperms (flowering plants) about 90% are considered endemic (Wagner and Funk 1995). A more conservative estimate for their numbers is 850 endemic species, although the figure could conceivably be as high as 1000 (Groombridge 1992; Davis *et al.* 1995). Some lineages have radiated spectacularly. The silversword alliance comprises 28 endemic species, apparently of monophyletic origin (i.e. a single common ancestor), but placed in three genera: *Argyroxiphium*, *Dubautia*, and *Wilkesia* (Wagner and Funk 1995). The genus *Cyrtandra* (Gesneriaceae) is represented by over 100 species, and although they may not be monophyletic, their radiation is no less impressive (Otte 1989; Wagner and Funk 1995). The Macaronesian flora provides the best developed Atlantic equivalent to Hawaii, with some exquisite examples of largely monophyletic radiations. For instance, there are more than 60 monophyletic endemic species of Crassulaceae, distributed across the genera *Aeonium*, *Greenovia*, *Aichryson*, and *Monanthes*; there are more than 20 species representing monophyletic taxa within 4 genera of Asteraceae, namely

Pericallis, Cheirolophus, Argyranthemum, and *Sonchus*;
and there are almost equally impressive radiations
within the genera *Echium* (Boraginaceae), *Lotus*
(Fabaceae), *Sideritis* (Lamiaceae), and *Limonium*
(Plumbaginaceae) (Sziemer 2000; Izquierdo *et al.*
2004).

Smaller oceanic islands, even in the tropics (e.g.
the Cook Islands), have much smaller floras and
typically lower proportions of endemics. This is due
in part to the reduced variety of habitats and to the
wide mixing of the typically sea-dispersed strandline species that dominate their floras. Another feature of the data which must be borne in mind when
examining statistics for island endemism is that in
some areas there are numerous species which are
endemic to islands, but which are shared between
different geographical entities. Thus, in the Lesser
Antilles (the Leeward and Windward Islands), the
shared or regional island endemics form a considerable proportion of the endemic flora as a whole.

Endemic animals

The following constitutes a selection of some of the
animal groups that are sufficiently well-known to
be placed into either a regional or global context
with some degree of confidence.

Land snails
There may be as many as 30 000–35 000 land snail
species in the world (Groombridge 1992), of which
a substantial proportion occur on islands (Wallace
1902). Some of the better-known of island snail
faunas are quantified in Table 3.3. Although only
eight archipelagos are listed in the table, they
account for approximately 7.7–9.0% of the world's
land snails by present estimates (some regions of
the world remain poorly known and the current
picture could change) (Groombridge 1992). Studies
in the Pacific suggest that, once again, it is the
larger, higher oceanic islands (15–40 km², >400 m
ASL) that typically have both most species and most
endemics, whereas low-lying atolls have neither
high richness nor high degrees of endemism. On
most islands with high snail diversity, the snails
appear to be concentrated in the interiors, especially mountainous regions with 'primary' forest

Table 3.3 Land snail species richness and endemism for selected
islands for which the data are complete enough for the proportion of
endemics to be estimated (from Groombridge 1992, Table 14.3)

Island or archipelago	Number of species	Number of endemic species	% Endemic
Hawaiian Islands	c.1000	c.1000	99.9
Japan	492	487	99
Madagascar	380	361	95
New Caledonia	300	c.299	99
Madeira	237	171	88
Canary Islands	181	141	77.9
Mascarene Islands	145	127	87.6
Rapa	>105	>105	100?

cover. For those islands with good data, a marked
positive correlation exists between numbers of
endemic plant species and endemic molluscs, but
not between molluscs and birds (Groombridge
1992).

Insects
No attempt will be made to cover the insects in a
systematic fashion, but the following statistics
from Hawaii and the Canaries give an idea. In the
family Drosophilidae (fruit flies), 511 species are
currently named and described for Hawaii, with
another 250–300 awaiting description (Wagner and
Funk 1995). Given the pattern of discovery, it is
estimated that there may be as many as 1000
endemic species. The radiation of tree crickets on
the archipelago has resulted in 3 endemic genera
and some 68 species. The orthotyline plant bug
genus *Sarona* is another endemic Hawaiian genus,
with 40 known species. Moving across to the
Atlantic Ocean, the Canarian insect fauna is
increasingly well studied, and comprises some
5700 species and 500 subspecies, of which almost
2200 species and 370 subspecies are considered
endemic (Izquierdo *et al.* 2004). Among them,
perhaps the most explosive example of radiation is
offered by the weevil genus *Laparocerus*, comprising some 65 monophyletic species, with approximately another 40 waiting to be described and
may be as many more waiting to be discovered
(Antonio Machado, personal communication).

Unfortunately, with the exception of a few well-studied islands, few groups of insects are sufficiently well known for broad biogeographic patterns to be established. The Lepidoptera provide an exception. The tropical Pacific butterfly fauna consists of 285 species, of which 157 (55%) occur on continental land masses, i.e. they have not speciated after arrival on the islands. One hundred are endemic to a single island/archipelago and 28 are regional endemics, i.e. they are found on more than one archipelago but not on the mainlands (Adler and Dudley 1994). Interestingly, butterfly speciation in the Pacific archipelagos is primarily a result of limited interarchipelago speciation. Intra-archipelago speciation appears important only in the Bismarcks and Solomons, which contain, respectively, 36 and 35 species endemic at the archipelago level. Of the other 24 archipelagos in the survey, only New Caledonia makes double figures, with 11 endemic species, and the others contribute just 18 species between them. The Bismarcks and Solomons are the two largest archipelagos in land area and the closest to continental source areas. Butterflies are generally specialized herbivores, their larvae feeding only on a narrow range of plants, or even a single species. On the smaller and more remote archipelagos, the related host plants simply may not be available for the evolution of new plant associations, thereby impeding the formation of new butterfly species (Adler and Dudley 1994).

Lizards

In their survey of the 27 oceanic archipelagos and isolated islands in the tropical Pacific Ocean for which detailed herpetological data exist, Adler *et al.* (1995) record 100 species of skinks in 23 genera, of which 66 species are endemic to a single island/archipelago and a further 13 are regional island endemics. Of the 23 genera, 9 are endemic to the islands of the tropical Pacific. A few species are widespread; for example, *Emoia cyanura* occurs on 24 of 27 archipelagos in the survey. The Bismarcks, Solomons, and New Caledonia appear particularly rich in skinks, but most archipelagos contain fewer than 10 species. Hawaii has only 3 species, none of which is endemic, in stark contrast to the birds,

insects, and plants. Some Atlantic competition in the biodiversity statistics game is provided again by the Canaries, which feature the endemic *Gallotia* lizards, an interesting monophyletic group composed of 7 extant species of varying sizes, plus at least one more extinct species (Izquierdo *et al.* 2004). Another important group of island lizards are the Caribbean anolines, which are typically small, arboreal insectivores. This is one of the largest and better-studied vertebrate genera: about 300 species of anoline lizards have been described, half of which occur on Caribbean islands (Losos 1994).

Birds

The examples of adaptive radiation best known to every student of biology from high-school days are probably the Galápagos finches and the Hawaiian honeycreepers. Therefore it will come as no surprise that islands are important for bird biodiversity. In excess of 1750 species are confined to islands, representing about 17% of the world's species; of these 402, or 23%, are classified as threatened, considerably in excess of the 11% of bird species worldwide (Johnson and Stattersfield 1990; Groombridge 1992, p. 245).

Adler (1994) has examined the pattern of bird species diversity and endemism for 14 tropical Indian Ocean archipelagos. Probably the most famous island endemics of all, the dodos of Mauritius, Rodrigues, and (possibly) Réunion, although extinct, are included as members of this data set, which incorporates both extant species and extinct species known only from subfossils (i.e. recently extinct species). Using stepwise linear regression it was found that numbers of each of the following were related positively to the number of large islands and total land area: total species number; number of continental species; and number of regional endemics. In addition, less remote and low-lying islands tended to have more continental species (of the low islands, only Aldabra had its own endemics), and higher islands tended to have more local endemics and proportionally fewer continental species.

Similar results were obtained for the birds of the tropical Pacific, but the Indian Ocean avifauna contained fewer species, had lower proportions of endemics, and had several families that were much more poorly represented, despite being diverse in the mainland source areas (Adler 1992). The generally small size of the Indian Ocean islands may be significant in these differences. The total land area is only 7 767 km², compared to 165 975 km² for the Pacific study, and there are also fewer island archipelagos in the Indian Ocean and they tend to be less isolated. The poor representation of certain families, notably the ducks and kingfishers, might be due to a shortage of suitable habitats, such as freshwater streams and lakes. Intra-archipelago speciation has been rare in the Indian Ocean avifauna, possibly occurring once or twice in the Comoros and a few times in the Mascarenes. Adler (1994) suggests that there may not be sufficient numbers of large islands to have promoted intraarchipelago speciation in the Indian Ocean. In contrast, it has been common within several Pacific archipelagos, accounting for most of the endemic avifauna of Hawaii, being important in the Bismarcks and Solomons, and occurring at least once in New Caledonia, Fiji, the Society Islands, Marquesas, Cooks, Tuamotus, and Carolines. In conclusion, Adler (1994) echoes Williamson's (1981) remark in respect of bird speciation on islands that 'geography is all important'.

Mammals

With the exception of bats, native mammals are not a feature of the most isolated oceanic islands, but less isolated islands include many that either once hosted or still do host interesting assemblages of native and often of endemic mammals. Perhaps the most impressive example is the Philippines, an archipelago of 7000 mostly true oceanic islands, featuring 170 species of mammals in 84 genera, of which 24 (29%) genera and 111 (64%) species are endemic (Heaney *et al.* 2005). Particularly impressive radiations have occurred in fruit bats and murid rodents, with patterns of endemism in the non-volant mammals clearly tied to Ice Age bathymetry, i.e. to the configuration of the islands as they were during the major low sea-level stands of the Pleistocene.

In the tropical Pacific, Carvajal and Adler (2005) note that a number of species of marsupials and rodents occur on the Solomons and Bismarcks, neither archipelago being particularly remote from New Guinea. But for the more remote islands, beyond the reach of non-volant terrestrial mammals (Table 3.1), mammal representation is restricted to bats, of which flying foxes are arguably the most important group, although other types of bat occur on islands. There are believed to be 161–174 species of flying fox (Chiroptera: Pteropodidae) and they are distributed throughout the Old World Tropics (but not the New World). In number of species, range, and population sizes, flying foxes have been the most successful group of native mammals to colonize the islands of the Pacific. The Pacific land region consists of about 25 000 islands, most of them very small. In addition to occupying small islands, most flying foxes have restricted distributions. Thirty eight of the 55 island species occupy land areas of less than 50 000 km² and 22 of these species occupy land areas of less than 10 000 km²; 35 occur only on a single island or group of small islands (Wilson and Graham 1992).

Comparisons between taxa at the regional scale

Within the figures cited above, those from G. H. Adler and his colleagues provide a common methodological approach and enable comparisons of patterns of endemism of different taxa at the regional scale (Table 3.4). The studies are based on 26–30 tropical Pacific Ocean islands or island archipelagos, plus in one case 14 Indian Ocean islands/archipelagos. The data include all known extant species plus others known only from sub-fossils.

Both birds and butterflies are capable of active dispersal over long distances and have colonized virtually every archipelago and major island within the tropical Pacific. Birds have a higher frequency of endemism than butterflies. Adler and Dudley (1994) consider that birds have superior dispersal ability, and that this might be anticipated to have maintained higher rates of gene flow than in butterflies, contrary to the higher degree of endemism. One plausible explanation is that the lower rate of endemism in butterflies might be a consequence of the constraints of the co-evolutionary ties with

Table 3.4 Degree of species endemism among tropical Pacific and Indian Ocean island faunas. (From Adler 1992, 1994; Adler and Dudley 1994; Adler *et al.* 1995; Carvajal and Adler 2005.)

Group	Total number	Continental	Regional endemics	Local endemics
Pacific Ocean butterflies	285	157 (55%)	28 (10%)	100 (35%)
Pacific Ocean skinks	100	21 (21%)	13 (13%)	66 (66%)
Pacific Ocean birds	592	145 (25%)	59 (10%)	388 (65%)
Pacific Ocean mammals	106	42 (40%)	7 (6%)	57 (54%)
Indian Ocean birds	139	60 (43%)	10 (7%)	69 (50%)

Continental, species also occurring on continental land masses; **regional endemics**, species occurring on more than one archipelago within the region but not on continents or elsewhere; **local endemics**, species restricted to a single archipelago or island. NB: Carvajal and Adler (2005) mistakenly give 292 as the figure for local endemic Pacific Ocean birds. The correct figure is as given above (G.H. Adler, personal communication, 2005).

particular host plants required by butterfly larvae. This may limit their potential for rapid evolutionary change on islands (see above). On the other hand, the capacity of lepidopterans to reach moderately remote oceanic islands should not be too lightly dismissed, as evidenced by Holloway's (1996) long-term light-trap study on Norfolk Island, which recorded 38 species of non-resident Macrolepidoptera over a 12 year period. Norfolk Island is 676 km from New Caledonia, 772 km from New Zealand, and 1368 km from the source of most of the migrants, Australia. Most of the arrivals appear to be correlated with favourable synoptic situations, such as the passage of frontal systems over the region.

Lizards are also widely distributed in the tropical Pacific. The proportions of skinks in the three categories of endemic are remarkably similar to the equivalent figures for Pacific birds (Table 3.4). In general, in each of skinks, geckos, birds, and butterflies, the proportion of species endemic to an archipelago is best explained by reference to the number of large (high) islands and to total land area. However, scrutiny of the data on an archipelago-by-archipelago basis reveals greater differences. For instance, 100% of New Caledonia's skinks are endemic to the Pacific Ocean islands, and as many as 93% are endemic to the New Caledonian islands themselves. The equivalent figures for birds (including subfossils) are 47% and 33%, respectively (Adler *et al.* 1995). These data contrast with those for Hawaii, on which most birds are endemic, but where there are no endemic lizards. Adler *et al.* (1995) suggest that these differences are explicable in relation to the differing dispersal abilities of lizards and birds. Birds, being better dispersers, reached Hawaii relatively early and have radiated spectacularly, whereas the three species of skink may only have colonized fairly recently and possibly with human assistance. New Caledonia, in contrast, may not be sufficiently isolated to have allowed such a degree of avifaunal endemism to have developed.

Numbers of birds and butterflies are remarkably similar on the less remote Pacific archipelagos east of New Guinea, but on the more remote archipelagos the numbers of bird species are far greater than those of butterflies (Adler *et al.* 1995). This is reflected statistically by archipelago area being the most important geographical variable in explaining bird species richness, whereas for butterflies isolation is the more important variable. Another pattern noted by Adler *et al.* (1995) is that in Pacific butterflies, Pacific birds, and Antillean butterflies the number of endemic species increases more steeply with island area than does the number of more widely distributed species. This observation supports the idea that a larger island area provides the persistence and variety of habitats most conducive to radiation on isolated islands (see further discussion in Chapter 9).

The proportions of Pacific island mammals in each of the endemism categories falls closer to the figures for birds and skinks than those for butterflies. However, the 5 marsupial species make it no further than the Bismarcks, and the 18 species of native rodents no further than the Bismarck or Solomon islands where 14 of the 18 species are endemics (Carvajal and Adler 2005). The remaining native mammal species indicated in Table 3.4 are all bats. Variation in richness in the mammal data appeared explicable as a function of a combination of intraarchipelago speciation in archipelagos of large islands, and interarchipelago speciation, particularly among more isolated archipelagos.

To summarize, while different taxa have radiated to different degrees on particular islands or archipelagos, it appears to be the case that the greatest degree of endemism is found towards the extremity of the dispersal range of each taxon (the 'radiation zone' *sensu* MacArthur and Wilson 1967). Islands that are large, high, and remote typically have the highest proportion of endemics. In total, islands account for significant slices of the global biodiversity cake.

3.5 Cryptic and extinct island endemics: a cautionary note

The foregoing account has taken little note of the 'state of health' of island endemics, and some of the assessments reviewed have included many highly endangered species and others already believed to be extinct. Before moving on from the geography of endemism to an examination of the theories that may account for the patterns of island ecology and evolution, it is important to consider both the problem of rare cryptic forms and the extent of the losses already suffered. Otherwise, there may be a danger of misinterpreting evolutionary patterns that are just the more resistant fragments of formerly rather different tapestries (Pregill and Olsen 1981; Pregill 1986). The role of humans in reducing island endemics to threatened status or extinction is important enough to warrant a separate chapter (Chapter 11) but for the moment these examples can serve to place a health warning on attempts to explain current suites of endemic species on islands.

Given the vast number of islands, especially in the tropics, and the general tendency for diversity to concentrate in less apparent taxa (beetles, and the like), it is understood that many undiscovered species remain for collectors and taxonomists among these relatively cryptic taxa. Only slightly more surprising is the number of vertebrates scientifically discovered only in recent years on under-researched islands such as the Philippines (cf. Heaney *et al.* 2005). Even in birds, probably the best-studied vertebrate group, 'new' species have been described in very recent times in remote islands, e.g. the flightless rail, *Gallirallus rovianae*, endemic to the Solomon Islands and first collected in 1977 (Diamond 1991a). Moreover, in the Canaries, which might be expected to be very well known scientifically, two quite sizeable endemic lizards new to science but of cryptic habits were discovered during the 1990s, *Gallotia intermedia* on Tenerife (Hernández *et al.* 2000) and *Gallotia gomerana* on La Gomera (Valido *et al.* 2000). Following the first discovery of *G. intermedia* in the Teno peninsular of Tenerife, a further population was found in another location adjacent to a teeming tourist resort.

Turning to already extinct island species, Schüle (1993) summarizes recent studies of vertebrate endemism for the Mediterranean islands, noting that although the small-animal faunas have shown good persistence, the larger terrestrial vertebrates of the early part of the Quaternary appear to have included endemic ungulates, carnivores, giant rodents, tortoises, and flightless swans, not one of which has survived to the present day. He regards it as plausible that some of the losses could be attributed to the arrival of new species driving older endemic species to extinction by competition. However, it is increasingly clear that some of this flux can be attributed to climatic change, sea-level change, and other environmental forcing factors (e.g. occasionally volcanism) that exposed the existing island assemblages both to substantial environmental change and to new colonists (Marra 2005). From the Late Pleistocene onwards, and especially in the Holocene, the extinctions—which included species of pigmy elephants, hippos and cervids, flightless swans, tortoises, and rodents—are often attributable to the arrival of humans (Schüle 1993; Blondel and Aronson 1999), although Marra (2005) cautions that we still have an incomplete understanding of the relative timing of human arrival and faunal turnover across the islands of the Mediterranean. Thus, on the one hand, the loss of *Myotragus balearicus* on the Balearic islands appears attributable to human settlement, which predates the extinction event, while the extinction of hippo from Cyprus, previously attributed to human hunting, has recently been queried because clear evidence of temporal overlap of humans with the hippo is lacking.

Estimates of the numbers of Hawaiian birds that were lost following Polynesian colonization *c.*1500 years ago, but before European contact, now stands at a minimum of 40 species (Olson and James 1982). Each of the 16 Polynesian islands that has yielded in excess of 300 prehistoric bird bones approaches or exceeds 20 extirpated species (i.e. lost from those particular islands, but not necessarily globally extinct), pointing to the likelihood of far more extinct species yet to be catalogued. About 85% of bird extinctions during historical times have occurred on islands (Steadman 1997*a*), a rate of loss 40 times greater than for continental species (Johnson and Stattersfield 1990).

The cataloguing of insular losses is clearly far from complete, and although we have some basis for estimating how many species of birds may be involved, for most taxa such calculations remain to be made (Milberg and Tyrberg 1993). Morgan and Woods (1986) appraise the problem for West Indian mammals. They calculate an extinction rate for the period since human colonization of one species every 122 years, and they emphasize that this is a minimum estimate. The mammals of the West Indian islands today are thus a highly impoverished subset of species that escaped the general decimation of the fauna. Morgan and Woods caution that biogeographical analyses that fail to incorporate the extinct forms are unlikely to prove reliable.

It is unsurprising, but significant biogeographically, that in addition to species disappearing altogether, a number of species have gone extinct from particular islands during the past 20 000 years, while still surviving on others. This is particularly evident in the bats, which supply the majority of persisting species of mammal, but is also a pattern found among lizards and birds. Many such losses

occurred as a result of climatic changes and the associated sea-level changes and habitat alterations at the end of the Pleistocene, and could be considered a 'natural' part of the biogeographical patterning of the region; others were doubtless due to human activities, including hunting, habitat alteration and the introduction of exotics. Interpretation of the present-day biogeography of the region thus requires knowledge not only of past faunas but also of the extent to which humans have been involved in their alteration.

The difficulty of this task is illustrated by a study from Henderson Island. In his investigation of 42 213 bird bones, 31% of which were identifiable, Wragg (1995) found that 12% of the fossil bird species recorded were accounted for by just 0.05% of the total number of identifiable bones, indicating them to be of uncertain status, quite possibly vagrants. Consequently, biogeographical studies that rely simply on a list of fossil birds might, exceptionally, assign resident status to temporary inhabitants. A different form of bias probably occurs in other studies because of the large mesh sizes (6 or 13 mm) commonly used in sieving soil samples. Larger-boned species such as seabirds, pigeons, and rails are usually found, but small passerines and hummingbirds are much less likely to be recovered (Milberg and Tyrberg 1993).

As yet, we have only a fragmentary picture of past losses, biased towards groups for which the fossil record has been revealing, notably vertebrates and especially birds. Much is still unknown. What, for instance, has been the impact of the losses from New Zealand of the giant moas on the dynamics of forests they formerly browsed? How might the loss of hundreds of populations and a few entire species of Pacific seabirds have influenced marine food webs, in which seabirds are top consumers? (Steadman 1997*a*). Many island fruit bats and land birds (including many now extinct) have undoubtedly had crucial roles as pollinators and dispersers of plants, and thus their extirpation must be anticipated to have important repercussions for other taxa, many of which may be ongoing. That is, unless there is more functional redundancy within oceanic island biotic assemblages than most authors currently recognize (compare Rainey *et al.* 1995; Vitousek *et al.* 1995; Olesen *et al.* 2002).

It has been observed that the rapidity of human-induced extinctions of island birds has been influenced by the nature of the island. The species of high, rugged islands, and/or those with small or impermanent human populations, have probably fared better than those of low relief and dense, permanent settlement. This is exemplified by the finding that a larger percentage of the indigenous avifauna survives on large Melanesian islands than on small Melanesian islands, or on Polynesian or Micronesian islands. This may be due to the buffering effects that steep terrain, cold and wet montane climates and human diseases have had on human impact (Steadman 1997*a*; Diamond 2005).

Steadman (1997*a*) has pointed out that the Kingdom of Tonga did not qualify in a recent attempt to identify the Endemic Bird Areas of the world because of its depauperate modern avifauna, yet bones from just one of the islands of Tonga, 'Eua, indicate that at least 27 species of land birds lived there before humans arrived, around 3000 BP. Forest frugivores/granivores have declined from 12 to 4, insectivores from 6 to 3, nectarivores from 4 to 1, omnivores from 3 to none, and predators from 2 to 1. As is consistent with predation by humans, rats, dogs, and pigs, the losses have been more complete for ground-dwelling species. It is highly likely that the means of pollination and seed dispersal, not just of the plants of 'Eua but of islands across the Pacific, have over the past few thousand years been diminished greatly as a consequence of this pattern of loss, replicated, as it has been, across the Pacific. That the attrition of endemic species from oceanic islands may have so altered distributional patterns, compositional structure, and ecological processes serves as a caution on our attempts in the following chapters to generalize biogeographic and evolutionary patterns across islands.

3.6 Summary

This chapter set out to establish the biogeographical and biodiversity significance of island biotas, particularly remote island biotas, and to provide an indication of the ways in which island assemblages are distinctive. In global terms, for a variety of taxa, islands make a contribution to biodiversity out of proportion to their land area, and in this sense collectively they can be thought of as 'hotspots'. Some of course, are considerably hotter in this regard than others. Their high biodiversity value is widely acknowledged, as is the threat to island biodiversity, and island thus typically feature prominently in global and regional conservation prioritization schemes.

Accompanying their high contribution to global diversity (through the possession of high proportions of island endemic forms), islands are, however, typically species poor for their area in comparison to areas of mainland. The more isolated the island, and the less the topographic relief, the greater the impoverishment. Remote island biotas are typically disharmonic (filtered) assemblages in relation to source areas. This can take the form of a climatic filter and thus a sort of biome shift on the island, but more interestingly involves the sampling of only a biased subset of mainland taxa. Such biases may take the form of the absence of non-volant mammals, the relative lack of representation within a particular family of plants, and so forth, and are interpretable in terms of the relative dispersal powers of the lineages and taxa under consideration. An island may be a remote isolate (oceanic in character) for a taxon with poor ocean-going powers, but be linked by fairly frequent population flows (continental in character) to a source area for a highly dispersive taxon.

Traditional zoogeographic and phytogeographic analyses have enabled biogeographers to identify unidirectional dispersal filters (so-called sweepstake dispersal routes), filter effects between different biogeographic regions, and cases where islands have been subject to the influence of multiple source areas. We appraise the issues involved in placing oceanic islands into traditional phytogeographical and zoogeographical regionalism schemes, showing that remote islands typically receive colonists from multiple sources and via multiple routes.

Historical biogeography has been for some time embroiled in debates between those advocating a prominent role for long distance transoceanic dispersal and those largely dismissive of the biogeographical significance of such processes, favouring tectonic scenarios of barrier creation, involving the splitting of ranges (vicariance events). Work reviewed in this and the previous chapter on the environmental history of islands and on molecular phylogenies, argues for both sets of processes playing a role, but in particular provides overwhelming evidence for long-distance dispersal being a biogeographically significant process.

In some cases, island forms may have changed less than the mainland lineages from which they sprung, or have persisted while the mainland population has failed, in which case they may be termed palaeoendemics. However, where novelty has arisen predominantly in the island context, they are called neoendemics, although it should be recognized that this is to divide a continuum. Recently, molecular phylogenetic analyses have begun to transform our understanding of island lineages and their relationships to continental sister clades, as illustrated herein by extensive reference to the complex colonization and evolutionary history of the Atlantic islands of Macaronesia.

In this chapter we have avoided discussion of the mechanisms of evolutionary change on islands, although noting some correlates with high proportions of endemics in island biotas. With increasing isolation, island size, and topographic variety, the number and proportion of endemic species increases. Endemism within a taxon appears to be at its greatest in regions near the edge of the effective dispersal range. Thus, although a high proportion of its plants and birds are endemic, Hawaii has only three skinks, none of which is endemic. Their colonization may perhaps have been human-assisted and fairly recent: Hawaii is thus beyond the effective dispersal range for skinks. In contrast, the less remote islands of the Pacific Ocean evidence a high proportion of endemism of skinks and, on still less remote islands, of rodents. Ancient 'continental' island fragments (e.g. Madagascar) can also be rich in endemics, although this can reflect varying contributions from (1) stranded 'vicariant' forms and (2) the operation of long-distance dispersal over long periods of time (e.g. particularly significant for the New Zealand flora).

Some lineages have done particularly well on oceanic islands, land snails and fruit bats among them. Particular genera, such as the Hawaiian *Drosophila* (fruit flies) and *Cyrtandra* (a plant example) have radiated spectacularly. Terrestrial mammals have made a good showing only on less remote islands, often those insufficiently remote to have escaped the attention of *Homo sapiens* at an early stage, and many have thus failed to survive to historical times. New discoveries continue to be made of living (extant) and fossil (extinct) island forms and this serves as a cautionary note, such data must be appraised carefully before we can attempt to explain current biogeographical patterns in terms of evolutionary theory.

PART II

Island Ecology

Volcanic islands can reach considerable heights, and then are subject to process of erosion, dissection and sometimes catastrophic landslides. In the distance, the youthful peak of El Teide (3718 m) is less than 1300 years old, whilst the foreground looks across the much older collapse feature of the Orotava valley. In the foreground, a stand of Canarian pine forest is disappearing into the cloud band that forms daily in association with the trade wind inversion, and which is a key regulator of water balance and of vegetation zonation on the island of Tenerife.

CHAPTER 4

Species numbers games: the macroecology of island biotas

Area is the devil's own variable.

(Anon.)

Coastal islands are notorious for their accumulation, at all seasons, of a staggering variety of migrant, stray, and sexually inadequate laggard birds. In the absence of specific information on reproductive activity, it is therefore unwarranted to assume tacitly that a bird species, even if observed in the breeding season, is resident.

(Lynch and Johnson 1974, p. 372)

Having established the biogeographical significance of islands, we now move on to the theories that have been developed from their study. We take our lead from Robert H. MacArthur and Edward O. Wilson's (1967) classic monograph *The theory of island biogeography*, by working from a consideration of ecological patterns and processes, through island evolutionary processes to the emergent evolutionary outcomes.

In this first *ecological* chapter, we focus on a research programme within ecological biogeography in which MacArthur and Wilson provided the seminal influence. This programme centres around species numbers as the principal response variable, and is concerned with: the nature of species–area relationships; the factors determining the number of species found on an island; and the rates of species turnover in isolates. MacArthur and Wilson's (1963, 1967) contribution represented a bold attempt to reformulate (island) biogeography around fundamental processes and principles, relating distributional pattern to population ecological processes, and at its heart lay a very simple, but powerful concept. Their **dynamic, equilibrium** model postulates that the number of species of a given taxon found on an island will be the product of opposing forces leading respectively to the gain and loss of species and resulting in a continual turnover of the species present on each island through time (Box 4.1). Although first proposed as a zoogeographic model (MacArthur and Wilson 1963), the model was promoted in their 1967 monograph as a general model covering all taxa and all forms of island.

Their theory was developed in large measure to account for apparent regularities in the form of species–area relationships, and invokes varying rates of species gain and loss as a function of isolation and area respectively. This they captured in the famous graphical model, in which immigration rate declines exponentially and extinction rate rises exponentially as an initially empty island fills up towards its equilibrial richness value (shown by

---

Box 4.1 The first formal statement of the dynamic equilibrium model

'We start with the statement that

$$\Delta s = M + G - D$$

where s is the number of species on an island. M is the number of species successfully immigrating to the island per year. G is the number of new species being added per year by local speciation (not including immigrant species that mainly diverge to species level without multiplying), and D is the number of species dying out per year.'
From MacArthur and Wilson (1963, p. 378)

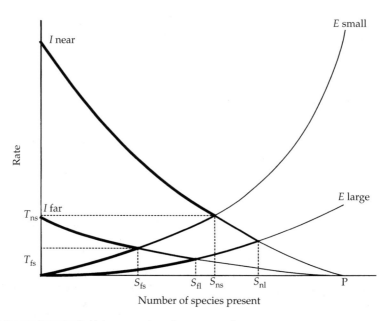

Figure 4.1 A version of MacArthur and Wilson's (1963, 1967) equilibrium model of island biogeography (EMIB), showing how immigration rates are postulated to vary as a function of distance, and extinction rate as a function of island area. The model predicts different values for S (species number), which can be read off the ordinate and for turnover rate (T) (i.e. I or E, as they are identical). Each combination of island area and isolation should produce a unique combination of S and T. To prevent clutter, only two values for T are shown.

the intersection). The immigration rate curve flattens with increasing isolation, and extinction rate with increasing area, thereby generating a family of curves providing unique combinations of richness and turnover for each combination of area and isolation (Fig. 4.1). Evolution is subsumed in the model with the assumption that as an island becomes ever more isolated, new forms are increasingly likely to arise as a result of *in situ* radiation rather than immigration. This elegantly simple core model will be termed here the **equilibrium model of island biogeography (EMIB)**, although as others have stressed it is really the *dynamic equilibrial* nature of the model that distinguishes it.

At the outset, it is important to emphasize that *The theory of island biogeography* contained a lot more than this simple model, indeed it was really just the starting point within the 1967 volume, and the authors went on to develop a series of increasingly sophisticated models of modifying effects, and of properties of effective island colonists, end-

ing the book by feeding these ideas into evolutionary models. As we discuss in a later chapter, at the time of their collaboration, Wilson had already published seminal papers on character displacement and the taxon cycle. So although it is fair to criticize the EMIB for its simplicity (which is also its strength), it is inaccurate to claim, as some have, that their *theory* gave no consideration to speciation and evolution.

MacArthur and Wilson's ideas have proved pivotal to debate in island ecology (and conservation biology) for several reasons. One is that their theory has focused attention on two of the most pervasive, fundamental themes of the subject:

- Do ecological systems operate in an essentially balanced, equilibrial state, or do they exhibit non-equilibrium modes of behaviour?
- To the extent that they approximate equilibrial behaviour, how inherently dynamic are they over time?

These themes can be found in subfields of biogeography as widely spaced as diversity theory, succession theory, and savanna ecology, to name but a few. The island laboratories once again have been a testing ground for developing and refining ideas of broader relevance (Brown 1981, 1986), and the EMIB was the starting point.

Hence, in the present chapter, we review the elements that MacArthur and Wilson drew together in their ecological model, and how well the EMIB has fared in the light of empirical testing. We go on to review recent developments in our understanding of the dynamics of species numbers in isolates, some of which build on, and others of which differ from the MacArthur–Wilson dynamic model. As will become evident, this chapter pays little attention to the compositional structure of island biotas; rather we are dealing in the emergent statistical properties, or **macroecology** (Brown 1995) of island assemblages. The regularities of form in these macroecological properties point to the existence of underlying mechanisms or rules governing ecological systems: the efforts reviewed in the present chapter have been concerned with identifying what they might be.

4.1 The development of the equilibrium theory of island biogeography

Several ingredients were combined in the dynamic equilibrium model (see account by Wilson 1995). The first is the long-known observation that species number increases in predictable fashion in relation to island area (e.g. Arrhenius 1921)—termed here the **island species–area relationship (ISAR)** (Box 4.2). For instance, the zoogeographer Darlington (1957) offered the approximation that, for the herpetofauna of the West Indies, 'division of island area by ten divides the fauna by two'. A second key ingredient was that in examining data for Pacific birds, MacArthur and Wilson (1963) noted that distance between the islands and the primary source area (i.e. island isolation) appeared to explain a lot of the variation. The third ingredient was the notion of the turnover of species on islands, a force that Wilson (1959, 1961) had previously invoked on evolutionary timescales in the taxon cycle (Chapter 9), and which other biologists had described as occurring on ecological timescales, for instance in analyses of plant (Docters van Leeuwen 1936) and animal

---

Box 4.2 Typology of species-number curves and relationships: an attempt to demystify

It is important to distinguish the four frequently used macroecological tools discussed in this box. Within the wider literature, a range of acronyms have been used for them. We use the acronyms SAC, SAD, and ISAR for convenience, but the reader should beware: the 'A' stands for three different terms.

Species abundance distributions: SADs
The species abundance distribution refers to how the number of individuals is apportioned across the species present in a community. SADs can be displayed graphically in several ways, e.g. as a plot of the frequency of each species against its rank importance in the community, or as a plot of the number of species of increasing classes of

abundance (see Magurran 2004). SADs are not a particularly key tool in island species-numbers studies, but assumptions concerning the form of the SAD are important to theoretical mechanistic arguments linking patterns of richness to species turnover (see text).

Species accumulation curves: SACs
The species accumulation (or collectors) curve is a graph of the *cumulative* number of species (*y* axis) with increasing sampling effort (*x* axis), where sampling effort could be variously formulated, e.g. (1) recording additional individuals within the chosen sampling area, (2) additional sample plots of a particular type from within a general locality, (3) additional contiguous sampling plots.

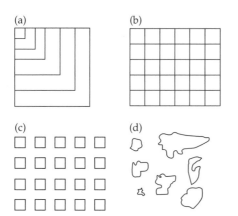

Classification by Scheiner (2003) of different types of sampling layout used in constructing species–area curves: (a) strictly nested quadrats, (b) a contiguous grid of quadrats, (c) a regular but non-continuous grid, and (d) a set of areas of varying size, often islands.

Researchers use SACs to estimate and compare the species richness of different localities (Magurran 2004). SACs based on design (3) are frequently used in plant ecological research to determine an appropriate minimum plot size to use for sampling vegetation. As sampling effort is represented by area, these plots are in fact often called species–area curves.

Species–area curves

The term species–area curve has been applied to two very different types of plot of species number versus area. First, it has been used for species accumulation curves where the *y* axis represents the *cumulative* number of species and where the *x* axis often (but not in all studies) represents contiguous areas (as above). Second, it has been used for plots of the species number *per sample*, where the *x* axis often (but not in all

studies) represents spatially separated areas, such as islands. The use of the same term for such very different phenomena has been and remains the source of considerable confusion (e.g. compare Scheiner 2004a,b with Gray *et al*. 2004). To save us from future confusion, Gray *et al*. (2004) argue for restricting the use of the term species–area curve to plots of numbers of species per sample against sample area, but the dual use of the term seems too deeply embedded for this proposal to succeed. Moreover, it is easy to find yourself using the same acronym for species accumulation and species–area curves. Our solution here is to avoid the use of the term species–area curve when a more precise term can be used: and we hope that the key distinctions between different types of analysis can then be more readily grasped. Just remember in reading the wider literature to check what type of analysis is being done under the heading of species–area curve or species–area relationship. Scheiner's (2003) scheme shown in the figure provides a useful starting point in appreciating differences between analyses.

Island species–area relationships: ISARs

Similar to the species–area curve, different authors may mean different things when they talk about the species–area relationship (SAR, or sometimes SPAR). In the present work we attempt to restrict the use of the term to refer to analyses of the number of species found on each of a set of distinct islands or habitat islands in relation to the area of each unit. Thus, the only permitted sampling design is the non-nested type (d) in the graphic. To make this explicit we have added the term island, to create the acronym ISAR.

(Dammerman 1948) colonization of the Krakatau islands. MacArthur and Wilson linked these observations to work (by Preston 1948, 1962 and others) on the distribution of individuals within an assemblage, i.e. the **Species–Abundance Distribution** (or **SAD**; Box 4.2), which had established the

prevalence of rare species in virtually all ecological assemblages.

MacArthur and Wilson's theory combines all these elements. In essence, from among the rare species typical of ecological systems, those species present in an island system at very low densities would have little chance of persisting and should frequently go locally extinct. And, the further they had to travel from the mainland, the less frequently they would recolonize. Hence the ISARs should vary in relation to isolation, as a function of varying rates of immigration to and extinction from the islands: producing predictable patterns of richness and turnover. There should also be systematic variation in SADs accompanying the pattern in ISARs; although sampling problems in detecting the form of the tail of SADs (and analytical problems in comparing them) make this harder to assess.

It has been noted that K.W. Dammerman (1948), in his monograph on Krakatau, had described several of the key elements of the eventual equilibrium theory, although he had failed to put them into a mathematical framework (Thornton 1992). Another biologist, Eugene Munroe, went one better than Dammerman. In his studies of the butterflies of the West Indies, Munroe presented a formulaic version of the equilibrium model, similar to that independently developed by MacArthur and Wilson in their 1963 paper (Brown and Lomolino 1989). Munroe's formula appeared only in his doctoral thesis (1948) and in the abstract of a conference paper published in 1953. Neither Dammerman nor Munroe went on to develop the ideas beyond a fairly rudimentary form.

In the following sections we explore, in a little more depth, the components MacArthur and Wilson brought together in their theory.

Island species–area relationships (ISARs)

Building on previous work, MacArthur and Wilson (1963, 1967), concluded that for a particular taxon and within any given region of relatively uniform climate, the ISAR (Box 4.2) can often be approximated by the **power function model**:

$$S = cA^z$$

where S is the number of species of a given taxon on an island, A is the area, and c and z are constants determined empirically from the data, which thus vary from system to system.

MacArthur and Wilson, like others before and since, were interested in determining species–area relationships in the most tractable fashion for further analysis, which is to find the best transformation that linearizes the relationship. Following Arrhenius (e.g. 1921) they favoured logarithmic transformations of both axes (sometimes termed **Arrhenius plots**). The equation thus becomes:

$$\log S = \log c + z \log A$$

which enables the parameters c and z to be determined using simple least squares (linear) regression. In this equation, z describes the slope of the log–log relationship and log c describes its intercept. Thus, a low z value (slope) means that there is less sensitivity to island area than for a system of high z value, while c values reflect the overall biotic richness of the study system, and thus vary with taxon, climate and biogeographical region. Values of z are easy to compare between study systems, but the parameter c changes with different scales of measurement used (e.g. km^2 versus miles^2), requiring conversion of data to a common measurement scale before they can be analysed for comparative purposes.

From the data at their disposal, MacArthur and Wilson (1967) found that in most cases z falls between 0.20 and 0.35 for islands, but that if you take non-isolated sample areas on continents (or within large islands), z values tend to vary between 0.12 and 0.17. Thus, the slope of the log–log plot of the species–area curve appeared to be steeper for islands, or, in the simplest terms, any reduction in island area lowers the diversity more than a similar reduction of sample area from a contiguous mainland habitat. Wilson's (1961) own data for Melanesian ants (Fig. 4.2) was one of the data sets used in this analysis, although there is a problem in the construction of this and other such comparisons, as we explain later.

Although it is possible to find many examples that broadly support this island–mainland distinction (e.g. Begon *et al.* 1986, Table 20.1; Rosenzweig 1995), there are also exceptions. Williamson (1988) reviews surveys providing the following slope

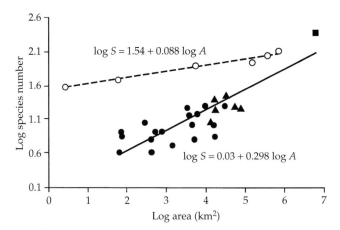

Figure 4.2 Species–area relationships for ponerine and cerapachyine ants in Melanesia. Solid dots represent islands; open circles, cumulative areas of New Guinea up to and including the whole island; triangles, archipelagos (not used in the regression); and the square, all of South-East Asia. (Adapted from Wilson 1961.)

ranges: ordinary islands, 0.05–1.132; habitat islands, 0.09–0.957; and mainland samples −0.276 to 0.925. It is important to establish the reliability of MacArthur and Wilson's generalizations on z, and whether the apparently errant values reported by Williamson can be explained away. More generally, the simple formulation $S = cA^z$ in practice has provided a wealth of opportunity for confusion and controversy and we therefore return to the topic and update what we have learnt since MacArthur and Wilson, below.

Species abundance distributions

In most plant and animal communities there are few species of many individuals and many more species of few individuals. This is one of the most robust patterns in community ecology. There are two foundational theories concerning the distribution of abundance, they differ in relation to the abundance of the rarer species. Fisher *et al.* (1943) suggested that the largest class of species is of those that are individually rarest. This gives rise to the **logarithmic series of abundance**. Preston (1948, 1962) developed the alternative theory that species more typically fit a **log–normal series of abundance**, i.e. that the most numerous species are

those of middling abundance. He argued that insufficient sampling commonly had given rise to the apparent fit of the logarithmic model. This can be illustrated diagrammatically for a hypothetical community if the abundances of individual species are plotted on a log scale against number of species on an arithmetic scale (Fig. 4.3). If the sample is small, then the sparsest species of the log–normal distribution will not be sampled, and the abundance distribution will be that shown to the right of the **veil line** drawn at point a (Preston 1948). On increasing the sampling effort, more of the rarer species of the system will be sampled, pushing the veil line to position b. By analogy, a small area veils'—or rather excludes—the existence of all the species whose total abundance falls below a critical minimum.

To put this into a more familiar ecological context, imagine an area delimited for study within a continuous woodland habitat, containing many birds, but just one or two pairs of a particular bird species—the rare end of the distribution. Through chance effects, such as harsh weather or predation, the species might be lost locally, but be replaced almost immediately by other individuals from the surrounding woodland taking over the territory. Repeat this thought experiment, but for an isolated island: the

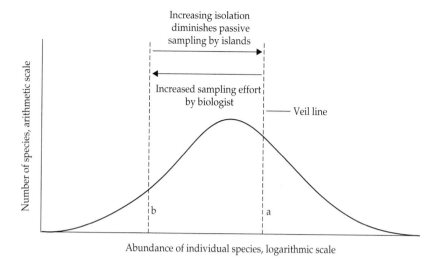

Figure 4.3 Preston's log–normal species–abundance relationship. The portion of this hypothetical abundance distribution that is sampled may be a function of either active sampling effort (by the biologist) or passive sampling effort (by an island). Only the portion to the right of the veil line will be sampled.

species would be lost and might not be replaced by immigration for some years (e.g. Paine 1985). Average this across the whole island assemblage and the island will, at any one time, fail to 'sample' a significant number of (principally) the rarer birds found in the mainland habitat patch (i.e. they will be absent). Preston showed that a log–normal series of abundance should give rise to z values of approximately 0.263, towards the low end of the range of values then available from islands and above those of continental patches. These differences therefore pointed to a role for isolation via population migration, and, in short, Preston's theorizing, combined with his analysis of island data, led him inevitably to a focus on turnover in his 1962 article, foreshadowing by a few months several of the elements of MacArthur and Wilson's (1963) paper.

For present purposes these points of detail are not critical, yet it remains of compelling interest to theoreticians to determine the best species abundance distribution (SAD) model, because SADs are foundational to much ecological theory (e.g. Hubbell 2001). One crude way of analysing this indirectly is by examination of the form of the ISAR, as on theoretical grounds it is expected that a plot of S versus log A should produce a straight line where the loga-

rithmic series applies, whereas a log S versus log A plot should produce a straight line where the log–normal series applies. By reference to the argument in the previous paragraph, it might be anticipated that the former situation would be found to apply for large samples (or areas) and the latter for small samples (or areas). While noting exceptions, both Williamson (1981) and Rosenzweig (1995) conclude that empirical studies from islands show a fair degree of support for Preston's model: the species–area plot is more usually linear on a log–log plot than any other simple transformation. Although this has a wider significance, for the moment it is sufficient to note that the position taken on species–abundance patterns by Preston, and by MacArthur and Wilson, was entirely reasonable.

The distance effect

MacArthur and Wilson (1963, 1967) predicted that ISARs would become steeper with increasing geographic isolation. Unfortunately, the increased impoverishment of island biotas with increasing isolation is confounded with variations in other properties of islands, particularly their area. The distance effect has turned out to be difficult to test;

moreover, those tests that have been conducted have proved equivocal. At the outset, MacArthur and Wilson (1963, 1967) produced a way of quantifying the effect for Pacific birds. First, they showed that there was a fairly tight, straight-line log–log ISAR for the closely grouped Sunda Islands, but that an equivalent analysis for 26 islands and archipelagos from Melanesia, Micronesia, and Polynesia had both a much greater scatter and a tendency away from a straight-line relationship. Secondly, they reasoned that a line drawn through the points for near islands could be taken to represent a 'saturation curve', and the degree of impoverishment of islands below the line could then be estimated as a function of area by comparison with the saturation curve. In general, the degree of impoverishment, when corrected for area, was found to increase from islands 'near' to New Guinea (less than 805 km) to intermediate, to far islands (greater than 3200 km).

Williamson (1988) suggests that this demonstration is invalid because the example used archipelagos rather than single islands, and included archipelagos of different climate or deriving their avifaunas from different sources. He reports that, in fact, the line tends to be flatter the more distant the archipelago, such that (with some caveats) slopes of 0.28, 0.22, 0.18, 0.09, and 0.05 can be derived for birds in the progressively more isolated archipelagos of the East Indies, New Guinea, New Britain, Solomons, and New Hebrides (Vanuatu) archipelagos, respectively. It appears that the assumptions made in analysing species–area curves in relation to isolation, e.g. whether to lump separate islands into archipelagos, whether to calculate isolation within or between archipelagos, whether to include islands of rather different climates, etc., are critical to the outcome of the analysis (compare Itow 1988; Williamson 1988; Rosenzweig 1995). In any case, the discovery since the 1960s of numerous 'new' bird species from subfossil remains on many of the islands of the Pacific (Chapter 3; Steadman 1997*a*) renders this particular part of the MacArthur and Wilson analysis problematic in hindsight, as several points on their graph are below their 'natural' level. Again, we re-evaluate the distance effect below.

Turnover, the core model (EMIB), and its immediate derivatives

It was these observations—of ISARs, SADs and distance—which were combined in the dynamic equilibrium theory, through the mechanism of the **turnover of species** on islands. The theory postulates that there are two ways in which islands gain species, by immigration or by evolution of new forms, and that these means of increasing species number will be balanced in the equilibrium condition by processes leading to the local loss of species from the island in question. Before immersing ourselves in the detail of the model, it is important to take note of the working definitions of the key terms involved (Box 4.3).

It is a reasonable generalization that immigration rate diminishes as a simple function of increasing distance, hence lowering the equilibrium point for more remote islands. This is only partially countered by increased speciation rates on remote islands, because of the relatively slow pace of phylogenesis. For the moment we will set aside the role of *in situ* speciation and concentrate solely on immigration. The local loss of a species population may be accomplished either by out-migration or by the death of the last representatives on an island, in either case leading to the local extinction (extirpation) of the species. The greater resource base of larger islands should mean that extinction rates for a given species richness are lower for larger islands than for smaller. Thus the EMIB postulates that the number of species found on an island represents a dynamic balance between immigration (*I*) and extinction (*E*), with immigration varying with distance from source pool and extinction with island area. This was presented in simple, accessible, diagrammatic form (Fig. 4.1)—possibly one of the keys to the broad uptake of the theory (Brown and Lomolino 1989).

MacArthur and Wilson (1967) also considered the effects of slightly more complicated configurations, for example where chains of islands ('stepping stones') are found strung out from a mainland, producing alterations in the expected immigration functions and thus in equilibrium values. However, the basic ideas may be most

---

Box 4.3 Definitions of terms involved in analyses of species turnover

MacArthur and Wilson (1967, pp. 185–191) developed the following formulations:

Immigration. The process of arrival of a propagule on an island not occupied by the species. The fact of an immigration implies nothing concerning the subsequent duration of the propagule or its descendants.

Immigration rate. Number of new species arriving on an island per unit time.

Propagule. The minimal number of individuals of a species capable of successfully colonizing a habitable island. A single mated female, an adult female and a male, or a whole social group may be propagules, provided they are the minimal unit required.

Colonization. The relatively lengthy persistence of an immigrant species on an island, especially where breeding and population increase are accomplished.

Colonization curve. The change through time of numbers of species found together on an island.

Extinction. The total disappearance of a species from an island (does not preclude recolonization).

Extinction rate. Number of species on an island that become extinct per unit time.

Turnover rate. The number of species eliminated and replaced per unit time.

MacArthur and Wilson's definitions are fine in theory, but in practice difficult to apply in empirical work. Very often available survey data do not allow researchers to distinguish between a species that has immigrated and one that is observed in transit or in insufficient numbers or condition to found a population. Similarly, total disappearance of a relatively inconspicuous plant or animal from a large, topographically complex island may be more apparent than real.

---

easily grasped without reference to such complicating features.

Figure 4.1 illustrates immigration and extinction as hollow curves, for the following reasons. When the number of species on an island is small relative to the mainland source pool, P, a high proportion of propagules arriving on an island will be of species not resident on the island immediately before that point. As the island assemblage gets larger, fewer arrivals will be of species not already present, and so the immigration rate (as defined in Box 4.2) has to decline as S nears that of the total mainland pool. The hollow form of the extinction curve recognizes that some species are more likely to die out than others, and that in accord with our knowledge of SADs, the more species there are in a sample or isolate, the rarer on average each will be, and so the more likely each is to die out. At equilibrium, immigration and extinction rates should be approximately in balance and thus equal the rate of species turnover ($I = E = T$).

Islands of different size or different isolation may have the same turnover, or the same species number, but they cannot have both the same turnover and species number. Figure 4.1 thus shows how the theory predicts different combinations of S and T as a function of area and isolation. This reading of the figure uses it as a representation of space. The diagram may also be understood as a representation of change through time for a particular island, commencing with initial colonization by the first inhabitants, when immigration rate is high, species number low, and extinction rate low. The curves may be followed towards the equilibrium condition when species number is high and stable, immigration rate has declined, and extinction rate has risen to meet it. (However, note that the visual appearance of the curves would be different if expressed in normal units of time: as shown in Fig 4.4 the increase in species richness per annum slows greatly as I and E converge.) If we take the case of a small, near island, species number at equilibrium will be S_{ns} and the rate of turnover at equilibrium will be T_{ns}. The portion of the I and E curves to the right-hand side of their intersection can, of course, be disregarded, as once they have met, neither rate should vary further to any significant degree; the island has reached its dynamic equilibrium point.

The equation for equilibrium is:

$$S_{t+1} = S_t + I + V - E$$

where S is the number of species at time t, S_{t+1} is the number at time $t + 1$, I denotes additions through immigration, V denotes additions through evolution (where applicable), and E denotes losses by extinction (it will be noted that as species are integers, continuous representations are approximations). Integration of the rates of I and E as they vary over time produces the colonization curve (Box 4.2). This rises steeply initially, but ever more slowly as eventual equilibrium is neared (Fig. 4.4).

MacArthur and Wilson (1967) recognized that biologists can rarely, if ever, be certain of recording all immigration and extinction events in real-world systems. But they provided a little-used way around this, reasoning that the colonization curve might be used to back-calculate indirect estimates of what the real I and E rates might be at two stages of the colonization process. The first point is near the beginning of colonization, where I may be assumed to be close to the rate of colonization and E to be insignificant (although both assumptions appear rather bold when successional factors are considered: Rey 1985; Chapter 5). The second point is near the equilibrium condition. In this case they developed a proof for derivation of the extinction rate at equilibrium as a function of the species numbers on other similar islands already at equilibrium and the time taken to reach 90% of the equilibrial number of species on the target island. They

detailed a number of studies in apparent support of this line of reasoning. The first of which was for bird recolonization of the Krakatau islands, which appeared to have reached a very dynamic equilibrium by the 1920s (MacArthur and Wilson 1963). This example is still cited in support of the theory (e.g. Rosenzweig 1995), but as MacArthur and Wilson (1963, p. 384) speculated might be the case, this was in fact a premature conclusion; indeed data had already been published which showed a further increase in species number (Hoogerwerf 1953).

The above constitutes the simplest manifestation of their theorizing, the basic dynamic equilibrium model (the EMIB), which occupies the second and third chapters of MacArthur and Wilson's (1967) monograph. It provides an essentially stochastic macroecological model of biological processes on islands, in which the properties of individual species get little attention. Its great virtues were first, that it was a dynamic model invoking universal ecological and population processes, and second, its apparent testability. The predictive nature of the equilibrium theory is important. Although it may have heuristic value without it, the contribution of the theory to biogeography was in large part to point the way towards a more rigorous scientific approach to the geography of nature. As Brown and Gibson (1983, p. 449) put it:

. . . like any good theory, the model goes beyond what is already known to make additional predictions that can only be tested with new observations and experiments. Specifically, it predicts the following order of turnover rates at equilibrium: $T_{SN} > T_{SF} \equiv T_{LN} > T_{LF}$.

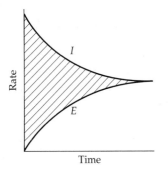
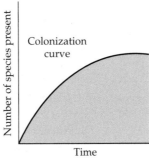

Figure 4.4 Integration of immigration and extinction curves (left) should, theoretically, produce the colonization curve as shown (right). (Redrawn from MacArthur and Wilson 1967, Fig. 20.)

The next element of their theory sketched out, in mathematical terms, a portrait of the biological attributes of the superior colonist and how island environments should 'select' for different species types at early and late stages of the colonization process. They borrowed from the standard notation of population models (r, the intrinsic rate of population increase; K, the carrying capacity), to describe early colonists as being **r-selected**—highly mobile, opportunistic colonizers capable of rapid growth, maturity and population increase, and later colonists as **K-selected**—slower dispersing and growing, but with greater ability to sustain their populations in resource-limited systems approaching the environmental carrying capacity. These ideas influenced Jared Diamond who in his work on island assembly rules introduced the terms **supertramps**, **tramps**, and **sedentary species** for the continuum from highly r-selected (equivalent to pioneer in the succession literature) to K-selected (late successional) species (see Chapter 5).

Another chapter considered additional effects of stepping stones, allowing enhanced dispersal routes and thus altering patterns of biotic exchange between different biogeographic regions. The later chapters also incorporated Wilson's ideas concerning niche shifts and the taxon cycle and radiation zones. MacArthur and Wilson (1967) thus built up their analyses from a highly simplified island ecological model, adding increasing evolutionary detail through the book. In parallel, there is a movement from an essentially stochastic first approximation to increasingly deterministic ideas. As they wrote on p. 121, 'A closer examination of the composition and behavior of resident species should often reveal the causes of exclusion, so that random processes in colonization need not be invoked.'

Lynch and Johnson (1974, p. 371), in their critical analysis of problems inherent in testing the EMIB, argued that for the theory to apply

' . . . turnover should reflect the stochastic nature of an equilibrium condition (i.e. the turnover should not be attributable to some systematic bias such as ecological succession, human disturbance of habitats, introduction of exotic species, etc.)'.

This is a fair assessment in relation to the simplest version of the EMIB, but as we have shown, it is unnecessarily restrictive in relation to the more extensive overall body of MacArthur and Wilson's theory, in which some structure is acknowledged and even predicted to occur. This is indicative of a general problem in assessing MacArthur and Wilson's theory: there is scope to define its limits in rather varied ways. Hence, although individual components and propositions may be tested and refuted, when it comes to assessing their legacy in the round, it is more a matter of assessing whether their assertion (1967, p. 65) that ' . . . only equilibrium models are likely to lead to new knowledge concerning the dynamics of immigration and extinction . . . ' has been borne out, and whether the research programme they were largely responsible for initiating continues to pose good questions and produce interesting insights.

4.2 Competing explanations for systematic variation in island species–area relationships

Attempts to evaluate and build on MacArthur and Wilson's theory were initially concerned, first, with the form of ISARs, and second, with the occurrence of the equilibrial dynamic at the heart of the theory (Gilbert 1980; Williamson 1988). In the following sections we consider both aspects. In the dynamic model (Fig. 4.1), both area and isolation are predicted to influence island richness, but typically the analytical approach has been to compute the ISARs for islands of varying isolation and to compare their slopes and intersects. Thus, the standard approach to the problem has been to treat area as the primary explanatory variable.

MacArthur and Wilson's EMIB is not the only explanatory model competing to explain patterns of variation in island species richness (e.g. Connor and McCoy 1979; Kelly *et al.* 1989). Here we list some alternative ideas that have been proposed either as alternatives to or as modifiers of the EMIB: some being of potentially wide application, others of more limited focus.

- **Random placement**. If individuals are distributed at random, larger samples will contain more species. An island can be regarded as a sample of such a random community, without reference to particular patterns of turnover. Connor and McCoy (1979) term this **passive sampling**, and advocate its use as a null hypothesis against all alternatives (debated in Sugihara 1981; Connor *et al.* 1983).

• The **equilibrium model (EMIB)** postulates the number of species on an island as a dynamic equilibrium between immigration and extinction, dependent on island isolation and area. Only this hypothesis involves a constant turnover of species.

• **Habitat diversity**. The number of species may be a function of the number of habitats: the larger the island, the larger the number of habitats.

• **Incidence functions**. Some species can occur only on large islands because they need large territories, others only on small islands where they can escape from competition (cf. Diamond 1974).

• **Species-energy theory**. According to this theory, the capacity for richness is considered a function of the resource base of the island, where the latter may be estimated, for example, using total primary productivity multiplied by area (Wright 1983). This mechanism may account for variation in species richness but is essentially neutral on the issue of turnover.

• **Small-island effect**. According to this theory, certain species cannot occur on islands below a certain size (e.g. Whitehead and Jones 1969). This effect may be more apparent in marine islands than non-marine isolates.

• **Small-island habitat effect**. In contrast to the previous idea, it has been suggested that in some systems, small islands may be different in character because of their smallness, so that they actually possess habitats not possessed by larger islands and thus sample an extra little 'pool' of species. Alternatively, they may just have a greater diversity of habitats than anticipated from their area, e.g. via telescoping of altitudinal zones (Chapter 2).

• The **disturbance hypothesis** postulates that small islands or 'habitat islands' suffer greater disturbance, and disturbance removes species or makes sites less suitable for a portion of the species pool (e.g. McGuinness 1984, and for a contrary example Wardle *et al.* 1997). Disturbance might also open up sites to invasion by new members.

The above list illustrates that simply finding a significant relationship between species and area is not conclusive in evaluating the EMIB. Kelly *et al.* (1989) made use of a rather underexploited feature of the EMIB: that given knowledge of the species pool, it should be possible to predict the proportion of species in common between islands, or *between equal-sized sample areas* on the islands. Kelly *et al.* (1989) have used the latter approach to evaluate the first three of the above hypotheses for 23 islands within Lake Manapouri, New Zealand. It had previously been established that there was a statistically significant ISAR for the whole flora, with some indication that habitat diversity might be the cause. In their study, they sampled for richness of vascular plant species in two vegetation types (beech forest and manuka scrub) with a fixed quadrat size. This should eliminate the effect of island area on observed species richness if either the random placement or the habitat diversity hypothesis is correct, but should leave an effect of island area if the equilibrium hypothesis is correct. The equilibrium hypothesis failed this particular test. As their sampling design was also intended to eliminate the random placement and habitat diversity hypotheses, they suggested that other ideas, such as incidence functions or the small-island effect, might have explanatory value.

Rosenzweig (1995) has expressed some doubts as to the meaning of this refutation, arguing that the samples were too small, that the restriction to single habitat types was inappropriate, and that the islands were insufficiently isolated. The plots used by Kelly *et al.* (1989) were of 100 m², which is not only a fairly standard size for use in New Zealand forests of this character (J. B. Wilson, personal communication), but is also respectable in relation to islands varying from 242 m² to 2.675 km² in total area. The restriction of sampling to particular forest types also seems a reasonable procedure for distinguishing between the three hypotheses as stated above. The point about isolation, although it does not invalidate the test, is a fair one. Unfortunately for advocates of the EMIB, similar problems apply to many other island ecological studies. Arguably, this study exemplifies the point we develop later, that the EMIB holds relevance to far more restricted geographical circumstances than many of its advocates have envisaged (Haila 1990).

4.3 Island species numbers and ISARs: what have we learnt?

Area and habitat diversity

Two facets of the relationship between area and species number have commonly been recognized: first, a direct area or area *per se* effect, and second, a habitat diversity effect. Although some authors have argued to the contrary, MacArthur and Wilson's (1967) theory is not really an 'area *per se*' argument. They wrote: 'Our ultimate theory of species diversity may not mention area because it seldom exerts a direct effect on a species' presence. More often area allows a large enough sample of habitats, which in turn control species occurrence.' (p. 8).

The **area *per se* effect** refers to the idea that given a uniform habitat, very small amounts of space can support only a limited number of individuals, and thus (through the properties of the SAD), will not be able to sustain all the species capable of occupying the habitat. As the area increases, more individuals can accumulate, sufficient for breeding populations of more species, and richness thus increases. The range in area over which richness will continue to rise will vary depending on the ecological characteristics of the system. It may be very small for lichens growing on rock outcrops, or very large for top carnivores. Gibson's (1986) data for habitat islands of grassland plants within an English woodland indicated that an island biogeographical effect of this sort, if present in the herb flora, was detectable only in patches of less than 0.1 ha. The **habitat diversity effect** is that as area accumulates, so typically does the variability of habitat, such that species specializing in e.g. fresh water, boggy, upland, or low pH habitats are added as the size and complexity of the island increases.

Numerous studies have since been published which quantify, via regression, values for z, and the varying explanatory values of area, isolation, altitude, island age, and various other measures of habitat diversity. Comparative studies of different taxonomic groups from the same islands typically return different explanatory models. For example, multiple regression analyses of Lesser Antillean data selected habitat diversity and maximum elevation for herptiles, isolation for bats, and isolation, area, and habitat diversity for birds (Morand 2000). Area is very often the most important explanatory variable because (via area *per se* and habitat diversity mechanisms) area is a prime determinant of the resource base available to biological communities. With larger islands or reserves, it becomes increasingly difficult to separate the area *per se* effect from the habitat diversity effect (but see Rafe *et al.* 1985; Martin *et al.* 1995). Depending on the size of the species pool, we can nonetheless appreciate that with increasing area of a uniform habitat there may come a point where there is sufficient area to accommodate breeding populations of a large array of species, but where the lack of particular microhabitats prevents all species from colonizing, e.g. the Krakatau islands lack mangrove habitats and thus mangrove species, although propagules are frequently cast up on the beaches.

Very often, habitat diversity may be expected to accrue in fairly simple fashion with increasing island area. However, as shown in many regression studies, particular habitat properties can account for significant amounts of variance in ISARs. Moreover, some studies have found evidence of thresholds whereby a degree of non-linearity is introduced into the ISAR over particular ranges of area. An early example is Whitehead and Jones' (1969) study of plant data from Kapingamarangi atoll (a Polynesian outlier in the west Pacific), in which they argued for threshold effects as follows. The smallest islands, up to 100 m^2 area, are effectively all strandline habitats: their flora is therefore restricted to the limited subset of strandline species. A little larger, and a lens of fresh water can be maintained, and the next set of plants can colonize, those intolerant of salt water. A little larger still and the island would be liable to human occupancy and would be topped up with introduced species. Such modifications to the form of ISARs indicate a greater complexity than captured in the simplest version of the EMIB, but are not necessarily inconsistent with equilibrium theories.

Several authors have developed more formal approaches to address the habitat effects that concerned Whitehead and Jones. One such study is Buckley's (1982, 1985) **habitat-unit model of island biogeography**. The standard method used by most authors has simply been to include indices of habitat diversity, along with area, isolation, altitude, and other geographical factors, in multiple regressions of species number (e.g. Ricklefs and Lovette 1999; Morand 2000). Buckley took a different approach, dividing islands into definable habitats and floristic elements, and relating species richness to area and isolation independently for each unit. His analysis was based on a series of small islets in inshore waters off Perth, Western Australia. He distinguished three terrain units: limestone, white calcareous sands, and red sands, and constructed separate species–area relationships for each. The values were then summed for the whole island. He found a significantly better fit for his habitat-unit model in predictions of actual species richness than derived from whole-island regression (Buckley 1982). This is unsurprising given the increased complexity of the analysis. He subsequently deployed the approach in a study of 61 small islets on bare hypersaline mudflats in tropical Australia, consisting of 23 shell, 19 silt, and 19 mixed-composition substrates (Buckley 1985). The flora comprised three elements: ridge species, salt-flat species, and mangrove species. In this second study, area was found to be the primary determinant of total species number. Moreover, elevation above mudflat was a better predictor of plant species richness than was substrate composition, probably through its effect on the soil salinity profile (cf. Whitehead and Jones 1969). Underlying these results, different floristic elements were found to behave differently as, not surprisingly, both island area and maximum elevation above mudflats were more important determinants of numbers of ridge species than of salt-flat species.

A similar approach to the habitat-unit model has been developed by Deshaye and Morrisset (1988). Working on plants in a hemiarctic archipelago in northern Quebec, they found that habitats worked

as passive samplers of their respective species pools, but that at the whole-island level species number was controlled by both area and habitat diversity. They forcibly expressed the view that data on habitat effects are crucial to interpretations of insular species–area relationships (cf. Kohn and Walsh 1994). In response to this need, Triantis et al. (2003) have proposed a simple approach by which to incorporate habitat diversity within species number modelling for island data sets. They introduce the term **choros** (K), apparently an ancient Greek word that describes dimensional space. K is calculated by simply multiplying the area of the island with the number of different habitat types present. Species richness is then expressed as a power function of the choros: i.e. $S = cK^z$, which is of course directly analagous to the more familiar $S = cA^z$. They compared the performance of both these models for 22 data sets of varying taxa (e.g. lizards, birds, snails, butterflies, beetles, plants), finding significantly improved fits in all but two cases. They further suggested that improvements in fit might be particularly evident when dealing with smaller islands, and the so-called 'small-island effect' (below). There are two points to be made concerning this approach, the first of which is the likely sensitivity of the outcome to how habitats are defined (Triantis et al. 2005). Secondly, that the choros model includes an additional parameter (the calculation is really $S = c(H*A)^z$ (where H is habitat number and A is area), and so an improvement in model fit is in many respects unremarkable. The approach may well prove useful in developing more powerful fits to particular data sets, but the problems of standardization of K across studies means that it will be less easy to incorporate in comparative analyses of multiple data sets.

Area is not always the first variable in the model

Not all studies identify area as the primary variable. Connor and McCoy (1979) speculate that non-significant correlation coefficients between species number and area are probably published less often than discovered, because they may be perceived by authors or reviewers to be uninteresting. One example of no relationship is provided by Dunn and Loehle (1988) who examined a series of upland and lowland forest isolates in south-eastern Wisconsin and for each data set reported slopes not differing statistically from zero. Their isolates were not real islands, nor was there an enormous range of areas in their data set. Factors other than size of isolate, such as disturbance and forest-edge effects, were argued to be influences on plant species number. Their result was consistent with the indirect nature of the control exerted by area, but also with the obvious point that meaningful depiction of species–area effects requires the selection of a wide range of island areas (Connor and McCoy 1979).

Power's (1972) study of bird and plant species on the California islands incorporated 16 islands or island groups with a broad range (0.5–347 km^2) of areas. The results were presented in a path diagram showing the relationships among variables as indicated by stepwise multiple regression (Fig. 4.5). Although area was an important determinant of plant species number, it was a relatively poor predictor of bird species numbers. The role of climate in mediating plant species richness was indicated by the role of latitude. Power suggested that near-coastal islands with richer and structurally more complex floras tended to support a larger avifauna. The analysis thus indicates the importance of hierarchical relationships between plants and birds in determining island species numbers, a line of argument which may also be developed in a successional context (Bush and Whittaker 1991).

Distance and species numbers

The depauperate nature of isolated archipelagos has long been recognized (Chapter 3). Several studies have shown that isolation explains a significant amount of the variation in species number once area has been accounted for in stepwise multiple regression (e.g. Fig. 4.5; Power 1972; Simpson 1974). Problems arise from the practical difficulties of measuring the isolation of a target island because of the existence of other islands (the configuration of which varies through time: Chapter 2), which may serve as networks for species movements across oceans, and because of the multicollinearity often found with other environmental variables (e.g. Kalmar and Currie 2006). Moreover, the effective isolation of islands and habitat islands is not determined solely by the distance from a designated source pool. Rather, for real islands it varies in relation to wind and ocean currents (Spencer-Smith *et al.* 1988; Cook and Crisp 2005), and for habitat islands it also depends on the characteristics of the landscape matrix in which they are embedded (Watson *et al.* 2005).

Several authors have made use of the power function model as indirect evidence for the presence or absence of equilibrium, and in particular cases, it may be possible to construct highly plausible

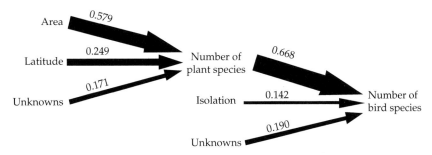

Figure 4.5 Path diagram showing relationships among variables in explanation of species numbers of plants and birds on the California islands (modified from Power 1972). The coefficient associated with each path is the proportion of the variation in the variable at the end of the path explained by the independent variable at the beginning of the path, while holding constant the variation accounted for by other contributing variables.

scenarios by such means. Studies of non-volant mammals on isolated mountain tops in western North America by J. H. Brown and colleagues provide an illustration (e.g. Brown 1971; Brown and Gibson 1983). The mountains of the Great Basin have been isolated from comparable habitat since the onset of the Holocene (c.8000–10 000 years). Brown assumed that immigration across the intervening desert habitats did not occur. It follows that the system dynamics must be dominated by extinction events, i.e. they are 'relaxing' towards lower diversity, ultimately to zero species. Small islands would have had higher extinction rates and have already lost most of their species, whereas large islands still retain most of theirs (Brown 1971). This explained the much steeper z values found for the species–area curve for mammals than for birds. For birds, the isolation involved was not sufficient to prevent saturation of the habitats. Thus, the mammals provided a non-equilibrium pattern and the birds an equilibrium pattern, although not quite to the EMIB model, as it appeared that on the rare occasion that a species went extinct, it tended to be replaced not by a random draw from the mainland pool, but by a new population of the same species (Brown and Gibson 1983).

It is possible to test such models by independent lines of evidence, if for example, subfossil remains can be found of species once found on the mountains but which have become extinct since isolation. Some such evidence was found in the Great Basin. The approach has been applied to a number of other North American mountain data sets, for instance by Lomolino et al. (1989), again showing how palaeoecological data can be combined with reasoning based on species–area relationships to derive plausible models of the dynamics of the insular systems. In this case, however, the montane isolates in the American South-West (Arizona, New Mexico, and southern Utah and Colorado) were argued to be influenced by post-Pleistocene immigrations as well as extinctions.

Connor and McCoy (1979) caution against such procedures, arguing that the slope and intercept parameters of ISARs should be viewed simply as fitted constants, devoid of specific biological meaning, and that it is important to examine the assumptions involved in such studies critically. For example, Grayson and Livingston (1993) observed one of the mountain species, the Nuttall's cottontail (*Sylvilagus nuttallii*) within the desert 'sea', suggesting that immigration across these desert oceans may occasionally be possible. Lomolino and Davis (1997) provide a re-analysis of data for both the Great Basin and the American South-west archipelagos. Their approach is predominantly statistical. They conclude that although the most extreme desert oceans (southern Great Basin) represent an effective barrier to dispersal, in less extreme cases (the American South-West) they represent immigration filters, allowing intermountain movements of at least some forest mammals on ecological timescales. The scenarios constructed are thus more complex than the original, invoking both equilibrium and non-equilibrium systems and varying contributions for immigration and extinction. Although plausible, these deductions ultimately demand verification by independent lines of evidence.

Species–area relationships in remote archipelagos

It was recognized in the EMIB that species number could increase through two means, immigration (I) and evolution (V). In ecological time, only immigration occurs at measurable rates, but in remote islands the evolution of new forms may be more rapid than the immigration rate from mainlands; the inclusion of V was thus necessary for the theory to extend to such systems. Wilson (1969) subsequently published a modified 'colonization' curve, showing how species number might rise above the value first achieved via a process of ecological assortment (e.g. successional effects) and, over long timespans, rise again via evolutionary additions. No means of quantifying the timescale or amplitude of the adjustment was made, making the adjusted theory difficult to falsify.

Several authors have applied equilibrium thinking to evolutionary biotas on remote islands (e.g. MacArthur and Wilson 1967; Juvig and Austring 1979). If the islands within a remote archipelago exchange species moderately frequently, species–area relationships may be detectable at the archipelago level. Moreover, taxa capable of rapid rates of speciation may perhaps so rapidly saturate even geologically young islands that a clear positive correlation could result between species richness and island area. This appears to apply to the highly volant Hawaiian *Plagithmysus* beetles, but not to the flightless Hawaiian *Rhyncogonus* weevils, which instead show a pattern of increasing species number with increasing island age, and a decline with island area (the largest Hawaiian island being the youngest) (Paulay 1994). Juvig and Austring (1979) have undertaken an analysis of species–area relationships within the Hawaiian honeycreeper-finches, which produced results they claimed were consistent with equilibrium interpretations. However, their analysis was undertaken before the extent of the anthropogenic species extinctions became clear. The full distribution of this family before Polynesian colonization is unknown, casting doubt upon interpretations of such species–area calculations (but see the attempt by James (1995) to correct for extinctions).

In addition, the history of introductions to remote islands suggests that, at least for particular taxa (e.g. plants), the total number of species on such islands can rise significantly, suggesting that the islands in question are effectively unsaturated and therefore non-equilibrial (e.g. Groombridge 1992, p. 150). This is perhaps a weak line of argument because carrying capacities have undoubtedly been changed by human alterations of habitats and this, in tandem with the anthropogenic increase in immigration rate, will set a new equilibrium point for the island in question (Sax *et al.* 2002). However, this does not mean that a new dynamic equilibrium will be achieved. Gardner (1986) has shown that the lizards of the Seychelles do not produce a significant ISAR, and that there are unfilled niches on some islands. The patterns of distribution of endemic and more widespread species were found to be best explained by reference to historical events, such as periods of lower sea level in the Pleistocene, and to non-equilibrium hypotheses (see also Lawlor 1986).

The applicability of equilibrium ideas to remote archipelagos may ultimately depend on the effective response rate of particular taxa and the scale at which they interact with the island environment. It is not sufficient to consider demonstration of an ISAR as being proof of dynamic equilibrium in those island groups in which evolutionary phenomena are dominant over ecological phenomena (cf. Williamson 1981, 1988; Lawlor 1986).

Scale effects and the shape of species–area relationships

Scale

As a number of authors have pointed out, it is important to take account of scale factors in analysing species–area relationships (e.g. Martin 1981; Lomolino 1986; Rosenzweig 1995; Sfenthourakis 1996). The details as to which variables best explain variations in S in a given limited data set hold fascination principally in their specific context. They are difficult to generalize. This is in large part because each group of islands (real or habitat) has its own unique spatial configuration and range of environmental conditions. The relative significance of such key factors as area and isolation will depend in part on how great the range (how many orders of magnitude) the sample islands encompass for each variable. Studies of plant diversity where maximum distance is of the order of 500 m, but where elevational range encompasses distinct habitats, vertically arranged, and where area varies by several orders of magnitude, are unlikely to show a significant isolation effect, but are likely to show both area and habitat effects. This is a description of Buckley's (1982) Perth islets data set, in which the purpose of the study required that isolation be effectively controlled. However, it is not just a matter of how great the range is but what the upper and lower limits of the variables are relative to the ecology of the taxon under consideration. This can be appreciated in terms of habitat—the addition of a patch of mangrove habitat will not do the same for plant diversity on a tropical island as the addition of an area of montane cloud-forest.

In terms of isolation, the effective mobility of different taxa (e.g. birds versus terrestrial mammals) and of different ecological guilds (e.g. sea-dispersed vs bird-dispersed plants) can differ greatly, which may also serve to shape the form of species–area comparisons across archipelagos. For a moderately dispersive group, it might be assumed that very small distances have no impact on colonization and thus on species richness. At the other extreme, beyond a certain limit, in the order of hundreds of kilometres, the group might be absent, and thus further degrees of isolation again have no impact. Similarly, Martin (1981) has shown that where the range of areas spans small island sizes, ISAR slopes are liable to be lower than for larger areas. It is thus not surprising that comparative studies show a range of forms for species–area effects between taxa and islands, with different explanatory variables to the fore (e.g. Sfenthourakis 1996). The collective value of such studies is the promise of refinement of, for instance, the dispersal limits for particular taxa, and thus of insights into the dynamics of particular insular systems (e.g. Lomolino and Davis 1997).

What shape is the species–area relationship?

Interest in the form of the species–area relationship is enduring amongst ecologists. Recently, Rosenzweig (1995) has argued that there are four different 'scales' of SPARs (the point scale, islands, intraprovince, interprovince). But, as we show in Box 4.4, it is not only the scale that varies between studies, but the spatial organization of the sample units and the nature of the response variable itself. Hence, the first thing to recognize in attempts to understand the form of the species–area relationship is that the literature encompasses apples (species accumulation curves: SACs) and oranges (island species–area relationships: ISARs). Whilst SACs take a variety of forms, the distinction between them and ISARs is fundamental (Box 4.4, Gray *et al.* 2004). To reiterate, an ISAR is simply the relationship for a set of islands between the number of species occurring on each island and the area of each island.

So, what is the shape of the ISAR? A trite answer would be 'what shape do you want it to be?' As set out above, the conventional approach has been to find a simple transformation of the data that more or less linearizes the relationship, as straight-line fits make the relationship tractable for further analyses. Tjørve (2003) has perhaps controversially argued that our approach to model fitting for SPARs should 'be based on the recognition of biology rather than statistics' (p. 833), i.e. that we should examine the fit of theoretical (mechanistic)

Box 4.4 Combining species accumulation curves and island species–area relationships: apples and oranges?

In Box 4.2 we set out a simple typology of macroecological tools used in analysing species diversity phenomena in relation to area. Although there are several ways in which areas of mainland may be sampled (graphics a, b, and c in Box 4.2) in order to construct a curve of cumulative species number with increased sampling effort (an SAC), there is only one way to construct an ISAR, and that is to plot the number of species per island versus area of that island. The ISAR approach can also be used for habitat islands, i.e. isolated patches of a distinct type of habitat that are surrounded by a strongly contrasting 'sea' of other habitats. In principle, it can also be adopted for comparing scattered sample plots of increasing area from within an unfragmented continuous ecosystem, but typically, published analyses use spatially nested sampling systems instead. Figure 4.2, derived from Wilson (1961), is a classic example. The mainland New Guinea data set represents cumulative areas of New Guinea: it is thus a nested sampling design, and hence species number is of necessity cumulative. According to the typology in Box 4.2, this makes it essentially a SAC on a regional scale, as distinct from the island data also plotted in Fig. 4.2, which we would term an ISAR. By plotting an ISAR for the island data and comparing them with an SAC for the mainland, might we be comparing apples and oranges?

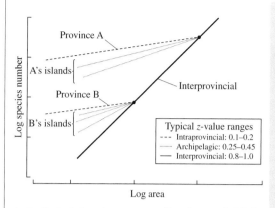

An idealized set of species–area curves (from Rosenzweig 2004).

Rosenzweig (1995, 2004) has developed an alternative typology of species–area curves, stressing differences in scale (see figure), but which also spans these different forms of construct. To show how our terms relate to his scheme, see the table below. Rosenzweig (1995) fashions a powerful argument based on the re-analysis of numerous data sets, and with important implications for conservation biogeography: hence it is important to understand his typology. In contrast to our present treatment, he uses a single term, species–area curve or SPAR, to refer to a sequence of SACs and ISARs applied at different 'scales'. The first level is local species accumulation curves; next comes species–area relationships for a set (archipelago) of islands; next, the construction of a species accumulation curve within a mainland on a regional scale; and finally, plots of overall richness of two or more regions/provinces of different area. This switching between different types of relationship within his analysis is deliberate and he comments on the implications of the nested design. He writes (Rosenzweig 1995, p. 10) in relation to intraprovincial SPARS:

'you must remember to keep your subplots contiguous when you group them to measure the diversity of larger areas. This is called the nested design. If you do not keep them contiguous, but amass them from scattered subplots, the result will have a steeper slope.'

Moreover, being a SAC, the form of such a relationship is typically curved (even in log–log

plots), initially rising swiftly and the rate of increase slowing as more and more of the species pool are included in the cumulative total. Hence we see that the three types of species–area curve generalized in Rosenzweig's scale diagram, are each constructed in quite different ways, with two being island or island-type species–area plots and one a SAC.

Returning to Fig. 4.2, the mainland data represent a regional SAC and the island data an ISAR. Of necessity the former is a smooth plot of a single data series, whereas the island data can include several islands of identical or near identical area, but varying species number, producing scatter around the line of best fit. If we wished to make a more direct comparison between the two data series, it would mean calculating an SAC for the island data, adding islands by order of increasing area, and tallying cumulative richness versus cumulative area. Even then, the spatial structure of the two data series would differ. Thus, to some extent, Wilson and Rosenzweig are comparing apples and oranges, but in the absence (as far as we are aware) of comparative analyses based on multiple mainland vs. island data sets, the implications of the differing structures of regional SACs and ISARS remain obscure.

The figure shows three of the four forms of species–area curves and how they relate to each other in respect of slopes and intersects. Rosenzweig's (1995, 2004) assertion that larger provinces will necessarily contain more species than smaller provinces, thus producing positive slopes for the interprovincial SPAR, is dependent on careful selection of provinces being compared so as to eliminate variation attributable to other factors (e.g. climate).

How the alternative typologies of species numbers analyses compare
Rosenzweig's scheme is as given in the figure, in which each SPAR type has a characteristic range of slope values, and which generalizes all but his point scale SPAR into a combined graphical model.

Our usage	Rosenzweig's scheme
Local SAC	The point scale SPAR
ISAR	Archipelagic SPAR
Regional SAC	Intraprovincial SPAR
Inter-Regional ISAR	Interprovincial SPAR

models against empirical data sets rather than setting out to find the best straight-line fit through the data.

Lomolino (2000c) has also questioned the approach of using data transformation in search of the best linear fit. He argues that there may be good ecological reasons to posit more complex scale-dependent relationships, and that untransformed SPARs should exhibit a sigmoidal form, with a phase across low values of area, where species numbers scarcely increase, followed by a rapid increase with area and a subsequent flattening as the number of species approaches the richness of the species pool (Fig. 4.6).

At the left-hand end of the plot, the existence of a 'small island' effect appears well established (Lomolino 2000c, 2002; Lomolino and Weiser 2001) although this is not to say it occurs in all data sets (below). Lomolino and Weiser (2001) provide a formal analysis of a large sample of ISARs, comparing the results of conventional regression analyses with those obtained from breakpoint regression. These analyses show that the small-island effect is comparatively common in island data sets. The threshold area tended to be highest for species groups with relatively high resource requirements and low dispersal abilities, and also to be comparatively high for more isolated archipelagoes. But with regard to the right-hand end, Williamson *et al.* (2001, 2002) contend that there is no evidence of the upper asymptote whereby the log–log ISAR flattens off across the largest islands. As Williamson and colleagues note, just as the species pool from which isolated biotas are derived contains a finite number of species, so too does the area, i.e. any particular ocean system contains a largest island. The shape of the ISAR as it approaches maximum area might be of a variety of forms, but they contend, empirically fails to show departure from a straight line relationship at the right hand end when plotted in log–log space. Recent analyses by Gentile and Argano (2005) of sets of terrestrial isopod data from Mediterranean islands in which sigmoid models were evaluated against linear, semi-logarithmic,

and logarithmic models tend to support this view. Despite the occurrence of a small-island effect, a sigmoid model failed to provide the best representation of the data. This suggests that the evolutionary effect posited by Lomolino on large islands in practice has an indistinguishable impact on the form of ISARs, which is essentially as suggested by MacArthur and Wilson (1967).

It is noteworthy that the form and biological meaning of ISARS and other forms of species–area plots remain the subject of so much debate and confusion. One important reason is that empirical data sets each have their own unique combination of variation in contributory environmental variables, of area, elevation, climate, habitat type, isolation, disturbance histories, etc. This introduces both structure (e.g. Kalmar and Currie 2006) and some noise into ISARs, which after all merely consider one independent variable: area. Comparison of published regression models, in which richness has been analysed against an array of independent variables, not surprisingly shows a diversity of answers as to which factors enter the models in which order, and with what particular relationship to richness. Each of these variables has relevance, but does not necessarily have significance over all scales within the empirical data gathered. Perhaps unsurprisingly, few analyses have attempted to tackle the possibility of threshold responses in respect of the complex array of potentially important, interacting variables: even for area, formal testing for threshold effects and more complex models of the form of the relationship has been only a recent and fairly controversial development. What we can say with respect to simple data transformations, is that conventional analyses have shown that sometimes semi-log and sometimes log-log plots of ISAR data provide the best fit. The generally preferred approach of log–log plotting of ISAR data typically provides a good fit over most of the range in area, with most evidence for systematic divergence being of the form of the small-island effect.

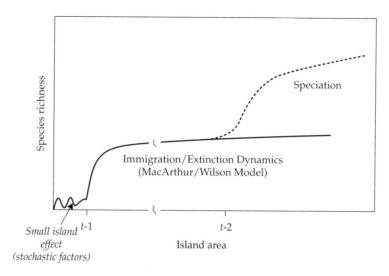

Figure 4.6 Might the island species–area relationship be scale-dependent? This model, taken from Lomolino (2000c), Fig. 6, examines the idea of thresholds in an arithmetic plot of species richness versus area. Starting at the left-hand end, there is little change in species number until a critical threshold is reached, but beyond area $t-1$, species number increases rapidly, as a function of the immigration/extinction dynamics of the EMIB. With islands larger than area $t-2$, species number shows a further increase, as it is afforced by *in situ* speciation. The 'small-island effect' is shown by many island data sets, but evidence for the existence of the second threshold and the upward curve towards the right-hand end of the plot is lacking. See discussion in text.

Species–energy theory—a step towards a more complete island species richness model?

The term **species-energy theory** appears to have been coined by Wright (1983), who modified the ISAR approach by replacing island area with a measure of the total energy available to a particular trophic level. He did so both in a graphical model equivalent to Fig 4.1 and operationally in the regression of island species numbers. His approach was thus very similar to that later adopted by Triantis *et al.* (2003) in the choros model, i.e. he replaced area in the species–area regression with the product of energy and island area. As with area, available energy does not estimate directly the variety of resource types present on an island, but it is likely to be correlated with it. Wright argued that, particularly for sets of islands of variable per-unit-area resource productivity, the species–energy model should work better than the original.

Wright provided two empirical demonstrations, first for angiosperms on 24 islands worldwide,

ranging in size from Jamaica (12 000 km^2) to the island continent of Australia (7 705 000 km^2), and second for land (including freshwater) bird species on an overlapping set of 28 islands, varying from Sumba (Indonesia) (11 000 km^2) to Australia. For angiosperms he used total actual evapotranspiration (total AET), a commonly used, although not ideal, parameter (see O'Brien 1993, Field *et al.* 2005), which estimates the amount of water used to meet the environmental energy demand, and found that the model $S = 123(\text{total AET})^{0.62}$ explained 70% of the variation in the logarithm of species number. For birds, he used, total net primary production (total NPP), and the model $S = 358(\text{total NPP})^{0.47}$ accounted for 80% of the variation in log species number. In both cases, the original species–area model explained significantly less variance. Wright's data were drawn from a set of islands ranging from the equator (e.g. New Guinea) to the arctic island of Spitsbergen (78° N), and thus were a far more climatically and biogeographically heterogeneous set than would normally be considered. They were thus well chosen to demonstrate how bringing climate-based energy availability into analyses of island species numbers may improve model fits.

Studies on gamma (regional) scale patterning of diversity within large landmasses have shown that remarkably simple climate-based models can account for the first-order pattern of species richness regionally and globally (O'Brien 1993, 1998; Hawkins *et al.* 2003, Field *et al.* 2005). These studies imply that there is a causal connection over time and space between climate-based water-energy dynamics, photosynthesis, subsequent biological activity, and species richness of a wide variety of taxa—and that it is worth giving more attention to climate and energy flows in island systems. However, there are also good reasons why a simple, universally applicable formulation may prove elusive.

On theoretical grounds, it is to be expected that spatial scale will be important to the applicability of species–energy theory (Whittaker *et al.* 2001). We know that the species richness of two samples will be dependent on how large an area is sampled. Comparison of a 1 m² patch of chalk grassland with a whole hillside of the same habitat will inevitably find the whole hillside to be richer. It follows that in searching for causal explanations of diversity patterns between differing locations, area should be held constant. On coarse spatial scales (comparing areas of tens to hundreds of square kilometres), climate is one of the more prominent sets of variables influencing plant species richness. At much finer scales, factors such as geology, slope, and grazing regime are likely to be proportionally more important. That is to say, different theories of diversity may hold relevance to different (albeit overlapping) spatial scales (Wright *et al.* 1993; Whittaker *et al.* 2001). Island species–energy theory builds both energy and area in to the ordinate, as island position on the energy axis is corrected for total island area. Thus it is likely that where island sizes span several orders of magnitude, deviations from the best-fit relationship may have different causes for different portions of the island size continuum.

Wright (1983) argues that species–energy theory is consistent with several long-established patterns. For instance, the energy requirements higher up the trophic pyramid are such that smaller areas may be unable to support top predators, and large-bodied animals are often among the first to be lost from areas newly turned into isolates. Wylie and Currie (1993) have applied the species–energy approach to mammals (excluding bats) on land-bridge islands from across the world. They made use of a slightly different set of explanatory variables, but the approach was otherwise very similar. They found, as Wright had for birds and angiosperms, that island energy explained more of the variation in mammal species richness than did area. However, the improvement was much less marked for mammals. The range of island sizes in the mammals data set was 0.4 km² to 741 300 km², and it is possible that the greater range, and in particular, the extension of range down to much smaller islands, may have given rise to a stronger relationship to area in their data. Latitude was also found to have a stronger effect on mammal species number than Wright had found. Of 100 log–log species–area regressions published in Connor and McCoy (1979), the average explanatory value of area was less than 50%. In the three data sets analysed by the species–energy method, the figures ranged from 70 to 90%.

It is a curious fact that the early impact of Wright's (1983) paper was far more evident in the literature on species-richness gradients within and across continents, than in island biogeography. However, a recent paper by Kalmar and Currie (2006), which models island bird richness globally, marks a key step forward in the application of these ideas within island biogeography. Kalmar and Currie compiled a data set of land-bird species richness for 346 marine islands, ranging from 10 ha to 800 000 km², and representative of global variation in climate, topography, and isolation. Using a fifth-root transformation of richness data (close to a logarithmic transformation) and a variety of simple transformations of explanatory variables, they found that as much as 85–90% of the variation in bird species numbers on islands globally can be accounted for by contemporary environmental variables. Their explanatory variables included area, elevation, various isolation metrics, mean annual temperature, and annual precipitation, and they also examined whether faunal region entered their models. Starting with the individual variables, they find constraining (triangular) relationships between richness as the dependent variable and area, temperature, precipitation, and (inversely) distance from the nearest continent, indicating that these variables may each set upper limits to richness, but that richness frequently falls below these limits because of the effects of the other limiting variables. Using correlation and general linear modelling, they show first, that the interaction terms between area and precipitation, and between area and temperature, are crucial to pushing the explained variance above 50%, and second, that the slope of the ISAR increases with rise in mean annual temperature, and with precipitation. Although their simpler models are easiest to interpret and readily reach R^2 values in the range of 0.64–0.70, their final multiple regression model based on quite complex manipulations of area, precipitation, isolation, temperature, and elevation attained a cumulative R^2 of 0.877. Halting the model after the inclusion, in this order, of 'the area–precipitation interaction', 'distance to continent', and 'temperature', provides a model explain 82% of the variation, which given that the data encompass global variation in island environments, and a large range of island areas, is a remarkably good fit. Faunal region had only a minor signal in the data, accounted for by lower than expected diversity in the Sub-Antarctic/Antarctic.

One of the most intriguing findings of their analyses relates to the role of isolation. Although they find that isolation has similar explanatory power to climate and area, the interactions of the explanatory variables tell a different story. According to MacArthur and Wilson (1967), the slope (z values) of ISARs should vary in relation to the difficulty of species dispersing to islands, i.e. more isolated sets of islands should produce steeper ISAR slopes. By contrast, in these global analyses, Kalmar and Currie (2006) report strong area–climate interactions, which are consistent with Wright's species-energy ideas, but they do not find area–isolation interactions. This is particularly intriguing given that their isolation metrics alone are capable of explaining just under 50% of the variation in their data set, and together with area 54%.

These seemingly contradictory findings may indicate that by selecting a globally representative set of islands, they have assembled a data set in which the variation in ISAR slope due to climate has largely swamped the intraregion influence expected as a function of island isolation. So, it is not that isolation has no correlation with richness of island birds, but it is hard to disentangle the independent effect of isolation from variation in area (with which distance is negatively correlated), topography, and climate and the further complications of stepping-stone effects that effectively alter isolation from continents. As Kalmar and Currie are quick to point out, these findings sit uneasily with the prominent role asserted for isolation in MacArthur and Wilson's dynamic island model.

David Lack (1969, 1976) argued that given the ability of birds to colonize remote islands comparatively early in their existence, and then to speciate on them, the apparent impoverishment of remote islands might be less a reflection of immigration–extinction dynamics than of a relative paucity of resources within their food chains. Kalmar and Currie's analyses seem to lend some support to this argument.

The issue remains, to what extent do the patterns discussed above reflect dynamic equilibria of the form proposed in equilibrium theory? To address this, it is essential, as Gilbert (1980) advocated, to give close consideration to species turnover.

4.4 Turnover

According to equilibrium thinking, it is a reasonable assumption that most islands are at, or close to, their equilibrium species number most of the time. The EMIB postulates that this should be a dynamic equilibrium, approached by means of monotonic alterations in rates of immigration and extinction. Each of these postulates requires examination. Before doing so, it is important to consider the nature of the evidence involved. Whether or not these postulates are well founded, there is another important issue that must be considered, which is: when turnover does occur at measurable rates on ecological timescales, is it homogeneous (across all species) or heterogeneous (involving particular subsets of species)?

This question is of fundamental importance to applications of island thinking to conservation (Chapter 10).

Pseudoturnover and cryptoturnover

If we are to assess the rate and nature of turnover of species over time, we have to be able to measure it accurately. Two forms of error have been described. **Cryptoturnover** refers to species becoming extinct and re-immigrating between surveys (or vice versa), thus depressing turnover rate estimates: a problem that will increase with the interval between surveys (Simberloff 1976). A different form of error, **pseudoturnover**, occurs when censuses are incomplete, and information on breeding status inadequate, leading to species appearing to turn over when they have actually been residents throughout, or alternatively when they have never properly colonized. For both these reasons, estimates of turnover (and of I and E) are dependent on census interval (Diamond and May 1977; Whittaker *et al.* 1989) as well as on the thoroughness of each survey (Lynch and Johnson 1974; Whittaker *et al.* 2000).

In assessing turnover, it is important to recognize the distinction between colonization rates and immigration rates (Box 4.3) and to realize that it is exceedingly difficult to measure *real* immigration and extinction rates (Sauer 1969; Abbott 1983). Most studies of faunal or floral build-up are based on lists of species found at different points in time. The rates presented are perforce observed rates of *changes in lists*: despite commonly being labelled immigration and extinction rates (e.g. Whittaker *et al.* 1989, 2000), they are really rather different from the terms as originally defined.

For the extinction of a species to occur, the species must first be present. Unfortunately, if we accept that most extinctions are likely to be of the rarer species, which are hard to detect, it follows that it will be difficult to determine which have really colonized in the first place. Only 'proper' colonists should be counted in determining the extinction rate. Examination of the definitions of terms (Box 4.3) reveals a certain fuzziness at the centre of this business. The distinction between immigration and colonization is indeterminate, as it relies on what constitutes a propagule for a particular taxon (actually it may vary from species to species) and by what is meant by 'relatively lengthy persistence'. Lynch and Johnson (1974) argue that persistence (for a colonist) or absence (for an extinction event) through at least one breeding cycle should be the minimum requirement. In practice, most island biogeographical studies have lacked adequate data on these phenomena, relying principally on lists of species recorded for different islands or different points in time, without adequate knowledge of breeding status, etc. These problems have plagued island turnover studies, and have led to some pointed exchanges in the literature.

The study that prompted Lynch and Johnson's critique and which purported to demonstrate turnover at equilibrium was that of Diamond (1969), who censused the birds of the Channel Islands, off the coast of southern California, 50 years on from an earlier compilation. Diamond found that among the nine islands 20–60% of the species had turned over. However, the interpretation of this turnover was disputed by Lynch and Johnson (1974), who, having first pointed to problems of pseudoturnover, went on to argue that most of the thoroughly documented changes in the birds could be attributed to human influence. This included the loss of several birds of prey due to pesticide poisoning, and the spread into the islands of European sparrows and starlings—both part of a continent-wide process of range expansion. They concluded that the occurrence of natural turnover at equilibrium had not been established by the study. Notwithstanding these data quality problems, Hunt and Hunt (1974) pointed out an interesting feature in the data for one of the islands, Santa Barbara: the predatory land birds were represented by tiny populations, generally just one or two pairs, and this level of the trophic hierarchy thus underwent greater fluctuations in species representation than birds of lower trophic levels.

Nilsson and Nilsson (1985) conducted an interesting experiment by re-surveying the plants of a series of small islets within a Swedish lake. Even with consistent survey techniques, they achieved at best only 79% efficiency per survey. As different species are missed on different occasions, a significant proportion of the recorded turnover could be attributed to pseudoturnover. Having assessed the rate of pseudoturnover, they were then able to calculate a best estimate of real turnover. However, such procedures cannot be applied with any confidence to surveys by different teams many years apart, wherein the expertise, experience, special taxonomic interests or biases, methods, and time spent in surveying are scarcely known or quantifiable. If the islands in question are large or otherwise difficult to survey, the problems are multiplied. It follows that larger, richer islands will have greater rates of apparent turnover because the degree of error in surveying them is liable to be at least as great as for small islands, and will therefore involve larger numbers of species. In circumstances of this sort, resort has to be made to special pleading in order to justify rate estimates and theory verification (e.g. Rosenzweig 1995, pp. 250–8). Such data may tell interesting ecological stories, but scarcely allow unequivocal tests of the EMIB to be made (Whittaker *et al.* 1989; 2000)

In his 1980 review, Gilbert argued vigorously that turnover at equilibrium, a key postulate of the EMIB, and one that set it aside from other competing theories, had not been convincingly demonstrated; the best empirical evidence coming from Simberloff and Wilson's mangrove arthropod experiments (below). Simberloff (1976) had acknowledged the same point in an earlier article, and in the terms demanded by Lynch and Johnson (1974) one is hard pressed to find convincing evidence of stochastic, equilibrial turnover to this day. One paper that comes close is Manne *et al.*'s (1998) analyses of immigration and extinction rates for bird species on 13 small islands of the British Isles. They use a maximum likelihood method to test if immigration rate declines with increasing distance, and extinction rate declines with island size (both should be concave functions, as Fig 4.1). Although the relationships were found to be in the predicted direction, the analyses failed to support the conclusions statistically.

When is an island in equilibrium?

MacArthur and Wilson (1967) understood that measuring the real rates of I and E in the field is exceptionally difficult. They therefore suggested a means of testing turnover using the variance–mean ratio. They argued that if a series of islands of similar area and isolation were to be selected, it is reasonable to suppose that the variance of the number of species on different islands will vary with the number of species present, i.e. variance should be a function of the degree of saturation. At the earliest stages of colonization the variance should be close to 1, declining to about 0.5 as islands become saturated. This provides an approach to assessing whether a series of similar islands (i.e. similar in isolation, area, etc.) might be viewed as equilibrial (e.g. Brown and Dinsmore 1988), but does not constitute a particularly precise tool.

Simberloff (1983) argued that for equilibrium status to be judged it is necessary for researchers to spell out how much temporal variation they will accept as consistent with an equilibrium condition. Only by adopting a specific colonization model, and then testing it against the data, can the equilibrium hypothesis be considered falsifiable. He undertook a study using a simple Markov model, in which extinction and immigration probabilities were estimated independently for passerine birds of Skokholm island, land birds of the Farne Islands, and birds of Eastern Wood (a habitat island in southern England). By comparison of the simulations with the observed data series of between 25 and 35 years, he found a poor fit with the equilibrium model. The data did not show the regulatory tendencies expected if species interactions cause species richness to be continuously adjusted towards equilibrium. For an alternative statistical analysis of these and other data sets, reaching a similar conclusion of non-equilibrium dynamics, see Golinski and Boecklen (2006).

As defined by MacArthur and Wilson (1967) in their glossary (Box 4.3), I, E, and T rates concern change per unit time, and take no account of the size of source pools (although they do elsewhere in that publication). A number of analyses of these phenomena have used formulations in which rates are expressed as a function of the size of the biota on an island during the course of the study, and/or as a function of the size of the mainland pool (e.g. Thornton *et al.* 1990). Where P is unknown, or unknowable, then evaluation of the theory must follow one of the other routes provided by MacArthur and Wilson (1967). Particular formulations may be optimal for different contexts (e.g. early periods of build-up, or near-equilibrium condition) and arguably should be given a different nomenclature (see the useful discussion in Thornton *et al.* 1990).

The rescue effect and the effect of island area on immigration rate

The EMIB promises a workable, testable model because extinction rate is influenced by area (or its correlates) and immigration rate by isolation. If, however, it can be shown that area also influences immigration, and isolation influences extinction, then the model shown in Fig. 4.1 is in some jeopardy (Sauer 1969).

It will be recalled that immigration rate is the term given to the arrival of species not already present on an island. It is logical to assume that a near-shore island for which there is a high immigration rate will also continue to receive additional immigrants of the species which *are* present. This might be termed **supplementary immigration** to distinguish it from immigration proper. An island population declining towards extinction might be rescued by an infusion of new immigrants of the same species, thereby lowering extinction rates on near-shore islands compared with distant islands. This effect of supplementary immigration was termed the **rescue effect** by Brown and Kodric-Brown (1977), who censused the number of individuals and species of arthropods (mostly insects and spiders) on individual thistles growing in desert shrubland in south-east Arizona. Their findings conformed quite well with the predictions of EMIB, with the exception that turnover rates were higher on isolated plants than on those in close proximity to others. This was argued to be because of increased frequencies of immigration events for

species present on the thistles at time 1, either preventing extinctions or perhaps effecting recolonization before a subsequent survey at time 2. Either way, each census was likely to record the species as present on the less-isolated thistles. It has been pointed out that this study was based on rather a notional form of island (individual thistles) and was not based on breeding populations. Since the phrase was coined, the rescue effect has been invoked by a number of authors in explanation of their data (e.g. Hanski 1986; Adler and Wilson 1989); often, as in the original, this seems to be based on its good sense rather than on any direct proof (but see Lomolino 1984*a*, 1986).

Similarly, immigration rate may be affected by area. A large island presents a bigger target for random dispersers. A simple study of sea-dispersed propagules on 28 reef islands in Australia showed that the fraction of the dispersing propagule pool intercepted by each island is proportional to beach length (Buckley and Knedlhans 1986). Larger islands also present a greater range of habitats and attractions to purposeful dispersers, such as some birds, elephants swimming across a stretch of sea, or mammals crossing ice-covered rivers in winter (Johnson 1980; Lomolino 1990). Williamson (1981) points out that MacArthur and Wilson recognized the possible influence of area on immigration in their 1963 paper but discreetly sidestepped this issue in their 1967 monograph.

At least one study has reported both the above 'problems'. Toft and Schoener (1983) undertook a two-year study of 100 very small central Bahamian islands, in which they recorded numbers of individuals and species of diurnal orb-weaving spiders. They found that the extinction rate was related positively to species number and to distance, and was negatively related to area. Immigration rate was positively related to area and negatively related to species number and distance. In consequence, the turnover rate showed no strong relationship to any variable, thus the net result of these interactions is that the effects of area and distance on turnover tend towards the indeterminate. Put simply, the EMIB does not work for this particular insular system.

The path to equilibrium

One of the few rigorous tests of the process of colonization and development of equilibrium was provided by Simberloff and Wilson's classic experiments on mangrove islets in the Florida Keys (Simberloff and Wilson 1969, 1970; Simberloff 1976). These experiments involved, first, a full survey of all arthropods on individual isolated mangrove trees (11–18 m in diameter); secondly, the elimination of all animal life from islets; and thirdly, the monitoring of the recolonization. Although seemingly a satisfactory demonstration of the achievement of equilibrium and one in which turnover was occurring, there were some problems. For instance, certain species groups were regarded as not treating the trees as an island. Some of these groups were included in the analysis and others excluded. Simberloff also had problems determining which species were proper immigrants as opposed to transients (Williamson 1981). One interesting features was that a degree of overshoot occurred, suggesting that the islands could support more than their 'equilibrium' number of species while most species were rare, but as populations approached their carrying capacities, competition and predation eliminated the excess species. As Simberloff (1976, p. 576) described it: 'more highly co-adapted species sets find themselves by chance on an island and persist longer as sets'. This development was termed an **assortative equilibrium**.

Demonstration of equilibrial turnover demands the most stringent data. The clearest evidence inevitably comes from 'islands' which are relatively small and simple, and for organisms with fast life cycles. The mangrove 'island' experiments are the classic first test and they fit this description. These tiny patches of mangrove consist of a single habitat type (albeit not to an insect), the arthropods concerned have very short generation times, and can be expected to respond very rapidly to changing opportunities and to exhibit relatively little community development. Such features may be atypical of islands in general. It is also notable that although turnover rates were indeed high in the early stages of the experiment, a couple of years on,

once the assortative process was completed, and pseudoturnover or transient species removed from the analysis, the rates were found to have slowed greatly. This is not expected if area and isolation remain constant.

A similar study was conducted by Rey (1984, 1985). His islands were patches of the grass *Spartina alterniflora*, varying in size from 56 to 1023 m^2, but structurally simpler than the mangroves. Once again, fumigation was used and arthropod recolonization was monitored on a weekly basis. Initially, the colonization rate was slow because extinction rates were high. As the assemblages built up, populations persisted longer and extinction rates fell. Rey makes a useful distinction between two extremes in turnover patterns. If all species are participating equally in turnover, such that each species is involved somewhere in the archipelago in extinction and immigration events, he terms the pattern homogeneous turnover. If, in contrast, only a small subset of the species pool is involved in turnover, this would be heterogeneous turnover. If strongly heterogeneous, the turnover would be inconsistent with the EMIB. Rey's data were intermediate between these two extremes, neither totally homogeneous, nor excessively heterogeneous.

Why is heterogeneous turnover problematic for the EMIB? The most basic equilibrium model of MacArthur and Wilson (1967) contains the assumption that each species has a finite probability of becoming extinct at all times. The rate of extinction would then depend on how many species there were. This model assumes entirely homogeneous turnover. However, their favoured model took into account that the more species there are, the rarer each is (on average) and hence an increased number of species increases the likelihood of any given species dying out. This results in the curved function for extinction. This is clearly correct under stochastic birth and death processes, provided you assume that the more species there are on an island, the smaller, on average, each population is. As already established, this provides us with the hollow forms of immigration and extinction curve and the expectation that equilibrium species number should be approached by a colonization curve of negative exponential form. It also generates the expectation that, on the whole, the rarest species on a particular island are most likely to become extinct. This view, although basically stochastic, is at least part way along the axis towards a model of structured turnover. Now, if, in general, turnover at equilibrium involves a subset of fugitive populations, while another, larger group of species, mostly of larger populations, is scarcely or not at all involved in turnover, then the turnover is highly heterogeneous. Both Williamson (1981, 1983, 1989a,b) and Schoener and Spiller (1987) have concluded that most empirical data point to turnover mostly being of such fugitive, or ephemeral species. Whether this is consistent with the EMIB or not is an arguable point: at best, if this really is an accurate description of turnover at equilibrium, then it suggests that the theory is 'true but trivial' (Williamson 1989a).

How is equilibrium approached? Can you assume that the more species an island has, the smaller the populations? Do the curves for immigration rate and extinction rate follow the trends described by MacArthur and Wilson of smooth decline and increase, respectively? Actually, as they put it themselves (1967, p. 22): 'these refinements in shape of the two curves are not essential to the basic theory. So long as the curves are monotonic, and regardless of their precise shape, several new inferences of general significance concerning equilibrial biotas can be drawn'. This acknowledges that it is the monotonicity of the rate changes which is critical. It is possible to test whether, indeed, immigration and extinction trends are monotonic. The answer in at least two cases appears to be 'no'. Rey's study illustrates this in a simple system. Observations from the Krakatau islands illustrate much the same point in a complex system.

The Krakatau islands were effectively sterilized in volcanic eruptions in 1883. Recolonization of the three islands in the group commenced shortly after. It started in a limited number of places and gradually claimed more of the island area. Niche space and carrying capacity increased as the system developed. Successional processes kicked in, and habitat space waxed and waned as grasslands developed and then diminished—overwhelmed by a covering of forest. Bush and Whittaker (1991)

demonstrated that, if the lists of species present are taken at face value, the trends in immigration and extinction from the lists (not the real, unmeasurable I and E on the islands) behave non-monotonically for birds, butterflies, and plants. Bush and Whittaker's (1991) bird data were later shown to be incomplete for the latest survey periods, but even after correction (Thornton et al. 1993), the non-monotonicity of trends remained (Bush and Whittaker 1993). It therefore appears that for complex islands, there is more ecological structure to the assemblage of an island ecosystem than is allowed for in the EMIB (see Chapter 5).

What causes extinctions?

What causes extinctions, and thus turnover, is one of the crunch issues connected with the equilibrium theory. MacArthur and Wilson (1967) were not explicit. There is no doubt that they regarded it as, in general, a function of population size, but to what extent does their theory assume it to be random (and homogeneous), as opposed to a deterministic outcome of species–species interactions, particularly competition? The 'narrow' answer is that the EMIB is essentially a stochastic formulation (above), but the 'broad view' answer is that their text acknowledges the importance of competition. For instance, the data for the recolonization of Krakatau by plants (up to 1932) failed to fit their expectations. One of the two explanations they offered invoked a successional pattern in which 'Later plant communities are dependent on earlier, pioneer communities for their successful establishment. Yet when they do become established, they do not wholly extirpate the pioneer communities . . .' (MacArthur and Wilson 1967, p. 50). This invocation of competitive effects is a community-wide, rather than one-on-one, form of competition, and is sometimes given the label **diffuse competition**. Wilson (1995, p. 265), writing retrospectively about his field reconnaissance before the mangrove islets experiments, recollects 'For ants the pattern was consistent with competitive exclusion. Below a certain island size, the colonization of some species appeared to preclude the establishment of others . . . '.

Rosenzweig (1995) invokes both competition and predation as reasons why extinction rates should rise with increasing size of the assemblage. The data he cites for a direct role of predation are not entirely convincing of a general effect, as they come (with one exception) from predators introduced by humans to remote islands, the biota of which had evolved without them. The exception was for data for orb spiders on very small islands, where they were preyed upon by lizards. The support in this case is not entirely unequivocal (Toft and Schoener 1983) and, indeed, Spiller and Schoener (1995) have shown that the effects of the lizards on spider densities within these islands vary significantly over time as a function of rainfall variations (data on turnover were not given in the latter study).

Additional evidence for predators causing rising extinction rates have come from studies by Thornton (1996) and his colleagues of the colonization of the newly formed island of Anak Krakatau (the fourth Krakatau island, which emerged from the sea in 1930). As the island began to support appreciable areas of vegetation in the 1980s, frugivorous and insectivorous birds were able to colonize in greater numbers, but following the establishment of raptors on the island, several species suffered considerable reductions. In two cases at least, their disappearance from the island was attributed to predation. This is an intriguing finding, as in other ecological contexts predators have often been invoked as preventing competitive exclusion and thereby enhancing diversity at lower trophic levels (Caswell 1978; Begon *et al.* 1986). In the mid-1980s Anak Krakatau was able to support only a single pair of oriental hobbies, which in 1989 were replaced by a pair of peregrine falcons, suggesting the position of top predator to be rather marginal. Anak Krakatau's avifauna is set in the context of three nearby islands, and individuals found on the island may well be only partially dependent on it for food supplies. For instance, Thornton (1996) has noted that the home range of the other avian predator on Anak Krakatau, the barn owl, is sufficient to take in all four Krakatau islands. In such cases, a predator may be able to remove one element of its *in situ* food supply while supplementing its diet on prey from other nearby islands (equally, new

arrivals or stragglers may be killed by resident predators on small islands, cf. Diamond 1974). In this fashion, Anak Krakatau may be functioning not as a closed but as an open system (cf. Caswell 1978).

The role of predators may also be scale dependent, such that development of the island's ecosystems to a greater biomass might diminish the ability of predators to cause the total loss of a prey species. This suggestion is consistent with Diamond's (1975a) observations of contrasts between Long Island and the more depauperate Ritter, both being islands recovering from volcanic disturbance. Caswell (1978) makes a similar suggestion on the basis of his non-equilibrium modelling of predator–prey relationships in open systems.

In order to demonstrate that the extinction curve does indeed have a concave, rising form, produced by species interactions, Rosenzweig (1995) had to resort to a complex line of reasoning, based on ISARS from two data sets, each of island archipelagos created at the end of the last ice age and presumed to be undergoing relaxation. This constitutes a rather indirect line of proof, reliant upon a series of assumptions being taken as read. It is intriguing that he had to work so hard to find support for such a basic element of the EMIB. The generality of the concave, rising trend of extinction does not seem to be founded on much empirical evidence yet. Data can also be found that, at face value, contradict this expectation.

There is no reason to assume that there is a single path to extinction. In the process of a species declining (perhaps in fluctuating fashion) to extinction, its population must, for some period, be small. The small size of the population may be a good predictor of extinction, but why is the population small in the first place? The explanation could lie in the trophic status of the organism; in predation, disease, or disaster; in resource shortage; or in competition with other species of its trophic level. It is, as Williamson (1981) notes, remarkable that the equilibrium theory was widely accepted with almost no evidence on the distribution of probabilities of going extinct and how these vary (cf. Pimm *et al.* 1988). It is arguable that, in many cases, habitat availability and stability, and food web connections (and continuous availability of resources), are more

han competition in determining extinc-
ilities (Whittaker 1992).

So, where does all this leave us in evaluating turnover on islands? The statement in Box 4.1 is of course a truism: the number of species on an island at a given point in time has to be a function of the number previously recorded and the numbers gained and lost in the interim, but the extent to which these rates behave according to the expectations of the EMIB remains hard to assess. Although we began in this chapter by discussing some of the core macroecological relationships within ecological systems, by focusing on species turnover, we have inevitably been drawn into discussing aspects of composition in the latter part of the chapter. We go on in the next chapter to look more formally at theories concerned with compositional pattern, and then pull together both sets of themes in the final chapter of this section of the book (Chapter 6).

4.5 Summary

Ecological systems have emergent statistical properties, detectable for example in regularities of form of the abundance and number of species, and which point to the existence of some fundamental governing ecological rules or processes. The study of systems in this way is termed macroecology. Island macroecology began long before the term was coined, in the early part of the twentieth century. It gained its core focus with the publication in 1967 of MacArthur and Wilson's equilibrium model of island biogeography (EMIB), which postulates that species number is the dynamic product of opposing rates of immigration and local extinction, as inverse functions respectively of isolation and area.

In this chapter, we first review the building blocks that lead to the formulation of the EMIB and then consider how it has fared alongside subsequent theoretical developments. Evaluation largely falls under two heads: spatial patterns and temporal patterns. We distinguish between several closely interrelated properties of ecological systems: species abundance distributions, species–area curves, species accumulation curves and island species–area relationships (ISARs). Variation in the form of ISARs was a central focus of MacArthur and Wilson's theory, and a substantial literature has since developed, demonstrating varying degrees of fit with the EMIB—implying that at the least a degree of modification is required. Variation in form may for instance be attributable to different variables depending on the magnitude of variation in independent variables (area, isolation, elevation, habitat diversity, etc) and their interaction. Yet, there is also the prospect that combined area-isolation-climate models can be developed that can account for the first order pattern of island species richness globally, at least for particular taxa.

The EMIB asserts that island richness patterns are dynamic outcomes—the product of perpetual turnover in membership and that once at equilibrium this process continues essentially unabated. On this score, the evidence appears at best equivocal, in part because of the enormous problems involved in accurately quantifying immigration and extinction rates and thus measuring turnover. Consideration of evidence on turnover draws us into discussion of compositional structure in the data, which receives much fuller consideration in the next chapter on island assembly theory.

CHAPTER 5

Community assembly and dynamics

... studies of island biogeography show that a community is not simply a collection of all those who somehow arrived at the habitat and are competent to withstand the physical conditions in it ... a community reflects both its applicant pool and its admission policies ... We need to formulate a new generation of population and community models with the transport processes explicitly built in ...

(Roughgarden 1989, pp. 217–19, with thanks to Thornton 1996)

In Chapter 4, we took a macroecological approach to island biotas: species were treated largely as exchangeable units. Such an approach ultimately has its limits. This chapter is therefore concerned with those aspects of island biogeographic theory dealing with compositional pattern and how it is assembled (mostly) in ecological time. By considering the much more complicated response variable of 'composition', it is possible to cast a different light on to some of the problems left over from the species numbers games (Worthen 1996). Pattern interpretation in this chapter will again invoke forces such as immigration, competition, and predation, but will also pay greater attention to species autecology, dispersal attributes, and succession. In essence then, whereas the previous chapter took a macroecological approach, this chapter is concerned with ecological biogeography.

Colonization and ecosystem development of a not-too-distant island arguably constitute just a special case of ecological succession, a process (realistically many processes) that in complex ecosystems, dominated by long-living organisms, may be evident over hundreds of years. We illustrate how succession can influence emergent patterns of species numbers through the example of the recolonization of the Krakatau islands, which is not only the best-studied 'natural experiment' in island recolonization, but was also the first test system used by MacArthur and Wilson (1967; see Wilson 1995). However, it is Jared Diamond's theory that provides the starting point for this chapter.

5.1 Island assembly theory

Some readers may have experienced the daily school assembly, the rules of which were that you all stood quietly in lines, class by class, and behaved yourselves while certain school rituals were observed. Of course, these behavioural rules were broken daily; furthermore, the precise order and number of individuals in each line varied, but on the whole the system approximated to the form laid down. The casual observer looking in might, however, have failed to recognize this on occasion. This analogy might apply to island ecosystems, but here detection of the rules, if they exist, requires a great deal more application. This particular branch of island ecology was given its impetus by an extensive series of studies of the avifauna of New Guinea and its surrounding islands by Diamond (e.g. 1972, 1974, 1975a). The impetus was lost for a while as the approach became embroiled in a technical dispute concerning the formulation and testing of hypotheses, and controversy concerning the relative significance of competition in pattern formation. Yet what Diamond (1975a) presented amounted to a fairly comprehensive island biogeographical theory in its own right, which we term **island assembly theory**.

Assembly rules

Diamond adopted the working hypothesis that through diffuse competition, the component species of a community are selected, and co-adjusted in their niches and abundances, so as to fit with each other and to resist invaders. In his studies he identified the following patterns or **assembly rules** (quoted from Diamond 1975a, p. 423):

1 If one considers all the combinations that can be formed from a group of related species, only certain . . . of these combinations exist in nature.
2 Permissible combinations resist invaders that would transform them into forbidden combinations.
3 A combination that is stable on a large or species-rich island may be unstable on a small or species-poor island.
4 On a small or species-poor island, a combination may resist invaders that would be incorporated on a larger or more species-rich island.
5 Some pairs of species never coexist, either by themselves or as part of a larger combination.
6 Some pairs of species that form an unstable combination by themselves may form part of a stable larger combination.
7 Conversely, some combinations that are composed entirely of stable subcombinations are themselves unstable.

These assembly rules were derived from several forms of distributional data (e.g. incidence functions, chequerboard distributions) for species and guilds, alongside data for recolonization of defaunated islands. The 'rules' were the descriptions and interpretations of these emergent patterns.

Incidence functions and tramps

Incidence functions are an essentially simple idea, reliant on the availability only of good data on species distributions across a series of isolates. They show the frequency of a species in relation to values of a potentially controlling variable, such as area or isolation (e.g. Watson *et al.* 2005). However, as originally employed by Diamond (1975a), they were **richness-ordered incidence functions**: plots of island species number, S, versus the incidence, J, of a given species on all islands of that value of S. Unless a huge number of islands is available, it is necessary to group together islands of a given range of values of S in order to obtain enough observations in each class. Figure 5.1 shows the functions calculated by Diamond for a subset of his bird species. The legend provides a labelling of the different distributional patterns, high-S species, A-tramp, and so on. In order to interpret these labels some background is needed.

Diamond's study area lies near the equator, with the predominant natural vegetation cover being rain forest. The New Guinea bird species pool consists of about 513 breeding non-marine species. Offshore, there are thousands of islands of varying sizes and degrees of isolation, for many of which bird data were available, in cases including instances of successful and unsuccessful colonization. Some land-bridge islands can be assumed to have gained much of their avifauna at times of lowered sea level, and were thus regarded as supersaturated. Others, having undergone disruption by volcanic action (e.g. Long and Ritter islands), were regarded as displaced below equilibrium. Others were regarded as equilibrial. Thus, within the 50 islands studied, although a significant species–area relationship was recorded, there were statistical deviations. These deviations appeared intelligible in relation to historical and prehistorical events, and were therefore interpreted in terms of long-term trajectories in avifauna size and composition.

Some species of birds, such as *Centropus violaceus*, *Micropsitta bruijnii*, and *Artamus insignis*, occur only on the largest, most species-rich islands. These were termed **high-S** (or **sedentary**) species (Diamond 1974), the precise usage being species found on the four richest of the Bismarck Islands (islands with 81–127 species) plus not more than one from the next richest category (islands with 43–80 species). At the other extreme, a small number of species are absent from the most species-rich islands and are

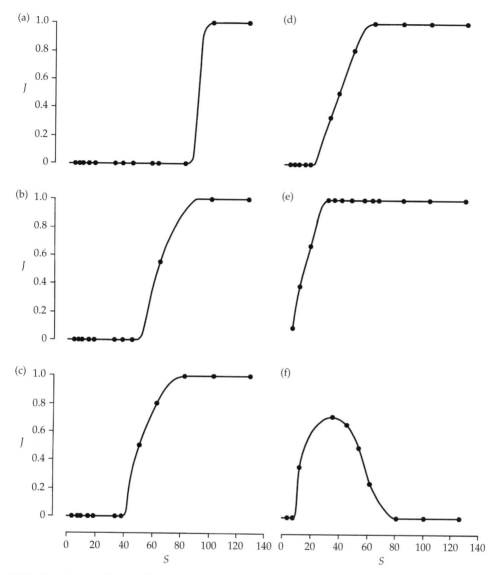

Figure 5.1 Incidence functions for birds of the Bismarck Archipelago. S=number of species on an island, but note that the points represent grouped data for a narrow range of values of S from between 3 and 13 islands per point, except the two largest values, which each represent a single island. The index $J = 1.0$ if the species occurs on all islands, and 0.0 if the species occurs on none. (a) *Centropus violaceus* (cuckoo), a high-S species; (b) *Diacaeum eximium* (berrypicker), an A-tramp; (c) *Pitta erythrogaster* (pitta), a B-tramp; (d) *Ptilinopus superbus* (pigeon), a C-tramp; (e) *Chalcophaps stephani* (pigeon) a D-tramp; (f) *Macropygia mackinlayi* (pigeon), a supertramp. See text for explanation of categories. (Redrawn from Diamond, J. M. (1975) In *Ecology and evolution of communities* (ed. M. L. Cody and J. M. Diamond). Copyright © by the President and Fellows of Harvard College. Reprinted by permission of Harvard University Press.)

concentrated in the smallest, or most remote, and most species poor islands. An example of this **supertramp** category is the pigeon *Macropygia mackinlayi*. In between, Diamond differentiated

arbitrarily four other categories of lesser **tramps**, for those species occurring on some or all of the most species-rich islands and, depending on the category of tramp, on other islands of varying size and species richness within the Bismarck Islands: A-tramps 3–9 islands; B-tramps 10–14 islands; C-tramps 15–19 islands; and D-tramps 20–35 islands. The smallest or most species-poor islands generally have representatives of both tramp and supertramp categories.

The conclusion reached from the incidence functions was that the island avifaunas represented highly non-random selections of the species pool. Rather, it appeared that species distributions were related to factors such as: competition; the absence of suitable habitat; island sizes being smaller than minimum territory requirements; seasonal or patchy food supply; historical factors such as land-bridge connections; disturbance and changes in carrying capacity; and interactions of these factors with population fluctuations.

The dynamics of island assembly

In addition to being found predominantly on the smaller islands, the supertramps were also characteristic of the recently recolonized volcanic islands (Diamond 1974), suggesting that these species are excellent colonists but poor competitors. The more widespread tramp species, being found in relatively high frequencies on all but the smallest islands, must, in contrast, be both capable colonists and good competitors. The species restricted to large islands might be absent from others because of poor dispersal or some form of intrinsic unsuitability to small islands. As this high-S set includes species of larger body size (e.g. herons), and species of large home ranges and strong dispersal potential (e.g. large hawks), the latter appeared important, at least for some of the species involved. Diamond concluded that the high-S and A-tramp categories actually included a fairly heterogeneous mix of species, some limited by dispersal and some by other factors. About 56% of the high-S species are endemic to the Bismarcks at the semispecies or species level, such that the patterns under discussion retain at least some signal from evolutionary timescales.

Some species distributions were interpretable in relation to habitat requirements. For example, as only five islands in the data set exceed 3500 feet

(1067 m) elevation, the 13 montane bird species are restricted to these 5 islands. Of the 5 islands, one is very small and another was recently defaunated by volcanic explosion: 12 of the 13 montane birds are lacking from both of these islands. Similar fits to habitat could be found for other Bismarck bird groups. The endemic species, which are mostly also high-S species, are mostly lowland forest or mountain species. By contrast, the tramp species confined to scarcer lowland habitats were not found to have differentiated beyond subspecies level. This was thought to be indicative of the greater frequency of dispersal needed to maintain their lineage in a scattered system of habitat patches across the archipelago. The supertramp species differ again, in that where they occur on an island, they tend to occur in a greater variety of habitats, and Diamond (1975a, p. 381) characterized their ecology as 'Breed, disperse, tolerate anything, specialize in nothing.' Thus, the lack of supertramps on an island is a product of competitive exclusion and not of dispersal constraints. The gradient between the supertramps and the high-S species can be viewed as equivalent to that between r-selected and K-selected species, or pioneering and late successional plant species. In each case, the gradient represents an approximation of a more complex reality, with varying degrees of explanatory power from one case study to another (Begon *et al.* 1986; and for a critical evaluation of r–K selection see Caswell 1989).

As will become apparent in Chapter 9, there are parallels between Diamond's ideas and E. O. Wilson's (1959, 1961) earlier ideas of taxon cycles. Both theories seek to explain distributional patterns by means of competitive effects, habitat relationships, and evolutionary considerations. Although we have placed them in separate sections of the book, they are both evolutionary ecological models, both are dynamic models, and both invoke a key role for competition. They differ principally in that the taxon cycle model focuses more on evolutionary change within the taxon as a biogeographical process, while the assembly rules ideas are concerned with identifying and explaining compositional regularities across a series of islands, invoking ecological forces such as ecological succession.

Diamond (1975a) himself suggested that the parallel was quite a close one, although he noted that while in the taxon cycle ants colonizing a species-poor island initially occupy lowland non-forest habitats and subsequently undergo niche shifts to occupy the interior, many colonizing birds already occupy lowland forests. This suggests that the superior average dispersability of the birds may be important to the differences in the patterns observed.

There can of course be great differences in dispersal abilities *within* taxa. This can be deduced for New Guinea birds by reference to those species absent from non-land-bridge islands. On the assumption that a dispersive species should somewhere have found one or more suitable islands in the non-land-bridge set, only 191 of the 325 lowland bird species have shown a capacity to cross water gaps greater than 8 km (5 miles) (Diamond 1975a). An example of an apparently poor disperser is the flycatcher *Monarcha telescophthalmus*, which occurs on all the large land-bridge islands around New Guinea, but no others. More direct evidence of dispersal abilities can be derived from the recolonization of islands such as Long, Ritter, and Krakatau, which are known to have been defaunated, or by direct observations of birds moving between islands (e.g. Diamond 1975a; Thornton 1996; Thornton *et al.* 2001). This observational information also played a part in the formulation of island assembly theory.

Chequerboard distributions

In developing island assembly theory, the incidence functions provided the first step in the reasoning. The second step relied upon a different form of distributional patterning, **chequerboard distributions**, found for pairs of congeneric bird species in the Bismarcks. A chequerboard (from the game chequers (US checkers), or draughts) is where two (or more) species have mutually exclusive but interdigitating distributions across a series of isolates, such that each island supports only one species. Several examples were given by Diamond (1975a). One example is provided by two species of flycatcher. *Pachycephala melanura dahli* is found on 18 islands, and its congener *P. pectoralis* on 11, but

they do not occur together on any island. Similar patterns were found for two species of cuckoo-doves of the genus *Macropygia* (Fig. 5.2). *Macropygia mackinlayi* has a weight range of 73–110 g and mean of 87 g, *M. nigrirostris* has a range of 73–97 g and mean of 86 g. Given these values, they might reasonably be expected to have similar resource requirements. The interpretation offered for these distributional patterns was that either chance determined first arrival, or that slight ecological advantages favoured one species over the other on a particular island. Once established, the resident would then be able to exclude the congener.

Chequerboards and incidence functions are different facets of the same distributional data sets. For instance, *Macropygia mackinlayi* was one of the supertramps identified by the analysis of incidence functions. Its congener occurs on all the larger land masses and only a few of the small islands. Again, Diamond invoked competitive exclusion to explain the failure to find both species resident on a single island. It will doubtless have occurred to the reader that as some islands lack both these species of cuckoo-dove, and as the identification of *M. mackinlayi* as a supertramp involves analysis of entire assemblages, not just other cuckoo-doves, there must be more to explaining the distributions of such congeners than simply their relationships to one another. The data are certainly consistent with a role for competitive interactions, but other factors and other biogeographic scenarios (e.g. Steadman 1997b) cannot be ruled out on the basis of Fig. 5.2. One means of testing the competition hypothesis would be provided in the event of one of the pair of congeners colonizing an island 'held' by the other. According to the interpretation of the incidence functions, there ought to be an asymmetry in their competitive interactions, and *M. mackinlayi* should generally be the loser. Such observations, unfortunately, are hard to come by.

Combination and compatibility—assembly rules for cuckoo-doves

Macropygia mackinlayi, although not overlapping with *M. nigrirostris*, does occur on islands with one of two larger species of cuckoo-dove: namely

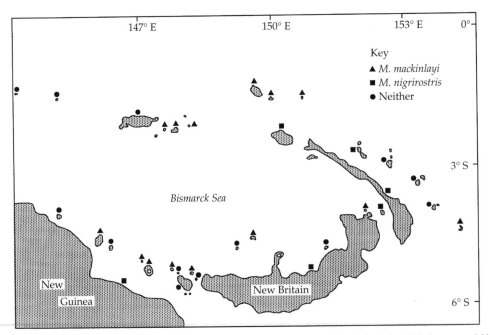

Figure 5.2 Checkerboard distribution of cuckoo-doves (*Macropygia*) in the Bismarck region. Of those islands for which data are available, most have one, but no island has both species. (Modified from Diamond, J. M. (1975) In *Ecology and evolution of communities* (ed. M. L. Cody and J. M. Diamond), Fig. 20. Copyright © by the President and Fellows of Harvard College. Reprinted by permission of Harvard University Press.)

M. amboinensis (average weight 149 g) or a representative of the *Reinwardtoena* superspecies (average weight 297 g). This suggests that a certain degree of niche difference, signified by a weight ratio of 1.5–2.0 between members of the same guild, may be necessary to enable coexistence (Diamond 1975*a*). Diamond undertook a more formal analysis of the species combinations to ensure that they were indeed non-random, and thus evidence for competitive effects in island assembly. There are 4 Bismarck cuckoo-dove species (represented by the letters AMNR), so 15 possible combinations, but only 6 combinations were actually observed on the 26 islands having one or more cuckoo-dove species. With such a small sample size relative to the number of permutations, it is not expected that each of the 15 possible combinations would occur. The analyses thus relied upon a modelling exercise designed to test whether the empirical data exhibited a significant degree of departure from a random processes of assembly.

The interpretation of the combinational patterns begins with the **incidence rules**, which determine that certain species combinations are 'forbidden' (i.e. do not occur). The second form of rule was termed a **compatibility rule**. The allopatric *M. mackinlayi* and *M. nigrirostris* exemplify this rule, as they are so ecologically similar in their resource requirements (if not in their incidence characteristics) that they are presumed to be unable to co-occur other than on the most temporary basis. Diamond postulated that the two are probably the product of fairly recent speciation from within a former superspecies. Thus the combinations MN, MNR, AMN, and AMNR cannot (i.e. do not) occur. The final rule type, **combination rules**, in effect mops up the unknowns. For instance, Diamond calculated on the basis of the incidence functions that the combination AMR should occur frequently on islands with a medium species number. It is allowed by the compatibility rules which were deduced, yet it was not found to occur within the data set. This was taken to imply additional combinational rules preventing apparently reasonable combinations occurring with the frequency to be expected by chance. As most changes to an island's avifauna are likely to be produced by single species additions or losses, recombinations requiring two or more gains and/or losses simultaneously (to avoid a 'forbidden' combination) might be anticipated to have low **transition probabilities**. Hence, some combinations might be viable, but difficult to assemble from the 'permitted' combinations.

Diamond undertook similar statistical analyses for the guilds of gleaning flycatchers, myzomelid sunbirds, and fruit-pigeons. The niche relationships among the species of *Ptilinopus* and *Ducula* fruit-pigeons provide a classic example of resource segregation. Diamond first established the form of community structure within the 'mainland' lowland forests of New Guinea. The larger pigeons forage preferentially on bigger fruits, providing one element of resource segregation. If large and small species feed within individual trees, the lighter pigeons are able to feed on the smaller, peripheral branches, better able to support their weight. Thus niche separation can be maintained within a guild of co-occurring species. Yet, within a single locality in New Guinea, no more than eight species will be encountered, forming a graded size series, each species weighing approximately 1.5 times the next smaller species. There are 8 size levels filled by combinations drawn from 18 species. On satellite islands off New Guinea, subsets are drawn in such a way that on smaller or more remote islands, size levels are emptied as the guild is impoverished according to a consistent pattern. Level 1 (the smallest birds) empties first, followed by levels 2 and 5, followed by level 8. It is intuitively sensible that the smallest birds should be excluded first as they will be restricted to the smallest fruit types, for which they would have to compete with slightly larger pigeons that can also make use of larger fruits. This form of structuring could also be identified on those islands regarded as being supersaturated, or undersaturated, suggesting that the time required to adopt such structure is shorter than that required for equilibration of species number

on islands. This observation resonates with subsequent analyses of ecosystem build-up within the flora and at least some elements of the fauna of Krakatau (e.g. Whittaker *et al.* 1989; Zann and Darjono 1992; Thornton 1996).

The above constitutes the bulk of Diamond's species assembly theory as it concerns us here. However, he also applied the ideas to non-insular communities, a context in which many of the recent developments in the formulation and testing of assembly rules have come (e.g. Wilson and Roxburgh 1994; Weiher and Keddy 1995; Wilson and Whittaker 1995). Diamond's studies incorporated distributional data at the island level in the first instance, but also involved: data on habitats in which particular species were observed or trapped; body weights; feeding relationships; observations of bird movements between islands; and historical records of species colonizations and extinctions, in cases providing apparent evidence of the unsustainability of certain combinations of species.

Criticisms, 'null' models, and responses

Considering how much evidence for prehistoric anthropogenic extinctions of island birds has emerged over the past decade, biogeographers today should hesitate to regard any distributional attribute of modern South Pacific birds as being uninfluenced by human activity.

(Steadman 1997*b*, p. 750)

Diamond's papers provide a coherent theory of the biogeography of the New Guinea island birds. This island assembly theory could be seen as a development of several of the lines of argument to be found in MacArthur and Wilson's (1967) *The theory of island biogeography*. Yet, the approach, and particularly the assembly rules, drew criticism from Daniel Simberloff and colleagues (Simberloff 1978; Connor and Simberloff 1979). The problems were sufficiently troublesome, or intractable, as to hamper subsequent progress, and the approach has arguably failed to receive the attention that its intrinsic interest warrants (Weiher and Keddy 1995).

One element of the logic employed by Diamond that might be criticized is the implicit acceptance of the equilibrium thinking at the heart of MacArthur and Wilson's (1967) monograph. Thus, for instance, he writes,

The fact that the larger land-bridge islands around New Guinea still have double their equilibrium species number [of birds], more than 10 000 years after the land bridges were severed, emphasizes for what long times many species can escape their inevitable doom on a large island' (Diamond 1975a, p. 370).

If it takes in excess of 10 000 years to reach equilibrium, the effect would appear to be a relatively weak one, potentially or actually to be overridden by other factors. The inevitability of their doom, the fate ultimately of all living things, as the fossil record reveals, is not necessarily in practice the consequence of their island systems moving closer to a *biotically regulated* equilibrium. Simberloff (1978) made a similar point when he argued that the statistical analyses of distribution were not uniquely interpretable as a function of diffuse competition.

A second critical paper was more strident. It claimed to show that 'every assembly rule is either tautological, trivial or a pattern expected were species distributed at random' (Connor and Simberloff 1979, p. 1132). The rules derived in Diamond's analyses were cast in terms of 'permitted' and 'forbidden' combinations (as rule 2, above) and 'resistance' to invasion. Thus, although they were derived empirically mostly from distributional data, competition was invoked within their formulation. One element of the critique concerned the great difficulty of demonstrating a particular role over time for competition in determining present-day distributional patterns (Law and Watkinson 1989; Schoener 1989). Simberloff (1978) argued that the role of competition had been overplayed by several authors, citing Diamond's studies in illustration. His argument focused largely on the methods by which non-random distributions are calculated. He argued that confidence limits were not given on the combinational rules, which were therefore difficult to evaluate, and that there were too few islands involved to allow rigorous statistical assessments. He argued that the **null hypothesis** should be formulated in terms of species being considered essentially as similar units, i.e. without differing dispersal abilities and ecologies, and that these species should be modelled as bombarding islands entirely at random. The different islands should be accepted as having different properties, however. While accepting that the first of these assumptions is biologically unrealistic, Simberloff's argument was that it provided a baseline by which to examine biogeographical distributions. When the data are not consistent with this null model, ecological phenomena, such as competition or dispersal differences, can then be examined for explanations. Simberloff also argued that the biological characteristics of each species should be examined for explanatory value as being a more parsimonious explanation than competition. Thus one species might be a superior colonist with respect to another because it has better dispersal ability and/or better persistence once immigration occurs (as a function of per capita birth and death rates).

Thus Simberloff, whose own formative experience was the mangrove islet study (a fast developing, near-source, dynamic system), adopted a position which is arguably closer to the stochastic, dynamic equilibrium model of island biogeography (EMIB) than did Diamond in his seminal study of New Guinea island avifaunas (a slow, more isolated, and less dynamic system with a strong evolutionary signal). This difference in view was recognized by Simberloff, who noted that botanists and invertebrate zoologists have most often sought distributional explanations in responses of individual species to physical factors, whereas vertebrate zoologists commonly invoke competition. The generality of competitive effects was in Simberloff's view largely unproven, and the reason was that appropriate statistical tests had yet to be applied to sufficient data sets—indeed such data sets were, by the nature of the statistical requirements, bound to be difficult to obtain. This line of argument was extended by Connor and Simberloff (1979) who contended that the patterns deduced by Diamond as the outcome of competition could be produced by random processes providing that the following three constraints were accepted: '(1) that each island has a given number of species, (2) that each species is found on a given number of islands, and (3) that each species is permitted to colonize islands constituting only a subset of island sizes' (Connor and Simberloff 1979, p. 1132). They based their claims on data for New Hebridean (Vanuatu) birds, West Indian birds, and West Indian bats, which they analysed for evidence of Diamond's assembly rules. The statistical part of their analysis failed to support the existence of assembly rules in their data.

The quote given earlier referred to tautologies and trivial features within the assembly rules. These arguments are difficult to summarize but amount to an attack on the line of reasoning used to construct the assembly rules. In illustration, the third rule 'A combination that is stable on a large or species-rich island may be unstable on a small or species-poor island' was argued to be reducible to 'A combination which is found on species-rich islands may not be found on species-poor islands' because, first, species number is used operationally by Diamond in lieu of island size and, secondly, because stability is largely dependent on the combination of species in question having been observed. As large islands contain on average more species, it would be expected by chance alone that some combinations found on rich islands will not occur on species-poor islands. By this line of reasoning, the third rule was deemed to be a trivial outcome of the definitions used and otherwise a probabilistic outcome.

Another important line of criticism concerned the chequerboard distributions, which they argued could represent ongoing cases of geographic speciation without re-invasion, i.e. they could be cases of divergence between allopatric or parapatric lineages, rather than being the outcome of competitive exclusion. The position they adopt at the end of their paper is clear. The assembly rules were flawed because at no time was a simple null hypothesis framed and tested, competitive effects were invoked without proof and where simpler explanations were available. Nonetheless, Connor and Simberloff stressed that they were not implying that island species are distributed randomly and that interspecific competition does

not occur, just that neither non-random distributions nor the primacy of competition in determining distributions were convincingly demonstrated by Diamond's analyses.

Not surprisingly, these criticisms drew a response, in the form of consecutive articles by Gilpin and Diamond (1982) and by Diamond and Gilpin (1983), and indeed generated a flurry of articles and comments within the literature of the late 1970s and early 1980s. Perhaps the core of Diamond and Gilpin's (1982) response is a challenge to the primacy of Connor and Simberloff's conception of a null hypothesis. Diamond and Gilpin (1982) take the reasonable position that biogeographical distributions may be influenced by many factors, including competition, predation, dispersal, habitat, climate and chance, and that the mix is liable to vary with the group of organisms considered and with spatial and temporal scale. Further, it follows that no single theory, hypothesis, or conceptual position has logical primacy, or special claim to be the most parsimonious null hypothesis (they signify this by placing 'null' in quotation marks throughout). Each should be regarded as a competing or contributory hypothesis to be evaluated on its merits. This is a pragmatic position adopted generally in the present book: different answers may be found to similar questions posed in different island biogeographical contexts. However, there is, of course, more to the debate than this. Diamond and Gilpin in fact took the opportunity to respond to a series of papers by Simberloff, Connor, Strong, and others, employing null hypotheses. The problem with null hypotheses, i.e. hypotheses of 'no effect', is the assumptions that are made in their formulation (Colwell and Winkler 1984; Shrader-Frechette and McCoy 1993). Diamond and Gilpin argue that in generating their null distributions Connor and Simberloff (1979) took the following inappropriate steps:

1 Whereas assembly rule theory posits competitive effects *within guilds*, they diluted their data by including data for the whole species pool instead of only for the target guild.
2 They incorporated hidden effects of competition into their constraints. The assumptions employed

in their null models build in occurrence frequencies and incidences of species which can be shown to be influenced by competition by reference to occurrence frequencies of members of the same guild in different archipelagos (Bismarck, Solomon, and New Hebrides). These data show that the fewer competing species of the same guild that share an archipelago, the higher the occurrence frequency of particular species, often by a factor of 10 or more.

3 They produced a statistical procedure incapable of recognizing a chequerboard distribution.

4 Simulation procedures used were inadequate to provide realistic simulations of the matrices.

This is an incomplete listing of the objections raised by Diamond and Gilpin (1982), but the general position they adopt is summed up as follows: 'how can one pretend that one's "null model" is everything-significant-except-X, when it was constructed by rearranging an observed database that may have been organized by X?' (Diamond and Gilpin 1982, p. 73). The problem thus afflicts the proponents on both side of the argument, and is one that has since reappeared in other applications of null models (papers in Strong *et al.* 1984; Weiher and Keddy 1995). All null models involve assumptions; the trick is to recognize what they are, the disagreements come over which are the most realistic and appropriate set to use (Box 5.1).

In their second paper, Gilpin and Diamond (1982) developed a test that they regarded as an improvement over the Monte Carlo modelling used by Connor and Simberloff, and affirmed their earlier conclusion that the distribution patterns were indeed non-random. Comparing results for the Bismarck Islands and the New Hebrides, they found non-random patterns in each system, but that the richer Bismarck Islands had clearer evidence of competitive effects. The New Hebrides have a smaller avifauna (56 compared with 151 species) and thus negative associations (as would be produced by competition) did not exceed a random expectation: neither were chequerboard distributions evident for the New Hebrides. In contrast, they re-affirmed the existence of negative associations and chequerboard distributions

amongst ecologically similar pairs of Bismarck species. They found that some pairs of species have more exclusive distributions than expected by chance, and invoked as likely controlling factors, competition, differing distributional strategies, and different geographical origins. Examples of the first two have been given above. In illustration of the third category, the hawk *Falco berigora* has spread from the west, whereas the parrot *Chalopsitta cardinalis* has spread from the east, but in each case to a limited degree. They are not members of the same guild, obviously, and their currently exclusive distributions relate to their differing geographical origins rather than competition. Other species pairs were found to have more coincident distributions than would be expected by chance alone, and this they interpreted in relation to shared habitat, single-island endemisms, shared distributional strategies, or shared geographical origins (Table 5.1).

As Gilpin and Diamond (1982) point out, much of the information about non-random co-occurrences is actually contained in the incidence functions—the first element in the whole analysis—which are much more intuitively accessible sources of information than are the more sophisticated modelling exercises. Equally, the chequerboard distributions provided unambiguous patterns, strongly suggestive of competitive effects: in such cases restricting analyses to members within a single guild is clearly necessary (Diamond 1975a). Thus, where observed within guilds of pigeons for which there is excellent evidence of colonizing ability across moderate

Table 5.1 Factors invoked in explanation of non-random co-occurrence of island birds by Diamond and Gilpin in their articles on assembly rules

Negative	Positive
Competition	Shared habitat
Differing distributional strategies	Shared distributional strategies
Differing distributional origins	Shared geographical origins
	Single-island endemics

Box 5.1 On null models in biogeography

As applied in statistical analysis, a **null hypothesis** is the hypothesis of no effect (in respect of a potential controlling variable), or of no difference (between two or more populations). The rejection of the null hypothesis leads to the inference that there is an effect or a difference. **Null models**, on the other hand, are computer simulations used to analyse patterns in nature, in an effort to achieve predictability and falsifiability in ecological biogeography. As typically applied within island biogeography, the approach taken is aimed to simulate random colonization of a set of islands by the members of a species pool, given particular constraints, such as how species number per island is constrained by island area. By repeating the random simulation a large number of times, it is possible to determine whether the observed pattern of species occurrence could have occurred by chance using a given level of significance.

The use of null models has been the subject of intense debate virtually wherever they have been applied within biogeography. Sometimes the arguments have been technical in nature, concerning how well the simulations achieve the intended goal of randomness or how well they match the intended constraints. More intractable is that it is very hard to reach an accord as to the correct way to set the constraints upon a system, i.e. which parameters are controlled and which are allowed to vary. Both sets of problems are exemplified by the assembly rules controversy discussed in the text. Here, the aim was to test for interspecific competition effects by building models that were neutral in respect of competition. But is this possible in the real world?

In practice, unlike for simpler forms of inferential statistics, we can rarely expect agreement on a single null formulation for use in biogeographical modelling. It is better to think of them not as 'null models' with the attached implications of a higher form of objectivity, but simply as **simulation models**, each involving a particular set of assumptions and constraints. Such models have value in exploring and testing out ideas, but always leave room for differences of interpretation.

For further reading and varying perspectives on the theme see, e.g. Gotelli and Graves (1996), Grant (1998), Weiher and Keddy (1999), Moulton *et al.* (2001), and Duncan and Blackburn (2002).

---

ocean gaps, and where an otherwise widespread species is absent from large, ecologically diverse islands offering a similar range of habitats to those occupied by the species elsewhere, but which happen to be occupied by the alternative species, competition is the obvious answer.

Other authors have since joined in the debate. Stone and Roberts (1990) re-analysed some of the data employed in Connor and Simberloff's (1979) article, but by means of a new form of test that they termed the C-score, a means of quantifying the degree of 'chequerboardedness'. Their analysis attempted to avoid some of the problems of Connor and Simberloff's methods while maintaining the constraints that they assumed. In contrast to Connor and Simberloff, their re-analyses led to rejection of random distributions for both the New Hebrides and the Antilles avifaunas (see also Roberts and Stone 1990). Thus, they found support for chequerboard distributions in both faunas, while stopping short of invoking competitive explanations for their findings. In a further analysis, Stone and Roberts (1992) examined the degree of negative and positive relationships within confamilial species of the New Hebridean avifauna. In this case they found evidence not for chequerboard tendencies, but instead for a greater degree of aggregation than expected by chance, i.e. confamilial species tended to occur together. This they interpreted as evidence for similarities of ecology within a family overriding competition between confamilial members. However, as they did not restrict analyses to particular ecological guilds, this 'intrafamily' result is both unsurprising and by no means contradictory to Diamond's (1975a) arguments.

For the record, Sanderson et al. (1998), Grant (1998), and Gotelli and Entsminger (2001) have continued the debate on how to construct null model simulations, again using the New Hebridean bird data as the subject matter. Although differing in the details, the overall picture is that some degree of non-randomness can indeed be detected.

Such debates are rarely settled by the triumph of one viewpoint to the complete exclusion of the other, but, as in this case, can sometimes result in the re-examination and refinement of the original models and positions. Connor and Simberloff's (1979) intervention led to increased attention to the problems of statistical analyses in island assembly structure, which was their stated intent. Even now, it would be a bold move to declare the debate over, but we offer the following conclusions. The original (Diamond 1975a) statements of island assembly theory laid too much emphasis on competition at the expense of the many other factors (Table 5.1) that could be involved in structuring species distributions across islands. Although there is room for debate about details of the assembly rules Diamond identified in his original analyses, the evidence is convincing that some non-randomness is involved. Moreover, within the avifaunal data (especially that from the Bismarcks), it is possible to find what appear to be compelling cases for competitive effects *within guilds* in particular cases.

At this point, we have to introduce a final note of caution: the patterns in distribution of the avifauna, and especially pigeons and doves may have been altered by humans. Steadman (1997b) provides a comprehensive review of prehistoric bone data for the extinction of columbids (pigeons and doves) from Polynesian islands. He notes that losses include the extinction of at least nine species in six genera, as well as the extirpation of numerous island populations of still extant species. If it were not for humans, a typically West Polynesian island would support at least six or seven species in four or five genera, instead of the observed one to six species in one to five genera. He comments that one

of Diamond's supertramp species, *Ducula pacifica*, which is found in both West and East Polynesia, appears to have expanded its range eastward after the arrival of people, which would be consistent with a response to human induced extirpation of populations of more competitive species. Having reviewed Polynesian evidence, Steadman notes that there is by contrast a paucity of prehistoric bone data from Melanesia. Some losses have, however, been documented, and Steadman considers it likely that many more await discovery. In time, it may turn out that some of the chequerboards and other assembly rule patterns of columbids at the heart of these debates have been structured not just by the 'natural' forces of competition, habitat controls, volcanic eruptions, and other historical processes, but also by human interference.

There has been something of a resurgence of interest in assembly rules in recent years (e.g. Weiher and Keddy 1995, Fox and Fox 2000). Gotelli and McCabe's (2002) meta-analysis of 96 previously published matrices found them to be highly non-random. As predicted by island assembly theory, there were fewer species combinations, more chequerboard species pairs, and less co-occurrence in real matrices than expected by chance. Although many of their data sets are not concerned with or are not limited to insular data and thus lie slightly outside the scope of this review, these findings are nonetheless significant in the present context.

Exploring incidence functions

One of the problems of evaluating island assembly theory as detailed above is that the biogeographical context of the island avifaunas under discussion is in each case quite complex. We now review some simpler systems, involving non-volant mammals on small, near-shore islands.

Hanski (1986) discusses occurrence and turnover of three species of shrews on islands in lakes in Finland, a system in which individual species populations could be studied in some detail. The largest species, *Sorex araneus*, was found to have a low extinction rate except on islands of less than 2 ha, on which demographic stochasticity becomes important. The smaller *S. caecutiens* and *S. minutus*, in contrast, were absent from many islands in the 2–10 ha range. Temporary food shortages, or 'energy crises' might be more problematic for the smaller species than for the larger *S. araneus*, thus potentially explaining their lower frequency of occurrence as a function of environmental stochasticity. Arguably, interspecific competition might also have a role to play, but Peltonen and Hanski (1991) subsequently reported no evidence for a significant effect of competition on extinction rates. Both studies conclude that the body size of species has a bearing on their incidence. As the larger species do not appear to be affected by interactions with the smaller species within the guild, they predict that they should occur consistently on islands greater than a minimum size set by demographic stochasticity. Small species should have a more erratic distribution, because of their increased susceptibility to environmental stochasticity. This interpretation combines elements of the incidence function and small-island effect hypotheses introduced in Chapter 4. Indeed, Hanski (1992) shows how, given knowledge of minimum population sizes, a simple model of incidence functions can generate realistic estimates of colonization and extinction rates for this system.

Incidence functions as originally devised are relatively crude univariate tools. Adler and Wilson (1985) provided a more sophisticated multivariate approach, using multiple logistic regression to estimate the probabilities of individual species occurrences. Their system consisted of records of 9 species of small terrestrial mammals for 33 coastal islands off Massachusetts, and their 11 explanatory variables provided various measures of island area, length, isolation, habitat, and source pools. Principal components analysis was used to generate two composite environmental variables, termed 'size' and 'isolation'. These composite variables were then used together with a categorical variable representing the dominant habitat type of each island ('domhab'), in multiple logistic regressions. For all but one species, statistically significant functions were obtained. In four cases (species of *Scalopus, Peromyscus, Microtus,* and *Zapus*), occurrence on islands was positively related to increasing island size, and in four cases (*Sorex, Blarina,*

Tamias, and *Clethrionomys*), to decreasing island isolation. The habitat variable, domhab, was also included in the functions of two species (*Peromyscus* and *Microtus*). Their study thus showed another means of quantifying relationships between species occurrence and a suite of environmental variables: their probability functions allowing the relative significance of different variables to be tested within a common framework. One other feature of interest was that they also used live-trapping over a 2–4 year period to show that the most widely distributed species across the islands are also those which reach the highest population densities within them.

The importance of both isolation and of constancy of resource availability were shown by Adler and Seamon's (1991) study of the population fluctuations in just a single species, the spiny rat *Proechimys semispinosus*, on islands in Gatun Lake, Panama. No other species of mouse- or rat-like rodents were captured on the islands. Their data revealed that colonization and extinction of the rat occurs regularly to and from the islands. Larger islands with year-round fruit production, regardless of isolation, have persistent populations of rats. Small, isolated islands only rarely have rats, because they lack a year-round food source and because immigration is rare. Small, near islands frequently have rats when fruits are present. These observations thus reveal how in some circumstances (of size, isolation, etc.) fluctuations in island carrying capacities influence species composition and turnover.

A particular land or aquatic system may provide dispersal filters which differentiate between different members of a source biota, and indeed different species within a single taxon (Watson *et al.* 2005). For instance, as shown by Lomolino (1986), particular functional groups of small mammals may be differentially affected by the same immigration filters. In regions with season ice cover, small coastal, inshore, or riverine islands may have their populations supplemented by mammals moving across the snow-covered ice in winter. In some cases, such movements have been recorded by tracking studies. In his studies of mammal movements across the snow-covered St Lawrence River, Lomolino found that larger, winter-active species were good colonizers, whereas the smaller species and hibernators seldom travelled across the ice, and if they did so, then only for limited distances. In related studies of incidences of species across 8 mainland and 19 insular sites, it was found that species that were known to utilize the ice cover had significantly higher insular rankings. Furthermore, within a particular functional guild, such as the insectivores, the smaller species were found to have lower insular rankings.

Although these data are strongly suggestive of immigration effects, species interactions may also be operative. It is a logical corollary of the EMIB that in addition to species numbers being a result of interactive effects of immigration and extinction, so also should species composition. Lomolino (1986) seeks to draw a distinction between combined (additive) effects of factors influencing immigration and extinction, and interactive or **compensatory effects**. The latter appears to be, in essence, merely another name for the rescue effect of Brown and Kodric-Brown (1977), whereby supplementary immigration of species present but at low numbers on an island, enables their survival through to a subsequent census. Thus, where immigration rates in the strict sense are high, so too will be rates of supplementary immigration. As the distinction between the two forms of population movement is effectively abandoned in this study, it might be termed instead the **arrival rate**. Species may thus have a high incidence on islands either where low arrival rates (poor dispersers and/or distant islands) are compensated for by low extinction rates (good survivors and/or large islands) or where high extinction rates are compensated for by high arrival rates. Thus species are common on those islands where their rate of arrival is high relative to their rate of loss. Of itself this is a trivial observation, but when combined with other ecological features it provides another option for understanding species compositional patterns, and one that stresses the dynamism inherent in the EMIB.

Figure 5.3 summarizes five hypothetical insular patterns for the distribution of particular species. First, a species may be distributed randomly (panel b) within an archipelago. Secondly, it may have minimum area requirements and not be dispersal-limited within the archipelago, in which case it will be present on virtually all islands over a critical threshold size (panel c). Thirdly, if it is a poor disperser which has low resource requirements relative to insular carrying capacities, it may occur only on the least isolated islands (panel d). Fourthly, it may depend on both island size and isolation, but not exhibit compensatory effects, thus resulting in the block pattern (panel e). Finally, it may exhibit the compensatory relationships between immigration and area and thus show the diagonal pattern of panel (f) of the diagram. Lomolino (1986) postulated that compensatory effects should be more evident for archipelagos with a large range of area and isolation relative to the resource requirements and vagility (mobility) of the fauna in question. If it is further assumed that within an ecologically similar group of species, larger species have both greater resource requirements and greater vagility than smaller species, an interesting prediction arises from this compensatory model. On the least isolated archipelagos, the smaller species will have a higher incidence on the smaller islands. But, as isolation of a hypothetical archipelago increases, the low persistence of the smaller species combined with their poorer vagility means that larger species should be the more frequent inhabitors of smaller classes of islands, i.e. their incidence relationships 'flip over' as a function of isolation. In short, if recurrent arrivals and losses are important in shaping the composition of the islands in question, the incidence of a species as a function of area will also vary as a function of immigration.

Lomolino developed a form of multiple discriminant analysis which enabled him to distinguish between the block effects and compensatory effects of Fig. 5.3. He applied this procedure to 10 species of mammals, each occupying at least 2 of 19 islands studied in the Thousand Islands region of the St Lawrence River, New York State. Circularity of argument in using distributional patterns to infer causation could be avoided as independent lines of evidence on species movements, body size, and other features of their autecology were available.

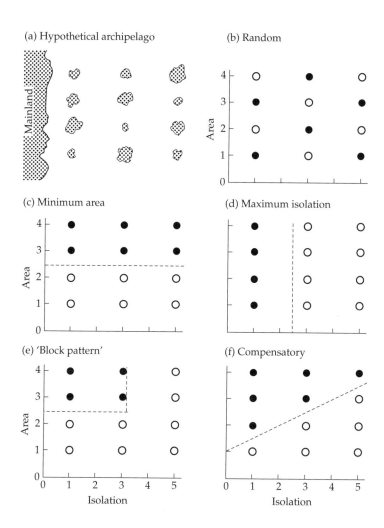

Figure 5.3 Patterns of insular occurrence of a species on islands of a hypothetical archipelago (a). Presence and absence in (b)–(f) are indicated by filled and open circles, respectively. The units for isolation and area are arbitrary. (Redrawn from Lomolino 1986, Fig. 1.)

It was found that none of the 10 species exhibited a maximum isolation effect (Fig. 5.3d), which was not surprising given the relative lack of variation in isolation in the data set. Only one species, *Microtus pennsylvanicus*, a small but dispersive generalist herbivore, showed no area or isolation effect. Six species exhibited minimum area effects (Fig. 5.3c) and the remaining three species exhibited compensatory effects (Fig. 5.3f). *Blarina brevicauda* is an example of the latter group, which is consistent with independent evidence that it is a relatively poor disperser. Lomolino (1986) applied his discriminant analyses to additional data sets for the islands of Lake Michigan, Great Basin mountain tops, and

islands in the Bass Straits. He found general support for the model he had developed, the relevance of scale of isolation to the incidence patterns, and in some cases, for patterns predicted by the compensatory model (see also Peltonen and Hanski 1991).

Microtus pennsylvanicus and *Blarina brevicauda* in the Thousand Islands region provide an illustration of another important effect structuring island communities, namely predation. The carnivorous shrew *Blarina brevicauda* preys mainly on immature stages of the vole *Microtus pennsylvanicus*. In the absence of the shrew on more remote islands, *Microtus* was found to undergo ecological release, occurring in habitats atypical of the species

(Lomolino 1984a). When introduced into islands, *Blarina* brought about drastic declines of *Microtus* densities, restricting it to its optimal habitat and, in at least one case, causing its extinction. As the vole is the better disperser, the two species exhibit a negative distributional relationship across the islands studied. Lomolino's studies thus demonstrate the significance both of recurrent arrivals and losses and of ecological controls such as predation in structuring the mammalian communities of relatively poorly isolated islands. As the islands in question undergo seasonal variation in their immigration filters, such that at times they are more akin to habitat islands than real islands, care must be taken in transferring such findings to other systems with different scales of isolation. The reliability of incidence functions is explored further in Box 5.2.

Linking island-assembly patterns to habitat factors

Haila and Järvinen (1983) and Järvinen and Haila (1984) examined the bird communities of the Åland islands, off Finland in the Baltic. There are thousands of islands in this group, ranging from only a few square metres to the main island of 970 km². Isolation of the islands tends to be slight, the main island being only 30 km from Sweden, and 70 km, via a steppingstone chain, from Finland. Twenty species of land birds found in the same latitudes in Finland and/or Sweden do not occur on the main island. These form three groups:

- species for which habitat is available on Åland, and for which absence is believed to be due to lack of over-water dispersal (e.g. green woodpecker, marsh tit, and nuthatch)
- species for which historical explanations can be offered (e.g. collared turtle dove and tree sparrow have recently expanded their mainland distributions but have yet to reach the island)
- species lacking because of absence of suitable habitat (e.g. great grey shrike).

They found that species numbers and density of the bird communities tended to be similar where habitats were of highly similar structure, and tended to differ when habitats were not directly comparable. They also noted that small, uninhabited islands in the archipelago lack those species of birds that are associated either directly or indirectly with humans. Their overall conclusion was that habitat factors provided explanations for the assembly of the bird communities of the Åland islands.

Graves and Gotelli (1983) used a null modelling approach in analysing the avifaunas of former land-bridge islands: Coiba, San José, Rey, Aruba, Margarita, Trinidad, and Tobago. They found in general that the avifaunas of these islands could be viewed as a random subset of the mainland 'habitat' pool when judged at the family level. They used families because they felt unable to assign species to guilds, and because species within a family are usually ecologically and morphologically similar. While acknowledging that this is not ideal—families do not necessarily represent units of interspecific competition—they contend that non-randomness of island avifaunas might reasonably be expected to be detectable at the family level. Using a simple model assuming all source-pool species to be equiprobable colonists, they found that out of 40 families, 3 are unusually common on land-bridge islands: pigeons (Columbidae), flycatchers (Tyrannidae), and American warblers (Parulidae), and only one, the puffbirds (Bucconidae), was present less often than expected. Thus, although in absolute terms many species and families are absent from the islands, the proportional representation of most families is consistent with mainland source pools. Comparison of the whole pool of species with the habitat pool showed that, as expected, the habitat pool is a superior predictor of species richness in each family. This implies that it is important when predicting species losses in fragmented landscapes to take account of patterns of habitat change that may unfold within fragments after isolation. Examination of the mainland range of the species in the pools showed that those species with widespread mainland ranges are disproportionately common on islands. This indicates either that they have persisted better since sea levels rose to produce the present degree of island isolation, or that

Box 5.2   How consistent are species incidence functions?

Incidence functions are simple analyses of the frequency of occurrence of individual species in relation to one or sometimes two properties of a set of islands: typically species richness, area, and isolation. Bierdermann (2003) provides an interesting analysis of area–incidence relationships of 50 species of vertebrates and invertebrates in which he demonstrates that area requirements increase essentially linearly with increasing body size on a log–log scale. Bierdermann rightly cautioned that it should not be assumed that species incidence functions are invariant. In illustration, Hinsley *et al.* (1994) have shown for birds of English woodlands that they may vary through time in relation to density independent mortality (extremes of weather). It might also be the case for habitat islands that they vary across the range of a species or as a function of the properties of the landscape in which the fragments are embedded.

Watson *et al.* (2005) tested this idea for three fragmented woodland landscapes in south-eastern Australia, all sufficiently close that climate and the regional species pool can be assumed similar. The landscapes were: (1) an agricultural landscape, used for both pastoral and arable farming; (2) a peri-urban landscape featuring pastures, hobby farms, and small urban areas; and (3) an urban landscape, namely the city of Canberra. They used logistic regression to establish sensitivity of species incidence to area, and to isolation in each landscape separately. Results demonstrated considerable variability in the incidence functions between the three landscapes and among species.

They may be categorized as follows:
8 species showed no sensitivity to area or isolation in any landscape
4 species showed minimum area thresholds in all landscapes
4 species showed minimum area thresholds only in the urban landscape
1 species showed a minimum area threshold but only in peri-urban and urban landscapes
3 species showed maximum isolation thresholds only in the agricultural landscape
6 species showed area effects in all landscapes and isolation effects (according to the

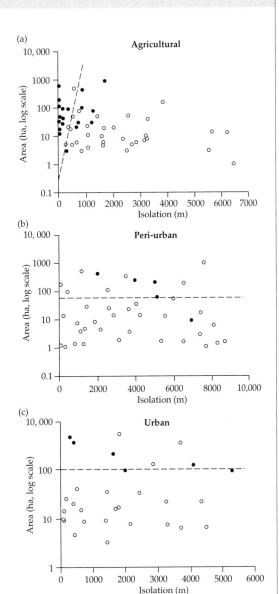

Box Figure 1. Area-isolation incidence plots for the eastern yellow robin *Eopsaltria australis* in three fragmented woodland landscapes in south-eastern Australia. Open circles are unoccupied remnants, closed circles are occupied remnants. The species displays a compensatory pattern in the agricultural landscape (a); and minimum area patterns in both the peri-urban (b) and urban (c) landscapes. Dashed lines represent the threshold where the probability of occurrence reaches 50%, as determined by logistic regression (From Watson *et al.* 2005).

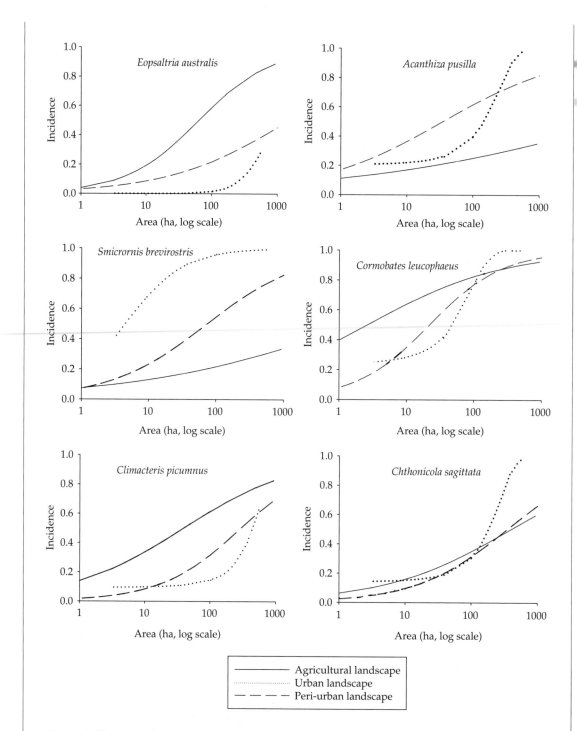

Box Figure 2. Incidence curves for six area-sensitive woodland species in the same three landscapes shown in the previous figure. Incidence functions were developed using logistic regression (From Whittaker *et al.* 2005, after Watson 2004).

compensatory pattern) only in the agricultural landscape

1 species (brown thornbill, *Azanthiza pusilla*) showed an isolation threshold in the agricultural landscape and area thresholds in the other two landscapes

None of the species was found to show a block area/isolation pattern

The variability in area/isolation is illustrated graphically for eastern yellow robin (Figure 1). Even where the species showed a common type of response between landscapes, the minimum area thresholds were found to vary significantly (Figure 2). These results show that species incidence functions can vary substantially, even within the same ecoclimatic region. As the woodland remnants themselves were similar in character across the study system, Watson *et al.* (2005) concluded that the differences in incidence functions between the landscapes were being determined by the nature of the matrix of non-woodland habitats in which the woodland fragments were embedded.

There are many ways in which matrix habitats might impact on the incidence of woodland birds within a patchy landscape, and it is not possible to know which were operative in this system. Nor is it possible to claim that the patterns shown, e.g. by the eastern yellow robin, would be replicated in another suite of urban, peri-urban, and agricultural catchments elsewhere within the range of the species. These findings do, however, imply that although species incidence functions may show some emergent tendencies (e.g. as reported for body size by Bierdermann 2003), they are not consistent properties within the range of a species.

widespread species tend to be better colonists, or possibly a combination of the two.

Anthropogenic experiments in island assembly: evidence of competitive effects?

It is difficult to assess the importance of competition in the process of community assembly on islands. One reason for this is the difficulty of observing immigration events. Human intervention in island systems has, however, inadvertently created opportunities to test for island assembly structure, and in cases provides evidence indicative of competition as a structuring force.

One such example is a study by Badano *et al.* (2005), who use the chequerboard index or *C*-score to analyse ant community structure in islands of the Cabra Corral dam, Argentina. They found that newly created islands showed a random pattern of species co-occurrence, but that on older (forest remnant) islands species showed a significant tendency to co-occur less often than expect by chance. This, they argued, is a consequence of competitive sorting on the older islands. Contrasting results were obtained in another study of newly created (forest remnant) islands from within a South American reservoir, this time an analysis of forest-interior birds nesting on islands within Lake Guri, Venezuela by Feeley (2003). He tested several different assembly rule models, finding strong evidence of nestedness (below) related to habitat specificity, such that specialist species tend to be absent from small, species-poor islands. This indicates that the communities have been shaped by structured losses of species rather than through the effects of interspecific or interguild competition, although there was some limited evidence of competitive limitations in the degree of size similarity permissible in co-occurring species.

The purposeful introduction of land birds to remote oceanic islands provides another form of inadvertent experiment on the role of competitive effects in structuring insular avifaunas. Such introductions have been for various purposes, including as game birds, as song birds, and to 'beautify' the native avifauna. Very often the native avifauna has been depleted concurrent with or before the introductions, with the exotic species to varying degrees replacing the lost species (Sax *et al.* 2002). Although the natural immigration rate of land birds to such islands in the absence of human transformation of the landscape can be assumed to be negligible, high rates of introductions of species by people have been documented in several archipelagos

(Lockwood and Moulton 1994). Only a proportion of introduced species succeeds in establishing and maintaining populations over lengthy periods. In tests of data for introductions of passeriform birds to the Hawaiian islands, Tahiti, and Bermuda, it has been found that the surviving species exhibit what is termed morphological overdispersion, i.e. they are more different from one another in characteristics that are believed to relate to niche separation than chance alone would predict (Lockwood and Moulton 1994). Similar results have been reported for game birds (galliforms) for the Hawaiian islands and New Zealand (Moulton *et al.* 2001). These results appear to support the hypothesis that interspecific competition shaped the composition of these anthropogenic communities.

Like other island assembly rules the morphological overdispersion pattern require careful scrutiny to ensure that it is not an artefact of, for example: (1) the biogeographic region of origin of the introduced species, (2) the ecological or taxonomic structure of the introduced species, or (3) the numbers of introduced birds (e.g. see Blackburn and Duncan 2001; Cassey *et al.* 2005a). Of these factors, the introduction effort has been highlighted as crucial in a number of data sets (Cassey *et al.* 2005a,b). On the other hand, in their analysis of game bird introductions, Moulton *et al.* (2001) argue that evidence for the introduction effort hypothesis in their data set is weak, citing examples of game birds failing because they were introduced into inappropriate environments in rather indiscriminate fashion: in such circumstances mere weight of numbers may not ensure survival. Hence, although it is certainly the case that greater numbers of released birds provides a greater likelihood of initial success of a species establishing, they argue that the emergent morphological pattern of the assembled community of exotic species shows evidence of competitive structuring.

Just when you think it might be safe to conclude in favour of competitive effects, once again the case has been disputed, by reference to the data derived from the New Zealand game-bird introductions. Duncan and Blackburn (2002) slay the beautiful hypothesis with the traditional ugly fact. They argue that competition among morphologically similar species could not have been responsible for the failure of most game-bird introductions to New Zealand because most species were released at widely separated locations or at different times, and would never have encountered other morphologically similar introduced game birds, or if they did would not have done so at the necessary densities to influence the outcome. They conclude that the significant pattern of morphological overdispersion in this instance must therefore result from some other cause, such as, for example, greater effort being expended on introductions of morphologically distinct species. Interactions with introduced predators, and with other anthropogenic changes in New Zealand further complicates attribution of the overdispersion pattern (Duncan *et al.* 2003). Accepting the criticisms of the New Zealand requires that the Hawaiian data stand alone, and taken in isolation the Hawaiian galliform data fail to show overdispersion, and so, according to Duncan and Blackburn (2002) the case falls. From this we may conclude that the most predictable feature of island assembly rules is that any claim of evidence of competitive effects will be contested: at last some evidence of competition, even if only amongst biologists!

5.2 Nestedness

If island biotas are in essence randomly drawn from a regional species pool, as per the simplest form of the EMIB, then an archipelago of islands should exhibit differences in composition from one island to the next and the degree of overlap should be predictable on the basis of a 'null' model. Diamond's assembly rules seek to describe departures from such an expectation, another form of departure is where island biotas exhibit nestedness. A **nested distribution** describes the situation where smaller insular species assemblages constitute subsets of the species found at all other sites possessing a larger number of species (Patterson and Atmar 1986). Diamond (1975a) intended his assembly rules approach to be deployed on groups of ecologically related species, or guilds, whereas the analyses of nestedness (like incidence functions) provide descriptive tools that can be applied to either narrow or broad groups of species. It might thus be possible for chequerboard distributions and the like to be detectable within a particular guild of birds, while the avifauna analysed as a whole exhibits a tendency towards nestedness.

Tendencies towards nestedness were recognized by early island biogeographers, such as Darlington (1957), but it was not until the 1980s that statistical tests were developed for formal detection of the pattern. When the first indices were developed by Patterson and Atmar (1986), they found that insular systems are commonly nested (Patterson 1990; Blake 1991; Simberloff and Martin 1991). Patterson and Atmar (1986) took the view that nestedness was most likely to occur in extinction-dominated systems and developed their first metric on this presumption. However, other authors argued that immigration might also produce nestedness patterns, requiring different metrics for quantifying nestedness. This led to much debate about how best to measure nestedness, a matter that is most likely still unresolved (e.g. Wright and Reeves 1992; Cook and Quinn 1995; Lomolino 1996; Brualdi and Sanderson, 1999; Wright *et al.* 1998; Rodríguez-Gironés and Santamaría 2006). Although the choice of metric is important at a detailed level, the general proposition that nestedness is a common feature of ecological assemblages (both insular and non-insular) is robust to the choice of metric (Wright *et al.* 1998).

The nestedness concept, as applied by Patterson and Atmar and as generally followed by others, is based on ordering the data matrix by the size of fauna or flora, i.e. it is **richness-ordered nestedness**. Some authors, however, have used the same term 'nestedness' when ordering the data matrix not by species richness but by island area, which we might term **area-ordered nestedness**, or even by island isolation, i.e. **distance-ordered nestedness** (e.g. see Roughgarden 1989; Lomolino and Davis 1997). The extent to which island assemblages are found to be nested when ordered in these different ways may be used to infer processes responsible for structuring them. However, whichever metric is used, and whichever variable is used for ordering the data matrix, inferences of causation ideally require independent lines of verification.

As Wright *et al.* (1998, p. 16) state 'nestedness is fundamentally *ordered composition* . . . Any factor that favours the 'assembly', or disassembly . . . of species communities from a *common pool* in a *consistent order* will produce nested structure' [italics in original]. They identify four factors or filters that may be important contributors in generating nested structure: passive sampling, habitat nestedness, distance, and area.

● So called **passive sampling** of a pool of species by a recipient island may produce nested structure if species are drawn randomly from the pool with the constraint that the availability of propagules is itself strongly non-random, assuming for instance a log-normal species abundance distribution for the pool community. Simulation models on this basis generate highly nested model communities, generally in fact more strongly nested than observed. Where passive sampling simulations do compare well with observed data sets, this may lead to the inference that differential immigration has structured the assemblages. However, Wright *et al.* (1998) argue that the simulation is not dependent on a particular mechanism of community assembly, and that other scenarios are conceivable. One might be that abundant species from within the pool are regionally 'successful' species with a low probability of going extinct from isolates.

● For **habitat nestedness** to produce biotic nestedness requires, first, a strong fidelity of species to particular micro-habitats, and second, for islands to acquire additional habitats in a fairly fixed order. Given which, we would anticipate that it will be more readily detectable in plant data sets (as plants often show high habitat affinity), as exemplified by Honnay *et al.*'s (1999) study of plants in forest fragments in Belgium.

● If nestedness is clearly related to **distance** from mainland sources, this suggests that there are predictable limits to species' dispersal abilities, such that the system is made up by islands 'sampling' a series of species' isolation-incidence functions (as Fig. 5.4). In many case studies, where available data are limited to a single time-point, it may be safest simply to label this pattern a 'distance' effect (Wright *et al.* 1998), but there are cases where the early stages of succession on a series of islands allow researchers to examine nestedness during the assembly process, and thus establish more directly the predictability of the colonization process. A good example is Kadmon's (1995) study of seven islands created by the filling of the Clarks Hill Reservoir, Georgia, USA. The islands were logged and cleared of woody plants before their separation from the mainland. Thirty-four years later, species number was found to be inversely related to island isolation. The woody plant floras were found to be nested when ranked by richness, or isolation, but not by area. These results point to the significance of differential colonization in structuring the data set. Only a small proportion of the species included in the analysis contributed to the observed nestedness, principally being those that lacked adaptations for long-range dispersal. Wind-dispersed species showed no evidence for nestedness. This study thus pointed to the significance of plant dispersal attributes in structuring the rebuilding of insular assemblages, a feature also of the recolonization data from Krakatau, as we shall shortly see.

• Nestedness can also result from differential **area** requirements across species, again as may be

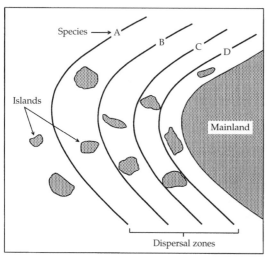

Figure 5.4 Hypothetical scenario for the production of nested subsets via differential colonization. From Patterson and Atmar (1986, Fig. 4), who contended, nonetheless, that extinction is a more potent structural determinant in the majority of insular systems.

revealed in the individual species' area-incidence functions of the component species. Where nestedness within a former land-bridge island data set is strongly area-related but is unrelated or only weakly related to isolation within a data set, this is taken to indicate a strong extinction signal in the structuring of assemblages. This rationale was invoked in classic early studies of nestedness of mammals on land-bridge islands (e.g. Patterson and Atmar 1986). This is on the grounds that before they became islands, the land-bridge islands were part of the mainland and can be assumed to have had a full complement of the mainland pool, with extinction then the dominant process in the sorting of the island biotas. Similarly, extinction-led sorting of remnant mountain-top 'island' has been invoked for the nestedness of mammal assemblages in the south-western USA (Patterson and Atmar 1986).

Given that nestedness appears to be so prevalent, it is worth considering factors that may constrain it:

• First, consider the implications of supertramp distributions: species that tend to occur preferentially on species poor islands, and to disappear in species rich assemblages. Such patterns, along with certain other of Diamond's assembly rule patterns (above) will form exceptions to any general system-wide tendency to nestedness.

• Second, nestedness must perforce be diminished as the geographic extent of the study system is expanded to incorporate different species pools, i.e. biogeographical turnover in species composition.

• Third, Wright *et al.* (1998) suggest that where the sample system involves mostly very small patches, the degree of nestedness may be diminished by high habitat heterogeneity between patches (i.e. the opposite of habitat nestedness).

It is now well established that nestedness is a common but not universal feature of many insular biotas. In an applied context, it has also been shown that within a given landscape, different taxa can respond in vastly different ways to fragmentation. For instance, Fischer and Lindenmayer (2005), examining data from fragmented forest habitats in south-eastern Australia, found that whereas birds showed a strong degree of nestedness, the arboreal marsupials were non-nested. In this case, their explanation for the nestedness of the avifauna invoked selective extinction as the dominant process.

Table 5.2 provides an interesting classification of nestedness patterns by Patterson (1990) using one particular metric. Opinions have varied as to the balance of explanatory power among the factors generating nestedness (Pattterson and Atmar 1986; Patterson 1990; Simberloff and Martin 1991; Cook and Quinn 1995; Nores 1995), but Patterson's conclusions appear to be broadly supported in a more recent multimetric, multisystem analysis by Wright *et al.* (1998), who have reaffirmed that land-bridge archipelagos are typically strongly nested, and

immigration experiments least nested. Their comparative analysis adds further support to the argument that differential extinction is a powerful force in producing nested structure, but also points to the realization that different types of system may be dancing to different tunes: sometimes immigration is dominant, sometimes extinction, sometimes habitat nestedness. The relevance of nestedness within conservation biogeography is examined in Chapter 10.

5.3 Successional island ecology: first elements

Consideration of nestedness in this chapter has led to the point where questions concerning the

Table 5.2 Summary of significant nested subset structure (according to Patterson 1990), in assemblages affected over various time periods by: mainly extinction; mainly colonization; or both processes in concert. The study was based on the nestedness index, N, of Patterson and Atmar (1986), which has been argued to be suboptimal for the analysis of systems affected by colonization (see discussion in Cook and Quinn 1995).

Processes	Nested	Non-nested
Holocene changes		
Extinction dominated	Southern Rocky Mts mammals	
	New Zealand land-bridge birds	
	Baja land-bridge mammals	
	Baja land-bridge herptiles	
	Great Basin mammals	
Colonization dominated	Baja oceanic birds	New Zealand oceanic birds
		Baja oceanic mammals
		Baja oceanic herptiles
Both processes	Baja land-bridge birds	
	Great Basin birds	
Long-term changes		
Extinction dominated	Australian wheat-belt mammals	Australian wheat-belt lizards
	North American park mammals	
	São Paulo sedentary birds	
	Bass Strait mammals	
Colonization dominated		All São Paulo birds
Both processes	Penobscot Bay mammals	
Short-term changes		
Extinction dominated		Mangrove area reduction
Colonization dominated	Dispersing shrews and voles	Breeding shrews and voles
	Weeds of young lots	Weeds of old lots
Both processes		Mangrove defaunation

colonization process require more detailed attention. Both island biogeography and succession theory go back a long way. Biologists have been studying them in combination since the nineteenth century (e.g. references in Whittaker *et al.* 1989), although not necessarily within the same frameworks we use today. Ridley may well have been influenced by the early Krakatau data when he wrote in the preface to his classic text on plant dispersal (Ridley 1930, p. xii):

An island rises out of the sea: within a year some plants appear on it, first those that have sea-borne seeds or rhizomes, then wind-borne seeds, then those borne on the feet and plumage of wandering seafowl, and when the vegetation is tall enough, come land birds bringing seeds of the baccate or drupaceous fruits which they had eaten before their flight.

This provides the first elements of a dispersal-structured or **successional model of island assembly**.

Successional effects were discussed in relation to equilibrium theory by MacArthur and Wilson (1967), but in a fairly simplified fashion and with little empirical data. Following their work, one of the more interesting contributions was provided by the data for the two volcanically disturbed islands from Diamond's (1974) study of New Guinea islands. Ritter was sterilized in 1888, and Long was devastated about three centuries before Diamond's survey of the two islands in 1972. Vegetation recovery on each island was found to be incomplete: Ritter supported *Pandanus* (screw-pines) up to 12 m high on its gentler slopes, but was bare in steeper areas, succession being retarded by landslides, strong prevailing winds, and erosion. Long's recovery was more advanced, but its forests remained comparatively open and savanna-like. The species–area plot in Fig. 5.5 reveals that both Ritter (4 species) and Long (43 species) remained depauperate (Diamond 1974). Observations of failed arrivals of other species on Ritter, including feathers at the plucking perch of a resident peregrine falcon, provided further support for a habitat/successional limitation explanation. Diamond compared his data with a survey from 1933, calculating by the means explained in Chapter 4

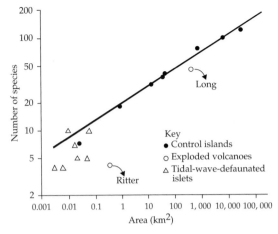

Figure 5.5 The relationship between log species number and log-island area for resident, non-marine, lowland bird species on the Bismarck Islands (redrawn from Diamond 1974). The line of best fit is for the so-called control islands. The explosively defaunated islands of Long and Ritter remain significantly ($P<0.001$) below the line of best fit, which Diamond interpreted as demonstration of incomplete succession and failure to attain equilibrium in the period since their volcanic disturbance.

that its species number was approximately 75% of the equilibrium value. He found that species numbers had increased only slightly in the interim, with a degree of turnover occurring—a result he attributed to the slow pace of forest succession 'arresting' the progress of the Long avifauna towards its equilibrium value. As turnover was not greater than on older islands he termed it a quasi-steady state. This hierarchically determined pattern provides an interesting parallel with subsequent studies of the build-up of bird numbers from the Krakatau group (below).

Diamond's line of argument was of much broader biogeographical scope, as we have seen, but the successional dynamics represented by Ritter and Long held a key place in the overall theory of the dynamics of the New Guinea islands. Diamond's reasoning was as follows. Small islands, on which extinction rates should be high, will have a high incidence of the best dispersing species, i.e. they have a high incidence of tramps. Larger defaunated islands will first be colonized by the *r*-selected supertramps. Other, less dispersive tramps arrive subsequently and eventually crowd out the less competitive supertramps. Before their loss, however, the island acts to supply the surrounding area with its surplus population, and the species will therefore survive by having found another defaunated island before being excluded from the first. Diamond calculated that for some supertramps the availability of Long island will have enabled the quadrupling of the entire population of the species. Long and Ritter thus hold considerable interest. In Long's case, this has been added to recently by studies of the emergent island Motmot, which has formed in the lake within Long Island, providing a form of nested colonization 'experiment' (see Thornton *et al*. 2001).

5.4 Krakatau—succession, dispersal structure, and hierarchies

Background

Krakatau was the first site used by MacArthur and Wilson to test their predictions of an approach to a dynamic equilibrium. The natural recolonization (Box 5.3) of the islands by birds between 1883 and 1930 appeared consistent with the EMIB: species number reached an asymptote and further turnover occurred (MacArthur and Wilson 1963, 1967). However, the plant data from the islands revealed a poorer fit, as species numbers continued to rise. This was postulated to be due to one of two effects: either that the pool of plant species was sufficiently large to prevent a depletion effect on the immigration rate, or successional replacement of the pioneer communities was incomplete, thereby leading to a reduction in extinction rate for a time. The latter effect was favoured, and illustrated in diagrammatic form, but the EMIB was not modified formally to account for the data.

Community succession

Although the first food chains to establish may well have been of detritivores, and microorganisms, consideration of the real business of community assembly begins with the higher plants. Partial survey data and descriptions of the plant communities are available from often brief excursions in 1886,

1896/97, 1905–08, most years between 1919 and 1934, 1951, 1979, 1982/83, and 1989–95 (Whittaker *et al*. 1989, 1992a; 2000). The system must be understood both in terms of successional processes and in relation to constraints on arrival and colonization. The broad patterns of succession will be described first.

The coastal communities established swiftly, most of the flora being typical of the sea-dispersed Indo-Pacific strand flora. In 1886, 10 of the 24 species of higher plants were strandline species. By 1897, it was possible to recognize a *pes-caprae* formation (named after *Ipomoea pes-caprae*) of strandline creepers, backed in places by establishing patches of coastal woodlands, of two characteristic types. First, patches which were to become representative of the so-called *Barringtonia* association (more generally typified by the tree *Terminalia catappa*), another vegetation type typical for the region, and secondly, stands of *Casuarina equisetifolia*. The latter is a sea- and wind-dispersed pioneering tree species, which also occupies some precipitous locations further from the sea. In the coastal areas it typically lasts for just one generation, as it fails to establish under a closed forest. It is notable that as early as 1897 the coastal vegetation types were already recognizable as similar to those of many other sites in the region (Ernst 1908). The coastal communities continued to accrue species over the following two decades, but have since exhibited relatively little directional compositional turnover, apart from the loss in many areas of *Casuarina*. They may thus be described as almost an 'autosuccession' (Schmitt and Whittaker 1998).

In the interiors, a much more complex sequence of communities has unfolded. To varying degrees this can be understood in relation to the considerable differences in habitats between and within islands, to differential landfall of plant species within the group, and to the dynamics of the physical environment, especially the disruption originating from the new volcanic island, Anak Krakatau. In 1886, most of the cover in the interior was supplied by 10 species of ferns, accompanied by 2 species of grass and 2 members of the Asteraceae: all 14 species being wind-dispersed. By 1897, the interiors had become clothed in a dense

Box 5.3 The Krakatau eruption of 1883: a dramatic start to a 'natural experiment'

The Krakatau group has undergone repeated phases of volcanic activity. In 1883 it consisted of three islands, arranged in a caldera which resulted from prehistoric eruptive activity. The largest island of the group became active in May of that year, commencing a sequence that ended on 27 August in exceptionally destructive eruptions. Two-thirds of the island disappeared, vast quantities of ejecta were thrown into the atmosphere, and a series of huge waves (tsunami) generated in the collapse resulted in an estimated 36 000 human casualties in the coastal settlements of Java and Sumatra flanking either side of the Sunda Strait some 40 km or so from Krakatau. The meteorological and climatological effects of the eruption were observable across the globe and the islands excited considerable and lasting scientific interest from a number of disciplines (Thornton 1996). Each of the three Krakatau islands was entirely stripped of all vegetation. The main island, now known as Rakata, lost the majority of its land area, but all three islands also gained extensive areas of new land resulting from the emplacement on to the pre-existing solid rock bases of great thicknesses of pyroclastic deposits. Estimates of ash depths are of the order of 60–80 m and to this day almost all of the area of the three islands remains mantled in these unconsolidated ashes, with relatively little solid geology exposed at the surface. No evidence for any surviving plant or animal life was found by the scientific team led by Verbeek later that year, and in May 1884 the only life spotted by visiting scientists was a spider. The first signs of plant life, a 'few blades of grass', were detected in September 1884.

The efficacy of the destruction has been hotly debated (Backer 1929; Docters van Leeuwen 1936). It is conceivable that some viable plant propagules might somehow have survived to be uncovered by later erosion of the ash mantle (Whittaker et al. 1995). However, there is no evidence for survival, and indeed the densest populations of early plant colonists were located on terra nova. Any viable rain forest propagules uncovered by erosion of the ash would in any case have emerged into a vastly different and hostile environment, and their successful establishment would thus have been entirely remarkable. The islands can be taken to have been as near completely sterilized as makes no practical difference. The fauna and flora have thus colonized since 1883 from an array of potential source areas, the closest of which, the island of Sebesi, is 12 km distant. All the nearest land areas, including Sebesi, were also badly impacted by the 1883 eruptions.

Here was a group of three islands, each mantled in sterile ashes and in time receiving plant and animal colonists. The potential of the islands as a natural experiment in dispersal efficacy and ecosystem recovery was appreciated, although, unfortunately, only botanists seized the opportunity at an early stage, with surveys in 1886 and 1897. Since then, numerous scientific teams have worked on Krakatau, to varying ends, but along the way accumulating a remarkable (if imperfect) record of the arrival, succession and turnover of plant species and many groups of animals (Whittaker et al. 1989; Thornton 1996; Whittaker et al. 2000).

It is important to appreciate the dynamism of the platform upon which ecosystem assembly has taken place. The early pace of erosion of the ash mantle must have been dramatic. It created a deeply dissected 'badlands' topography which remains geomorphologically highly dynamic. The extensive new territories around the coasts were also subject to extreme rates of attrition, and steep cliffs formed rapidly around much of the shoreline. Shallow shelving beaches, which provide the most favourable points for the colonization of many plant and some animal species, are restricted in their extent. In 1927, after nearly 46 years of inactivity, a new island began to form in the centre of the 1883 caldera, finally establishing a permanent presence, Anak Krakatau, in 1930. Through intermittent activity it grew by the mid-1990s to an island of over 280 m in height and 2 km in diameter (Thornton et al. 1994). Over this period it caused significant and widespread damage to the developing forests of two of the three older islands, Panjang and Sertung, but only indirectly impacted on Rakata (Whittaker et al. 1992b; Schmitt and Whittaker 1998).

grassland, dominated by *Saccharum spontaneum* and *Imperata cylindrica*, interspersed with small clusters of young pioneer trees. Ferns dominated only the higher regions of Rakata and the balance of species had also shifted in favour of the flowering plants. By 1906, the woodland species had increased considerably in the interiors, although remaining patchy, and the fern communities were gradually receding upwards. The grasslands were tall and dense and so difficult to penetrate that proper exploration of the interiors was greatly hampered. The woodlands of the lowlands continued their rapid development, with *Ficus* spp., *Macaranga tanarius*, and other animal-dispersed trees to the fore.

Forest closure took place over most of the interior of each island during the 1920s, such that by 1930 very little open habitat remained. This key phase of system development was fortunately the subject of detailed investigations by Docters van Leeuwen (1936). As the forests developed, habitat space for forest-dependent ferns, orchids, and other epiphytic plants became available, and their numbers increased rapidly in response. Conversely, the pioneering and grassland habitats were reduced, species populations shrank, and some species disappeared. Rakata is a high island, *c.*735 m, and altitudinal differentiation of forest composition was evident as early as 1921. The highest altitudes thereafter followed a differing successional pathway, in which the shrub *Cyrtandra sulcata* was for many years a key component. Although most vegetation changes appear to have been faster in the lowlands, spreading up the mountain of Rakata, one key species, the wind-dispersed pioneering tree *Neonauclea calycina*, first established a stronghold in the upper reaches, before spreading downwards. By 1951 it had become the principal canopy tree of Rakata from just below the summit down to the near-coastal lowlands. At the close of the century it was clearly in decline in the lowlands, as the forests developed a more patchy character, with a number of other large canopy and subcanopy species varying in importance from place to place within the interiors.

The patterns of development on the much lower islands of Panjang and Sertung were broadly similar up to *c.*1930, although differences in the presence and abundance of particular forest species

were noted. Since 1930, both islands have received substantial quantities (typically in excess of 1 m depth) of volcanic ashes over the whole of their land areas. Historical records and studies of ash stratigraphies demonstrate that some falls of ash have been very light, but *c.* 1932/35 and 1952/53, and possibly on other occasions, the impact has been highly destructive (Whittaker *et al.* 1992*b*). For instance, Docters van Leeuwen (1936) described how in March 1931 the most disturbed forests of Sertung resembled 'a European wood in winter', and how grasses re-invaded (possibly resprouted) within the stricken woodlands. Since then, forests dominated to a considerable extent by the animal-dispersed trees *Timonius compressicaulis* and *Dysoxylum gaudichaudianum* have become characteristic of large areas of both islands. The interplay of slight environmental differences between islands, differential patterns of landfall, and disturbance episodes from the volcano, has produced a shifting successional matrix across each island, influencing diversity patterns at fine and coarse scales in complex ways that are difficult to attribute in other than general terms to the array of potential causal variables (Schmitt and Whittaker 1998).

Although different forest types have been recognized on the Krakatau islands, these 'communities' are not discrete and forest successional pathways are in practice more complex than simple summary successional schema (e.g. Whittaker *et al.* 1989, Fig. 15) might be taken to imply. It is notable that of all the vegetation types of the Krakatau islands, only the *pes-caprae* formation and *Barringtonia* association were assigned with any confidence to phytosociological types by the earlier plant scientists. As Docters van Leeuwen (1936, pp. 262–3) put it, 'all other associations in the Krakatau islands are of a temporary nature: they change or are crowded out.' Although the coastal systems are similar to those of other locations in the region, those currently recognized from the interior by Whittaker *et al.* (1989) lack documented regional analogues of which we are aware. The interior forests of the Krakatau islands continue to accrue new species of higher plants, and the balance of species in the canopy is undoubtedly in a state of flux, with strong directional shifts in the importance of particular species being evident over the period since 1979 (Bush *et al.* 1992; Whittaker *et al.* 1998). The business of forest succession is ongoing.

A dispersal-structured model of island recolonization

The community-level changes have been underpinned by trends within the colonization data

(Whittaker *et al.* 1989, 1992*a*). Figure 5.6 simplifies these trends into three time-slices, restricting the treatment to the island of Rakata, for which the best data (least confounded by volcanic disturbance) are available. The figures are for all species recorded from the island, whether still present or not (Whittaker and Jones 1994*a*). Phase 1 represents the pioneering stage during which wind-dispersed pioneers, first ferns, and then grasses and composites

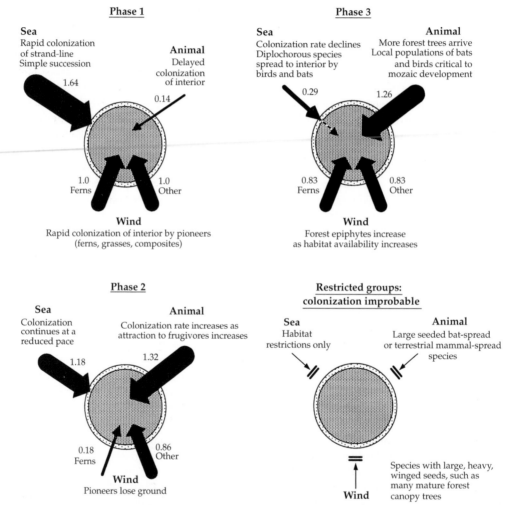

Figure 5.6 Plant recolonization of Rakata Island (Krakatau group) since sterilization in 1883. The three phases correspond with survey periods and represent convenient subdivisions of the successional process. Phase 1, 1883–1897; phase 2, 1898–1919; phase 3, 1920–1989. Arrow widths are proportional to the increase in cumulative species number in species/year (these values are also given by each arrow). The flora is subdivided into the primary dispersal categories, i.e. the means by which each species is considered most likely to have colonized: animal-dispersed (zoochorous), wind-dispersed (anemochorous), and sea-dispersed (thalassochorous). The model distinguishes between strand-line (outer circle) and interior habitats and, in the fourth figure, identifies constraints on further colonization. (From Whittaker and Jones 1994*a*.)

(Asteraceae), initiated the colonization of the interior. As emphasized above, colonization of the strandlines was rapid during this phase, gradually diminishing thereafter.

Phase 2 represents the period during which the extensive grasslands waxed and waned, as, increasingly, animal-dispersed trees and shrubs spread out from their initial clumps to form woodlands. The interiors have been filled almost exclusively by species which are primarily either wind- or animal-dispersed. The relative balance between these two groups shifts dramatically between phase 1 and 2. Few ferns colonized during phase 2 despite the efficiency of their dispersal system (microscopic spores). The interpretation offered for this is that there are actually relatively few ferns in the regional species pool which typify such extreme, seasonally droughted pioneer sites, and they arrived very quickly during phase 1. Wind-dispersed early successional flowering plants also continued to accrue during phase 2, including Asteraceae and terrestrial orchids, but very few trees and shrubs. The vast majority of arboreal species are animal-dispersed, and after a very slow start, in which only two species of animal-dispersed species were found by 1897, their rate of arrival took off in the early 1900s.

Phase 3 runs from the point when forest closure became extensive, to the 1989 datum. During this phase, the colonization rate of sea-dispersed species slackens, which reflects the relatively limited source pool, but also the limited array of coastal habitats available. In the interiors, it was only when forest habitats became available that the second wave of ferns, those of forest interiors, could build in numbers. Hence their rate of discovery increased rapidly in the period of forest formation, peaking around 1920. The ease of dispersal of ferns, with their microscopic propagules, can be demonstrated by comparison of the size of the Krakatau flora (all islands 1883–1994) with the native flora of West Java. The ratio for spermatophytes is 1:10.1 and for ferns 1:4.2, indicating Krakatau to have a remarkably rich fern flora (Whittaker *et al.* 1997). Wind-dispersed flowering plants colonizing the newly available forests of phase 3 were predominantly orchids, many of which are epiphytic, together with a mix of other epiphytic and climbing herbaceous species. The single largest group recorded for phase 3, however, is the animal-dispersed set, mostly being trees and shrubs.

More detailed analyses of dispersal mechanisms suggest further structural features in the data. For instance, wind-dispersed species can be split into those with dust-like propagules, plumed seeds, and winged seeds or fruits. The larger, winged propagules have limited dispersive ability and, although significant in the regional pool of forest canopy trees, only one or two species of this category have colonized Krakatau (Whittaker *et al.* 1997): they are thus noted in Fig. 5.6 as improbable colonists. Colonization of the animal-dispersed species of Krakatau, with the exception of a few human-introduced cases, can be attributed to the actions of frugivorous birds and bats. Whittaker and Jones's (1994*b*) best estimates of their relative roles are given in Table 5.3. Fruit bats swallow only the smallest seeds (<1 or 2 mm) and it was assumed that they would be unlikely to carry larger seeds in their mouths or claws over the many kilometres of

Table 5.3 Estimates of the numbers of plant species found on Krakatau between 1886 and 1992 for which birds and bats have a dispersal role. Under the heading of animal dispersal two modes of transport may be distinguished: those seeds which are eaten, swallowed and which pass through the gut or which are eventually spat out, and those transported by external attachment to the animal. The data set include all four Krakatau islands and all records (i.e. including some species which have not maintained a presence) (source: Whittaker and Jones 1994*b*.)

Dispersal mode	Number of species
Animal, gut passage (bird and/or bat)*	124
Animal, attached externally (bird)	10
Human introduction, but spread by feeding birds and/or bats	15
Sea-colonist, then spread by animals	24
Total animal introduction and/or spread	173

* Of these 124 species, 50 were considered to be bird-dispersed, 31 could potentially have colonized either by bird or bat transport, and 31 were considered to have arrived by bird but then to have the potential to be bat-spread within the islands. These estimates should be regarded as first approximations (e.g. see further work on bat-dispersal on Krakatau, by Shilton 1999).

sea separating Krakatau from source areas. Thus, small-seeded animal-dispersed species might be introduced either by birds or bats, but larger-seeded species can be considered strictly bird-assisted colonists. Beyond a certain threshold seed size it is likely that only the largest fruit-pigeons can effect an introduction. However, once a species has established on an island, a wider range of dispersal agents may be involved in local dissemination.

Where plant species have two rather contrasting dispersal vectors, they are termed diplochores. Twenty-four sea-colonist, but potentially animal-spread species occur on Krakatau, of which 14 are bat-spread, 4 may be bat- or bird-spread, and 6 are bird-spread. This mixed group of diplochores includes several of the earliest colonists of the Krakatau strandlines. They likely had an important role in kick-starting the succession of animal-dispersed species. As shown, restrictively animal-dispersed species were laggardly colonists. There cannot be much advantage in visiting a barren island if you happen to be a frugivorous bird or bat. Once the fringing plant communities of sea-dispersed species established and began to fruit, however, the islands would have provided a suitable food supply in the form of the fruit of the diplochorous species. From the early accounts we know that many of the first true animal-dispersed plants were concentrated around the coastal fringes. Admittedly the interiors were less well explored, and certainly pockets of trees quickly became dotted throughout the interiors, yet we can be fairly confident that the first points of landfall, and the first roosting points for frugivores and the first food supply, were provided by the coastal vegetation.

The first frugivores observed on the islands were birds, six species being observed in 1908 (the first bird survey), whereas it was not until the next zoological survey in 1919 that bats were noted. Birds may thus have been the more important group in terms both of simply the number of animal-dispersed colonists and being earlier in arrival (although survey data are too poor to be conclusive). However, bats have had a differing role to birds. First, they have spread some species that birds do not transport. For instance, the important coastal and near-coastal tree *Terminalia catappa* only invaded the island interiors after the arrival of the bats. Secondly, Whittaker and Jones (1994b) suggested that they might be more important to the early seeding of open habitats because of differences in their foraging behaviour (in part mediated by interactions with predators). More recent detailed work by Shilton (1999; Shilton *et al.* 1999) supports these ideas, and also crucially demonstrates the ability of bats to retain viable seed in the gut for over 12 hours between the final feed of one night and the start of feeding on a subsequent night, by which time they may easily have moved from mainland Java or Sumatra to the Krakatau islands (or vice versa). Although their role in long-distance seed dispersal seems limited to only a few small-seeded plant genera with the right types of fruit (especially *Ficus* species), we are now convinced that their role in kick-starting forest succession was more important than earlier researchers (e.g. Docters van Leeuwen 1936) recognized.

The point of detailing these patterns and processes here is to illustrate that on real islands (in the sea and some kilometres in area) succession is complex and demonstrates hierarchical interdependency across trophic levels. In order to understand how compositional patterns develop in the vegetation and flora, it is necessary to consider the ecological attributes of the plant species, their habitat relationships, and their hierarchical ecological links with animals (Bush and Whittaker 1991). Animals effect their dispersal, in cases their pollination, and of course also act as seed predators. In turn the vegetation provides the habitat and food resources of the animals; no fruit supply means no resident frugivores.

Colonization and turnover— the dynamics of species lists

Given the basic elements of vegetation succession and colonization patterns, broken down into plant dispersal types, it is possible to interpret much of the pattern in the rates of immigration and turnover of both plants and animals of the Krakatau islands. What follows here flows on from the consideration of turnover in Chapter 4, but it is placed here because of the structural features evident within it. An important caution applies to these studies. The Krakatau islands are large, topographically complex, and sufficiently difficult to explore that some parts have never, or only rarely, been penetrated by scientists. For both plants and animals, surveys span over 100 years of expeditions, have involved considerable turnover in personnel (and thus in expertise and methods), differences in intensity and differing areas of search. They have also taken place at irregular intervals. For these reasons, we must be cautious in interpreting changes in rates of immigration and extinction. The rates calculated are, in any case, not strictly as defined in the EMIB, rather they represent 'arrival in the lists' and 'departure from the lists'.

Whittaker *et al.* (1989) presented an analysis of higher plant colonization for 1883–1983, which illustrates the smoothing effect of calculating immigration rates over greater intervals of time, ignoring the intervening survey data (Fig. 5.7). The repeat surveys of Docters van Leeuwen (1936) in 1919 and 1922 produce a pronounced spike in the immigration data (another occurs for the same reasons at the end of the series), but the peak for the 1920s remains even when this survey frequency effect is removed. The low point, *c.*1951, reflects survey deficiency. Bush and Whittaker (1991) therefore calculated 'immigration' and 'extinction' rates for plants using survey data grouped into adjacent surveys, and ignoring the 1951 datum. Reasoning that for most plants, especially large forest trees, temporary disappearance from the lists was more likely to reflect survey inefficiency than genuine turnover, they provided upper and lower estimates: first, the recorded turnover and, secondly, assuming the minimum turnover allowable from the data (Fig. 5.8) (see also Whittaker *et al.* 1989, 1992*a*). The overall trend in species richness is also shown in this figure, and although the lack of adequate surveys between 1934 and 1979 makes the precise shape of the curve unknowable, additional exploration between 1989 and 1997 continued to turn up additional species records. Providing regional source pools are not destroyed by human action, it seems probable from our analyses of

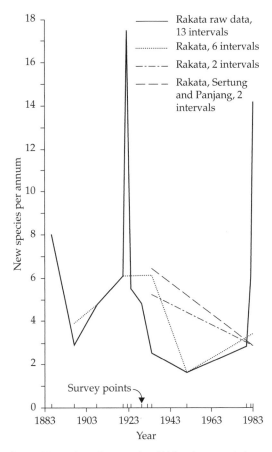

Figure 5.7 Numbers of new species of higher plants recorded on Krakatau over a variety of intervals, expressed as an annual rate. The figures exclude re-invasions. (Redrawn and corrected from Whittaker *et al.* 1989, Fig. 16.)

forest dynamics (e.g. Schmitt and Whittaker 1998; Whittaker *et al.* 1998) that species number can continue to rise for some time to come.

Evident in the analyses of 'immigration' and 'extinction' rates is that discovery of new species on Rakata peaked during the period of forest formation and closure, and the rate of loss of species from the lists also has a peak, lagging behind the 'immigration' peak, and reflecting the loss of early successional habitats. Further survey work between 1983 and 1994 allowed Whittaker *et al.* (2000) to re-examine the patterns of species loss in the light of an improved knowledge of the present day flora. Their approach was to focus on the 325 species

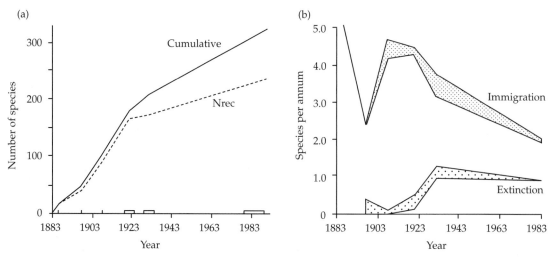

Figure 5.8 Plant recolonization data for Rakata Island (Krakatau) 1883–1989 (redrawn from Bush and Whittaker 1991). Surveys are grouped into survey periods, and the mean year of each period is used to estimate rates. Shaded lines represent the difference in rate according to whether minimum turnover is assumed or if 'recorded' immigration/extinction is assumed. (a) Cumulative species totals and number of species recorded at each survey (Nrec), curves for higher plants. Nrec figures are on the basis of the assumption of minimum turnover, i.e. species are only counted as going extinct if they fail to be found in all subsequent surveys. This assumes that temporary absences are artefacts of sampling rather than the true extinction and subsequent re-invasion of the species. (b) Immigration/ extinction curves for spermatophyte species. Rates are calculated as follows: (i) assuming recorded turnover, i.e. that all extinction and immigration records are real (immigration rate = number of species not recorded at time 1 but recorded at time 2/time 2 − time 1; extinction rate = number of species recorded at time 1 not found at time 2/time 2 − time 1); (ii) assuming minimum turnover (immigration rate = number of species not previously recorded, but found at time 2/time 2 − time 1; extinction rate = number of species recorded at time 1 not found at or after time 2/time 2 − time 1).

found within the first 50 years of recolonization (1883–1934), and to examine their persistence in the contemporary flora using first the 1979–1983 datum adopted in their earlier work and second the more exhaustive 1979–1994 data. The improved survey data reduced the number of 'extinct' species by one-third to 94, of which some 41 species could be regarded as ephemerals that never really established breeding populations in the first place, i.e. the true number of extinctions could be 50 or fewer. Although the estimates of extinction rate were thus shown to be unreliable, structural features in the extinction data were rather more robust. Losses relate to (1) the original abundance of a species as recorded in the 1883–1934 period (Fig 5.9a), (2) the number of islands on which a species occurred (another indicator of abundance) (Fig. 5.9b), and to a lesser degree (3) the primary dispersal mode of the species (indicative of the autoecology of the species). In short, exactly as we might anticipate, it

is the ephemeral (not proper 'colonists') and the rare which have failed to persist.

Animal-dispersed species have the lowest overall rate of loss, whilst among sea-dispersed species there is an interesting contrast between core members of the regional strandline flora, which were found on all three islands and have shown high persistence, and those species found initially on only one island, the majority of which have disappeared having either never established populations in the first place, or having been swept away by coastal erosion (Fig 5.9b). To sum up, the first 110 years of the succession have involved huge transformations in the environment and habitat types, and in the character of the forest, but even so, few of the species that established themselves in the first 50 years have subsequently failed to persist. Those that have failed are a statistically non-random subset made up mostly of the initially rare species.

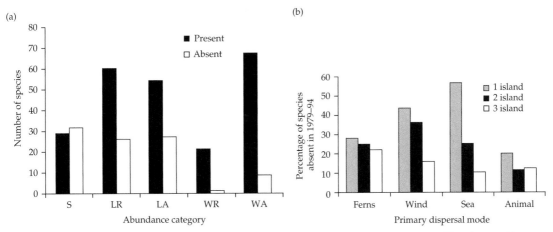

Figure 5.9 Persisting and failing species from the 1883–1934 flora of Krakatau (Rakata, Sertung, and Panjang islands), as determined by survey data for presence on the islands between 1979 and 1994. (a) Presence/absence in relation to abundance categories derived from descriptive accounts from earlier botanical work, where S = singleton records; LR = localized and rare; LA = localized but abundant; WR = widespread but rare (low density); WA = widespread and abundant. (b) Losses from the flora in relation to primary dispersal mode and the number of islands on which each species was recorded in the 1883–1934 period. Wind, sea, and animal refer to spermatophytes only; ferns are all wind-dispersed, in the form of spores. (Source: Whittaker *et al.* 2000, Figures 2 and 3). Unsurprisingly, initially rare species have a lower rate of persistence than initially widespread species. Further scrutiny of the raw data allows exclusion of a number of the extinctions documented in this figure and a lower overall estimate of extinction rate in the flora. Such an adjustment, if made to Fig. 5.8b, would induce a steeper negative trend to the extinction curve than shown there.

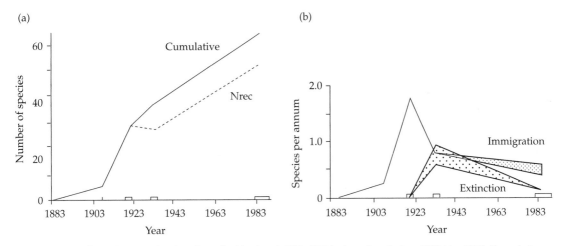

Figure 5.10 Butterfly recolonization data for Rakata Island (Krakatau) 1883–1989 (redrawn from Bush and Whittaker 1991). The peaks in immigration and extinction appear to correlate with habitat succession as grassland gave way to forest habitat on the islands. (a) Cumulative total and number of species recorded at each survey (Nrec), as Fig. 5.10. (b) Immigration and extinction curves, as Fig. 5.8.

The data quality problems seen for the plants undoubtedly afflict even the best of the animal data sets for Krakatau, but are less well understood. Here we take them essentially at face value. The butterfly data (Fig. 5.10) and land-bird data (Fig. 5.11) show similar trends coincident with forest closure, but contrasting trends towards the end of the data series, with butterflies seemingly climbing more steeply and birds approaching an asymptote. In the first two decades, the poverty of the vegetation arguably presented a limited array of opportunities for butterfly species, many of which

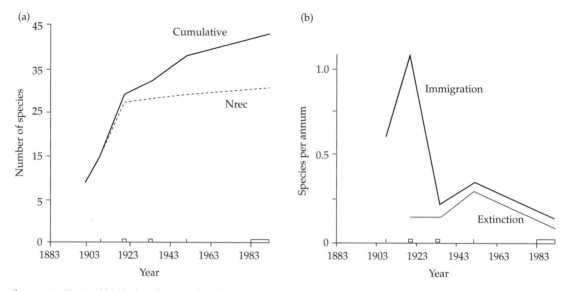

Figure 5.11 'Resident' land-bird recolonization data for Krakatau (redrawn from Thornton *et al.* 1993). (a) Cumulative total and number of species recorded at each survey (Nrec), assuming minimal turnover. (b) Immigration and extinction curves, as species per year for inter-survey periods, as Fig. 5.8.

require the presence of fairly specific food plants. The rate of arrival appears to have peaked during the period of most rapid floral and habitat diversification. Extinction also peaked following forest closure. Bush and Whittaker (1991) attribute four of the losses to succession, as open habitats were lost, other losses being possibly due to sampling deficiencies and to the inclusion of likely migratory species in the calculations. Thornton *et al.* (1993) took this view in replotting the data, thus finding a lower peak in the extinction rate, although the general trends in rates were the same.

Figure 5.11 shows the trends in the bird data as calculated by Thornton *et al.* (1993). These data are similar to the butterfly data but have an additional survey point, for 1951. This survey, by Hoogerwerf (1953), was missed by MacArthur and Wilson (1967) who suggested from the three survey collations for Rakata and Sertung up to 1934 that equilibrium might already have been reached. The data in Fig. 5.11 show that numbers subsequently rose slightly but at a much reduced rate (Thornton *et al.* 1993). There is evidence in these data of successional effects involving different trophic levels. The data also indicate a degree of turnover, the precise

figure involved being highly sensitive to assumptions made as to breeding status of birds (Thornton *et al.* 1993). The colonization curve is now fairly flat, but only future survey data can show if the birds of Rakata have achieved a dynamic equilibrium. Rawlinson's analyses of reptile data also indicate both an important role for abiotic factors and a low turnover rate. There have been only two extinctions of stabilized reptile populations, one due to habitat changes (canopy closure and coastal erosion) and the other to catastrophic alteration of habitat by eruptions of Anak Krakatau. No evidence was found of an approach to a dynamic equilibrium.

Thus, for Krakatau, the fit of the data (Figs 5.8, 5.10, 5.11) to the expectations of the EMIB appears to be poor, the best relationship being for birds, although even here the fit is imperfect (Fig. 5.11b). MacArthur and Wilson's (1967) stated position was that the precise shapes of the I and E curves could be modified within the confines of the EMIB, providing that they remained monotonic, 'When a new set of curves must be derived for a new situation, the model loses much of its virtue . . . ' (p. 64). Departure from monotonicity suggests that the intersects may be unstable or perhaps that there may be alternative stable states. These simple analyses of the species lists from Krakatau clearly show non-monotonic episodes in the early phases of island successions. This implies that succession is imposing structure that swamps the general stochastic processes represented in the EMIB. Much of the turnover is successional, and quite a lot involves species which may well not have established breeding populations. If the turnover attributable to the loss and gain of habitats and to succession is set to one side, undoubtedly some turnover remains. Precisely how much is real, in the sense of extinction of established populations, is very hard to quantify. Overall, turnover can be judged to be heterogeneous rather than homogeneous *sensu* Rey (1985).

The degree of organization in the Krakatau assembly process

Is the assembly of the Krakatau system governed by chance, or is it deterministic and predictable in its behaviour? It is difficult to provide definitive answers to this loosely phrased question because of the deficiencies of the historical data and because there are not enough other similar 'experiments' to compare it with. However, some interesting insights have emerged from studies of the new island, Anak Krakatau.

The emergence and development of the new island has modified the conditions for arrival, mortality, and survival of plants and animals within the island group as a whole (Box 5.3; Bush and Whittaker 1991, 1993). It has impinged upon developments on the other islands, not least through its volcanic activity. Its own history of environmental development has been distinctive, and its colonists can be assumed to have been drawn predominantly from within the archipelago, rather than from the more distant source pools of the post-1883 period. Indeed, some populations of animals move between the islands in foraging. As an experiment in island colonization, it thus cannot be viewed as a proper replicate. Nevertheless, it is interesting to note that there has been a bias in the early patterns of species assembly on Anak Krakatau towards those species which colonized the archipelago in

the early post-1883 phase (Thornton *et al.* 1992). This applies to several groups of plants and animals (e.g. birds, reptiles, and bats). The exception to the pattern was provided by the butterfly data set. This was probably due to the dependence of butterfly species on the availability of particular habitats and food plants, which differed sufficiently between old and new runs of the experiment to buck the trends.

The recolonization of Anak Krakatau has not been an uninterrupted process. Partomihardjo *et al.* (1992) have analysed the succession of floras following, first, the island's appearance in the centre of the Krakatau caldera in 1930, and, subsequently, following wipe-outs of the vegetation in 1932/3, 1939, 1953, and damaging, but not entirely destructive eruptions around 1972 (Thornton and Walsh 1992; Whittaker and Bush 1993). Surveys have been carried out only intermittently, but we have in essence three or four runs of the assembly experiment on Anak Krakatau, for each of which a species list is available. This rather small-scale natural experiment has demonstrated a strong degree of repetition, a temporal nestedness, with a core of constant species, which make it back each time and are added to (Whittaker *et al.* 1989; Partomihardjo *et al.* 1992). The first assemblage was really just a seedling flora, a few members of which were not found in the second assemblage. Yet, of 32 species identified from these two surveys, 30 recolonized subsequently. The eruptions in 1972 are presumed to have severely reduced the 'third' flora, yet all but 2 of its 43 species were recorded between 1979 and 1991.

If the Anak Krakatau flora is broken down into arbitrary functional guilds, there are basically three sets of species:

1 *Strandline species.* These are the largest set, representing 58 of 125 spermatophytes found between 1989 and 1991. Propagules of these species are sea-dispersed, and are produced in large numbers locally on the other islands. This is a very consistent set. Forty-two of these species have been found on all four Krakatau islands.

2 *Pioneers of interior habitats.* These species are wind-dispersed ferns and grasses, with a few

composites. Again they are locally available, are very dispersive, and produce large numbers of propagules. They are adapted to a highly dispersive life style and are very capable of 'finding out' such open environments, on which their survival depends (they are presumed to be not good competitors). As with the strand species, they are in essence a subset of those found on the other islands post-1883.

3 The third set of species, the *second phase colonists of interior habitats*, only managed to become established in the most recent flora. It will be interesting to establish the extent to which it may repeat following any further wipeout. Comparisons with the older islands suggest that in relation to the sequence on the older islands this group of species is likely to be less predictable in colonization sequence because of the more variable patterns of seed production and dispersal (mostly by birds or bats) within the archipelago. For instance, some of the contemporarily most abundant tree species on the other islands (e.g. *Arthrophyllum javanicum*, *Timonius compressicaulis*, and *Dysoxylum gaudichaudianum*) were recorded on Anak Krakatau at an earlier stage of successional development than in the post-1883 sequence.

Whittaker and Jones (1994*a*) have formulated these and other observations from Krakatau into an informal 'rule table' in which the dispersal, successional, turnover, and compositional characteristics of five such functional plant guilds were tentatively put forward should the opportunity arise to evaluate them elsewhere. Within plant successions there may be both autogenic 'facilitation' effects and diffuse competitive effects between guilds, culminating in later successional communities shading out earlier ones. Yet much of the structure may be attributable to a form of relay floristics, mediated through dispersal attributes of the plants and dependent, for some guilds, on hierarchical links between plant and animal communities (Bush and Whittaker 1991; Whittaker 1992; Thornton 1996; and compare with Thornton *et al.* 2001).

Just as in other island systems, as well as there being structure as to which species assemble and in what sequence, there may be particular sets of

species which fail to colonize. Whittaker and Jones (1994*a*) highlight those types for which they regard Krakatau as probably undersampling the regional species pool (Fig. 5.6). More formal comparison of the Krakatau floras with other sites in the region has provided support for these observations (Whittaker *et al.* 1997). Later successional species with poor dispersal adaptations, those which have large, winged, wind-scattered propagules, those dispersed by terrestrial mammals, and large-seeded bat-fruits (lacking diplochory) remain deficient on Krakatau, whereas highly dispersive forms such as ferns and orchids have been 'oversampled'.

The Krakatau plant and animal recolonization data thus suggest a general trend in the degree of predictability through time. While the draw of early pioneers is fairly predictable as a function in large measure of their superior dispersability, precisely which later successional species happen to establish breeding populations is much less predictable. Thornton (1996) has made a further, interesting point regarding the elements of chance and determinism. He draws the analogy between community assembly and the construction of a jigsaw, arguing that the further on in the process a species joins the system the less influence it can have over subsequent events, as it operates within successively narrow bands set by what has gone before. Thus a late-joining species may be unpredictable in its identity, yet may be predicted to have relatively little impact on community trajectories. The parallel with the assembly rules of Diamond is clear. At early stages of the recovery process, a number of species combinations or community pathways are possible, but as the system becomes more complex, and the jigsaw more complete, the number of pieces that will fit in to a given gap will decline, eventually to just one. This form of analogy has some intuitive value for groups such as birds, but is harder to sustain for plants. Successional pathways are not as discrete and limited as such a picture implies. The islands are continually subject to the vagaries of a varied environment, ranging from volcanic eruptions, through storms and landslides, to droughts. In such a disturbed setting competitive replacement within natural plant communities simply cannot be detected on a one-on-one basis (if it ever can), and returning to the analogy, the jigsaw is continually being disturbed and reformulated in new ways, while perhaps never quite being completed.

To conclude, the island ecological rates for Krakatau contain a lot of sampling noise, through which clear structural features are evident, revealing island recolonization to be in essence a special case of succession. Rates of turnover have, in addition, been affected by the physical environmental dynamics of the system. Hierarchical features across trophic levels are evident and it has also been argued that, even within a single taxon, not all species additions have equivalent effects. The bird data from Anak Krakatau indicate that although the addition of another insectivore or frugivore may not perturb the existing community structure significantly, the addition of the first predatory birds caused a major community perturbation (Thornton 1996). Similarly, the appearance of the first animal-dispersed fruits, and the first birds, and the first bats, post-1883 must also have constituted important events. A more recent example is the crossing of a threshold population size of fig trees on Anak Krakatau, in providing for resident populations of fig wasps and of frugivores (Compton *et al.* 1994). Once again, we see that different taxa reveal a variety of trajectories of their colonization curves, and greater and lesser degrees of turnover.

5.5   Concluding observations

This chapter has reviewed a range of forms of structure in the composition and dynamics of island biotas. We have paid particular attention to successional processes, as exemplified first within the New Guinea birds studies of Jared Diamond and colleagues, and second, through the Krakatau case study. Notwithstanding the debates concerning the statistical tests of such patterns, the evidence for non-random patterns from a variety of empirical studies is overwhelming (Schoener and Schoener 1983; Weiher and Keddy 1995). Competition is merely one of the forces involved, but it does have a role, even if it remains difficult to demonstrate. Progress has been made, as the tools of analysis

of island assembly patterns, incidence functions and nestedness have been debated, refined and evaluated. In the very nature of the response variables, this body of work encompasses more complexity than the previous chapter and in some ways has been harder to advance. By encompassing more complexity, it is correspondingly harder to find generality. Yet, theories of island assembly may in the end be of as much or perhaps more practical value to conservationists (Chapter 10; Worthen 1996) than, for example, are attempts to refine geometric models of the relationship between coastline configuration and immigration rate. We hope thus to have shown how the study of ecological biogeography can shed some light on the species numbers games of the previous chapter. The aim of the next chapter is to provide a synthesis drawing some of these ideas together in a re-assessment of progress in island ecology.

5.6   Summary

Island biotas are not simply random draws from regional species pools. Instead, they typically exhibit compositional structure: some species, species combinations, or species types, are found more frequently, and some less frequently, than might be expected by chance. A comprehensive island assembly theory was formulated by Jared Diamond, in his studies of birds on islands in the Bismarck and other groups around New Guinea. His analyses were founded on several forms of distributional patterns. Incidence functions describe the frequency of a species as a function of island species richness (or sometimes island size). Those species found only on the most species-rich islands, which were also the larger, land-bridge islands, were termed high-S species. Supertramps were, in contrast, found only on small, remote, species-poor islands, including islands recovering from past disturbance. Between these two extremes are the 'lesser tramps', found on varying numbers of the islands. Chequerboard distributions, were also observed, whereby taxonomically and ecologically related species have mutually exclusive but interdigitating distributions across a series of islands. Within particular guilds of birds, Diamond found evidence for compatibility rules, i.e. that a certain degree of niche difference is necessary to enable coexistence. After these various effects were considered, it was apparent that some combinations of species were more frequent and others less frequent than expected, suggesting combination rules. In total, seven distinct assembly rules were derived empirically on the basis primarily of the distributional data. The interpretations of these patterns rested on the assumption that through competition the component species are selected and co-adjusted in their niches and abundances, thereby 'fitting' with each other to form relatively species-stable communities.

Diamond's analyses drew pointed criticism from Simberloff and colleagues, who in a twin-pronged attack questioned both the primacy given to competition and the extent to which the distributional patterns departed from a random expectation in the first place. On the first count, Diamond and Gilpin responded by emphasizing the extent to which factors other than competition (e.g. predation, dispersal, habitat controls, and chance) were in fact integral to the theory. On the second count, they contested the basis for the null models, demonstrating that the findings depend heavily on the biological assumptions that particular authors build in to their null models.

The debate has continued and reappeared in other guises. Competition is undoubtedly a force in shaping island biotas but it is extremely difficult to distinguish the precise extent of its effects. Subsequent studies using similar types of distributional 'tools' (e.g. incidence functions) have demonstrated several common features, often finding a role for island area, but also involving a range of habitat determinants. Some convincing evidence of a different form of assembly rule invoking competitive effects is provided by morphological overdispersion in assemblages of exotic species introduced to oceanic islands. These patterns have been reported for several groups of oceanic islands and for both passeriform and galliform bird groups.

Nested distributions are where smaller faunas (or floras) constitute subsets of the species found in all richer systems. Nestedness is attributable variously to 'passive sampling', nestedness of habitat, differential immigration (frequently using the surrogate of island isolation), and differential extinction (area). Statistical tests of the degree of nestedness have been developed and applied to many island data sets. Nestedness turns out to be a common feature of insular biotas, varying in importance depending on the nature of the islands and the taxon being considered.

Island recolonization constitutes in essence just a special case of ecological succession. This is illustrated by reference to the recolonization of Krakatau. In this natural experiment, dispersal attributes, hierarchical relationships across trophic levels, and habitat changes are each seen as significant in structuring island species assembly. Turnover from the species lists can be attributed largely to a combination of succession, habitat loss, in a few cases predation, and to the comings and goings of 'ephemerals', i.e. it is heterogeneous in nature. For plants, as might be expected from first principles, it is the rare species that most frequently fail. The fit with equilibrium theory appears poor: while some taxa or 'guilds' may have reached an asymptote, overall the system has yet to equilibrate, if indeed it is destined ever to do so.

CHAPTER 6

Scale and island ecological theory: towards a new synthesis

What appears patchy to a grasshopper may appear uniform to a gnu.

(Wright *et al.* 1998, p. 19)

In the previous two chapters we have reviewed island ecological theories from both macroecological and ecological–biogeographical perspectives, tracing the development of ideas following the appearance of the seminal work of MacArthur and Wilson (1963, 1967). We return again to their equilibrium theory at the outset of this chapter. The exploration of these ideas by ecologists has produced many insights, sometimes consistent with the original models, sometimes inconsistent. Departures from predicted behaviours are in many respects of greatest interest, as they show how the models need to be modified. As a research programme, this body of work is characterized by a general assumption that the systems can be characterized as dynamic and equilibrial: indeed we might consider this paradigmatic. Yet, as we know, this assumption may not be supported in all cases. Accordingly, we will consider the alternatives in the present chapter. In so doing we hope to reconcile the apparently conflicting lessons from 40 years of research and to suggest where island ecological theory might be heading.

Several themes run through this chapter. They include: the need for alternatives to the dynamic equilibrium model, recognizing the possibility of dynamic but non-equilibrium systems, and of largely static ecologies, which again may be equilibrial or non-equilibrial; the issue of scale dependency; and the need for models to accommodate hierarchical links within ecological systems. Building

in such complexities may lead to more realistic models, but we should not readily abandon the search for simple, unifying theories and models (Brown 1995), and therefore, we also consider some recent attempts to develop more satisfactory simple frameworks for island ecology.

6.1 Limitations of the dynamic equilibrium model of island biogeography: a reappraisal

. . . the equilibrium model and its derivatives suffer from extreme oversimplification by treating islands as functional units with no attention to internal habitat diversity and by treating species as functional units with no allowance for genetic or geographical diversity. This is not even good as a first approximation, because it filters out the interpretable signal instead of the random noise. The authors are in such a hurry to abandon the particulars of natural history for universal generalization that they lose the grand theme of natural history, the shaping of organic diversity by environmental selection . . .

(Sauer 1969, p. 590, on MacArthur and Wilson 1967)

Brown (1981) observed that the EMIB has three characteristics which he claimed any successful general theory of diversity must possess.

• First, it is an equilibrium model, thus historical factors, climatic change, successional processes, and the like are acknowledged, but side-stepped. The theory explains rather the ultimate limits, the theoretical patterns of diversity.

• Secondly, it confronts the problem of diversity directly, number of species being the primary currency of the model. It also takes account both of biological processes and some, at least, of the characteristics of the environment.

• Thirdly, it is empirically operational, making robust, qualitative predictions that can be tested (although, as we saw in chapter 4, the model turns out be much more difficult to test than once thought).

Brown notes that the fact that the EMIB has been repeatedly and unequivocally falsified does not diminish the contribution it has made to advance our understanding of diversity. In contrast to the judgement by Sauer (1969, above), he therefore characterizes it as a useful beginning:

The model implies that the determination of diversity is a very stochastic process. Species continually colonize and go extinct, the biota at equilibrium is constantly changing species composition, and all of these processes occur essentially at random. MacArthur and Wilson knew that this was not so, but ' . . . it was a useful simplifying assumption . . . ' (Brown 1981, p. 883).

Perhaps the key virtue of the model as a general model is that it is dynamic. The distinction between **historical** and **dynamic hypotheses** is based on the repeatability or probability of recurrence of a particular state or form. Whereas, historical, or time-bound, knowledge refers to the analysis of complex states having very small probabilities of being repeated (i.e. states of low recoverability), physical, or dynamic, or timeless, knowledge refers to the analysis of states having a high degree of probability of being repeated (Schumm 1991). The search for dynamic hypotheses of species richness can thus be likened to the search for the general laws of ecology, from which historical contingency supplies the deviations (Brown 1999; Whittaker *et al.* 2001). Hence, the abiding attraction of the MacArthur and Wilson model and of the dynamic, equilibrial framework they put in play.

Many of the problems with the essentially stochastic EMIB of MacArthur and Wilson have been discussed in the previous two chapters, as we have focused on the extent to which equilibrium models apply; our ability to predict island species numbers and identities, and on what forces structure island biotas. Table 6.1 summarizes some of the key issues. There is no doubting the lasting heuristic value of the dynamic equilibrium model, but more in question is the continuing widespread use of the model, its direct derivatives and assumptions in predicting species losses in relation to habitat loss and fragmentation: an issue we return to in Chapter 10.

Table 6.1 Some of the limitations identified in MacArthur and Wilson's equilibrium model of island biogeography (EMIB)

Author	Limitations of EMIB
Sauer (1969)	Theory ignores autoecology—but species are not interchangeable units (e.g. see Armstrong 1982 on the effects of rabbits introduced onto islands)
Brown and Kodric-Brown (1971)	Distance may affect extinction rate (E) (through the rescue effect), whereas E is treated as a function only of area in the EMIB
Lynch and Johnson (1974)	The data are rarely adequate for testing turnover
Hunt and Hunt (1974)	Turnover can be confounded by trophic-level effects
Simberloff (1976)	Re: mangrove data set, most turnover involves transients, i.e. is pseudoturnover
Gilbert (1980)	Turnover at equilibrium has not been demonstrated
Williamson (1981)	Area may affect immigration rate (I), whereas I is treated as a function only of isolation in the EMIB
Williamson (1981, 1989a,b)	Immigration, extinction, and species pool are each poorly defined. EMIB is imprecise on reasons for extinction, and most turnover is ecologically trivial (cf. Simberloff 1976)
Pregill and Olsen (1981)	Ignores historical data and role of environmental change
Haila (1990)	EMIB has a narrower domain of applicability than originally thought (may be operational on the population scale)
Bush and Whittaker (1991)	Ignores successional effects and pace, and the hierarchical links between taxa

There remain those, such as Rosenzweig (1995), who defend the theory and argue that it 'holds up well'. He contends that it is inaccessibility that defines 'islandness' in a biological sense, and that it follows from this that a biological definition of an island is preferable to physical definitions such as being surrounded by water. His definition is: 'An island is a self-contained region whose species originate entirely by immigration from outside the region' (p. 211). This is a restrictive definition that would seem to rule out numerous real islands that have high proportions of endemic species. Moreover, the term 'self-contained region' is also important. By it, Rosenzweig means that each species should be a **source species**, i.e. it has an average net rate of reproduction sufficient to maintain positive population. The alternative condition is that species are **sink species**, which are dependent on influx from outside to maintain their populations. That is, species which only maintain their presence through continual supplementary immigration should be discounted, or if they are in abundance presumably the study system should not be considered insular. This definition, if rigorously applied, would see a high proportion of study systems excluded from any test of the EMIB. This is, in practice, to define islandness by how well the EMIB applies to an island. We prefer to state the case the other way around: the equilibrium model holds only for a limited subset of islands. Can we generalize as to which?

6.2 Scale and the dynamics of island biotas

The processes invoked in the EMIB as determining how richness of an island may change from one point in time to the next are immigration, speciation, and local extinction (Box 4.1), which of course has to be correct. It is also reasonable to suppose a general tendency towards equilibrium values, but that this condition will be approximated to greater or lesser degrees for particular islands and archipelagos as other forces and factors come into play. The degree of fit is argued here to be broadly related to scale, and in particular the range of scale in area and isolation of the island system

considered. This line of thinking was developed in a seminal paper by Haila (1990), who argued that the key is to clarify what is mean by an 'island', relative to the ecological processes at work (note the parallels with Rosenzweig's thinking, above). The processes themselves form a continuum, with characteristic coupled space–time scales, as shown in Fig. 1.1. Haila distinguishes (1) the individual scale; (2) the population scale 1—dynamics; (3) the population scale 2—differentiation; and (4) the evolutionary scale. He illustrated this framework with examples taken from northern European archipelagos, although his data provide only indirect evidence of some of the processes at work.

• On the **individual scale**, a patch of land is an island if some crucial phase of the life cycle of individual organisms obligatorily takes places within its boundaries. Consider those birds which breed on an island but overwinter elsewhere, such that their life cycle is influenced by factors external to the island. Birds of prey, on the other hand, may include several small islands of an archipelago within their territory. In such circumstances individuals may be viewed as an integral part of the regional population, such that the insularity of the environment on this *individual* scale has few population or community-level consequences, other than through being inferior or superior places for reproduction.

• When islands are larger and more isolated, they may support populations that are dynamically independent of those on other land areas. At this point the relevance of the EMIB model may be apparent. To be an island on the **population dynamic scale**, two criteria are identified: first, the whole life cycle of the organisms concerned must be confined within the island and, secondly, the island population must be able to demonstrate independence from the mainland dynamics for several generations. For instance, there is no evidence that bird populations on single islands from the Åland archipelago are dynamically independent. Although thse islands are about 40 km from the Swedish mainland and 70 km from mainland Finland, and up to some tens of square kilometres in area, island population dynamics commonly occur in parallel with those of the mainland. Only for a few sedentary bird species on the largest and most isolated Baltic island is there evidence of dynamic independence.

• Should the criteria for independence expressed above be fulfilled for long periods, the island populations become increasingly independent as genetic units, and this allows for the second scale of population processes, **differentiation**. Exemplification of this comes from several studies, two cases being the various subspecies of wrens (*Troglodytes troglodytes*) on different North Atlantic archipelagos (Williamson 1981), and the studies of Berry (e.g. 1986) on the founder effect in small mammals on offshore islands around Britain. The relevance of the equilibrium theory would seem to be fairly slight in such circumstances.

• The natural progression in this scheme is to the **evolutionary scale**, by which Haila (1990) means processes leading to the divergence of species, i.e. taxonomic differentiation to a greater degree than indicated by the term 'subspecies' or 'variety' or by slight niche shifts. Empirical support for the proposition that remote islands typically reach and maintain the condition of dynamic equilibrium is patchy at best (Chapter 4).

Of these four scales, the EMIB is restricted mainly to the *population dynamics* scale, in that the other scales do not appear to satisfy the criteria necessary for its operation. The scales of Haila's framework grade one to another, such that particular studies may be at the interface, and the applicability of the theory uncertain. Even within a single taxon, such as birds, or bats on the Krakatau islands, some species populations may effectively be bound by the confines of their island, while others may have wider territories inclusive of the islands under study. It is not, then, that the theory holds for all islands within a certain space–time configuration, but that the effects of the processes it represents may be prominent within this conceptual territory, whereas elsewhere (in the remaining subject space) they are subsumed by other dominant processes. It also follows from these arguments that it should be considerably easier to develop models based on individual species (e.g. Lomolino 2000*a*,*b*) than whole-system models.

Haila's (1990) scale framework is predominantly a biological one. He makes reference to environmental dynamics, but not prominently. It will, however, be apparent that particular forms of environmental change and disturbance will also have characteristic scale patterns, which determine their relevance to interpretation of the differing scales of island ecological process. In short, it is not sufficient to assume ecological systems to be in equilibrium, this must be demonstrated (Weins 1984), and in systems for which abiotic forcing dominates, more complex, non-equilibrium models may be required.

Residency and hierarchical interdependency: further illustrations from Krakatau

One of the features imposed by the equilibrium paradigm has been the restriction of focus to species that are resident on islands. Although this has not always been adhered to, it does result, for example, in seabirds and (often) migrant birds being excluded from consideration in island ecological studies. Such species may be of huge significance to the ecology of an island and may also act to introduce other fauna and flora.

An illustration of the problems concerning notions of residency comes from the fruit bats of Krakatau. The largest flying fox species is *Pteropus vampyrus*. A colony of several hundred individuals was observed intermittently on the islands during the 1980s and 1990s. *Pteropus* have been observed to gather in large roosts on islands but fly to mainland sites in order to feed (and vice versa), and their nightly range can be as much as 70 km (Dammerman 1948; Tidemann *et al.* 1990). Within mainland areas *Pteropus* is also known to be highly mobile, exhibiting movements that may represent non-seasonal nomadism in response to variation in food availability, rather than regular seasonal migration. Krakatau may thus provide just one roost site of a number used by the same colony within a larger geographic area. Their intermittent use of it may reflect variation in food supply in the region, human interference at other roost sites, and avoidance of periodic ash falls on Krakatau. These animals and other, smaller fruit bats, act as important seed vectors for the introduction and/or spread of a particular subset of Krakatau plant species (Shilton 1999; Shilton *et al.* 1999).

The large fruit-pigeon, *Ducula bicolor*, is another important species of frugivore, which for island ecological purposes is considered as a Krakatau resident (Thornton *et al.* 1993), yet which exhibits a roaming behavioural pattern, in which large flocks move between offshore islands in the region (Dammerman 1948). The two species of Krakatau *Ducula* (the other is *D. aena*) are believed to have played crucial roles in introducing animal-dispersed (i.e. zoochorous) plants—particularly those with larger seeds—to Krakatau (Whittaker and Jones 1994*b*). The particular identities and timing of the trees introduced by such means may have important ramifications for the species of animals which may subsequently be supported within the islands.

All these species of frugivores have been highly significant to the structure inherent in the assembly process for Krakatau, the interior forests being predominantly composed of bird- or bat-dispersed plants. Arguably, the behaviours of these species exhibit characteristics of connectedness to other locations which are more akin to metapopulation models (Chapter 10) than to the notions of residency demanded by the EMIB. In short, for some animal taxa and some guilds, there are potentially complex hierarchical links between their turnover patterns and floral development and turnover. Other forms of hierarchical links may also be found, such as those linking predator–prey groups in an area. Although such observations are of explanatory value, it is, as we saw with the debates about island assembly rules, a considerable challenge to incorporate such processes into general predictive models.

Immigration and extinction curves are theorized in the EMIB to have smooth forms, the former falling and the latter rising to a point of intersection: the dynamic equilibrium. The patterns of arrival and disappearance from the Krakatau lists do not correspond with this expectation (Chapter 5), in part because of hierarchical, successional features evident in the data, which are simply absent from the model. The big early kinks in the rates relate to the key switch from open to closed habitats. It is conceivable, however, that this switch marks the end of significant autogenically derived (i.e. biotically driven) switches in trends. In which case, once beyond the early phases of succession, the EMIB may become more realistic, i.e. at some stage, an island must fill up with species and some form of equilibrium be established. However, given the complexity of succcessional processes in tropical lowland forests, Bush and Whittaker (1993) question whether it is reasonable to suppose that population processes can bring the system to equilibrium before the next major disturbance moves the system away from that condition again.

Forest succession is a slow process: the lifespan of individual canopy trees can exceed 300 years, and even early successional species may live for decades. If a true biotic equilibrium in the flora were to be established beyond doubt, it must be after a period of several generations of biotically mediated interactions, i.e. hundreds of years. However, such conditions are unlikely to be reached in very small patches because of varying forms of environmental variability and disturbance. Events such as hurricanes, volcanic eruptions, flooding, landslides, and other high-magnitude phenomena occur within continental and island landscapes, and typically have periodicities of less than several hundred years (Chapter 2).

For this argument to hold force for animal groups/taxa, there must be a strong pattern of dependency of animal groups on plants for habitat and food resources, such that their patterns of colonization, carrying capacity and turnover are tied to the dynamics of the plant communities. In turn, animals can have key roles in shaping plant succession, through roles as seed dispersers and pollinators (Bush and Whittaker, 1991, Elmqvist *et al.* 1994). If the degree of dependency is weak, it may be that merely the recovery of forest biomass and productivity, which is likely to be more swiftly accomplished (Whittaker *et al.* 1998), may be sufficient to allow equilibration of particular animal groups. In practice, some 120 years on from the initiation of the Krakatau experiment, it remains difficult to provide a definitive answer to this question.

6.3 Forms of equilibria and non-equilibria

We are not dealing with a well-stabilized situation of long standing. This may account for some of our difficulties in fitting existing situations into a theory which concerns itself only with final equilibrium.

(Preston 1962, p. 429)

We have argued that there is an important distinction between equilibrium and non-equilibrium conditions. There is an equally important distinction between dynamic and static models of island ecology. These notions provide for a variety of combinations, as shown in Fig. 6.1, in which we highlight four extremes: dynamic equilibrium, static equilibrium, dynamic non-equilibrium, and static non-equilibrium. Some example systems that appear to match the different conditions are provided in Table 6.2.

The **dynamic equilibrium hypothesis** corresponds to the EMIB, the properties of which have already been considered at length. The core EMIB model is essentially stochastic (i.e. occupying the extreme bottom left corner of Fig. 6.1), but MacArthur and Wilson (1967) recognized in their book that some structuring of turnover commonly occurred, hence their fuller theory could be assigned a position a little to the right within the

figure. By either view, their ideas are strongly associated with the dynamic equilibrium corner. Figure 6.1 recognizes a continuum from purely stochastic (homogeneous) turnover, through increasingly heterogeneous turnover to the ideas of habitat determinism with minimal turnover. This end of the axis describes the **static equilibrium hypothesis**, which may be characterized crudely as where the key controls of species presence/absence appear to be habitat controls (as Martin *et al.* 1995) and where species turnover measured on a timescale of generations is insignificant. This is not to deny compositional change over time, but that in the absence of human interference the turnover is unmeasurable, giving the appearance of stasis.

It is possible to reconcile ideas of habitat determinism with the occurrence of systematic variation in species richness related to area and isolation, providing that habitat structure itself is significantly influenced by the area and isolation of an island (Martin *et al.* 1995). Hence, once again, the pattern of turnover is central to distinguishing between alternative hypotheses. The emphasis on habitat controls has often been associated with David Lack, who based his analyses on studies of island birds (e.g. Lack 1969, 1976). He argued that the failure of birds to establish on islands comes

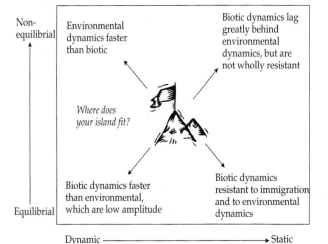

Figure 6.1 A representation of the conceptual extremes of island species turnover. The dynamic equilibrium condition corresponds to MacArthur and Wilson's (1963, 1967) theory; the static equilibrium equates to Lack's (e.g. 1969) ideas on island turnover of birds; the static non-equilibrium to Brown's (1971) work on non-equilibrium mountain tops; and the dynamic non-equilibrium to Bush and Whittaker's (1991, 1993) interpretations of Krakatau plant and butterfly data. Considering a single taxon, different positions in this diagram may correspond to different islands or archipelagos, and different taxa in the same island group may also correspond to different positions.

from a failure to find the appropriate niche space rather than from a failure to disperse to the island. He therefore played down the relevance of turnover. Others have supported this viewpoint insofar as saying that 'most genuine turnover of birds and mammals seems attributable to human effects' (Abbott 1983) or that turnover, where it does occur, mostly involves subsets of fugitive species and is thus ecologically trivial (Williamson 1981, 1988, 1989a). A similar conclusion was reached in a study of newly isolated Canadian woodlots. Once successional effects were removed from the analysis, turnover within the vascular flora, if occurring at all, was at a very slow pace (Weaver and Kellman 1981).

A similar distinction, between the dynamic and the static, can be made within non-equilibrium ideas. First, we may recognize disturbed islands: those in which equilibrium is reached only rarely because environmental dynamics outpace the response times of the biota (cf. the disturbance hypothesis of McGuinness 1984). The biotas may track changing equilibrium points through time, but always remain a step or two out of phase. This

idea has been given various labels (e.g. Heaney 1986; Bush and Whittaker 1993; Whittaker 1995), but within the scheme developed here it is identified as the **dynamic non-equilibrium hypothesis**. We may distinguish this notion from that introduced by Brown (1971) for relictual assemblages dominated by extinction. In Brown's system, although isolates may be losing species on a millennial timescale, and are thus in a non-equilibrium condition, on ecological timescales the system of isolates appears effectively static, i.e. they fit the **static non-equilibrium** condition.

We regard this diagrammatic model as being of heuristic value although, of itself, it lacks predictive capacity. Moreover, it is often going to be difficult to assign a study system to a particular condition such that ready agreement will be reached amongst ecologists that the data unequivocally support a particular interpretation (compare e.g. Bush and Whittaker 1991, 1993, and Thornton *et al.* 1993). See Box 6.1 for an illustrative discussion.

The difference between the dynamic non-equilibrium hypothesis and the dynamic equilibrium

Table 6.2 Exemplification of classification of studies of island richness and turnover as per Fig. 6.1, showing that different taxa from within a single island group (Krakatau) support different models, and that in different contexts, data from the same taxa (terrestrial vertebrates, birds, invertebrates, plants) have been interpreted as supporting different models. (After Whittaker 2004.)

Dynamic, non-equilibrial	Static non-equilibrial
1. Krakatau plants (Whittaker *et al.* 1989; Bush and Whittaker 1991)	7. Krakatau reptiles—several species introduced by people, only two species have been lost, both related to habitat losses (data in Rawlinson *et al.* 1992)
2. Krakatau butterflies (Bush and Whittaker 1991)	8. Great Basin mountain tops, North America, small mammals (Brown 1971)
3. Bahamas, plants on small islands (Morrison 2002*b*)	9. Lesser Antilles—small land birds, examined over an evolutionary time-frame (Ricklefs and Bermingham 2001)

Dynamic, equilibrial	Static, equilibrial
4. Krakatau birds are a reasonable fit, although showing clear successional structure in assembly and turnover (Bush and Whittaker 1991; Thornton *et al.*, 1993)	10. Krakatau terrestrial mammals—no recorded extinctions to date (Thornton 1996)
5. Mangrove islets of the Florida Keys, arthropods (Simberloff and Wilson, 1970)	11. Bahamas, ants on small islands (Morrison 2002*a*)
6. British Isles, birds on small islands (Manne *et al.*, 1998)	12. Oceanic island birds (Lack 1969, 1976 [but see Ricklefs and Bermingham 2001]; Walter 1998)
	13. Canadian woodlots plant data (if successional effects are removed from the analysis) (Weaver and Kellman 1981)

Box 6.1 Implications of anthropogenic changes in richness of oceanic islands for island theory

In Chapter 5 we considered how introductions of exotic species to oceanic islands provided inadvertent experiments in island assembly structure. Here we consider the implications of anthropogenic impacts on island species number. Sax et al. (2002) took the approach of estimating historical richness prior to human-induced extinctions and colonization events, and comparing these values with richness subsequent to these processes. They did so for vascular plants and for land birds on 11 and 21 oceanic islands/archipelagos, respectively. As shown in the figure, the number of naturalized exotic plant species greatly exceeds the number of extinctions (panel A), whereas for birds, exotic additions approximately match the number of extinctions from the avifaunas (panel C). Hence, in comparing current with historic richness estimates, humans have led to an approximate doubling in size of the floras (panel B), but little net change for birds (panel D). In terms of actual numbers, the number of extinctions varies between 0 and 71 for plants, and 1 and 64 for birds: in both cases, the Hawaiian islands providing the highest value.

These results necessarily involve a number of caveats concerning data quality, lumping of islands within archipelagos, and so forth. But, they are intriguing if viewed through the lens of the equilibrium paradigm. Taking a simplistic line of reasoning, we might argue as follows. The floras of the oceanic islands have approximately doubled in size, as naturalizations have vastly exceeded extinctions. Hence, the islands must originally have been below their intrinsic capacity for richness, i.e. they were non-equilibrial. The birds, in contrast, have seen a balance of extinctions and establishments, and must thus have been in equilibrium historically, and have now re-established their equilibria.

How might we explain away the apparent mismatch between the plant data and the equilibrium condition? Many of the islands in the study once possessed extensive areas of forest that have been cleared for agriculture and settlements. The removal of forest cover and its replacement by predominantly low-stature

vegetation has provided ideal conditions for the establishment of numerous species of (predominantly) herbaceous weeds, which have a much smaller modular size than the pre-existing tree flora, i.e. many more species can be fitted in. Moreover, the bombardment of the islands by introduced plants effectively means that the immigration rate has been radically altered, and in many cases, continues to be. For example, there is no sign of a slowing down of the pace of introductions and establishment of exotic species in the Galápagos (S. Henderson, personal communication). It is thus evident that we have simultaneously altered both immigration rate and carrying capacity. Hence, the enrichment of the floras may indicate not previous non-equilibrium, but that we are currently en route to some new anthropogenically determined enhanced equilibrium point.

Unfortunately, if we are to allow this argument to stand for plants then the same arguments should apply to birds, in which case their numbers should also have gone up. We now have to defend the lack of change in bird richness. Birds differ in having a much higher proportion of losses of native species (from a variety of causes, e.g. Blackburn et al. 2004) than the case for plants, and it turns out that the lost species have on average been replaced by exotics. As discussed in Chapter 5, not all introductions of birds to oceanic islands are successful, so perhaps for birds we see an equilibrium condition being re-established despite the hike upwards in immigration events. This in turn would imply that on average, the intrinsic carrying capacity of the island is determined by the resource base it represents rather than by the particular vegetation cover.

Have we squared the circle? Perhaps, but if so, it is by special pleading, and we have ended up with an argument that indicates that a substantial alteration to the immigration rate and to vegetation cover has had great impact on plant numbers but not on bird richness. Moreover, there is a great deal more to understanding the factors governing anthropogenic changes to island species numbers, whether taking a

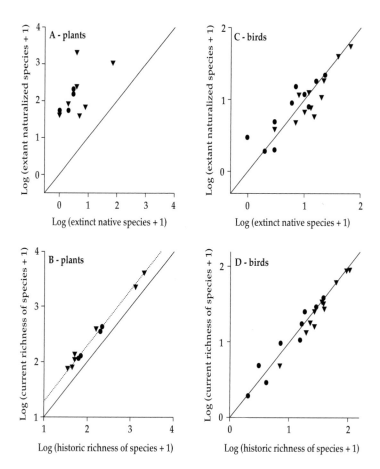

Species richness of vascular plants and land birds on oceanic islands. Triangles represent islands inhabited before European contact; circles represent those that were not. (From Sax *et al.* 2002, Fig. 1.)

macroecological modelling approach concerning factors like introduction effort and predation (e.g. Blackburn *et al.* 2004, Cassey *et al.* 2005*b*) or considering the detailed context of turnover for particular islands (Duncan and Blackburn 2002), such as the impact of disease organisms and their carriers on Hawaii (Pratt 2005).

We hope to have illustrated how difficult it is to bring such arguments about the equilibrium assumption to a clear conclusion. We are left, nonetheless, with some intriguing observations.

First, plant species numbers on oceanic islands appear to be capable of significant increases without *necessarily* involving high turnover (extinction) of native species. Second, the proportional relationships seen in the figure suggest that the slope (*z* value) of the ISARs (island species–area relationships) have been little changed despite the enrichment of the floras and the turnover involved in the avifaunas. Third, it seems the role of isolation is less significant for determining island land bird species numbers than predicted by the EMIB.

hypothesis (or EMIB) is of degree, relating to the causation of turnover and the biotic response times (Whittaker 1995). In fact, MacArthur and Wilson (1967) acknowledged the implications of environmental disturbances for particular island systems. For instance, they recognized the following phenomena: the occurrence of successional turnover; the influence of storms on immigration rates; and that much extinction, especially that resulting from storms, drought, and invasion of new competitors, is accompanied by a severe reduction in the carrying capacity of the environment. Nonetheless, they tended to downplay the general significance of disturbance, arguing that only equilibrium models are likely to lead to new knowledge concerning the dynamics of immigration and extinction. But, as others have noted, where turnover is the product principally of abiotic forcing, turnover patterns may be poorly predicted by equilibrium models (Caswell 1978; Heaney 1986; Whittaker 1995; Morrison 1997; Shepherd and Brantley 2005).

Figure 6.2 is Bush and Whittaker's (1993) attempt to generalize the Krakatau island ecological trends and to set out alternative trajectories, allowing for taxa of differing ecological roles and response times. It encapsulates the temporally bumpy rates associated with successional processes and provides for two possible scenarios whereby equilibrium might be reached. The data for rate changes indicate that equilibrium might be approached by a declining immigration rate, combined with two alternative extinction trends. In the first, extinction rate is low and turnover rather heterogeneous (interactive). In the second, it is essentially squeezed out of the system altogether (non-interactive). Given the problems of collecting adequate survey data, and the high degree of pseudoturnover which is a feature of empirical data from such large and complex systems, the two conditions may not be readily distinguishable in practice. Close examination of the Krakatau plant data demonstrated that a high proportion of apparent turnover involved species best regarded as ephemerals, which had not really colonized in the first place (Whittaker et al. 2000). The third form of projection recognizes that the islands experience environmental change, which may be dramatic and

destructive. The longer the period before equilibrium is attained, the greater the likelihood that an event of intermediate or high magnitude will cast the system away from equilibrium and into the perpetual or dynamic non-equilibrium condition of Fig. 6.1. In this projection, the biota chases perpetually moving environmental 'goalposts'.

The relevance of the three projections in Fig. 6.2 to particular Krakatau taxa depends on features of their ecology such as generation times and dispersal attributes, and their position within the biological and successional hierarchy (cf. Schoener 1986). Within the conceptual space of Fig. 6.1, different taxa or guilds (or the same taxa in different island contexts) may thus occupy different positions: the reptiles arguably towards the static non-equilibrium; the plants of Anak Krakatau towards the dynamic non-equilibrium; the coastal flora at the archipelago level towards the static equilibrial position; and the birds of Rakata, at least at times, relatively close to a dynamic equilibrium but with less homogeneous patterns of turnover than indicated for a strict EMIB position. The two diagrams are thus complementary, in showing the conceptual space (Fig. 6.1) and temporal features characteristic of different points within it for one exemplar system (Fig. 6.2).

Dynamic non-equilibrium, or multistate models have become prominent in several subfields of ecology in recent decades (e.g. savannah ecology), but thus far relatively few non-equilibrium models appear to have been developed in island ecology (see e.g. Caswell 1978; Weins 1984; Williamson 1988; Villa et al. 1992; Russell et al. 1995; Whittaker 1995). Incorporating environmental disturbance into island ecological models and modelling presents significant but not insurmountable difficulties. An example is provided by Villa et al. (1992). They constructed a model island consisting of a habitat map of cells containing individuals, a list of the species involved, and their life history characteristics. The model involved colonization of the island, and varying degrees of perturbing events that caused the mortality of individuals. The model was highly simplified, but nonetheless allowed the testing of ideas. One intuitively appealing result was that the global attainment of equilibrium within the simulations was strongly dependent on the rate of

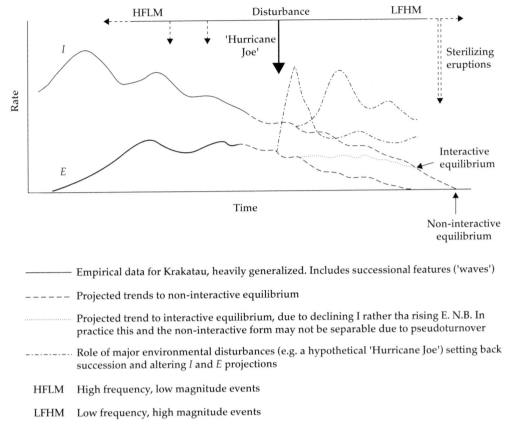

Figure 6.2 A model of island immigration and extinction incorporating disturbance. The figure provides a highly generalized representation of data for the recolonization of Krakatau and three hypothetical projections. (From Bush and Whittaker 1993.)

increase of the species involved. This demonstrated that equilibrium for slow-growing species (i.e. those with a low intrinsic rate of population increase) should be tested over longer periods, a point equally clear from Schoener's (1983) review of empirical data. Another interesting feature was that in the presence of disturbance, turnover was mostly accounted for by ephemeral populations (arguably trivial ecologically). This held even in equilibrial simulations, a finding again in concordance with empirical evidence, in the form of bird data from Skokholm (Williamson 1989a,b).

Figure 6.1 is thus offered as a restatement of island ecological theory, in which the 'narrow' view of EMIB is represented in the broader subject space

against other 'narrow' view alternatives. The four extreme positions on turnover are thus dynamic equilibrium, static equilibrium, dynamic non-equilibrium, and static non-equilibrium. In practice, they may be difficult to distinguish with the data available. Moreover, the response times, dispersal abilities, and trophic status of different taxa may mean that comparative studies of different taxa from the same set of islands adopt differing positions in this framework (Table 6.2; Bush and Whittaker 1993). Once again, following Haila's argument, the characteristics of the target taxon/group in relation to the spatial scale of the system will largely dictate the applicability of these labels to case study systems. Similarly, Crowell

(1986), in discussing the mammalian fauna of a set of land-bridge islands, aired the notion that it may be possible to recognize a continuum, from isolated land-bridge islands having non-equilibrial faunas, to those of intermediate isolation with both relict species and those in equilibrium, to islands so near, and/or so small, that all species are in equilibrium.

The options sketched out in Fig. 6.1 are also similar to the distinctions drawn about 30 years ago by Caswell (1978) (see also Weins 1984; Schoener 1986; Case and Cody 1987). The scale of Caswell's application is rather different from those generally under discussion here, but the parallel is apparent nonetheless. He set out to develop models of predator (or disturbance)-mediated coexistence. He distinguished between open systems (involving migration between 'cells') and closed systems (no migration), and between equilibrium and non-equilibrium systems, concluding that it would not do to suggest that communities as a whole are entirely equilibrium or entirely nonequilibrium systems. 'Perhaps a community consists of a core of dominant species, which interact strongly enough among themselves to arrive at equilibrium, surrounded by a larger set of nonequilibrium species playing out their roles against the backdrop of the equilibrium species' (Caswell 1978, pp. 149–50).

6.4   Temporal variation in island carrying capacities

The dynamic equilibrium paradigm anticipates that island species number will in general stabilize and show little temporal fluctuation. Here we focus on some of the mechanisms that may lead to fluctuations in richness through time, relating mostly to the idea that an island's carrying capacities may change, but also noting that both island area and degree of isolation may also change over time (Table 6.3).

The most obvious cause of change in isolation is alteration in relative sea level, but effective isolation may be influenced by more subtle environmental changes. Many accounts of species turnover rates have treated ocean waters as a constant barrier to dispersal. Scott's (1994) observations of an invasion of San Clemente island, 80 km from the Californian mainland, led him to question this assumption. In the late summer of 1984, 26 black-shouldered kites (*Elanus caeuleus*) took up temporary residence on the island for a period of several months, although previously only one or two kites had been recorded during 19 autumn/winter cycles of bird observation. Scott suggested that the 1984 irruption may have been a consequence of an unusual pattern of Catalina eddies—which are seasonal cyclonic winds—possibly linked to an El Niño event. Each pulse of kites arriving on the island coincided with the first or second day of a Catalina eddy. It appears that the eddy system is like a door to San Clemente island that irregularly opens and closes. The same El Niño influenced carrying capacity for Darwin's finches in the Galápagos (Chapter 9). From such simple empirical observations, it might be reasonable to postulate that fluctuations in the weather might produce variations in rates of immigration, extinction, and turnover.

The prevalence and implications of intense disturbance events

Abbott and Black (1980), in their study of vascular plant species on 76 aeolianite limestone islets, attributed slight changes in the turnover patterns to an increase in rainfall and to the passage of a cyclone. In discussing other small island studies, they note that ' . . . as most cays occur in tropical regions, they are subject to frequent cyclonic or hurricane disturbances which often cause waves temporarily to inundate cays and wash vegetation away' (p. 405) (see also Sauer 1969). Similar findings were reported by Buckley (1981, 1982) in his analyses of plants on sand cays and shingle islands on the northern Great Barrier Reef.

Periodic intense disturbance is a natural feature of many islands across the globe (Whittaker 1995), and has increasingly been afforced by human influences. In illustration, hurricanes are frequent, devastating events throughout the region 10–20° north and south of the equator (which covers a lot of islands) (e.g. Nunn 1994). For the Caribbean,

Table 6.3 How might the capacity for richness of an island vary through time? Left and right boxes provide hypothetical examples of how increase/decrease might arise from factors under the same general heading (Modified from Whittaker 2004).

Factors leading to an increase	Factors leading to a decrease
Primary succession, increasing biomass, system complexity and niche space, in part driven by species colonization, in part enabling further colonization	Late successional stages if passed through simultaneously across an isolate might see competitive exclusion of suites of earlier successional species (e.g. Wilson and Willis (1975), data from Barro Colorado Island, Panama)
Arrival of 'keystone species' (e.g. fig trees on Krakatau (Whittaker and Jones 1994a,b; Thornton 1996))	Arrival of a 'superbeast' (e.g., first vertebrate predator), may over-predate naive prey species
Humans may increase habitat diversity, introduce new species, manipulate ecosystems to raise productivity, etc.	Humans may clear habitat, introduce new pest/predator or highly competitive species, degrade habitats, etc.
Area/habitat gain, e.g., through coastal deposition, uplift, sea-level fall	Area/habitat loss, e.g., through coastal erosion, subsidence, sea-level increase
Moderate disturbances might open up niche space, e.g., for non-forest species on an otherwise forested island	Major disturbance (e.g. hurricane or volcano) may wipe out much standing biomass, massively reducing populations
Increasing climatic favourability, increasing biological activity, increased NPP	Climatic deterioration, lower NPP
Evolution, on remote islands, a similar effect to primary succession, primarily operating through hierarchical links between organisms at different trophic levels	*In situ* co-evolution is unlikely to generate reductions in richness, but as oceanic islands age they subside and erode (Chapter 2), such that in parallel with the processes of speciation, carrying capacity will first peak and then decline through time (Chapter 9)

between 1871 and 1964 an average of 4.6 hurricanes per year was recorded, the island of Puerto Rico having a return period of just 21 years (Walker *et al.* 1991*a*; Chapter 2). Hurricanes are high-energy weather systems, involving wind speeds in excess of 120 km/h, affecting paths tens of kilometres wide. They can have huge ecological impacts on island ecosystems (e.g. Walker *et al.* 1991*b*; Elmqvist *et al.* 1994).

During post-hurricane succession in the Caribbean, it has been found that the animal community may continue to change long after the initial 'greening-up' of the vegetation, through invasions and local extinctions as the vegetation varies through successional time (Waide 1991). Initially, nectarivorous and frugivorous birds suffer more after hurricanes than do populations of insectivorous or omnivorous species, a consequence of their differing degrees of dependence on re-establishment of normal physiognomic behaviour in the vegetation.

Turnover patterns in response to disturbances of this and other kinds can be anticipated on theoreti-

cal grounds to be dependent upon taxon choice (e.g. Diamond 1975*a*, p. 369). Schoener (1983) established empirically that percentage species turnover declines from 'lower' to 'higher' organisms, as follows (approximate figures): 1000%/year in protozoa, 100%/year in sessile marine organisms, 100–10%/year in terrestrial arthropods, and 10–1%/year in terrestrial vertebrates and vascular plants. Moreover, across this range of organisms, turnover declines approximately linearly with generation time. Given which, and restricting consideration only to islands subject to such environmental dynamics, the attainment of a dynamic (biotically balanced) equilibrium in certain *K*-selected (long-lived, only moderately dispersive) taxa may be postulated to be comparatively rare.

Variation in species number in the short and medium term

It might be argued that islands devastated by events such as volcanoes and hurricanes are extremes: Krakatau is just a special case. However, the Indian Ocean tsunami of 26 December 2004, the eruption of Montserrat, in the 1990s, and the recent increase in hurricane activity in the Caribbean suggest otherwise. In addition, less dramatic environmental variation may also perpetuate non-equilibrium conditions. For example, Lynch and Johnson (1974), who may have coined the term 'successional turnover' in its island ecological sense, argued that where turnover is successional in nature, such as produced after a fire, then the turnover is determined not primarily by area–distance effects, but by the timing, extent, and nature of changes occurring in the habitats. They regard use of an equilibrium model in such situations as inappropriate. They also argue that interpretation of faunal change is especially difficult if both equilibrium and non-equilibrium (e.g. successional) turnover are involved.

Extreme climatic events, such as freezes, have been recognized as important determinants of island avifaunas. An illustration is provided by Paine's (1985) observations of the extinction of the winter wren (*Troglodytes troglodytes*) from Tatoosh, a 5–6 ha island, 0.7 km from the north-western tip of the Olympic Peninsula (Washington, USA). The loss of the species followed an extremely cold period in December 1978, and it took 6 years to re-establish a breeding population, which was presumed to be because of the difficulty of a relatively sedentary species invading across a water gap. Abbott and Grant (1976) found evidence of non-equilibrium in land-bird faunas on islands around Australia and New Zealand, a result that did not appear to be attributable to sampling error or to humans. They suggested in explanation that islands at high latitudes (such as these) are generally subject to irregular climatic fluctuations and may not therefore have fixed faunal equilibria (cf. Russell *et al.* 1995). Abbott and Grant (1976) argued that the extent to which climate fluctuates about a long-term average value determines the extent to which species number does the same. 'In this sense, species number 'chases' and perhaps never reaches a periodically moving equilibrium value, hindered or helped by stochastic processes' (Abbott and Grant 1976, p. 525).

Another study that appears to demonstrate that equilibrium values change in response to modest, non-catastrophic changes in island environments on timescales of years and decades is that of Russell *et al.* (1995). They found that a non-equilibrium model provided improved predictions of observed turnover of birds for 13 small islands off the coast of Britain and Ireland. They suggested that turnover could be viewed as operating on three scales: first, year to year 'floaters' (trivial turnover); secondly, on a timescale varying between 10 and 60 years, an intrinsic component equivalent to that envisaged in the EMIB was observable; thirdly, most islands show a change in numbers over time due to so-called extrinsic factors, such as habitat alteration.

Long term non-equilibrium systems

Extending the time scale, the glacial/interglacial cycles of the Quaternary have involved high-magnitude effects, the boundaries of biomes being shifted by hundreds of kilometres, sea levels falling and rising by scores of metres. These changes have resulted in at least two types of relictual pattern studied by island biogeographers. The first is that of land-bridge islands, those formerly connected to larger land masses (their mainlands), which gained at least some of their biota by overland dispersal before becoming islands (e.g. Crowell 1986). The second well-studied context is where habitat 'islands' have become isolated in a similar fashion within continents (e.g. Brown 1971). In both contexts, data often support a non-equilibrium interpretation.

The EMIB would have it that slopes of species–area curves should be steeper (higher z values) for distant (oceanic) island faunas than for nearer (often land-bridge) islands in the same archipelago. This is due to the combined effects of extinction and low colonization rates on very isolated islands, and because the impact of both is greater on small islands (Fig. 6.3a). However, as Lawlor (1986) noted in a review of data for terrestrial mammals on islands, large remote islands tend not to attain the species richness predicted for them on the basis of their area. This is

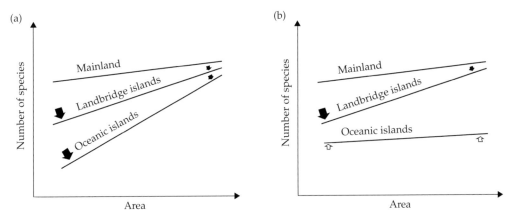

Figure 6.3 Equilibrium and non-equilibrium representations of island species–area relationships for a hypothetical archipelago, as summarized by Lawlor (1986). (a) By an equilibrium argument, the effects of recurrent colonization and extinction produce steepening of the species–area curves with increasing distance, as extinction (arrows) has greater relative impact on remoter islands, for which immigration is increasingly difficult. (b) In the non-equilibrium model, oceanic islands may be undersaturated, as immigration (open arrows) is too slow to 'fill' the islands (extinction is relatively unimportant), whereas land-bridge islands undergo 'relaxation', i.e. their richness patterns are extinction dominated (solid arrows).

because immigration is very slow to distant archipelagos, and is little influenced by island size in such remote contexts, thus producing relatively flat intraarchipelago ISARS. Land-bridge islands, in contrast, may have steeper species–area curves because of the predominance of extinction, as over long time-periods, members of the assemblage stranded by island formation suffer occasional attrition, supplemented by little immigration (Fig. 6.3b, Millien-Parra and Jaeger 1999).

In such cases, the fauna at the point of isolation has been termed **supersaturated**, i.e. it contains more species than at true equilibrium; thereafter, the dominance of extinction over immigration results (in theory) in **relaxation** of species number, declining towards eventual, hypothetical equilibrium. Classic illustrations of this effect have also come from non-volant mammals occupying montane habitats in the Great Basin of western North America. These studies have been selected for their illustrative value, and by no means all non-equilibrial systems are products of Quaternary climate change. The distinction between true oceanic islands and land-bridge islands is, as Lawlor (1986) has shown, a crucial one for understanding species–area relations of non-volant mammals

of remote islands, and analyses which fail to distinguish the two groups are thus likely to be misleading.

Implications for endemics?

If environmental fluctuations can be important to some small to moderate-sized near-continental islands, why do we not hear more of the adverse effects of natural disturbance in remote oceanic islands? Might they not be anticipated to be worst affected, being unable to replenish species stocks swiftly after such events as cyclones? Perhaps the effects of variability are largely 'built in' (evolutionarily) to their ecologies?

Miskelly (1990) reports widespread reproductive failure and delayed breeding by Snares Island snipe (*Coenocorypha aucklandica*) and black tit (*Petroica macrocephala dannefaerdi*), two endemic land birds, on the Snares Islands, in the New Zealand sub-Antarctic. Snipe also suffered unusually high adult mortality. He attributed these phenomena to the pronounced El Niño–Southern Oscillation (ENSO) event of 1982–83, which produced heavy rainfall and significantly lowered temperatures on the Snares Islands, thereby reducing the invertebrate food supply. Neither bird species invested much energy in reproduction during the ENSO event compared with subsequent years. The poor breeding seasons in this study contrast dramatically with the greatly enhanced reproductive success of land birds on the Galápagos, which benefited from improved plant growth and seed production enabled by the high rainfall (Gibbs and Grant 1987). Very heavy rainfall stimulated the production of plants and therefore the caterpillars feeding on them, important to finches in the breeding season. A significant increase took place in the population size of Darwin's ground finch on Daphne Major island. Gibbs and Grant (1987) concluded that rare events can have a major influence on key population processes, including processes of significance to evolutionary changes, in long-lived birds living in temporally varying environments (see also Grant 1986). These events did not involve species gains and losses, but warranted attention from ecologists because of the importance of the endemic species involved.

Whittaker (1995) argued that island endemics are unlikely to be lost very often in response to even large-scale disturbances, such as hurricanes or volcanic eruptions. First, this is because species and ecosystems have evolved within the context of the disturbance regimes. Secondly, and linked to this, they tend to occur on larger oceanic islands, which may well provide refuges unaffected by the perturbation somewhere within their land area. A recent illustration comes in the impact of the eruption of Chances peak in the 1990s on the endemic Montserrat oriole. The eruptions led to the loss of the species from a substantial part of the range, but sampling indicated that some 4000 individuals persisted in the less badly affected Centre Hills area (Arendt et al. 1999). By this line of thinking, small oceanic islands, such as cays and atolls, should support relatively impoverished systems with few endemics restricted to them.

Support for this general line of reasoning is provided by a study of the Laysan finch (*Telespiza cantans*), one of the endangered Hawaiian honeycreepers. It occurs on the small coral island of Laysan in the Hawaiian archipelago, although in prehistorical times it also occurred on Oahu and

Molokai, and a small population has recently been introduced to Pearl and Hermes reef. Laysan is only 397 ha in area and of that only 47% is vegetated. Morin (1992) found that weather conditions played a crucial role in regulating reproductive success. In 1986, a severe storm caused almost total mortality of eggs and chicks regardless of clutch size. Later that year, the number of fledglings per nest increased as clutch size increased. The species is omnivorous and eats some part of almost every plant on the island, as well as invertebrates, carrion, and eggs. It also has life-history parameters conditioned by long experience of the fluctuating conditions on the island. Reproduction can be spread over a long breeding season, with the potential for multiple broods, and the species has a long reproductive lifespan. Notwithstanding this flexibility of response, it was concluded that stochastic weather events and predation are the two key factors currently limiting the Laysan finch population, in contrast to most of the other honeycreepers of Hawaii, whose populations are threatened principally and acutely by introduced disease carried by exotic mosquitoes (Pratt 2005).

The South Pacific islands of Samoa have two flying fox (fruit bat) species, *Pteropus samoensis* (endemic to Samoa and Fiji) and *Pteropus tonganus*. Pierson et al. (1996) discuss the response of each of them to two severe cyclonic storms, Ofa, which struck in 1990, and Val, in 1991. Flying foxes are important elements of the ecosystem, possibly warranting keystone status, because of their roles as seed dispersers and plant pollinators. Activity patterns and foraging behaviour were disrupted for both species. Before the hurricanes *P. tonganus* was much the commoner, with a population on the small island of Tutuila of about 12 000, contrasting with fewer than 700 *P. samoensis* individuals, and it was thought that the more restricted endemic might be the more vulnerable to storm effects. However, the endemic species was found to have survival strategies not observed in the more common and widely distributed *P. tonganus*, and the latter experienced the more severe declines, of the order of 66–99% per colony. The numbers of *P. samoensis* seemed to be relatively unaffected. Its survival strategies included making greater use of leaves in the aftermath of the storm and also feeding on the fleshy bracts of a storm-resistant native liane. *Pteropus tonganus*, on the other hand, seems to have adopted the unfortunate behavioural response of entering villages to feed on fallen fruit, where they were killed by domestic animals or people.

Pierson *et al.* (1996) make the point that island species of the tropical cyclone belt have evolved with recurrent cyclones. It appears in this case that the species of narrowest geographical range actually has evolved more effective responses to cyclone events. Given that flying foxes can live for 20 years or so, individuals may experience several severe storms in a lifetime. Such events in themselves are unlikely to lead to extinction of the species, provided that other threats, such as habitat loss and hunting, are constrained. Where the two are combined, however, extinction becomes more likely (Christian *et al.* 1996). Although wind damage is typically patchy, it seems reasonable to suggest that areas of high topographic complexity, e.g. volcanic cones and deep valleys, are the most likely areas to retain patches with some foliage, and so should be given priority in reserve design.

6.5   Future directions

... in practice it is next to impossible to find a set of islands without environmental change or evidence of the severe impact of man's activities

(Case and Cody 1987, p. 408)

We have learnt a lot from studying island ecology under the ruling dynamic, equilibrial paradigm. We can continue to do so, whether focusing on improving macroecological models for island species number by incorporating energetics or by new approaches to nestedness and assembly structure. In this chapter we have reviewed evidence for departures from the dynamic equilibrium condition. Of course, some degree of variance can be accepted within the equilibrium paradigm, as systems of rapid response rates and small amplitude environmental fluctuations may swing rapidly back in to balance, such that most of the time an island could indeed be close to equilibrium. Application of dynamic, equilibrial models does

not demand a precise fixed value (Simberloff 1976). However, depending on the relative timescales of changes in the island environment and response by the biota, we contend that islands may variously be: always in equilibrium; in equilibrium most of the time but occasionally disturbed from equilibrium; or never in equilibrium, but always lagging behind changes in environment.

The emphasis on habitat determinism by Lack and others has been opposed by many supporters of dynamic island biogeography. They see it as returning the subject to a narrative form. The presence of many endemics on very isolated islands clearly demonstrates that even among a taxon of generally good dispersers, e.g. birds, there comes a point where isolation is sufficient to render immigration and colonization a rare and essentially probabilistic process. However, there is no need to draw such a stark distinction between dynamic and static views of islands. Studies reviewed in this chapter suggest that is possible to build bridges between these differing positions and that very often the determinants of species number and composition lie somewhere between the essentially stochastic, dynamic equilibrium and more 'static' behaviour, and require a more complex explanation of community organization on islands than provided for by the EMIB.

Conceivably, some island systems (large and remote) might respond to a severe El Niño event, for example, by exhibiting considerable flux in population sizes, biomass, and productivity, without much impact on species turnover rates. In other cases (small and near-shore islands), a severe cold snap in the winter might have little effect on biomass and productivity but might greatly impact on populations of small birds, leading to measurable alteration of species turnover rates. On the whole, there appears to have been insufficient attention paid to systems departing from the assumptions of the dynamic equilibrium paradigm. Similarly, as exemplified by Krakatau, where disturbances are severe, or environmental change is very considerable, there may be long-term ecological changes which unfold over the course of decades or centuries. In such cases, a successional model of island ecology can provide a more appropriate

Box 6.2 Lomolino's (2000) tripartite model of island biogeography

In a special issue on island biogeography published in the journal *Global Ecology and Biogeography* in 2000, Mark Lomolino (2000*a*) argued that given the acknowledged limitations of the equilibrium theory, a new island model is required, and that it should: (1) include feedbacks among system components (as Bush and Whittaker 1991), (2) acknowledge the role of evolution (as Heaney 2000), and (3) be species-based (Lomolino 2000*b*). He stated 'Species vary in many ways that affect immigration, extinction and speciation, and many biogeographic patterns derive from, not despite, differences among species. For example, the form of the species-isolation and species–area relationships may reflect very general patterns of variation in immigration abilities and resource requirements among species.' (Lomolino 2000*a*, p. 3).

Lomolino's graphical model (see figure) maintains the simplicity of MacArthur and Wilson's (1967) classic island biogeography scheme, but pulls out a third dimension, explicitly recognizing the role played by speciation in promoting species richness, as a function of the island's area and isolation. In many respects this graphic merely sketches out ideas contained within MacArthur and Wilson's (1967) monograph. As we saw, the first statement of their model explicitly includes speciation, and they went on to discuss how particular taxa typically show the best patterns of evolutionary radiation towards the periphery of their spread across an ocean's islands (they termed this the 'radiation zone').

The graphic provides a useful summary of some of the more important patterns and processes we have discussed in the island ecology section, while also pointing to the main process we have

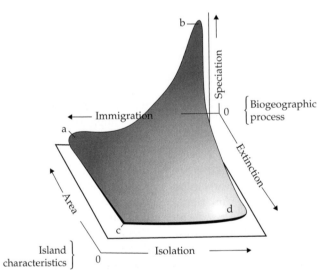

Lomolino's (2000*a*) tripartite model focuses on the relative importance of the three fundamental processes of biogeography: evolutionary change, extinction, and dispersal. He notes that: (1) immigration rates should increase not only with proximity to a source region but also with the vagility of the target species; (2) extinction rates should decrease with island area but increase with the target species' resource requirements (which may in turn be broadly related to body size, Bierdermann 2003); finally, (3) speciation rates should be more important where extinction and immigration rates are lowest, and therefore will be greater on the largest and most isolated islands, but will decrease with species vagility and resource requirements. The shading represents the relative levels of richness and the relative resistance to and resilience of insular biotas following disturbance. The biotic characteristics of islands in the four corners (a, b, c, and d) are set out in the table. (From Lomolino (2000*a*), Fig. 1.)

Island characteristics relating to the positions indicated in the figure

Island feature	(a) Large, near	(b) Large, far	(c) Small, near	(d) Small, far
Immigration	High	Low	High	Low
Speciation	Low	High	None	Low
Extinction	Low	Low	High	High
Species richness	High	High	Moderate to low	Low
Turnover	Low	Low	High	Low
Island type	Continental islands	Continental fragments or large oceanic islands	Small continental islands	Depauperate oceanic islands
Example	Newfoundland, Tasmania	Madagascar, Hawaii	Skokholm (Wales)	Easter Island, Tristan da Cunha

hitherto neglected: evolution. Moreover, although Lomolino's model recognizes the patterns of diversity and turnover, it doesn't assume equilibrium.

So, is the way ahead to take a species-based approach? Lomolino's argument that many biogeographic patterns derive from, not despite, differences among species echoes many of the points discussed throughout our examination of assembly rules, incidence functions, and the importance of species-level effects in the dynamics of island biotas. A species-based approach is in some respects more flexible, can be more relevant to the needs of conservationists in particular circumstances (e.g. modelling the incidence or potential incidence of an endangered species), and opens up a new set of hypotheses

for testing. However, we should also recall that species responses and impacts can vary between systems. For example, Watson et al. (2005) demonstrated variation in species incidence functions of woodland birds in the Canberra area (Australia), which appeared attributable to properties of the matrix in which the woodland systems were embedded (see Box 5.2). Similarly, many species considered benign and non-invasive within their native ranges, upon being introduced into an exotic island, turn into radical transformers of ecosystems (Chapter 10). This suggests that their pattern of incidence in one area may not accurately predict their incidence elsewhere. Hence, there are likely to be limits to the predictive power of species-based models, just as there are with system-level models.

---

representation of turnover and compositional patterns than do simpler stochastic models based only on parameters such as area and isolation.

These arguments amount to a case for more complex models of island ecology. Building in greater complexity is fine under two conditions: (1) you are sure of your starting assumptions, i.e. the paradigm you are working under, or (2) you are attempting to develop better understanding of detailed case study systems or areas. However, we should not abandon the arguably more important scientific task of searching out the simple explanation where it will suffice, and of finding new models of general applicability. Box 6.2 reviews one possible direction.

Consideration of scale has been another key theme of the last two chapters, as we have developed the argument that although some processes may be universal, their impact is detectable at particular scales of analysis, while at other scales other processes or confounding variables dominate the signal. The increasing availability and sharing of data should allow the scale sensitivity of island ecological models to be more thoroughly investigated, allowing us to determine the applicability of particular theories within the space/time/scale continuum.

Throughout the island ecology section of this book we have to a remarkable degree disregarded the role of humans, largely reflecting the approach of the research literature in the field. Yet, as clear from particular parts of the discussion, human societies have had profound impacts on island ecologies, nowhere more so than on some of the more remote oceanic islands. We are only beginning to understand how this may have impacted on island diversities. This caution serves to reinforce the take home message of this section that it is not enough to assume that a system is in equilibrium, any more than to assume that it is not. This should be a matter of empirical investigation, particularly if other theories are to be constructed on the basis of equilibrial assumptions (cf. Weins 1984). We need to remember this when we turn to the application of island theories to conservation problems in the final section of the book.

6.6   Summary

Although the equilibrium model of island biogeography (EMIB) has served to advance understanding of island ecology, the work reviewed in the previous two chapters demonstrates that the dynamic equilibrium framework is no longer sufficient to encompass the field. We argue in the present chapter that the EMIB has a narrower frame of reference than original envisioned, working for only a subset of island species data sets. This subset is largely determined by the scale parameters of the target system relative to the space usage and dispersal powers of the target taxon.

Haila distinguishes four coupled space-time scales, which he terms: (1) the individual scale, (2) the population dynamics scale, (3) the population differentiation scale, and (4) the evolutionary scale. Systems that correspond to the population dynamics scale are most likely to show dynamics in tune with the EMIB. We illustrate how even within a single taxonomic group, such as birds or mammals, different species can in practice have radically different spatial scales of interaction with an island system. Temporally, the dynamics of different components of an ecosystem also respond on varying wavelengths. For instance, whereas bird communities may respond fairly rapidly to changing island

carrying capacity, or disturbance events, the successional dynamics inherent in forested ecosystems imply that vegetation systems may take decades to hundreds of years to establish a meaningful equilibrium condition, always assuming that no further disturbance occurs. The importance of taking account of hierarchical relationships across trophic levels is also stressed, such that for instance, the equilibrium point for birds might change over time in response to longer term dynamics in the vegetation.

To accommodate these complexities, we present a framework in which the dynamic equilibrium (EMIB) model is placed alongside dynamic non-equilibrium, static-equilibrium, and static non-equilibrium concepts, and we discuss the applicability of these ideas to case study systems. The lack of long-term high resolution data series hampers our ability to distinguish between these competing hypotheses, but we go on to evaluate evidence for how and why longer term variation in island carrying capacities may occur. Natural disturbance phenomena, such as tsunami, hurricanes, volcanic eruptions, and even marked fluctuations in weather conditions linked to ENSO events, may all impact on island ecologies, with varying potential significance to island species numbers and composition.

Increasing the complexity of island ecological models to incorporate such factors will improve the realism of the models but at the cost of generality. An alternative approach is to develop species-based models, which may allow us to avoid some of the limitations of the EMIB, but it has yet to be established that species-based models have predictive power across the range of a species, or that they will lead to improved models for whole assemblages. Given the importance of understanding the emergent ecological behaviour of assemblages of species, we suggest that further efforts be devoted to exploring the scale sensitivity of island ecological models, the degree to which systems demonstrate equilibrium dynamics, and how hierarchical links across trophic levels influence emergent properties, such as species richness and turnover.

PART III

Island Evolution

The Macaronesian clade of *Echium* (Boraginaceae) consists of 28 endemic species (two on Madeira, 23 on the Canaries, and three on the Cape Verde islands), of which 26 exhibit woodiness. They derive from a single herbaceous continental ancestor. The photograph shows *Echium wildpretii* ssp. *trichosiphon*, endemic to the summit scrub of La Palma (Canary Islands) (with kind permission of Ángel Palomares).

CHAPTER 7

Arrival and change

Island forms of the same species [of birds] show every gradation from extremely small differences in average size to differences as great as those which separate some of the species. The differences tend to be greater the greater the degree of isolation. Some of them are adaptive and others seem non-adaptive.

(Lack 1947a, p. 162)

In the island ecology section we explored patterns and processes while treating species as standard and unchanging units of analysis. This is a useful simplifying assumption, but, as we know, species are not fixed, nor are they homogenous entities. In this section of the book we consider the special insights into evolutionary change that islands have provided.

We start from the point at which a new colonist arrives on an island, and then consider the micro-evolutionary processes that go to work on such populations and which result in a fascinating array of emergent island effects or syndromes. The changes with which we are concerned in this chapter are expressed in genotypic and phenotypic characters, sometimes exemplified by island endemic *species*, but frequently evidenced at subspecific level, i.e. they do not necessarily involve speciation and the recognition of distinct island and mainland species. The processes include a mix of essentially chance effects, resulting in the divergence of island lineages from their mainland source pools, and of directional evolutionary forces that appear to operate in a consistent fashion across many remote islands and taxa. These micro-evolutionary processes and patterns are the building blocks involved in the phenomena of phylogenesis and the emergent macro-scale patterns of island evolution described in the following two chapters.

7.1 Founder effects, genetic drift, and bottlenecks

Islands form in two fundamentally different ways: either by arising out of the sea, at which point they are empty of land organisms, or by once contiguous areas of land becoming separated from one another by incursions, as for example when sea levels rise at the termination of ice ages (Chapter 2). Thus classic oceanic islands start empty, whilst land-bridge islands have a full complement of species when they first become proper islands. However, some islands have such a complex or poorly known environmental history that this simple dichotomy can be difficult to apply, or to agree upon (e.g. see McGlone *et al.* 2001; McGlone 2005; McDowall 2005; Morrone 2005). Here we set these issues to one side, mostly taking examples from true oceanic islands, thus allowing us to examine colonization events associated with long-distance dispersal to islands that start (and remain for long periods) species-poor for their area, and which develop from a biased subset of the mainland species pool.

The **founder principle** (Mayr 1954) is that typically a species immigrating to a remote island will establish by means of a tiny founding population. This contains only a subset of the genetic variation in the source population and it subsequently receives no further infusion. This immediately provides a bias in the island population, on which other evolutionary processes then operate. Genetic variability can be increased after establishment by mutation and re-sorting. One important form of re-sorting is **genetic drift**, the chance alteration of allele frequencies from one generation to the next, which may be particularly important under sustained conditions of low population size. In addition to such non-selective drift there may also be distinctive selective features of the novel island environment. The outcome of these phenomena could, in theory, be a rapid shift to a new, co-adapted combination of alleles, hence contributing to reproductive isolation from the original parent population.

Although it is possible to demonstrate the occurrence of genetic drift, as Berry *et al.* (1992) have done in their study of a small population of house mice on Faray in the Orkneys, the contribution of founding effects and drift to island speciation remains the subject of keen debate (Barton 1989; Clarke and Grant 1996). It is particularly problematic that founding *events* (i.e. colonizations) have been theorized to produce a variety of rather different founding *effects*, and that the latter are difficult to isolate empirically from other microevolutionary processes. In fact, there are at least three distinguishable models of how founder effects may lead to speciation:

- Mayr's peripatric speciation
- Carson's founder-flush speciation
- Templeton's genetic transilience.

These theories place greater degrees of stress, respectively, on:

- the impact of the general loss of heterozygosity combined with random drift
- the role of variations in selection pressure (particularly its relaxation immediately after colonization)
- changes in a few major loci within the genome (see discussions in Templeton 1981; Barton and Charlesworth 1984; Clarke and Grant 1996).

The founding event represents a form of **bottleneck**, a term implying a sharp reduction in population size. Bottlenecks may occur at other points in the lifespan of a lineage, e.g. following catastrophic disturbance of a habitat, or in the context of species pushed to the brink of extinction by human action. Studies of the Hawaiian *Drosophila* indicate that founder events have occurred and at three levels: in the colonization of the archipelago, of its constituent islands, and of patches of forest habitat (*kipukas*) within islands (Carson 1983, 1992; Carson *et al.* 1990). Significantly, this work indicates that

bottlenecks produce a marked loss of heterozygosity only if they are sustained or are repeated at short intervals. In contrast, if a bottleneck lasts only one generation and recovery to a large population size is rapid, then increased genetic variance for quantitative traits can occur.

According to theories of founder-effect speciation, the narrowing of the gene pool during a bottleneck may serve to break up the genetic basis of some of the older adaptive complexes, liberating additive genetic variance in the new population (Carson 1992). Novel phenotypes might subsequently result from natural selection operating on the recombinational variability thus provided in the generations immediately following the bottleneck. According to a model proposed by Kaneshiro (e.g. 1989, 1995), an important element in this scenario is differential sexual selection. He found that a minority of male drosophilids perform the majority of matings, and that females vary greatly in their discrimination, and acceptance, of mates. In normal, large populations the most likely matings are between males that have a high mating ability (the studs), and females that are less discriminating in mate choice. This provides differential selection for opposing ends of the mating distribution and a stabilizing force maintaining a balanced polymorphism. In a bottleneck situation, however, there would be even stronger selection for reduced discrimination in females, and if this was maintained for several generations, it could result in a destabilization of the previous co-adapted genotypes.

In such bottleneck situations, with reduction in mate discrimination by females, the likelihood of a female accepting males of a related species increases, thus allowing hybridization to occur. The term given to the incorporation in this way of genes of one species into the gene pool of another via an interspecific hybrid is **introgressive hybridization**. With the development of molecular tools, hybridization between animal species (especially insects) has been discovered to occur naturally much more often than had previously been suspected (Grant and Grant 1994; Hanley *et al.* 1994; Kaneshiro 1995; Clarke and Grant 1996). (This presents a small but perhaps too often overlooked challenge to the assumption that lineage development occurs solely through branching (Herben *et al.* 2005).) Thus, although bottlenecks can serve to reduce the genetic base of a new colony, in some instances they may serve to provide *enhanced* genetic variability within a population, the very stuff on which natural selection may then go to work!

Laboratory studies of Hawaiian *Drosophila* have provided a number of important insights into founder effects, but recently these phenomena have been studied in an increasing range of island taxa (e.g. Clarke and Grant 1996). For example, Abdelkrim *et al.* (2005) describe how the use of complementary genetic techniques of analysis allows the detection of varying severities of founding events in initial colonization and subsequent interisland dispersal events for ship rat (*Rattus rattus*) in the Guadeloupe archipelago. Another Hawaiian example is provided by endemic members of the plant genus *Silene*. Populations appear to be much more polymorphic on the older island of Maui than on the younger island of Hawaii (i.e. the main or Big Island), suggesting that genetic variation had been lost in colonization of the latter (Westerbergh and Saura 1994). The process of restricted gene flow followed by genetic drift also seems to be repeated as younger volcanic terrain is colonized within the island of Hawaii.

Although some regard founder effects as a key part of the most likely mode of speciation in most Hawaiian Drosophilidae (e.g. Kaneshiro 1989, 1995), others dispute their significance. For instance, Barton (1989) observes that enzyme heterozygosity within Hawaiian drosophilids is similar to that of continental species, and queries whether, in practice, founder effects cause much loss in genetic variation. Barton (1989) regards other elements in the island laboratory as being of greater significance than founder effects, and he mentions specifically what he terms **faunal drift**, the random sampling of species reaching an 'empty' habitat, and thus providing a novel (disharmonic) biotic environment in which selection drives divergence into a variety of new niches (see also Vincek *et al.* 1997).

Sarre *et al.* (1990) examined morphological and genetic variation among seven island and three mainland populations of the sleepy lizard, *Trachydosaurus rugosus*, in South Australia. Their work supports the significance of random genetic drift and inbreeding in small isolates. All but two of their islands were isolated from the mainland between 6000 and 8500 years ago. Contrary to some other recent studies of vertebrates on South Australian islands, their allozyme electrophoresis analysis showed that heterozygosity levels did not vary significantly among the populations, i.e. there has been little erosion of heterozygosity within the insular populations. Nonetheless, the island populations exhibited reduced allelic diversity, with rare mainland alleles tending not to be present on the islands. This may have been due in part to bottlenecks occurring at some point or points since isolation of the populations. There was also a greater degree of genetic divergence shown between island populations than between mainland populations, and this they attributed to genetic drift on the islands.

Allozymes are different forms of an enzyme specified by allelic genes, i.e. by genes that occupy a particular locus. Allozyme electrophoresis has become a conventional method of measuring heterozygosity levels. It is thus of interest that two measures of developmental stability varied significantly among the island populations in a manner distinct from the results of the allozyme study (Sarre and Dearn 1991). The morphological measures used were: fluctuating asymmetry—the degree of random difference between left and right sides in bilateral structures (see Box 10.1); and percentage gross abnormalities, involving features such as missing ear openings, missing toes, club feet, and deformed head and labial scales. Three of the island populations stood out as having distinctly higher levels of these forms of developmental instability. This was interpreted as being due to the expression of deleterious genes (in turn due to inbreeding depression) or the genetic disruption of co-adapted genotypes. One important conclusion drawn from their study is that reduced levels of heterozygosity *per se* are unlikely to be the prime cause of the observed developmental instability. In short, the allozyme technique was failing to reflect the genetic basis for the deleterious morphological features. The relative contribution of environmental and genetic factors to developmental stability in the sleepy lizard cannot at this stage be determined.

In contrast to the emphasis given in this study to drift, Thorpe and Malhotra (1996) stress the importance of natural selection for current ecological conditions in their studies of Canarian and Lesser Antillean lizards. It may be significant that Thorpe and Malhotra's islands, unlike Sarre's, are large and contain a considerable range of habitats, i.e. the relative importance of stochastic processes versus directional selection may turn out to be strongly dependent on the environmental context.

Although opinions differ as to the overall significance of founding effects for the rapidity of evolutionary change and ultimately speciation on islands, it has increasingly become evident that founding effects can be detected by several different techniques, and for a range of plant and animal taxa (e.g. Clarke and Grant 1996; Drotz 2003; Hille *et al.* 2003; King *et al.* 2003). Moreover, in several well-specified case studies, it is possible to see a nested sequence of events, from the initial colonization of an archipelago, through interisland movements, down even to intraisland colonization events (Westerbergh and Saura 1994; Abdelkrim *et al.* 2005).

Implications of repeated founding events

What are the implications of such scenarios? First, island lineages are likely to contain less genetic diversity than the mainland source population. Second, because there is a general trend towards loss of dispersability once established on a remote island (discussed below), and because of the often significant interisland distances, the repetition of such founding events is likely to lead to further differentiation between populations within an archipelago, especially if there is significant environmental variation amongst the islands to add a selective component to lineage development. Available data tend to confirm these expectations (e.g. DeJoode and Wendel 1992; Frankham 1997; Hille *et al.* 2003). DeJoode and Wendel (1992) compiled data in the form of allozyme variability for approximately 60 Pacific island endemic plant taxa

(a mix of species and varieties), finding that they exhibit roughly half the variability previously reported for continental plant species, and two thirds of that shown by continental endemics of restricted range. Recent analyses of populations of silvereyes (*Zosterops lateralis*) that colonized islands in the southwest Pacific from mainland Australia after AD 1830 allow an unusually well-specified quantification of founder effects. The analyses indicated that single founder events had little impact on genetic diversity, partly because founding populations have apparently been quite large (24–200 or so), but that repeated island hopping was accompanied by a gradual decline in allelic diversity (Clegg *et al.* 2002; Grant 2002).

In general, differences in genetic variability among island and mainland populations have been found to be larger: (1) the smaller the size of the founder event, (2) the larger the differences in population sizes, (3) the smaller the immigration and dispersion rates, (4) the smaller the island size, and (5) the greater the distance from the continent (Frankham 1997). Indeed, analyses of a few tree species colonizing the Krakatau islands, which are isolated from mainland Java and Sumatra by only about 40 km, indicate no loss of genetic variation in the island populations, and suggest that for some species the island and mainland populations can be considered panmictic, i.e. as if involving random mating within a single extensive population (Parrish 2002). Moreover, similar results were found for two species of Krakatau fig wasps (Zavodna *et al.* 2005); despite the small size of these insects, such a distance may be scarcely a barrier to colonization and gene flow. As Grant (2002, p. 7819) commented about the silvereye study

. . . Perhaps it is more parsimonious to invoke single-immigration events for each island, but I think this is likely to be wrong. If it is wrong, it carries an important implication: except for the most isolated islands, repeated immigration may obscure or obliterate any founder-effect changes that take place following the initial colonization.

Reviewing allozyme variation in 69 Canarian endemic plant species, Francisco-Ortega *et al.* (2000) report an average species-level genetic diversity at allozyme loci of 0.186, which is more than twice as high as the mean reported for Pacific island endemics (0.064) by DeJoode and Wendel (1992). They suggest that this discrepancy might be explained through two non-exclusive arguments. The first argument is that the greater age of the Canarian archipelago (e.g. 20 million years for Fuerteventura; Anguita *et al.* 2002), means that there has been a longer period of time for genetic variation to accrue via mutation. However, phylogenetic studies of several endemic groups do not support great ages for the majority of endemic lineages, hardly exceeding 4–5 million years (Marrero and Francisco-Ortega 2001*a,b*). Moreover, there is also evidence of a considerable degree of interfertility among several endemic species belonging to radiating genera, and in cases even for interfertility between members of distinct monophyletic genera, which supports the supposition that they represent relatively recent derivatives from common ancestors (Francisco-Ortega and Santos-Guerra 2001). Second, the close proximity of the islands to Africa, 96 km today but only 60 km during glacial period sea-level minima (García-Talavera 1999), would have facilitated multiple introductions into the Canaries and thus genetic bottlenecks may have been less extreme than for some of the more remote Pacific islands. Although a monophyletic origin has been demonstrated for a large majority of Canarian endemics, this does not mean that there could not been several dispersal events prior to the radiation processes (Francisco-Ortega *et al.* 2000). These ideas are interesting but speculative: further work is clearly needed to establish how genetic diversity levels vary between and among archipelagos, and then to test what factors may be responsible.

It may be relevant to this discussion that the bulk of the genetic diversity displayed by the Canarian endemic plants appears to reflect interpopulation, rather than intrapopulation variation. This is consistent with (1) repeated founder events within the archipelago, and even within islands, (2) the apparent existence of dispersal limitations of pollen and seed among populations, and (3) the prevalence of autogamic (selffertilization) reproduction in many of these species. Moreover, when a species has been found

to hold a larger component of intrapopulation than interpopulation genetic variation, as was found for the Gomeran endemic shrub *Echium acanthocarpum*, the population disjunction has been interpreted in historical terms as constituting just the surviving remnants of a formerly continuous island population (Sosa 2001). If intraarchipelago events are crucial to explaining patterns of genetic variation within lineages, comparisons between archipelagos or different ocean basins may in some respects be confounded.

Finally, in this discussion we have assumed that bottlenecks are essentially stochastic, but conceptually, at least, we might recognize that they could sometimes be regarded as deterministic. Whereas stochastic bottlenecks are survived by a random subset of the population, deterministic bottlenecks are survived by a specific genetically 'pre-adapted' subset. A deterministic bottleneck will give rise to a greater subsequent degree of resistance to a new episode of the same disturbance, whilst a stochastic one won't. We can illustrate this idea with two hypothetical examples. First, in the event of the introduction of the myxoma virus to an insular rabbit population, a few individuals may survive to perpetuate the population, and these will tend to be those least affected by the disease: subsequently genetic resistance will spread through the population. Second, imagine that a volcanic eruption produces a lava flow that destroys entirely a plant community, except for those plants that happen to have germinated on small hillocks that protrude over the level of the lava. The survivors will dominate the seed rain in the next generation, but have not been selected for their resistance to being covered with lava.

In conclusion, what have we established in this section? First, the restriction of populations to an extremely small size, either on colonizing an island (the founding event), or subsequently in the history of a lineage (other bottlenecks) may, in theory, produce a variety of effects:

• loss of genetic diversity (heterozygosity)
• addition of novel genetic combinations
• the introgression of genotypic variation from related species (or varieties).

Thus, bottlenecks can variously diminish heterozygosity (a general indicator of fitness), or in contrast can serve to 'shake things up', generating a sort of mini-genetic revolution (*sensu* Carson and Templeton 1984), contributing to rapid phases of evolutionary change in island populations. Secondly, although the number of studies is growing, providing some support for the operation of these mechanisms, their relative contribution to island evolutionary change remains uncertain (Grant 2002). Thirdly, notwithstanding recent developments in molecular biology and genetics, it is hard to be sure how well the specific phenotypic properties upon which the success of a lineage depends are measured by tools such as allozyme electrophoresis (see papers in Clarke and Grant 1996). It is surely notable that in comparison to related continental lineages, island endemic forms often display far greater morphological differences than are reflected in estimates of average heterozygosity (e.g. Crawford *et al.* 1998), which implies that something about these island environments encourages rapid lineage diversification through throwing specific switches in the genetic make-up. That 'something' likely reflects the general impoverishment, disharmony, and environmental/biotic novelty of remote islands. While founder events and other largely stochastic bottlenecks are part of the story, the emergent patterns encourage evolutionary scientists to invoke significant roles for deterministic causes alongside or in preference to stochastic founding effect processes, as will be discussed later.

7.2 After the founding event: ecological responses to empty niche space

Do islands possess vacant niche space? The relationship between isolation and impoverishment may not necessarily take a simple linear or log–linear form (Chapter 4), but it is accepted that oceanic islands possess fewer species for their area than equivalent areas of continent. Yet, if this represents an equilibrium condition (whether dynamic or static), this could be interpreted as indicating that the islands involved are actually 'full' and that

the 'missing' ecological niches are lacking because there is no space for them. Although this argument might apply to older oceanic islands, surely there must be plenty of vacant niche space in the species poor and disharmonic early stages of the lifespan of an island? This is generally assumed to be the case, but the concept of empty niche space turns out to be another slippery one to pin down and quantify. This is partly because the term **ecological niche** has itself been variously defined. When first introduced it reflected the figurative usage of 'a place in the community', but developed into the description of the types of opportunities that occur for species to use resources and avoid predators (Schoener 1989). Hutchinson's (1957) usage was of a different form: the n-dimensional hypervolume of environmental axes within which a species occurs. A plant or animal has a certain range of an environmental resource, such as temperature, at which growth is optimal. Outside this optimal range it may survive but not thrive, until a point is reached, the limits of its species tolerance, beyond which the species cannot persist. Each resource gradient forms a part or dimension of the hypervolume.

In the island context, the term **empty niche** refers to the lack of representation of a particular mainland ecological guild or niche, providing evolutionary opportunities for colonizing taxa to exploit the vacancy: the idea being that a fundamental niche exists and it is just a matter of filling it. As the assembly rules debate shows, it can be quite difficult to determine if such a fixed form of architectural blueprint exists and so while the concept of the empty niche is of intuitive value, it can prove extremely difficult to apply.

The literature discussed in the following sections mostly refers to a form of niche theory that derives from the resource gradient idea. Typically, it does not encompass a multidimensional characterization of the niche, but is limited to a single or very limited set of characters, such as mobility, body size, or jaw size, each of which provides a surrogate measure for one dimension of the n-dimensional hypervolume of the organism. The ambiguities of the niche concept are avoided in many such studies by describing the distributions of species populations along resource spectra. As Schoener (1989) points out, much of this literature is, in essence, concerned with competition, chasing the elusive question (Law and Watkinson 1989) of the extent to which competing species can coexist. The answer remains elusive in the island evolutionary context because it is easier to detect functional changes (particularly morphological changes) in island forms, than it is to determine the role of interspecific competition in relation to other potential causes of such change. The best chance here is to examine contemporary processes, which may then be inferred to hold a wider historical relevance. The greatest methodological difficulties come when it is the 'ghost of competition past' that is being sought, i.e. when researchers are setting out to deduce past events from current pattern (Law and Watkinson 1989).

Notwithstanding these cautions, if we are to understand how adaptive radiations of organisms, such as the Galápagos finches or Hawaiian honeycreepers have come about on oceanic islands, we have to be able to build explanatory models that start with the arrival on a remote island of the first seed eater, the first nectar feeder, or even the first land-bird species. The founding population thus can exploit hitherto largely untapped resources: in short, empty niche space. As the island fills up with a mix of further colonists and of neoendemic species the availability of empty niche space must fluctuate, eventually declining again as the island ages and subsides.

Ecological–evolutionary responses to the differential occupancy on remote islands (as compared with continents) provide interesting insights, which we now attempt to review, starting with two general responses. First, the phenomenon of ecological release, i.e. of niche expansion by colonist species, and second, density compensation, the term given for the general pattern of increased average density of island species that goes hand in glove with the lower richness of island ecosystems.

Ecological release

Ecological release occurs when a species in colonizing an island thus encounters a novel biotic environment in which particular competitors or other interacting organisms, such as predators, are absent. The response can take perhaps two main forms: (1) loss of 'unnecessary' features (e.g. defensive traits, bold patterning), and (2) increase in variation in features such as beak morphology.

Lomolino (1984*a*) argues that ecological release attributable to the absence of predators may be more widespread than generally recognized and, conversely, that competitive release may be overestimated. He cites examples of Fijian fruit bats of the genus *Pteropus* being more diurnal in the absence of predatory eagles than those on less isolated Pacific islands, and of the meadow vole *Microtus pennsylvanicus* being essentially indiscriminant of habitat type on islands without one of its major predators, *Blarina brevicauda*.

Sadly, the loss of defensive traits commonly involves a lack of fear of, and ability to flee from, humans and terrestrial vertebrate predators: undoubtedly a contributing factor to the demise of island species such as the dodos and the many species of flightless rails extinguished from islands across the Pacific (Steadman 1997*a*). The Solomon Islands rail, *Gallirallus rovianae*, is not only flightless but also exemplifies another tendency: the loss of bold patterning (Diamond 1991*a*). The loss of elaborate or distinctive morphology and a return to simpler song types both appear common features of island endemics, especially where islands possess only one representative of a genus (Otte 1989). The general interpretation offered for these changes is that organisms released from competition with closely related species no longer require such accurate mating barriers, and so these features are gradually lost.

The second form of release in the absence of close competitors on an island is to free a colonist from constraining selective pressure, thereby allowing it to occupy not only different niches but also a wider array of niches than the continental ancestral form (Lack 1969; Cox and Ricklefs 1977). This form of ecological release, the increase in niche breadth, forms a key part in the reasoning of the competitive speciation model we consider later and is an important part of many scenarios for island evolution (e.g. adaptive radiation). Island finches provide classic examples of evolutionary increases in niche breadth within lineages. In illustration, granivorous finch species on remote islands such as Hawaii and the Galápagos are highly divergent in terms of beak sizes (and seed sizes utilized) compared with finches on continents and, plausibly, this can be related to an absence of competitor lineages on the islands (Schluter 1988).

Increases in niche breadth within an island lineage are not always morphologically apparent. The ancestor of the Darwin's finch of Cocos Island, *Pinaroloxias inornata*, colonized an island where 'empty' niche space and the lack of interspecific competition provided conditions for ecological release. This endemic finch has diversified behaviourally and, while showing little morphological variation, exhibits a stunning array of stable individual feeding behaviours spanning the range normally occupied by several families of birds (Werner and Sherry 1987). This intraspecific variability appears instead to originate and be maintained year-round behaviourally, possibly via observational learning.

Density compensation

Density compensation refers to higher than normal (i.e. than mainland) densities of a species on an island, linked to the lower overall richness of the island assemblage, and often suggested to indicate a form of competitive release. It was recognized by Crowell (1962) in a comparative study of birds on Bermuda with those of similar habitats on the North American mainland. This density adjustment may occur to an apparently excessive degree, in which case it is termed **density overcompensation** (Wright 1980). The occurrence and degree of density compensation is variable, and a single species may exhibit higher, lower, or comparable densities on different islands as compared with the mainland. This is, on reflection, unremarkable, as particular species experience different conditions on different islands, e.g. in relation to presence/absence of predators, the presence/absence of competitors, or the suitability of habitats.

To examine the generality of density compensation effects, we should consider entire guilds or entire assemblages rather than just subsets of species. Remote islands are typically considered

species poor for their area, and one corollary of this is that the species–abundance distribution is expected to be attenuated, with fewer very rare species able to sustain a presence than would be found in an average equivalent area of mainland. Where colonizing species find 'empty niche space' and undergo ecological release, this effect will be all the greater. It follows that the average population density of species occurring on islands should be greater than for a mainland assemblage, and that this effect should be more marked the lower the richness per unit area (Fig. 7.1).

As discussed in the island ecology section, variation in island species richness can best be accounted for by bringing additional variables, particularly climatic and habitat factors, into the regression models. Given which, we ought to take account statistically of the same variables in testing if overall density levels differ from those that would be anticipated as a function of the overall resource base of an island and the accompanying richness of species thus supported. In short, density compensation across an entire assemblage is expected as a simple corollary of lower richness, and this 'null model' should first be considered before reaching for other factors in explanation (Williamson 1981).

Even so, there is a considerable literature on density compensation, focusing on a variety of explanations for apparent cases of overcompensation. Two prominent ideas are: (1) a species being able to find more food per unit effort and thus attaining a higher abundance within the same habitat without a niche shift; and (2) niche enlargement, such that a species utilizes a wider range of habitats, foraging strata, foraging techniques, or dietary components (MacArthur *et al.* 1972). In turn, these mechanisms may be underpinned by the compositional character of the island assemblage, for example, the absence of top predators, or of large-bodied competitors. In the latter case, a similar bird biomass might be distributed amongst more numerous smaller-bodied bird species.

Blondel and Aronson (1999) provide an example (Table 7.1) of increased density through both mechanisms (1) and (2) among tits (*Parus* species) in Corsica. The island has three species, and the nearby

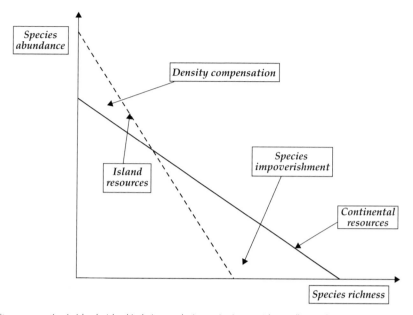

Figure 7.1 Density compensation in islands. Island isolation results in species impoverishment (lower obtuse triangle), which promotes the acquisition of available resources by the species present on the islands, allowing an increase in their density (upper obtuse triangle). From Olesen *et al.* (2002), slightly modified.

Table 7.1 Niche enlargement and density inflation in tit species (great tit *Parus major*, blue tit *P. caeruleus*, coal tit *P. ater* and crested tit *P. cristatus*) in matching habitat gradients on Corsica compared with the nearby French mainland. Values are breeding pairs/10 ha. Habitats follow a gradient of increasing structure complexity. Each species occupies on average more habitats and has higher population sizes on Corsica than on the nearby mainland. The crested tit (*P. cristatus*) is not present on Corsica (from Blondel and Aronson 1999)

Species	Low matorral	Medium matorral	High matorral	Low coppice	High coppice	Old oakwood
Mainland						
Great tit				2.2	3.1	3.2
Blue tit						11.5
Coal tit						0.2
Crested tit						1.8
Total mean				**2.2**	**3.1**	**16.7**
Corsica						
Great tit	1.6	1.7	2.5	3.6	2.6	4.7
Blue tit		0.5	0.2	3.3	7.9	14.2
Coal tit				1.2	2.1	4.1
Total mean	**1.6**	**2.2**	**2.7**	**8.1**	**12.6**	**23.0**

mainland has four. Comparable habitats have more tit species in Corsica than in the mainland, and the Corsican population densities of all tit species are greater.

Several other mechanisms have been suggested that may influence density patterns in addition to those mentioned (e.g. see MacArthur *et al.* 1972,

Wright 1980). It is hard, therefore, to generalize on the combination of factors involved: they are specific to particular species on particular islands. It is sufficient for present purposes to note that mainland and island species densities can vary substantially. The phenomena of ecological release and density inflation do not apply universally to all island populations, but there appears to be good evidence for their operation for many island lineages, providing selective environments that encourage genetic adjustments in the island populations. From first principles, we can anticipate that they will be particularly prevalent on remote islands at early stages of faunal and floral colonization. We go on to consider other niche shift trends in a later section.

7.3   Character displacement

Character displacement can, and frequently does, refer to size changes among island forms. What distinguishes this form of niche shift is the causal interpretation: that it is competition between two fairly similar varieties or species that brings about selection, in one or both, away from the region of resource overlap (Brown and Wilson 1956). Diamond *et al.* (1989, p. 675) define ecological character displacement as 'the effect of competition in causing two initially allopatric species to diverge from each other in some character upon attaining sympatry'. Darwin (1859) termed this 'divergence of character', and it has also been termed 'character coevolution' or 'coevolutionary divergence', although there is actually not one, but a suite of related theoretical ideas centred around this notion (Otte 1989).

Caribbean anoline lizards provide examples of character displacement, Schoener (1975) finding shifts in perch height and diameter consistent with competitive effects. Species of similar size were found to affect one another more than dissimilarly sized species, while larger species affected smaller ones more than the reverse (cf. Roughgarden and Pacala 1989). Furthermore, according to a distributional simulation using Monte Carlo analysis, the distribution of lizard species over microhabitat types on satellite islands of the Greater Antilles was also found to be consistent with competitive effects (Schoener 1988). However, the results of this analysis depended on the assumptions built into the null model—a general problem with null models in ecology and biogeography (Chapter 5).

There is also some evidence for competitive displacement in Hawaiian crickets. Otte (1989)

presents an array of observations, including data on song variation in *Laupala*, a genus of swordtail crickets. The songs were found to be hyperdispersed, the divergence in signals being greater between coexisting populations than between allopatric populations, suggesting that competitive interactions are important. Otte concluded that character displacement is probably extremely common in the Hawaiian crickets and, by extension, elsewhere, although it is difficult to demonstrate unequivocally. While the most convincing demonstrations of character displacement tend to be from systems involving few interacting species (cf. the anoline lizard studies), such effects may also be detectable, and certainly are relevant, where a somewhat larger network of interacting species is involved.

Given the difficulty in validating ideas of character displacement (Connell 1980), it would be ideal if an experiment were to be devised that controlled for other complicating factors. The principal problem is that character changes with time are generally not observed, but are merely inferred by comparison of the characteristics of populations in allopatry with those in sympatry. Such studies provide corroboration but not validation.

Diamond *et al.* (1989) have reported a natural experiment which comes close to solving this problem, in demonstrating divergence within populations in sympatry for less than three centuries. In the mid-seventeenth century, Long Island, off New Guinea, erupted violently and destructively, initiating a primary succession on both Long itself and the nearby islands of Tolokiwa and Crown. By analogy with the recolonization of Krakatau, it is reasonable to assume a delay of a few decades for re-vegetation to proceed to the point at which a forest bird assemblage would have developed. Thus the bird populations of the three islands can be taken to have been founded less than 300 years ago. The avifauna contains two different-sized species of honeyeaters of the genus *Myzomela*—*M. pammelaena* and *M. sclateri*. They are the only islands on which the two coexist. Allopatric populations on other islands can be identified as the probable sources for the founding populations of the Long group. Thus, the serendipitous 'experiment' has a maximum duration of about three centuries, for which two otherwise allopatric species have co-occurred. Furthermore, within the Long group, the two species occur together in all habitats and at all altitudes, often being found in the same flowering tree. They are thus sympatric even at the finest intraisland scale. Diamond *et al.* (1989) found that the Long populations are significantly more divergent from each other than are the allopatric source populations. The larger species, *M. pammelaena*, is bigger still on Long than in the source populations, and the smaller *M. sclateri* is even smaller. Comparing samples of the two species from allopatric populations, the weight ratio was found to lie between 1.24 and 1.43, but for the sympatric Long island populations the value was 1.52.

The 300-year time span may be unremarkable by analogy with similar niche shifts in other circumstances. Conant (1988) found that Laysan finches (*Telespyza cantans*) introduced to Pearl and Hermes Reef underwent measurable shifts in bill shape within 18 years, in adjusting to their local food supply. Similarly, significant change in diaspore size in *Lactuca muralis* in Barkley Sound took place within just five plant generations (see below; Cody and Overton 1996).

7.4 Sex on islands

There are many ways in which plants and animals reproduce, the simplest dichotomy being between sexual and asexual modes. What works best in the island laboratory? How does the possession of a particular breeding system affect the chances of colonization or speciation on islands? Does hybridization create new forms and thus engender diversification, or does it restrict the tendency to radiation by pulling back two diverging forms from the brink of irrevocable parting?

Dioecy and outcrossing

Plants exhibiting sexual reproduction may as a simple generalization be classified as monoecious (male and female flowers on the same plant), hermaphrodite (bisexual flowers), or dioecious (plants that, generally, bear either male or female flowers

throughout their lifespan). At least theoretically, it should be more difficult for dioecious species to colonize a remote island, given the apparent need to have both sexes present for success (Ehrendorfer 1979; Sohmer and Gustafson 1993). In practice, early comparison by Bawa (1980, 1982) suggested that islands might have relatively high proportions of dioecious species. He noted that 28% of the Hawaiian plant species are dioecious, compared to a figure of 9% for a well-specified tropical mainland flora from Panama. Since then, the Hawaiian figure has been revised downwards, to 15% for species that are strictly dioecious, or to 21% if a broader grouping of sexually dimorphic species is used. This is still the highest incidence of dioecy of any well-specified flora (Sakai *et al.* 1995a). However, it is by no means clear that Hawaii is representative: indeed dioecious species are reportedly scarce in both the Galápagos (McMullen 1987) and the Canaries, where only 2% of the endemic flora is dioecious (Francisco-Ortega *et al.* 2000).

The high proportion of dimorphic species in the Hawaiian endemic flora is intriguing. Sakai *et al.* (1995a) find that dimorphy is frequent in part because many colonists were already dioecious, but that autochthonous evolution of dioecy has occurred in at least 12 lineages, including several species-rich lineages. Interestingly, one-third of currently dioecious species derive from monomorphic colonists. Further study of other island floras will be needed to establish how representative these patterns are of other oceanic island floras.

Baker (1955) argued that hermaphroditic plants with self-compatible reproduction systems, able to establish (potentially) from a single founder individual, would have better chances of island colonization than dimorphic or obligated out-crossers. From the data available it appears that the floras of Hawaii (Carr *et al.* 1986), New Zealand (Webb and Kelly 1993) and the Galápagos (McMullen 1987) do have smaller proportions of self-incompatible taxa than nearby continents, consistent with Baker's rule. Moreover, in some species, such as *Fragaria chiloensis*, dioecious in America but hermaphroditic in Hawaii, or *Coprosma pumila*, dioecious in New Zealand but monoecious in Macquarie (Ehrendorfer 1979), there is evidence that 'continental' allogamy (out-crossing) has been substituted by insular autogamy (self-compatability). We might anticipate that any initial advantage in autogamy at the point of colonization should be of limited duration in relation to subsequent lineage development. Interestingly, 237 species of Canarian endemic plants (almost half the endemic flora) exhibit features (such as unisexual flowers, heteromorphic styles or self-incompatibilty) promoting out-breeding (Francisco-Ortega *et al.* 2000).

Patterns of dioecy and self-crossing vs. outcrossing in island floras currently remain poorly understood. As they became better quantified, such properties may shed interesting light on the evolutionary assemblages of island floras in terms of what they reveal about the importance of pollinator mutualisms (below), and may ultimately help us to understand why some lineages have radiated on islands while others are absent or have failed to radiate (Sakai *et al.* 1995*b*; Barrett 1996).

Loss of flower attractiveness

Island plants often display a loss of flower showiness, such that they tend to have small, actinomorphic (radially symmetrical), white or green non-showy flowers with simple bowl-shaped corollas, instead of the large, tubular, zygomorphic (bilaterally symmetrical), bright-coloured flowers frequently displayed on the continents (Carlquist 1974). This trend, is evidenced in a comparison of Australian and New Zealand floras, and probably reflects the generalistic and promiscuous nature of the island pollinators (Barrett 1996), as well as their smaller sizes when compared to the continental pollinators. Inoue and Kawahara (1996) were able to demonstrate a direct correlation between the flower size of *Campanula* species and of their pollinators. Flowers were larger on the mainland, where they were visited by larger pollinators (bumblebees and megachilid bees) lacking from the islands. Pollination in the island populations was carried out by smaller pollen-collecting halictid bees, which were shown experimentally to be indifferent to flower size, in contrast to the preference for larger flowers shown by the nectar-feeding megachilid and bumblebees on the mainlands.

Anemophily

Some island plants appear to have shifted from ancestral animal pollination to wind pollination (anemophily). Examples include *Rhetinodendron* in the Juan Fernández islands, *Plocama* and *Phyllis* in the Canaries, and *Coprosma* in New Zealand (Ehrendorfer 1979), and it has been suggested as a general island trend (see e.g. Carlquist 1974, Ehrendorfer 1979). Possible benefits include: (1) gaining independence from impoverished pollinator services (e.g. Goodwillie 1999; Traveset 2001), (2) windy conditions favouring anemophily, and (3) outcrossing fitness benefits yielded by wind pollination (Barrett 1996). However, as a definitive analysis of the prevalence of anemophily has yet to be carried out, it would be hasty to conclude in favour of these mechanisms. Preliminary data for the incidence of anemophily in the floras of well-studied archipelagos yield values varying from 20% for Hawaii (Sakai *et al.* 1995*a*,*b*), to 29% for New Zealand and 34% for the Juan Fernández islands (Ehrendorfer 1979). In contrast, McMullen (1987) stressed the paucity of wind pollination within the native flora of the Galápagos. Furthermore, anemophily may be linked to other important features, such as woodiness or dioecy, thus not of itself indicating selective advantage.

Parthenogenesis

Reproductive mode may also be tied up with colonization probabilities in certain animal taxa. According to Hanley *et al.* (1994), **parthenogenetic** (asexually reproducing) lizard species are relatively common on islands. They suggest that this may be because of: an enhanced colonization ability; the opportunity to flourish in a biotically less diverse environment (in which the relative genetic inflexibility that might be associated with parthenogenesis is not too disadvantageous); or the opportunity to escape from hybridization or competition with their sexual relatives. Their study examined the coexistence of parthenogenetic and sexual forms of the gecko *Lepidodactylus* on Takapoto atoll in French Polynesia. The two forms of *Lepidodactylus*, although closely related and hybridizing to a limited extent (through the activities of males of the sexual form), are considered separate species. Their coexistence appears to be stable in the short term at least, as the proportions of their two populations were similar in 1993 to a previous survey in 1986. Hanley *et al.* note that the asexuals are actually extremely heterozygous relative to the sexual population. This may seem odd at first glance, but apparently all parthenogenetic vertebrates with a known origin have arisen from a hybridization event, in large part accounting for their higher levels of heterozygosity. In this case, far from the asexual form suffering from competition with the sexual, it is the parthenogenetic form that has the wider distribution within the atoll (perhaps indicative of hybrid vigour), with different clones having overlapping but different habitat preferences, and with the sexual form more or less restricted to the lagoon beach. The coexistence of the two forms appears to be facilitated by the specialization of the *less* successful sexual taxon to beach habitats.

Hybridization

According to figures cited by Grant and Grant (1994), speciation through hybridization may account for as many as 40% of plant species (all allopolyploids arise in this way), but is considered to be much rarer in animals. However, cross-breeding in animals is not so rare, and there have been numerous studies of hybrid zones, wherein hybridization takes place between two recognized species. When successful, it provides a form of gene flow between them (cf. Kaneshiro 1995, above). Linking the themes of pollination and hybridization, Parrish *et al.* (2003) report evidence of hybridization of dioecious fig species on islands in the Sunda Strait region of Indonesia (including the Krakatau islands), which indicates a breakdown in pollinator specificity on the islands (as hypothesized by Janzen 1979), possibly linked to temporary shortages of the matching fig wasp pollinator species. Hybridization is also considered to be common within the Hawaiian flora. Indeed, even in the species-rich genus *Cyrtandra*, which was once thought not to hybridize much (if at all), recent

studies suggest it to be quite common (Sohmer and Gustafson 1993).

Grant and Grant (1994) studied the morphological consequences of hybridization in a group of three interbreeding species of Darwin's finches on Daphne Major (Galápagos) between 1976 and 1992. *Geospiza fortis* bred with *G. scandens* and *G. fuliginosa*. The hybrids in turn backcrossed with one of the recognized species. Interbreeding was always rare, occurring at an incidence of less than 5% of the populations, but nonetheless was sufficiently frequent to provide new additive genetic variance into the finch populations two to three orders of magnitude greater than that introduced by mutation. During the course of their studies Grant and Grant (1996*b*) noted a higher survival rate among hybrids following the exceptionally severe El Niño event of 1982–83. This event led to an enduring change in the habitat and plant composition of the island, which appears to have opened up the food niches exploited by the hybrids, and provided for most of the back-crossing observed during the study. Their work thus incidentally demonstrates that significant fluctuations in climate can have important ecological and evolutionary consequences.

7.5   Peculiarities of pollination and dispersal networks on islands

We have established that islands tend to be species poor and disharmonic, that island colonists go through founding events and subsequent expansion in numbers, and that evolutionary ecological mechanisms subsequently include density compensation, niche expansion, and niche shifts. To provide another example of density compensation or over-compensation, densities of anoline lizards as great as two individuals per square metre have been measured on some Caribbean islands (Bennett and Gorman 1979). Such high densities must, it is reasoned, lead to a greater relative level of intraspecific competition for resources than is usual. In such circumstances, we can expect to see selection for the exploitation of new food resources, such as nectar, pollen, fruits, or seeds. Here we review evidence for the distinctive ways in which island pollination and dispersal networks have developed.

We highlight two features. First, because of the impoverishment in species, island pollination and dispersal networks tend to feature endemic super-generalist species, both among the animals and plants (Olesen *et al.* 2002). Secondly, pollination and/or dispersal relationships can appear bizarre in relation to those found on continents, because the functions are carried out, in the absence of more common continental pollinators or dispersers, by groups of animals not frequently involved in such functions (Traveset 2001).

The emergence of endemic super-generalists

Traditionally, pollination studies focused on single plant species or, at most, on guilds of related plant species and the insects pollinating them (Traveset and Santamaría 2004). This reflects formerly dominant ideas that pollination mutualisms were the result of highly co-evolved interspecific relationships. More recent community-level work has shown that pollination processes are typically more generalized in nature than previously thought, with many insect and plant species interacting, and thus yielding weak pairwise relationships. However, the connectance level (the percentage of all possible animal–plant interactions within a network that are actually established) has been found to diminish in species-rich communities (Olesen and Jordano 2002). On islands, where species-poor communities prevail, the existence of pollination networks with a high degree of connectance is thus to be expected. Studies within the Azores, the Canaries, and Mauritius show connectance levels varying between 20 and 30% (Olesen *et al.* 2002). The exception to this range was provided by the Macaronesian laurel forests, a community dominated by palaeoendemic tree species, where the value obtained was 9%, indicating a higher level of specialization.

Recent work suggests that island networks often involve a few super-generalized animal pollinators serving many plant species, and a few super-generalized plant species that are visited by many pollinators (Olesen and Jordano 2002). *Bombus canariensis*, the endemic Canarian white-tailed bumblebee, provides us with one such example. It occurs from the coast up to 2000 m and is known to pollinate 88 plant species (Hohmann *et al.* 1993). In a laurel forest studied on La Gomera, *Bombus canariensis* was found to visit 48% of the plant species participating in the network. Similarly, the endemic Canarian bee *Anthophora allouardii* visits at least 46% of the plant species of a subdesert scrub community on Tenerife (Olesen *et al.* 2002). An even greater degree of generalism is displayed by an endemic *Halictus* bee in Flores (Azores) (60%), by the endemic day gecko *Phelsuma ornata* from Ile aux Aigrettes (Mauritius) (71%), and by the endemic bee *Xylocopa darwini* from the Galápagos (77%).

Among plants, the Macaronesian islands again provide examples of endemic plant species with generalized pollinator relationships. These include *Aeonium holochrysum*, a member of the coastal subdesert scrub, visited by 80% of the known community of pollinators, and *Azorina vidalii*, an Azorean endemic shrub, visited by 75% of the pollinator community. Conversely, *Lavandula buchii*, an endemic plant restricted to the old Teno and Anaga massifs of Tenerife, is pollinated by only 10 insect species (Delgado 2000), in comparison with its continental congener *Lavandula latifolia* from Southern Spain, which is pollinated by 80 species (Herrera 1989). This illustrates that a more comprehensive comparative analysis is needed to establish if the apparent tendency towards generalist species is a general feature of remote islands.

It has been suggested that one reason for the success of many introduced species on islands is the formation of 'invader complexes', whereby introduced species preferentially establish interactions with other introduced species (D'Antonio and Dudley 1993). However, Olesen *et al.* (2002) find little support for the invader complex concept in pollination networks analysed on the Azores and Mauritius. Rather, endemic supergeneralist species tend to include newcomers in their networks, either in the form of endemic insects pollinating exotic plants, or by endemic plants being pollinated by exotic insects. In these cases, the endemic super-generalists may be pivotal to the successful establishment of alien species that have left behind their continental networks (Olesen *et al.* 2002).

Unusual pollinators

The absence of many common continental pollinators (bees, wasps, humming-birds, etc.) or dispersers (some guilds of land-bird, non-volant mammals) from remote islands has provided opportunities for animals that do not usually perform these functions to acquire or extend these roles. The best illustration of this island feature comes from recent work on lizards, which have adopted roles as pollinators almost exclusively on islands, including New Zealand, Mauritius, the Balearic islands, and Tasmania. Olesen and Valido (2003) report that of 37 lizard species now known to act as pollinators, only two species are from mainland areas, both in fact peninsulas (Baja California and Florida).

The first published observations revealing lizards as pollinators derive from Madeira in the 1970s, when *Teira dugesii* was observed drinking nectar and licking the stigmatic lobes for pollen on both native and introduced ornamental plant species (Elvers 1977). Several species of Canarian endemic lizards (*Gallotia* spp.) are now known to feed on nectar of *Euphorbia, Echium,* and *Isoplexis.* Similarly, on the Balearic islands, *Podarcis lilfordi,* an endemic lizard, is known to visit some 20 plant species, including the flowers of the native *Euphorbia dendroides,* for its highly concentrated nectar (Traveset and Sáez 1997). Of special interest is the role of *Hoplodactylus* geckos in New Zealand and of *Phelsuma* day geckos as pollinators in Indian Ocean islands. These geckos are known to visit many plant species (Whitaker 1987; Olesen *et al.* 2002), transporting pollen over considerable distances (Nyhagen *et al.* 2001). The introduction of exotic predators (e.g. rats, mongooses, monkeys) has the potential to disrupt these pollinator services, potentially leading to extinction of native plant species.

Unusual dispersal agents

Again it appears that lizards have a disposition to act as dispersal agents on islands, a role performed less extensively by mainland lizard species. Of some 200 lizard species reported to have a role in seed dispersal, two thirds were island species (Olesen and Valido 2003). Examples include 18 New Zealand lizard species, among them the gecko *Hoplodactylus maculatus* and the skinks *Oligosoma grande* and *Cyclodina alani* (Whitaker 1987). On the Balearic islands, *Podarcis lilfordi* and *P. pityusensis* are known to feed on the fruits of more than 25 species (Pérez-Mellado and Traveset 1999). All the endemic Canarian lacertids (seven *Gallotia* species) and skinks (three *Chalcides* species) are known to feed on fruit. *Gallotia* lizards eat fruits of at least 40 native plant species, which is slightly more than the half of the Canarian plant species bearing fleshy fruits, and also feed on the fruits of at least 11 introduced species (Valido and Nogales 1994).

Another 'unusual' plant dispersal agent comes in the form of the giant tortoises (*Geochelone* spp.) of the Galápagos (Racine and Downhower 1974) and Aldabra atoll (Hnatiuk 1978). Today, the Aldabran species (*Geochelone gigantea*) has been introduced on several islets off Mauritius where goats and rabbits have recently been eradicated, as a functional substitute for extinct Mauritian tortoises (*Geochelone triserrata* and *G. inepta*), which are considered to have been the original dispersers of the endemic Ile aux Aigrettes ebony (*Diospyros egrettarum*). After a pilot study in a fenced enclosure, viable fruits of the ebony were found dispersed in tortoise faeces away from the parent trees (Zavaleta *et al.* 2001).

7.6   Niche shifts and syndromes

Thus far in this chapter we have considered some of the filters that might prevent lineages successfully establishing on islands and the implications of small population sizes and of the possession of particular life-cycle traits connected to breeding systems. We have also begun, inevitably, to consider emergent patterns of changing biological and niche characteristics. Here, we continue this process under the heading of some of the more prominent island syndromes, starting with the loss of dispersal powers.

The loss of dispersal powers

Given that oceanic islands are so hard to reach, it may appear paradoxical that so many flightless forms occur on them. For a distant island popping out of the sea for the first time, there will inevitably be a strong bias towards groups that are highly dispersive, as these will, by definition, have a greater chance of colonizing. This was reflected in the biogeographical patterns discussed in Chapter 3, of filter effects, the gradual loss of taxa with increasing distance from continental source areas, and of disharmony. The relevance of these effects is broad, and wide ranging, from the evolutionary opportunities thus presented on islands like Hawaii, to the ecological structuring in less remote islands such as Krakatau (Chapter 5). The island biogeography of remote islands thus bears the imprint of dispersal filtering of faunas and floras as its first organizational principle.

Once on an isolated island, however, there may be a strong selective force against dispersal ability, particularly in insects and birds. Examples given by Williamson (1981) of flightless forms include the 20 endemic species of beetle on Tristan de Cuhna, all but two of which have reduced wings. A number of ideas have emerged in explaining such tendencies. One put forward by Darwin (letter to J. D. Hooker, 7 March 1855; Roff 1991) is that highly dispersive propagules are more liable to be lost from the gene pool, for instance by a highly dispersive plant or insect being blown off to sea, so selecting for less-dispersive forms among the island population. A second idea, applicable to flightless birds such as the famously extinct dodos, is that the lack of predators removes the selective advantage of flight in ground-feeding species. Indeed, flight might be disadvantageous in that the energy used in maintaining and using flight muscles is effectively wasted. However, flightless species within generally volant groups can also be found on continents, so before attempting to test ideas concerning loss of dispersal, it might be prudent to confirm that the incidence of flightlessness on islands is higher than expected by chance alone.

Roff (1991, 1994) has undertaken just such analyses for insects and for birds. Darwin's original

observations on insects were based, of course, on relatively few data, and in his analysis of the much larger data sets now available Roff (1991) found that statistically there is no evidence of an association between oceanic islands and insect flightlessness. He argues that the large size of most oceanic islands makes flightlessness of little positive selective value: only a small fraction of the population is likely to be lost from such sizeable land masses (see also Ashmole and Ashmole 1988). It should be stressed, however, that Roff's analyses are based on a dichotomy between volant and non-volant forms. He does not take into account island forms that have evolved a reduced flight capability or behaviour without actually losing the ability altogether.

In his second article, on birds, Roff (1994) points out two important factors that complicate the analysis. First, in the case of most bird groups, the small sample size limits the potential for formal statistical analysis. Secondly, the phylogenetic constraints on flightlessness must be considered. The condition may have evolved once, followed by subsequent speciation within a flightless lineage, or it may have evolved several times, in extreme cases once for each flightless species. For example, among important flightless bird groups, the 18 known penguin species are considered to be monophyletic, having speciated from a common flightless ancestor, whereas the 41 ratite species (ostriches, etc.) represent between one and five evolutionary transitions to flightless forms, depending on whose phylogeny one accepts. In contrast, flightlessness in rails has most probably evolved separately for each of the 17 (or more) flightless species, among the 122 (or more) recognized rail species. Truly flightless rails are only found on islands and they represent about half of the island rail species (Roff 1994). More ratites occur on islands than on continents, but the taxonomic distribution does not indicate that flightlessness is more likely to evolve on islands in this group. Examples of flightlessness from other groups of birds include the following island forms: the New Caledonian kagu (*Rhynochetos jubatus*), the two extinct dodos (*Raphus cucullatus* in Mauritius and *Pezophas solitaria* in Rodrigues), New Zealand's kakapo (*Strigops habroptilus*, the only flightless parrot), the Auckland Islands teal (*Anas auck-landica*), the Galápagos cormorant (*Phalacrocorax harrisi*), the three Hawaiian ibises (*Apteribis*), and four Hawaiian ducks (*Anas*); these 13 species representing at least eight evolutionary transitions. Taking all these factors into account, Roff concluded that there is an insular tendency to flightlessness, although this could only be formally shown for the rail group. A separate analysis by McCall (1998) records that flightlessness has evolved independently in at least eleven extant bird families, eight of which currently contain both volant and non-volant species. They noted that the volant species in these families tend to be characterized by relatively short wings. As short-winged species pay a relatively high energetic cost to fly, their findings support the hypothesis that the energetic costs of flight provides a key factor in selecting for flightlessness in these families.

On the topic of flightless rails, Diamond (1991*a*) reported a newly discovered species, *Gallirallus rovianae*, which is believed to be either flightless or weak-flying. It was collected in 1977 from New Georgia, Solomon Islands. The ancestral form is volant and more boldly patterned than the new species. Diamond regards *G. rovianae* and its relatives on other oceanic islands as cases of convergent evolution, whereby 11 groups of rails have independently evolved weak-flying or flightless forms and also, in several cases, reduction in boldly patterned plumage, on widely scattered oceanic islands. These weak- or non-flying rails do not fare well in contact with people, and as many as half of those extant at the time of European discovery have subsequently become extinct. Moreover, subfossil remains of other extinct species have been found on most of those Pacific islands that have been explored by palaeontologists, and it is likely that there are many more extinct forms awaiting discovery (Steadman 1997*a*). There is thus clearly a strong selection pressure for reduced flying ability in rails, and this appears to be related to the energetic and weight burden of unnecessary flight muscle on islands lacking terrestrial mammalian predators.

From the plant world, the genus *Fitchia* (in the Asteraceae), endemic to the Polynesian islands of the south Pacific, exemplifies reduction in dispersability (Carlquist 1974; Williamson 1981). Its closest continental relatives are herbs with spiked fruits transported exozoically (i.e. by external attachment to an animal), but in *Fitchia* the spikes are relict, the fruits are much larger, the plants are trees, and their fruits drop passively to the forest floor. Endemic Pacific island species of *Bidens* (also in the Asteraceae) illustrate various steps in the loss of barbs and hairs and other features favouring dispersiveness, these changes being accompanied by ecological shifts from coast to interior and, in some cases, wet upland forest habitats (Carlquist 1974; Ehrendorfer 1979). It might be argued that these examples suffer from a certain circularity of reasoning, with traits being used to infer the reasons for their selection. To allay such concerns, a demonstration is needed of such changes actually occurring. Once more, the Asteraceae provide the evidence.

Cody and Overton (1996) monitored populations of *Hypochaeris radicata* and *Lactuca muralis*, both wind-dispersed members of the Asteraceae, on 200 near-shore islands in Barkley Sound (Canada). The islands ranged from a few square metres in area up to about 1 km^2. Significant shifts in diaspore morphology occurred within 8–10 years of population establishment, the equivalent in these largely biennial plants of no more than five generations. The diaspore in these species consists of two parts, a tiny seed with a covering (the achene), which is surrounded by or connected to a much larger ball of fluff (the pappus). The clearest findings were for *Lactuca muralis*, the species with the largest sample size. Founding individuals had significantly smaller seeds (by about 15%) than typical of mainland populations, illustrative of a form of a nonrandom founder effect termed **immigrant selection** (Brown and Lomolino 1998, p. 436). Seed sizes then returned to mainland values after about 8 years, but pappus volumes decreased below mainland values within about 6 years. If the pappus is thought of as a parachute and the seed the payload, then the ratio between the two indicates the dispersability of the diaspore (Diamond 1996). Figure 7.2 provides a schematic representation of the trends in dispersability. Cody and Overton were able to quantify these phenomena because

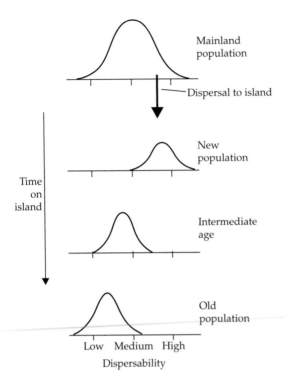

Figure 7.2 A schematic diagram of the evolution of reduced dispersability of wind-dispersed plants on islands, as suggested by the changes recorded in *Lactuca muralis* (Asteraceae) on near-shore islands in British Columbia, Canada. Maximal dispersability characterizes the youngest island populations, as the founding event(s) will be from the upper end of the dispersal range of the mainland population. Thereafter selection acts to decrease dispersability of propagules on islands as more dispersive diaspores are lost to sea. (Simplified from Cody and Overton 1996, Fig. 1.).

they were working on a micro-scale version of the oceanic island examples usually cited, i.e. their populations were very small, occupying tiny areas, and their plants were short-lived—ideal for monitoring short-term evolutionary change.

The development of woodiness in herbaceous plant lineages

Several plant families with predominantly herbaceous species in the continents are characterized by having woody relatives on islands. Outstanding examples include the silversword alliance and lobelioids in Hawaii, tree lettuces in Juan Fernández and Macaronesia, tree sunflowers in

St Helena and Galápagos and buglosses and daisies in Macaronesia (Givnish 1998).

Whether island woodiness is a derived or basal feature has long been debated. Advocates for the former view include both Darwin (1859) and Carlquist (1974, 1995). Those arguing that it is a basal, relictual characteristic include Bramwell (1972), Sunding (1979), and Cronk (1992). There are cases of woody island plants that are considered relictual, for instance the Atlantic island laurel-forest tree species discussed in Chapter 3, but these species are unexceptional in being woody forms. For predominantly herbaceous taxa represented on islands by woody forms, analyses of molecular phylogenies all point to woodiness being a derived characteristic, having developed from originally herbaceous colonists (Kim *et al.* 1996; Panero *et al.* 1999; Emerson 2002).

Hawaii provides good examples. Most colonists were weedy, herbaceous, or shrubby, with only a minority of tree-like plants: subsequently, woodiness has developed in the insular lineages (Carlquist 1995). The Amaranthaceae and Chenopodiaceae provide classic cases, in which the endemic species are only woody (or in cases suffrutescent—woody only at the base of the stem), whereas in continental source areas they are principally herbaceous (Carlquist 1974; Sohmer and Gustafson 1993; Wagner and Funk 1995). For the Atlantic islands, analyses of nuclear rDNA of woody Macaronesian *Sonchus* and five related genera have likewise shown them to have been derived from continental herbaceous ancestors, most probably from a single founding species. During the adaptive radiation of the lineage, it has undertaken a limited number of interisland transfers, has diversified in to differing habitats, and has developed considerable morphological diversification, including the development of the woody lineage (Kim *et al.* 1996). Similarly, the Macaronesian clade of *Echium* (Boraginaceae), consisting of 29 endemic species (2 on Madeira, 24 on the Canaries, and 3 on the Cape Verde islands), of which 26 exhibit woodiness, derive from a single herbaceous continental ancestor (Fig. 7.3; Böhle *et al.* 1996). The acquisition of woodiness

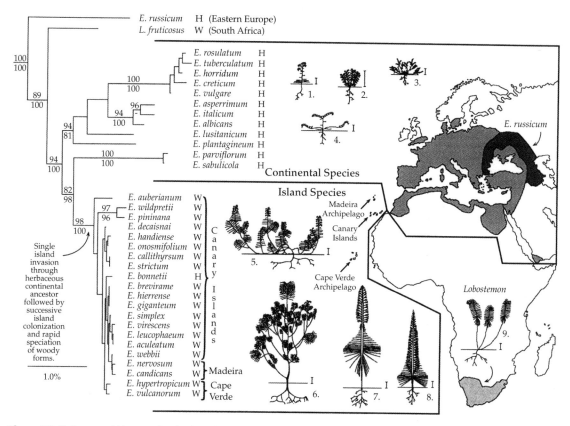

Figure 7.3 Phylogeny and biogeography of *Echium* showing that the woody species of Macaronesia are part of a monophyletic archipelagic radiation (source: Böhle *et al.* 1996). Altogether there are 29 endemic species of *Echium* in Macaronesia. The numbers refer to *Echium* species that exemplify a growth form or habit: (1) *Echium plantagineum*, (2) *E. humile*, (3) *E. rosulatum*, (4) *E. parviflorum*, (5) *E. onosmifolium*, (6) *E. decaisnei*, (7) *E. simplex*, (8) *E. wildpretii*, (9) *Lobostemon regulareflorus*. The technical details of the phylogeny are as follows. It was based on analyses of 70 kb of non-coding DNA determined from both chloroplasts and nuclear genomes. Numbers above branches indicate the number of times the branch was found in 100 bootstrap neighbour-joining replicates, using the Kimura distance, whereas the numbers below branches indicate the number of times the branch was found in 100 bootstrap parsimony replicates.

from herbaceous ancestors has been repeated several times in some lineages, as in the Macaronesian endemic genus *Pericallis* (Panero *et al.* 1999).

Several reasons have been suggested for the development of lignification on islands. As reviewed by Givnish (1998), they include the following.

1 The **competition hypothesis** (Darwin 1859). Herbaceous colonists would obtain a competitive advantage by growing taller and developing into shrubs/trees.

2 The **longevity hypothesis** (Wallace 1878). Woodiness might allow extended lifespans, increasing the chance of sexual reproduction where pollinators are scarce.

3 The **moderate insular climate hypothesis** (Carlquist 1974). In comparison with the closest continents, islands typically have milder (and moist) climates, promoting the development of herbs to tree-like growth forms.

4 The **promotion of sexual out-crossing hypothesis** (Böhle *et al.* 1996). Evolution of the woody habit accompanies increased longevity, which provides greater likelihood of cross-pollination in environments such as islands where pollination services may be unreliable. This hypothesis is an extension of hypothesis (2).

5 The **taxon cycling hypothesis** (Givnish 1998). Successful island colonizers are usually weedy species that established themselves easily in open/marginal habitats, where they can grow rapidly. Increased specialization and selection in competitive environments, with scarce resources, may have precipitated a shift toward persistence through woodiness (Givnish 1998, Panero *et al.* 1999). This hypothesis is an extension of hypothesis (1).

It is too early to conclude which of the above might provide the best explanation, if indeed there is a single answer.

Size shifts in island species and the island rule

It presents another apparent paradox that two conflicting traits are viewed as common in island forms: **gigantism**, examples of which include the Galápagos and Indian Ocean tortoises (Darwin 1845; Arnold 1979), and **nanism** (dwarf forms), illustrated by island subspecies of ducks, which are commonly smaller than mainland species (Lack 1970). Of course, size variations within taxa are not restricted to islands, and so there is the danger of inferring a general rule from a few striking cases. For instance, the famous Aldabran and Galápagos giant tortoises (*Geochelone* spp.) reach up to 130 cm in length. However, giant tortoise are readily eliminated by humans, and other giant *Geochelone* occur, or occurred, in South America, Madagascar, the Seychelles, the Mascarenes, and on the African mainland. It is not certain whether large size evolved after reaching islands or whether it merely bestowed the capabilities of dispersing to islands as remote as the Galápagos (Arnold 1979).

Unusual size has been reported for many island taxa, including plants and insects, but it is amongst vertebrates that the most striking general patterns are found, for there appears to be a graded trend from gigantism in small species to dwarfism in larger species. This has been termed the **island rule** (van Valen 1973, Lomolino 1985, 2005), and has been found to hold for mammalian orders, non-volant mammals, bats, passerine birds, snakes and turtles (Fig 7.4, Box 7.1). How well particular taxa fit this island rule depends, as ever, on the assumptions involved in the analysis (e.g. compare Lomolino 2005 and Meiri *et al.* 2006), but the synthesis provided by Lomolino (2005) provides both a compelling body of evidence and a persuasive interpretation of apparent exceptions. Analyses of data for continental mammal species have suggested that there is an optimum body size for energy acquisition at about 1 kg body weight (Damuth 1993). Linking this observation to the island rule, it would seem that islands favour the evolution of a body size most advantageous for exploiting resources in particular dietary categories, and that this must in turn be linked in some way to the constrained area of islands, and the absence of the full array of competitors and predators found within a continent.

The body size at which equivalence in body size between island and continental populations is reached varies between taxa. Lomolino (2005) interprets these values as indicating the optimal size for a particular body architectural plan (bauplan) and ecological resource exploitation strategy (Fig. 7.5). Whereas in species-rich mainland communities, interspecific interactions (especially predation and competition) facilitate evolutionary divergence from the optimal size, on islands selection favours change towards the optimum body size. For instance, islands often lack terrestrial vertebrate predators, but do have birds of prey. This switch in predation regime might favour larger rodents, which would then become dominant in the population through intraspecific competitive advantage (Reyment 1983, Adler and Levins 1994). At the other end of the scale, carnivores (Box 7.1) and artiodactyls (even-toed ruminants) typically become smaller, suggesting small size to be advantageous in restricted insular environments of limited food resources (Lomolino 1985).

There is considerable variability around the trend line (Fig 7.4), some of which may be explained by variables such as island isolation and area. For example, body size of the tricoloured squirrel (*Callosciurus prevosti*) decreases with increasing isolation, and increases with island area (Heaney 1978). Some size changes can only be understood

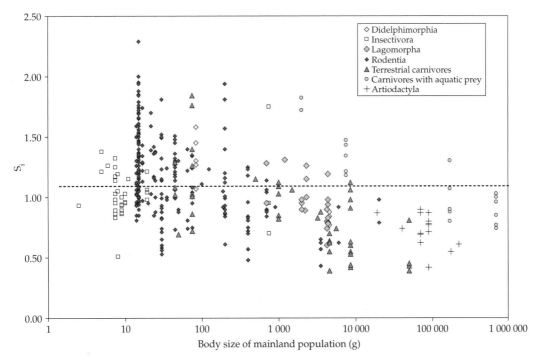

Figure 7.4 Demonstration of the island rule: body size trends for populations of insular mammals. S_i = relative size of insular forms expressed as a proportion of body mass of their mainland relative. (Redrawn from Fig. 1 of Lomolino 2005).

Box 7.1 Dwarf crocodiles and hominids

The dwarf crocodiles of Tagant (Mauritania)
Seventy years after the last reports of their existence and six years after the IUCN listed them as extirpated, relict populations of the Nile crocodile (*Crocodylus niloticus*) have been rediscovered in wetlands in southern Mauritania (Shine *et al.* 2001). These populations appear to have been isolated by climatic change associated with the Pleistocene/Holocene transition, and have persisted in association with seasonal ponds, known as gueltas. In summer, these gueltas can be smaller than 1 ha. Today it is thought that there may be small populations of these dwarf crocodiles in some 30 or so of these Mauritanian wetlands, feeding mainly on the guelta fishes, in competition with human settlers. In these small 'islands' the species has undergone dwarfism, resulting in adults of no more than 2–3 m body length, just half that of adult Nile crocodiles.

Dwarfism in human islanders?
Considerable press interest followed the discovery of the remains of six individuals belonging to what has been claimed to be an extinct species of human (*Homo floresiensis*) from Liang Bua Cave in the island of Flores, Indonesia (Brown *et al.* 2004; Morwood *et al.* 2004). The adults were only 1 m tall, weighing just 25 kg and with brains smaller than one third that of modern humans. They are thought to have reached Flores c.50 000 years BP, persisting alongside modern humans until finally becoming extinct about 18 000 years BP.

Where did they come from? Although *Homo erectus* were considered to have reached Java as early as 1.5 Ma, they were until fairly recently thought not to have crossed Wallace's line (the 25 km gap of open ocean dividing Bali from Lombok). Finds of 0.8 Ma stone tools in Flores might indicate that they did reach Flores, although no bones have yet been found. Thus *H. floresiensis* may have evolved from ancient *H. erectus*. Human dwarfism due to insularity has not previously been reported, but this change in body size is consistent with the island rule, as discussed in the text.

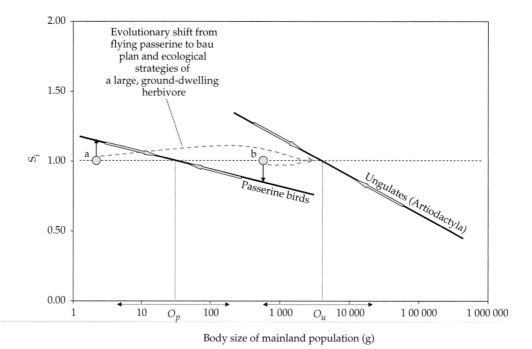

Body size of mainland population (g)

Figure 7.5 Lomolino's (2005, Fig. 6) graphic of the island rule shows the rule for two different bauplans and a possible transition between them. In essence, small forms tend to get larger (arrow at point a) and larger forms get smaller (arrow at point b) but following different lines, matching separate optima, dependent on the fundamental body plan and ecological role. Most size changes are gradual but occasionally major evolutionary transformations occur, involving shifts in bauplan and ecological role in response to the distinctive island opportunities and selective forces, e.g. the flight loss and gigantism of the now extinct moa of New Zealand. An evolutionary transition from a passerine bird bauplan to that typical of a large ungulate (large non-volant herbivore) would involve a transition from the left-hand line, with optimum O_p (passerine) to the right-hand line, with optimum O_u (ungulate).

within Lomolino's model as a result of a shift to another bauplan, i.e. to another niche (implying both size and function change). As an example, New Zealand moas were giant flightless birds that, in the absence of terrestrial mammals, essentially acquired the bauplan of ungulates (Fig. 7.5; Lomolino 2005).

It is also possible that there might be a form of non-random founding effect (termed immigrant selection by Brown and Lomolino 1998) involved in some size shifts. For instance, in active dispersal events, the founder group might comprise significantly larger individuals from within the mainland population, better able to survive a long-distance journey to an island. Or, conversely, significantly smaller plant propagules may be better predisposed to arrive on a new island if a passive dispersal event has to occur (e.g. Cody and Overton 1996,

above). However, it is hard to see how immigrant selection would assist in generating the size reductions in larger mammals. Support instead for the overwhelming role of selection comes from a recent study of anthropogenically fragmented forest in Denmark, where body size changes consistent with the island rule were reported within just 175 years (Schmidt and Jensen 2003, cited in Lomolino 2005).

Fossil and subfossil remains provide further exemplification of the island rule. For instance, the Mediterranean islands once featured an endemic fauna of dwarf hippopotami, elephants and deer, and giant rodents (the so called HEDR fauna), which survived from the Pliocene to the sub-Recent period (Reyment 1983). Schüle (1993) argued that for animals like elephants, which start large, islands provided insufficient resources to sustain such large body weights. Even infrequent scarcity could provide strong selection pressure towards more agile, smaller forms. Some of the best evidence for nanism seems to be for fossils of the Cretan form of dwarf elephant, which show a wide size range, interpreted as representing stages in the development of the dwarf form. 'The flatland, lumbering ancestral form, with good swimming capabilities, had been irrevocably transformed into a clever climber of rocky slopes' (Reyment 1983, p. 302). The prize for the greatest degree of dwarfing goes to the Maltan *Elephas falconeri*, which was barely more than a metre in height (Lister 1993) and only a quarter the size of the putative ancestor, *E. namadicus*.

These remarkable decreases in size may have occurred quite quickly. Vartanyan *et al.* (1993) reported remains of the woolly mammoth, which appear to have persisted on Wrangel island, 200 km off the coast of Siberia, when the island became separated from the mainland c.12 000 BP. The mammoth became extinct from its continental range by around 9500 BP (either because of climate change, or humans, or a combination of both), but it persisted on Wrangel until some point between 7000 and 4000 BP. Not only did the population thus attain relictual status, but between 12 000 and 7000 BP it appeared (on the basis of fossil teeth) to shrink in body size by at least 30%. However, size changes need not be so rapid, nor follow so soon after colonization. Sicilian elephants probably took between 250 000 and 600 000 years to shrink to about half the dimensions of the ancestral *Elephas antiquus* and about 1 million years to reach one quarter of the ancestral size.

All Pleistocene ungulates, other mammals of any considerable size, flightless birds, and giant tortoises on the Mediterranean islands were almost certainly wiped out by the first humans to arrive. This is, however, only recognizable in the archaeological record on Cyprus and Sardinia. It appears that the extant wild ungulates on the Mediterranean islands were all introduced by humans.

Although the island rule now appears to be clearly established, there is evidence of Holocene size reductions in lizards from small islands in the Caribbean (Pregill 1986). In particular cases, the evidence is strongly supportive of some form of

co-evolutionary competitive pressure brought about by formerly allopatric taxa coming into sympatry (cf. Roughgarden and Pacala 1989). In other cases, the activities of humans, such as habitat alteration and the introduction of terrestrial predators, may be responsible not only for the extinction of larger lizards from within island assemblages but also for size reductions (Pregill 1986). In illustration, in the Virgin Islands, populations of the St Croix ground lizard (*Amereiva polyps*) have declined steeply over the past century, and body size of mature adults has also decreased. The ecology of these islands has been greatly affected by humans, and in particular, by the introduction of the mongoose (Pregill 1986). Associations between human colonization and the extinctions of larger species from within insular lizard assemblages are also known from the fossil records of the Canaries, Mascarenes, and Galápagos archipelagos. Indeed, Pregill (1986) concludes that insular lizards that draw our attention because of their exceptional size may have been rather ordinary in prehistory. Fossils of extinct large frogs and snakes are also known from the Caribbean. Humans have also selectively eliminated larger vertebrates across the globe during the late Quaternary: so this form of filtering is far from being just an island pattern.

Food chain links may be important in selecting for both smaller and larger forms. An interesting example is Forsman's (1991) study of variation in head size in adder *Vipera berus* populations on islands in the Baltic Sea. In accord with the general trend for smaller vertebrates noted above, it was found that relative head length of adders was smallest in the mainland population. An interesting further feature of this study was that, within the island set, relative head length increased on the islands with increasing body size of the main prey, the field vole, *Microtus agrestis*. This was interpreted as the outcome of stabilizing selection for head size within each population, i.e. an evolutionary response to the variation in body size of the main prey species and the small number of alternative prey species available on islands. Schwaner and Sarre (1988) also linked large body size in a snake species to food resources. In this case, the Australian tiger snakes (*Notechis ater serventyi*) on

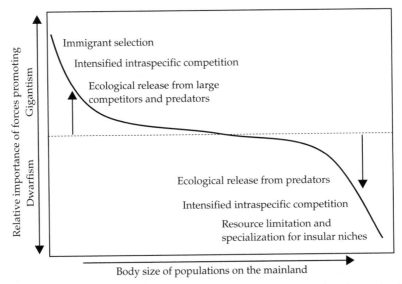

Figure 7.6 Factors hypothesized to be important in generating the island rule in island vertebrates and how they vary in relation to body size. (Redrawn from Fig. 9 of Lomolino 2005.)

Chappell Island in the Bass Strait eat an abundant, high-quality resource, which requires little effort to locate, but which is highly seasonal: chicks of mutton birds. Here, larger body size might enable greater fat storage, thus enhancing survival of individuals through long periods of fasting.

The size shifts and associated morphological changes, like other 'micro-evolutionary' processes/ patterns reviewed in the present chapter, sometimes lead to island populations being recognized as subspecies, or full species, but do not necessarily do so. The island rule is, as Lomolino (2005) writes, an emergent pattern, and although numerous particular mechanisms might be involved, varying from taxon to taxon, the generality of the pattern (occurring in many taxa, all over the globe) suggests that a relatively powerful underlying selective force must apply, and that this in turn must relate to island disharmony. Of the various factors discussed, those Lomolino holds to be most important are summarized in Figure 7.6.

Changes in fecundity and behaviour

Vertebrates colonizing islands also frequently display changes in behaviour (Sorensen 1977; Stamps and Buechner 1985). For birds we have already discussed examples of changes in feeding niche, loss of flight, loss of defensive instincts on remote islands, and a general tendency to follow the island rule. To provide an example of a behavioural change, Lack observed that birds on the Orkney isles select a wider range of nesting sites than on the British mainland (Table 7.2). Lack (1947b) also noted that island birds frequently have reduced fecundity, expressed as smaller clutch sizes than their continental relatives. The same trend has been reported for island lizards and mammals (Stamps and Buechner 1985), often combined with the production of larger offspring. Further trends that have been claimed for insular animals include a diminution of sexual dimorphism, a tendency towards melanism, and the relaxation of territoriality (Stamps and Buechner 1985; Wiggins et al. 1998).

Small clutch sizes

Bird clutch sizes are known (1) to increase with latitude, and (2) to decrease on islands when compared with continental areas at the same latitude (Lack 1947b; Ricklefs 1980; Blondel 1985). Furthermore, clutch sizes decrease with decreasing island area and with increasing isolation within the same archipelago (Higuchi 1976), and at least for some related bird species, clutch size also decreases

Table 7.2 Alterations in nesting sites in some Orkney birds first noted by David Lack (after Williamson 1981, Table 6.4). In all cases the normal mode is also found in Orkney

Species	Normal mode	Orkney mode
Fulmarus glacialis (fulmar)	On cliffs	On flat ground and sand dunes
Columba palumbus (woodpigeon)	In trees	In heather
Turdus philomelos (song thrush)	In bushes and trees	In walls and ditches
Turdus merula (blackbird)	Woods and bushy places	Rocky and wet moorland
Anthus spinoletta (rock pipit)	Sea cliffs	Out of sight of sea
Carduelis cannabina (linnet)	Bushes and scrub	Cultivated land without bushes; reedy marshes

with increasing endemism level of the island focal species. For instance, the Canarian endemic blue chaffinch (*Fringilla teydea*) has a mean clutch size of 2.0 eggs (Martín and Lorenzo 2001), the endemic subspecies of chaffinch *Fringilla coelebs canariensis* a mean clutch size of 2.6 (Martín and Lorenzo 2001), whereas the northwest African subspecies of chaffinch (*F. c. africana*) shows a value of 3.9 (Cramp and Perrins 1994). Canarian endemic laurel-forest pigeons (*Columba bollii* and *C. junoniae*) lay one egg, instead of the two eggs laid by the native but non-endemic rock dove (*C. livia*). The Canarian kestrel (*Falco tinnunculus canariensis*) has a mean clutch size of 4.41 eggs, whereas Moroccan kestrels, at the same latitude, have an average clutch size of 4.80.

Several hypotheses, more or less interconnected, have been suggested in explanation of the decrease in fecundity of island birds.

• **The energy reallocation (Cody) hypothesis.** This is a refinement of David Lack's hypothesis on the evolutionary significance of clutch size, and has been especially advocated by Cody (e.g. 1971). It posits that greater environmental predictability on islands will lower mortality, resulting in lower population fluctuations than in mainland regions. This might result in selection for smaller clutch size, with reinvestment of the saved energy in other components of fitness, such as better quality of young or increased longevity for parents through better foraging efficiency, predator avoidance or competitive ability (Blondel 1985).

• **The resource predictability (Ashmole) hypothesis.** First postulated by Ashmole (1963) and supported by Ricklefs (1980), this hypothesis predicts

that clutch size is determined by the differences in food resources between the non-breeding and the breeding season, increasing the clutch size in direct proportion to seasonality. In stable island environments, where seasonality is buffered by the ocean, there is less seasonal fluctuation in food supply, thus favouring a diminution of the clutch size.

• **The density compensation hypothesis.** Island isolation precludes the arrival of many mainland competitors, which drives density compensation in the island species (MacArthur *et al.* 1972), resulting in higher population sizes and thus in higher intraspecific competitiveness for resources, favouring higher investment in fewer young (Krebs 1970).

• **The reduced predation and parasitism hypothesis.** Island isolation prevents many groups of continental predators and parasites arriving, thus the insular bird populations have higher survival rates, and competitiveness, which results in a diminution in clutch sizes.

Relaxation in territoriality
Insular lizards, birds, and mammals often exhibit reduced situation-specific aggression toward conspecifics. This relaxation in aggressive behaviour can be expressed in the form of: (1) reduced territory sizes, (2) increased territorial overlap with neighbours, (3) acceptance of subordinates on the territory, (4) reduced aggressiveness, or (5) abandonment of territorial defence (Stamps and Buechner 1985). These changes are often associated with unusually high densities, niche expansion, low fecundity and the production of few, competitive offspring. Two main non-exclusive hypotheses

have been suggested to explain this relaxation in territoriality:

- **Resource hypothesis**. This hypothesis suggests that territorial behaviour is primarily adjusted to resource densities, and that resources on islands are, because of the lack of competitors resulting from isolation, more abundant than in the mainland. More resources will result in reduced territory sizes, increased territorial overlap, and changes in the vigour of territorial defence on islands (Stamps and Buechner 1985).

- **Defence hypothesis**. In addition to any effects of resources, the buffer effect exerted by the ocean on island weather will result in milder climates and this, together with the absence of predators, will result in the survival of more conspecific individuals without territory. These 'have-nots' will intrude and contend with those that own territories. This will elevate the costs of defence for owners of insular territories, selecting for reduced territories, increased territorial overlap, and acceptance of subordinates. If defence costs became exaggerated, animals might benefit by expending less energy in territory defence and reallocating their resources into producing fewer but more competitive young. Thus, this hypothesis links the observed trend of small clutch sizes of insular birds to territory size via the mechanism of defence costs (Stamps and Buechner 1985).

As with many of the phenomena discussed in this chapter, these ideas are interesting, but at present we cannot conclude which of them might hold greater explanatory power across different islands.

The island syndrome in rodents

Although the various evolutionary changes in ecological niche in island populations have mostly been considered separately so far, it will be apparent to the reader that many of the changes evolve together. A proper understanding of individual changes requires models that incorporate several such phenomena simultaneously. Adler and Levins (1994) set out to construct such a model for island rodents, synthesizing results from a range of empirical studies, principally of mice and voles. They noted that island populations of rodents tend to evolve higher and more stable densities, better survival, increased body mass and reduced aggressiveness, reproductive output, and dispersal. It is striking that island rodent populations of different species and from disparate geographic areas, often demonstrate similar sets of patterns. Adler and Levins termed these collective island-mainland differences the **island syndrome**. They summarize the main traits and some of the more likely explanations for them in tabular form (Table 7.3).

Table 7.3 Short-term and long-term changes in island rodents and proposed explanations (from Adler and Levins 1994, Table 2)

Island trait	Proposed explanation
Reduced dispersal	Immediate constraint (short-term response) and natural selection against dispersers (long-term response)
Reduced aggression	Initially, reduced population turnover, greater familiarity with neighbours, and kin recognition. Long-term directional selection for reduced aggression
Crowding effect	Isolation ('fence effect' resulting from reduced dispersal) and reduced number of mortality agents such as predation, both of which result in crowding of individuals and consequently higher population densities
Greater individual body size	Initially, a reaction norm as a response to higher density. Long-term directional selection for increased body size in response to increased interspecific competition
Lower reproductive output per individual	Initially, a reaction norm as a response to increased density. Long-term directional selection in response to decreased mortality
Greater life expectancy (higher survival probabilities for individuals)	Reduced number of mortality agents such as predation

The higher densities of rodent populations indicated by the crowding effect are an important part of this analysis. As already highlighted, because islands tend to have fewer species on them than equivalent continental areas, population densities will differ between islands and mainlands, in effect as a statistical artefact. Williamson (1981) therefore distinguishes the notions of **density compensation** and **density stasis**. Density compensation is where island communities have the same total population density but distributed over fewer species, i.e. with large population sizes per species (above). Density statis is where the overall population of the community on the island is less than that of the reference mainland system, such that population sizes per species are the same as the mainland. Distinguishing between the two conditions is tricky and necessitates an appropriate experimental design.

According to Adler and Levins (1994) a number of studies of island rodents have avoided the pitfalls identified by Williamson, demonstrating that higher average densities of rodent populations do indeed occur in some cases on islands. They cite several examples of experimental studies using fenced enclosures, which have shown that rodent populations in such circumstances can reach abnormally high densities and often destroy the food supply (the 'fence effect' of Table 7.3.), thereby indicating dispersal to be an important regulator of population densities under normal circumstances in which there is no (complete) fence. As rodent populations on small islands lack an adequate dispersal sink, the fence effect may be invoked as a force providing selective influence on the population.

One of the mainstays of island *ecological* analysis in recent decades has been the study of the relative significance of island area and isolation (Chapter 4). Adler and Levins' (1994) scheme extends to dealing with how these factors may work in an island *evolutionary* context (see also Adler *et al.* 1986). They hypothesize that average population density might, in general, increase with increasing isolation and hence with reduced dispersal, but that this fence effect might decline with an increase of island area to the point where it disappears altogether as islands more closely resemble a mainland. Thus the island syndrome might be expected to occur only in islands which are: (1) sufficiently isolated, (2) not too large so as to resemble mainland, and (3) not so small that they cannot support persistent populations. Figure 7.7 illustrates the differing effects that area and isolation are envisaged to have. Effects which may be expressed largely as reaction norms in the short term i.e. within the range of phenotypes of the founding population, may, under sustained selection pressure, produce rapid evolutionary change and locally adapted island populations that differ from mainland populations. The effect of isolation is a direct one, on dispersal, whereas that of area is less direct. Larger islands and mainlands typically have more predators, competitors, and habitat types, and because of this (and especially because of predation effects), densities are depressed compared with those of small islands.

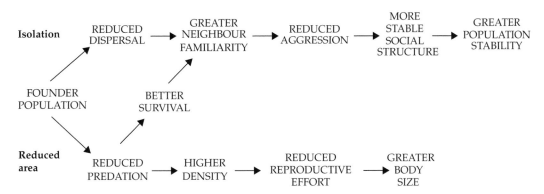

Figure 7.7 Schematic diagram showing the initial effects of island isolation and area on rodent populations. Long-term effects of insularity are directional selection for increased body size, reduced reproductive output, and reduced aggression (Redrawn from Adler and Levins 1994, Fig. 2).

Adler and Levins' (1994) model is based on studies of rodents, but there is no particular reason why it could not be extended to and tested on other systems.

7.7 Summary

This chapter has been concerned with what we might think of as micro-evolutionary change; the often small, incremental alterations in genotype, and in niche, which provide the building blocks of the more spectacular emergent patterns. We have considered essentially random processes associated with founding events and genetic drift, and we have discussed at greater length an array of island species traits or syndromes that appear to point to general selective forces operating on islands. We have also set out some of the many hypotheses put forward in explanation of these changes.

The founder principle is that small bridgehead populations bring with them a biased subset of the genetic variation of the source population. Such bottlenecks appear to be important in the rapid evolutionary divergence of mainland and island lineages, although the scale and significance of founder effects remains a contested issue. Analyses of reproductive behaviour and the sexual systems of colonists provide further insights into evolutionary change on islands. Hybridization appears to have been more important in island evolution than once realized.

Island forms have undergone a wide variety of forms of niche alteration, including shifts and expansions in feeding niche, alterations in nesting sites in birds, loss of defensive attributes, and loss of dispersal powers. One of the most striking patterns is the so-called island rule, of size increase in small vertebrates, and size reduction in large vertebrates. Each of these trends requires careful evaluation before a significant island effect is accepted. In some cases niche shifts can be due to character displacement caused by competitive interactions, or to ecological release in the absence of the usual array of competing species. Particular syndromes of traits can be identified which can be understood in relation to the geographical context of the island systems under examination. For example, island pollination networks seem to be characterized by the emergence of endemic super-generalists, which can be linked to the disharmonic nature of remote island ecosystems.

Although much of the discussion has been inconclusive, raising as many questions as answers, we hope that we have set out some of the more important evolutionary–ecological mechanisms, the tool kit from which island evolution has fashioned the peculiar features of island endemics. In the final section, we have begun to see how these mechanisms can be put together into more complex theories that might explain emergent patterns of island faunas and floras, including the most spectacular radiations of species and genera from single ancestral colonists. To understand these in more detail we have now to consider the process of speciation, what it entails, and how we can organize ideas about island speciation into alternative explanatory and descriptive frameworks. This we do in the next chapter.

CHAPTER 8

Speciation and the island condition

... it is the circumstance, that several of the islands possess their own species of the tortoise, mocking-thrush, finches, and numerous plants, these species having the same general habitats, occupying analogous situations, and obviously filling the same place in the natural economy of this archipelago, that strikes me with wonder. It may be suspected that some of these representative species, at least in the case of the tortoise and some of the birds, may here-after prove to be only well-marked races; but this would be of equally great interest to the philosophical naturalist...

(Charles Darwin in 1845, writing about the Galápagos archipelago. From Ridley 1994, p. 79)

Having covered the more important micro-evolutionary processes and outcomes in the previous chapter, the aim of this chapter is to provide the basic frameworks by which we understand macro-evolutionary patterns of change on islands. We thus start with the nature of the species unit, and then we outline a set of alternative frameworks within which we can organize ideas about island evolution: specifically these frameworks are distributional, locational, mechanistic, and phylogenetic. We reserve consideration of the most striking emergent patterns of island evolution to Chapter 9.

Evolutionary change does not necessarily imply speciation, and in the previous chapter we focused on the changes in niche that characterize island forms without paying much attention to whether they constitute 'good' species or merely varieties. In fact, the determination of the species unit, although of central importance in ecology and biogeography, is not always clear cut. As highlighted in the above quotation, Darwin was uncertain as to the taxonomic status of the members of that most emblematic group of island evolution, the Galápagos

finches. When he collected them he viewed the different kinds of finches as merely variants of a single species (Browne and Neve 1989). Moreover, the full significance of his Galápagos collection did not immediately strike Darwin. Rather, his key insights into species transmutation came some time after the voyage, as he was attempting to make sense of his observations and of the taxonomic judgements made of his collections. Although he was right to state that the Galápagos biota would prove of abiding interest whatever the taxonomic status of the members of the radiating lineages, it is important for the 'philosophical naturalist' to appreciate the nature of the species unit.

8.1 The species concept and its place in phylogeny

No one definition has as yet satisfied all naturalists; yet every naturalist knows vaguely what he means when he speaks of a species.

(Darwin 1859, p. 101)

... it will be seen that I look at the term species, as one arbitrarily given for the sake of convenience to a set of individuals closely resembling each other, and that it does not essentially differ from the term variety, which is given to less distinct and more fluctuating forms.

(Darwin 1859, p. 108)

If we are to study distributions of organisms, let alone speciation, it is a prerequisite that we have a currency, i.e. units that are comparable. Thus the most fundamental units of biogeography are the traditional taxonomic hierarchies by which the plant and animal kingdoms are rendered down to species level (and beyond). So, for example, a particular form of the buttercup can be described by the following nomenclature:

Kingdom   Plantae
Division   Spermatophyta
Order   Ranunculales
Family   Ranunculaceae
Genus   *Ranunculus*
Species   *bulbosus*
Subspecies   *bulbifer*

How is this taxonomy decided? Are the units actually comparable? The following is a fairly typical textbook definition of the species unit: 'groups of actually or potentially interbreeding natural populations which are reproductively isolated from other such populations' (Mayr 1942, p. 120). This is a form of the biological species concept—promoted notably by Ernst Mayr—which has been predominant for the past 60 years (Mallet 1995). Yet the question of 'reproductive isolation' (Table 8.1) is often an unknown quantity, as populations may be geographically separated (non-overlapping distributions are termed allopatric). In such circumstances they may exhibit sufficient morphological differentiation to be considered subspecies or even separate species, yet actually remain capable of interbreeding if placed into a common enclosure. Then again, species may be capable of interbreeding

Table 8.1 A classification of isolating mechanisms in animals (from Mayr 1963, slightly modified)

1. Mechanisms that prevent inter-specific crosses (pre-mating mechanisms)
   a. Potential mates do not meet (seasonal and habitat (ecological) isolation)
   b. Potential mates meet but do not mate (ethological isolation)
   c. Copulation attempted but no transfer of sperm takes place (mechanical isolation)
2. Mechanisms that reduce full success of inter-specific crosses (post-mating mechanisms)
   a. Sperm transfer takes place but egg is not fertilized (gametic mortality)
   b. Egg is fertilized but zygote dies (zygote mortality)
   c. Zygote produces an F_1 hybrid of reduced viability (hybrid inviability)
   d. F_1 hybrid zygote is fully viable but sterile, or produces deficient F_2 (hybrid sterility)

in the laboratory and have sympatric distributions (i.e. occur in the same geographic area) but remain reproductively isolated in the wild because of behavioural differences. Sometimes a rider is added to restrict the definition to those individuals that can successfully interbreed to produce viable offspring. Yet this rule too can be broken, as species which are recognized as 'good' may interbreed successfully in hybrid zones (Ridley 1996), an example being the native British oaks *Quercus petraea* and *Q. robur*. Studies of the genetics of British populations have shown that both maintain their integrity over extensive areas of sympatry, but form localized 'hybrid swarms' in disturbed or newly colonized habitats (White 1981). Apparently this is a stable situation, as Miocene fossil leaf impressions suggest that these taxa became differentiated no less than 10 million years BP. Some prefer to treat such partially interbreeding populations as 'semispecies' within 'superspecies' (White 1981). Hybridization between closely related island species appears to be fairly common (Silvertown *et al.* 2005), especially early on in the development of an adaptive radiation (Grant 1994).

For many groups, moreover, the taxonomy remains poorly known and subject to revision. For example, molecular analyses of the *Bufo margaritifera* complex of toads in Central and South America suggest that toads recognized by this name actually represent composites of morphologically cryptic species that are genetically unique (Haas *et al.* 1995). If the criterion of being reproductively isolated is a difficult one to employ on living organisms, it is even more so with many fossil organisms. Often the best that can be done is to determine whether the morphological gaps between specimens are comparable with those for living species that are reproductively isolated.

Thus, even the species unit, although central to so many biogeography purposes, has blurred edges (see also Otte and Endler 1989; Ridley 1996). The units below the species are a matter of considerable debate and strong opinions are held that they are either important ecologically and therefore matter or that they are arbitrary and the whim of the taxonomist (see discussion in Mallet 1995). The significance of this issue can be shown by reference to the

Lepidoptera: butterflies are thought to comprise about 17 500 full species, but the number of currently recognized subspecies approaches 100 000 (Groombridge 1992). These problems are general to taxonomy and systematics, but have particular force in relation to islands on which there is often no adequate way of testing the potential reproductive isolation of geographically isolated 'species', 'subspecies', or 'varieties' on different islands within an archipelago or ocean basin. Even in well-studied lineages the integrity of the species unit can be difficult to resolve. For example, although recent phylogenies based on mtDNA sequences (Sato *et al.* 1999) recognize 14 species of Galápagos finches (13 on the Galápagos plus the Cocos Island finch, which is descended from a Galápagos ancestor), Zink (2002) argues that the species-level taxonomy within the genera *Geospiza* (ground finches, 6 species) and *Camarhynchus* (tree finches, 4 species) should be considered unresolved. Zink raises the possibility that *Geospiza* and *Camarhynchus* are in effect each functioning as a single, highly variable species, which would mean reducing the number of recognized species from 14 to 6. This taxonomic uncertainty is a product of the degree of hybridization going on in this relatively youthful radiation (Grant and Grant 1996a). Elsewhere in this volume we follow general practice in recognizing 14 species.

In a review of the species concept, Mallet (1995) notes the prevalence, even at textbook level, of as many as seven or eight different notions of how to define species. This may be bewildering to many, and for the present purposes a way out is needed. This is not the place for a lengthy review of the problems of taxonomy and systematics; the intent has instead been to ensure that the reader appreciates that difficulties exist with the species concept and its application. The solution Mallet recommends is to return to Darwin's pragmatic usage of 'species' and 'varieties', but to make use of new knowledge from genetics as well as morphology in determining when the species label is appropriate. Darwin's position is made clear by the quote given above, and in the following passage from *On the origin of species*:

. . . 'varieties have the same general characters as species, for they cannot be distinguished from species, except,

firstly, by the discovery of intermediate linking forms . . . ; and except, secondly, by a certain amount of difference for two forms, if differing very little, are generally ranked as varieties . . . '

We now recognize that there is more to it than this, but even with modern genetic techniques it can be difficult to agree on what constitutes a 'good' species.

If species are in essence arbitrary units, or 'well-marked varieties', then it will be apparent that determining the point at which speciation has occurred within a radiating lineage is also problematic. 'That is like asking exactly when a child becomes an adult. I am content to know that initially there was one lineage and now there are more.' (Rosenzweig 1995, p. 87). The debates on this crucial matter will continue, and although some may not like it, the view taken here is a pragmatic one, following Darwin and Wallace: the precise level at which species are defined is arbitrary. As with other currencies, exchange rates can change as a function of the methods and assumptions used in the calculation.

At higher levels, species are grouped into genera, then families, and so forth. In flowering plants such groupings of species are based principally on the evolutionary affinities of floral structure and so it is possible to find many different growth forms and ecologies within a single family. Thus the Asteraceae (Compositae—the daisy family) all have a composite flower structure, but some are herbs and others are trees. In general, there tends to be less variation in functional characteristics within a genus than between genera (within the same family) and there is presumed to be a closer evolutionary relationship, with all members of a genus ultimately descended from a common ancestor, different from that of other genera within the same family. Some species are so distinct that they may be the only members of a genus or family, whereas others belong to extremely species-rich genera, e.g. figs (*Ficus*; Moraceae). Insular examples include, at the one extreme, *Lactoris fernandeziana*, the monotypic representative of the endemic family Lactoridaceae from the Juan Fernández islands (Davis *et al.* 1995); and at the other extreme, on the

Hawaiian islands, over 100 species of *Cyrtandra* of the widespread family Gesneriaceae (Otte 1989). In general, the greater the taxonomic distinction between organisms, the more ancient their differentiation, although calibrating the timescale is difficult, particularly for fossil organisms.

In recent decades the methods of traditional evolutionary systematics have been challenged by the development of phylogenetic systematics, or **cladistics**, which presumes to supply a more objective means of quantifying the relatedness of a taxonomic group. This involves scoring the degree to which different organisms share the same uniquely derived characteristics that other taxa outside the group do not possess. These characteristics are then used to construct a **cladogram**, or **phylogeny (phylogenetic tree)**, which is a branching sequence setting out the most parsimonious ('simplest') model for the relationships between taxa, i.e. it provides a hypothesis for the evolutionary development of a lineage. The data employed may be morphological or genetic (i.e. based on DNA sequences). The papers in Wagner and Funk (1995) provide examples of both as applied to Hawaiian plants and animals, as well as an excellent explanation of the methods of cladistics.

Historical biogeographers also make use of what are termed **area cladograms**. An area cladogram is constructed by first determining the phylogeny and then replacing the species names with the geographic location in which those species are found. This new tree in effect provides a series of hypotheses of the sequence of dispersal and vicariance (population subdivision by barrier formation) involved in the evolutionary development of the lineage, emphasizing the role played by geological/environmental history in lineage development.

Although the proponents of phylogenetic systematics are often forthright in praising the approach and condemning traditional systematics, it should be recognized that the decisions as to which characters to include in the phylogeny, and the rules used in its construction, are the choice of the user. The phylogeny thus constructed is therefore merely one approximation, albeit hopefully a basically reliable one, to the history of events that produced the lineage. Once such a phylogeny has

been constructed, it can then be used to explore the likely sequence of interisland and intraisland speciation events within archipelagos, or between islands and mainlands.

The most exciting opportunities have been opened up by advances in molecular biology and genetics and these methods have been seized upon in island evolutionary studies. Such studies benefit from use of the 'genetic clock' which, on the assumption of a more or less constant rate of accumulation of mutations and thus of genetic differences between isolates, allows estimates of the dates of lineage divergence (Box 8.1). An excellent illustration is provided by the studies of Thorpe *et al.* (1994) on the colonization sequence of the western Canarian lizard, *Gallotia galloti*, in which the direction and timing of colonization as postulated by genetic clock analyses (nuclear and mitochondrial DNA divergence) was shown to be entirely compatible with the independently derived geological data for the timing and sequence of island origins.

It should be noted that these molecular clocks may not always run to time. A key problem is exemplified by Clarke *et al.*'s (1996) study of two species of land snails (*Partula taeniata* and *P. suturalis*) on the island of Moorea (French Polynesia). It was found that although the two species exhibit significant morphological and ecological differences, they are genetically close. Clarke *et al.* attributed this closeness to 'molecular leakage' or 'introgression': the convergence of genetic structure through occasional hybridization. Such **introgressive hybridization** is now considered to be quite common in particular island lineages (Clarke and Grant 1996). Even low rates (as low as 1 in 100 000) may be enough to upset the phylogenetic trees and molecular clocks upon which scenarios of island evolution discussed in Chapter 9 now rest (Clarke *et al.* 1996; Zink 2002). For further insights into these and other methodological problems of biogeography and systematics, see Myers and Giller (1988), Thorpe *et al.* (1994), Ridley (1996), the exchange between Herben *et al.* (2005) and Silvertown *et al.* (2005); and for some reassurance about the use of molecular clocks, see Bromham and Woolfit (2004) and Box 8.1.

Box 8.1   Molecular clocks

Dating events in evolutionary biogeography presents considerable challenges because of the extended time periods involved, extending way beyond many conventional dating techniques such as radiocarbon isotopes. Traditionally, fossils are dated by means of the geological record and their place in the stratigraphic column, but this can be a very imprecise approach and doesn't provide a means of dating colonization and speciation events in extant taxa. **Molecular clocks** provide a solution to this problem based on the idea that proteins and DNA evolve at a more or less constant rate. The extent of molecular divergence is then used as a metric for the timing of events within the development of the lineage.

Geneticists have developed an impressive array of rather complicated analytical procedures to ensure that such molecular clocks are reasonable. Nonetheless, they are inevitably based on limited amounts of information and involve a number of important assumptions. They should be regarded as providing plausible hypotheses for the developmental sequence and timing of particular monophyletic lineages. Greater confidence can be placed in such molecular clocks if they can be independently calibrated.

In illustration, Carranza et al. (2006) have made use of a molecular clock in their study of the phylogeography of the lacertid lizard *Psammodromos algirus* in the Iberian peninsular and across the Strait of Gibraltar in North Africa. They sampled genomic DNA for three genes from 101 specimens of the subfamily Gallotiinae, mostly of the target species, but including specimens of *Gallotia* species from the Canaries. These specimens provide a means of 'rooting' the phylogenetic tree, and also a means of calibrating it. This is because it is reasonable to assume that *G. caesaris caesaris*, endemic to the island of El Hierro, commenced divergence from its nearest relative, *G. c. gomerae*, endemic to the island of La Gomera, shortly after the formation of El Hierro (approximately 1 Ma). The clock used by Carranza et al. indicated that diversification in the genus *Gallotia* began on the Canaries about 13 Ma, with a rather earlier data of 25 Ma for the first speciation events within the Iberian/North African genus *Psammodromus*.

This case study is illustrative of a growing number of studies that place the development of a monophyletic island lineage into context alongside related mainland lineages. This body of work has produced some interesting surprises, for instance showing evidence of successful back-colonization of island forms into mainland regions (e.g. Nicholson et al. 2005).

8.2   The geographical context of speciation events

Distributional context

In order to appreciate the full gamut of possibilities involved in speciation it is helpful, if slightly artificial, to distinguish the geography of speciation from the mechanism (Table 8.2; Box 8.2). The geographical context can be viewed either in distributional or locational terms, i.e. distinguishing the degree of overlap of populations involved in speciation events on the one hand, from the issue of where the evolutionary change is taking place.

The terms **sympatry** and **allopatry** denote, respectively, two populations (or species) which overlap in their distributions and two populations (or species) which have geographically separate distributions. In geographical terms, a new species (or subspecies) may, in theory, arise in one of three circumstances. If the new form arises within the same geographical area as the original, then it may be termed **sympatric speciation**, if it arises in a zone of contact (hybrid zone) between two species

Table 8.2 A simplified framework for speciation patterning

Form	Pattern
Distributional	
Sympatric	Overlapping
Allopatric	Separate
Parapatric	Contiguous
Locational (and historical)	
Neo-endemic	Change on island
Palaeo-endemic	Island form relict
Mechanistic	
Allopatric	Drift and selection
Polyploidy	Changes in chromosome number
Competitive speciation	Other sympatric mechanisms
Tree form (phylogeny)	
Anagenesis	Replacement of original
Anacladogenesis	Alongside original
Cladogenesis	Lines diverge and replace original

it may be termed **parapatric speciation**, and if it arises in a separate geographical area it may be termed **allopatric speciation**. In reality the distinction between these conditions may be blurred as: (1) the spatial scale we choose to describe the geography of the populations may not be concordant with the scale at which the members of the populations interact; and (2) we may be wholly ignorant of the past history of the distributional overlap of the populations such that, for instance, a currently sympatric species pair may well have speciated while geographically isolated from one another.

As evident from the debate concerning vicariance vs dispersalist explanations in biogeography (Chapter 3), many authors have assumed that allopatric speciation is by far the dominant mode of speciation. The demonstrable occurrence of sympatric forms of speciation exposes the weakness

---

Box 8.2   On archipelago speciation, allopatry and sympatry—a clarification

We provide this simple summary as a quick look-up and clarification of the ideas involved.

• **Allopatric speciation** is speciation occurring between two (or more) populations that are geographically isolated from each other. All island neo-endemics exemplify allopatry in the sense that they are isolated from the mainland source pool. However, within an archipelago, populations may also be allopatric at the inter-island scale, or even within islands, for example, isolated within forest islands surrounded by lava (*kipukas*, Hawaii).

• **Sympatric speciation** is a term applied to cases where speciation occurs without geographical separation of the populations involved, i.e. the opposite of allopatric speciation. In practice, the past roles of allopatry and sympatry can be difficult to distinguish at the intra-island scale, or even at the intra-archipelago scale where the islands are close to one another and the organisms involved have high dispersal abilities.

• **Parapatric speciation** refers to situations intermediate between the extreme cases of allopatry and sympatry, where the geographical segregation is incomplete, and two separating forms are contiguous in space with a hybrid zone in the area of contact/overlap.

• **Archipelago speciation** is a term sometimes used where speciation of island lineages has occurred mostly between islands rather than within them. This tends to imply an allopatric model, especially where inter-island distances are considerable, but may in fact involve both allopatric and sympatric phases. The taxon cycle model, as applied to the West Indian avifauna, is also an example of archipelago speciation. Sometimes, environmental differences across an archipelago such as the Canaries are great enough that **adaptive differences** are evident, but sometimes very little niche segregation is evident, in which case the archipelago speciation might be considered **non-adaptive.**

• **Are these terms operational?** Given the huge environmental changes involved in the life cycle of isolated volcanic islands, and archipelagos, it can be very difficult to be sure that current distributions indicate the circumstances in which speciation occurred. Indeed, taking the Galápagos finches as a classic case study of island evolution, it appears likely that this adaptive radiation has involved allopatric, sympatric and parapatric phases (Chapter 9). Similarly, the distinction between adaptive and non-adaptive changes may not always be clear cut: evolutionary change commonly involves a mix of stochastic and directional mechanisms.

of making such assumptions (Sauer 1990). The production of island neoendemics of course occurs in isolation from the original continental source population, and to that extent can be considered to have occurred allopatrically. However, much island speciation has occurred in the condition of island-scale sympatry (i.e. differentiation of two forms occurring on one island), but on the finer within-island scale, this still allows for differing degrees of population overlap (Keast and Miller 1996, p. 111).

Diehl and Bush (1989) suggest that populations utilizing different, spatially segregated habitats should nonetheless be considered sympatric when all individuals can move readily between habitats within the lifetime of an individual. This must be judged separately for different taxa as, for example, a within-island barrier for land snails may not be a barrier to bird populations (Diamond 1977), which in cases might move freely even between separate islands within an archipelago. Indeed, the distinction between allopatry and sympatry is something that varies substantially within higher taxa. In illustration, the Krakatau islands, isolated from the mainlands by about 40 km, are beyond the apparent colonization limits of certain plant families (Whittaker *et al.* 1997), yet for some species of plants and their insect pollinators, the island populations can be considered as freely interbreeding (panmictic) with populations on the mainland (Parrish 2002). Within large oceanic archipelagos it is probably commonplace for large radiations of monophyletic lineages to have involved a mix of sympatric and allopatric speciation events, as for example, appears to apply to the Hawaiian fruit flies *Drosophila* (Tauber and Tauber 1989) and *Sarona* (a genus of phytophagous insects) (Asquith 1995).

Locational and historical context—island or mainland change?

We have previously introduced the distinction between neoendemic forms (those that have evolved *in situ* on the island) and palaeoendemic forms (which have become endemic because of the extinction of the same or closely related forms

from mainland populations). This general idea was recognized by A. Engler as long ago as 1882 and is discussed in two papers by Cronk (1987, 1990). Cronk compared the two species of the endemic St Helenan genus *Trochetiopsis* (redwood and ebony) with other species in the same family (Sterculiaceae), based on the morphology of present-day pollen and ancestral pollen in sediments. From this he concluded that the ancestors of the present species probably arrived by long-distance dispersal during the Tertiary period, from African or Madagascan stock (Fig. 8.1). The lineage has since diverged into two forms, but the degree to which they differ from their mainland relatives is only partly explained by *in situ* change. Evolution and extinction elsewhere among the branch to which the ancestor belonged has meant that the rest of the family has effectively evolved away from the St Helena genus over the period since colonization.

The key point is that in attempting to understand evolution on islands, we cannot assume that over timescales of millions of years, evolution on the continents that supplied the initial island colonists has somehow stood still. We cannot therefore interpret all island/mainland differences, and in particular all insular endemism, as necessarily a function of evolutionary change on the island in question. 'There is no doubt that adaptive radiation occurs on islands, but this may be interpreted as the elaboration of a relict ground-plan, without requiring that ground-plan to have arisen by evolution on the island' Cronk (1992, p. 92).

Paulay (1994) comments that relict taxa are well known on large, old, continental fragment islands, the lemurs of Madagascar being classic examples. The primitive weevil family Aglycyderidae of both Pacific and Atlantic islands provides another case of ancient forms restricted to islands. Although short-lived oceanic islands may not be thought to be ideal for long-term survival of relict lineages, if the species involved have good dispersal abilities they may survive by island hopping. Thus three of the six families that comprise the most primitive order of pulmonate land snails are restricted to islands, yet representatives may be found on even the most remote Pacific islands, strongly suggesting their

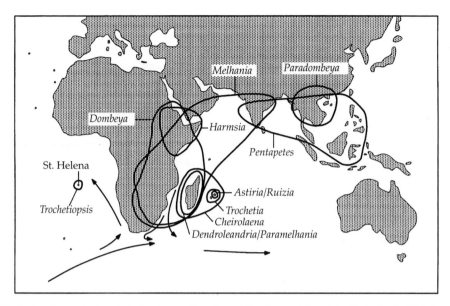

Figure 8.1 Distribution of genera in the subtribe Dombeyeae (Sterculiaceae). The closest relatives of the *Trochetiopsis* of St Helena are *Trochetia* (Mauritius) and *Dombeya*. The centre of generic diversity is in Madagascar and the Indian Ocean islands. According to Cronk, present-day ocean currents (arrows) differ from those predominant at the time the ancestral Dombeyeae stock colonized St Helena during the late Tertiary, when there was a stronger flow from the Indian Ocean to the South Atlantic. (Redrawn from Cronk 1990, Fig. 3, with permission of the New Phytologist Trust.)

distributional boundaries to be set by biological factors rather than an inability to reach mainland areas.

The labelling of particular species or traits as 'relictual' and the relative importance of palaeo- and neoendemism on islands has proven controversial (Chapter 2; Bramwell 1979; Cronk 1992; Elisens 1992; Carlquist 1995). One example discussed in Chapter 7 is the question of whether woodiness of island forms of normally herbaceous taxa could be considered relictual (basal) or a derived characteristic. The application of modern genetic techniques argues convincingly for the latter (Kim *et al.* 1996; Panero *et al.* 1999). In practice, as more modern genetically derived phylogenies become available it seems likely that the dichotomy between palaeo- and neoendemic forms will become increasingly irrelevant. Instead, we will be able to compare the complex, and often individualistic, patterns of lineage development across and between islands, with recent work even pointing to cases of back-colonization of island forms to mainlands (e.g. Carine *et al.* 2004; Nicholson *et al.* 2005).

8.3   Mechanisms of speciation

When examining the mechanisms of speciation rather than focusing on the geographical circumstances, it becomes clear that there are different processes at work. Rosenzweig (1995) summarizes these under the three headings: **geographical** or **allopatric speciation**, **competitive speciation**, and **polyploidy** (involving an increase in the chromosome number). Each form holds relevance to the island context, but as insularity defines the island condition, the allopatric form of speciation will be dealt with first.

Allopatric or geographical speciation

The essence of this model has been summarized by Rosenzweig (1995, p. 87):

• A geographical barrier restricts gene flow within a sexually reproducing population.
• The isolated subpopulations evolve separately for a time.

- They become unlike enough to be called different species.
- Often the barrier breaks down and the isolates overlap but do not interbreed (or they interbreed with reduced success).

The first step of this model suggests a vicariant event (barrier formation). This scenario applies to certain ancient continental fragment islands (e.g. Madagascar) and to former land-bridge islands. In such cases, the island population is assumed not to have experienced a bottleneck founding event. For true oceanic islands, the starting point is different: dispersal across a pre-existing barrier. The founding population may be quite small, and thus the founding event may involve a bottleneck effect (Chapter 7). In either scenario, considering a now remote island population, it develops in effective isolation from the mainland source population. Subsequent environmentally-determined bottlenecks, genetic drift effects and selective pressures in the novel island context then provide the engine for further differentiation from the mainland form.

The failure of the isolating barrier may not be a frequent feature for oceanic islands, but can happen, for instance, as a function of sea-level change providing 'stepping stones' between island and mainland. This can lead to **double invasions**, where a second invasion of the mainland form occurs notwithstanding the continued existence of an oceanic barrier. This was the hypothesis put forward to explain the occurrence of two species of chaffinches on the Canaries (Lack 1947a). The endemic blue chaffinch (*Fringilla teydea*) was considered to have evolved from an early colonization event, to a point where it was sufficiently distinct to survive alongside a later-arriving population of mainland common chaffinches (*F. coelebs*). More recent genetic work based on mitochondrial DNA has supported this basic idea, with an interesting twist, as it seems likely that the later colonizing *F. coelebs* reached the Canaries via the Azores and Madeira (Marshall and Baker 1999). As often the case with such data, the results and interpretation are not straightforward, but this was regarded as the most likely sequence, with

strong support also for a back-colonization event from the Canaries to Madeira. Hence, it seems likely that the sequence of population movements amongst the islands was rather more complex than implied by the double invasion model. Similar complexity of movements may be necessary to account for other Atlantic island phylogenies (e.g. Kvist *et al.* 2005).

Interestingly, available evidence shows that the greatest degree of radiation within the Macaronesian flora can be attributable to monophyletic genera, with paraphyletic lineages paradoxically producing fewer endemic species. Silvertown (2004) reported that 20 monophyletic endemic genera account for 269 endemic plant species, while 20 paraphyletic lineages account for just 38 species. Double or multiple invasions occur within the monophyletic lineages, but the movements occur at the intra-Macaronesia scale rather than in the sense of mainland–island colonization events. Silvertown attempts to explain this paradox in terms of a competitive mechanism involving niche pre-emption (see discussion in: Silvertown 2004; Saunders and Gibson 2005; Silvertown *et al.* 2005).

The barriers that exist between islands within an archipelago are undoubtedly important features of remote oceanic island groups. As Rosenzweig (1995, pp. 88–9) puts it 'Suppose propagules occasionally cross those barriers, but usually after enough time for speciation has passed. Then the region and its barriers act like a speciation machine, rapidly cranking out new species.' The radiations that characterize many taxa on the Hawaiian islands reflect this, with interpretations for the relationships among the 700 (or more) species of *Drosophila* involving a number of phases of interisland movements as a part of this most spectacular of radiations (e.g. Carson *et al.* 1970; Carson 1992). Radiation on archipelagos is examined further in the next chapter.

An illustration of the contrast between the archipelago context and the single remote island is provided by that most famous of examples of island evolution, Darwin's finches. The Geospizinae are recognized as a distinct subfamily, found in the Galápagos archipelago, which comprises nine islands larger than 50 km² (and a larger number of smaller islets), with predominantly desert-like conditions. Here they have radiated into 13 different species (Sato *et al.* 1999; but see Zink 2002). The family is found also on Cocos Island, a lone, forested island of about 47 km², which is, just, the nearest land to the Galápagos. Here there has been no possibility of the island hopping that appears to characterize archipelago speciation, and the group is represented by a single endemic species *Pinaroloxias inornata*, one of just four species of land bird (Williamson 1981; Werner and Sherry 1987). Analyses of mtDNA data indicate that the Cocos finch is probably a descendent of a Galápagos tree finch (Sato *et al.* 1999).

Yet, as Sauer (1990) cautions, within an archipelago, in addition to single species being present on different islands, quite commonly several sibling species occur on a single island, where they are mainly segregated by habitat. In many instances, isolation may have been only indirectly relevant as, by screening out competitors, isolation offers opportunities for adaptive radiation to those species that do arrive. Some, at least, of this radiation may take the form of the next class of mechanism, competitive speciation.

Competitive speciation

This is the term that Rosenzweig (1978, 1995) gives to cover a variety of related modes of sympatric speciation. The idea here is that a species expands its niche to occupy an unexploited ecological opportunity, followed by that species' sympatric break-up into two daughters, one in essence occupying the original niche, and the other the newly exploited one. The expansion happens because of a lack of competition with other species, a feature common to remote island faunas and floras. But the break-up into separate species happens because of increased competitive pressure between those portions of the population best able to exploit the two different niches. The operation of these processes must take place over the course of many generations, and observation of the complete process in the field is thus not an option. Rosenzweig illustrates the idea by means of a

'thought experiment', the essence of which, placed in the island context, can be rendered as follows.

A colonizing species of bird will have a particular feeding niche, determined morphologically and behaviourally, and perhaps most clearly expressed by features such as bill size and shape (Lack 1947*a*; Grant and Grant 1989). According to theory, such characters would be expected to have a unimodal distribution of values. The environment of the island contains empty niches, unexploited ecological opportunities, which for the purpose of our experiment are taken to be distributed bimodally, i.e. there are two resource peaks. The species, unconstrained by competition with the full range of mainland forms, expands from its original modal position somewhere along this resource space so that it occupies each of these niches. Individuals with genotypes that match best to one or the other of these niches become numerically dominant within the population, which has by this stage grown close to its carrying capacity. At this point, disruptive selection kicks in, such that the mid-range phenotypes lose fitness relative to those that are better suited to one or the other of the main feeding niches. Those individuals adapted primarily to a peak in the resource curve may also exploit the valleys either side of that peak. If they do so at all successfully, no opportunity remains for valley phenotypes in between the two peaks. They will be few in number in the population, and their offspring will have fewer resources to tap. Although they may also reach up to the peaks, they will not be effective competitors in those portions of the resource continuum. Valley genes will thus have little success and will be bred out, as in time members of each 'peak' population that can recognize others of their type (and breed with them alone) will be at a selective advantage. Therefore, isolating behavioural mechanisms develop and at this point the lineage can be viewed as having split.

If neither resource peak matches closely the original niche of the species, the divergence may result in two novel forms and the effective disappearance of the original species (termed **anagenesis**: see below). There may also be more than two resource peaks, thus accommodating a larger radiation.

This idea, of 'adaptive landscapes', was originally introduced by Sewall Wright (1932) and has been incorporated into a number of different evolutionary models (see e.g. Templeton 1981, Futuyma 1986, Otte and Endler 1989).

Intentionally, the account given here is a simple rendition of ideas culled from a much larger body of literature. Based on laboratory evidence, competitive speciation may occur in as few as 10–100 generations, not as fast as polyploidy (almost instantaneous), but faster than geographical speciation, which seems to require thousands or even hundreds of thousands of years (Rosenzweig 1995).

Grant and Grant (1989) recorded an early phase in the process of sympatric speciation in a study of a temporary reproductive subdivision within *Geospiza conirostris*, one of the Galápagos (Darwin's) finches occurring on Isla Genovesa. The population was partly subdivided ecologically for a brief period, but this division then collapsed again through random mating. All of the known sympatric species of ground finches differ by at least 15% in at least one bill dimension, suggesting that this may indicate a threshold in niche separation needed to sustain separate species. During the temporary subdivision, the two groups of males differed by a maximum of 6% in bill dimensions, and this, occurring as it did over but a brief period, was insufficient to foster any discrimination in a mating context. One of the dry-season food niches sustaining the division declined catastrophically and the division collapsed. Nonetheless, this study shows how sympatric division may be fostered in taxa in which polyploidy does not occur. Grant and Grant would have had to be extraordinarily lucky to come along just at the right time to record a sympatric split that 'stuck'. They therefore did not regard the failure of this event as evidence against the idea, which they concluded to have some relevance for vertebrates. Rather, they took it to indicate that in order to foster sympatric divergence leading to speciation, the niches or habitats to which different members of a population adapt must be markedly different and display a long-term persistence. This is more likely to occur on larger, higher islands.

How important is competitive speciation in the island context? This is difficult to judge as it cannot be decided merely by investigating the degree of geographical overlap between sibling species in ecological time. This is because, given the probabilistic nature of dispersal and the magnitude of past environmental change, populations currently isolated from one another may once have been sympatric and vice versa. Indeed, it may commonly be the case that **archipelago speciation**—diversification within archipelagos—involves a mix of allopatric and sympatric phases of subtly varying combination from one branch of a lineage to the next (cf. Clarke and Grant 1996). Over the lifespan of a large island, the same may also apply within a single island. The relative roles of allopatric and competitive speciation within the island context cannot therefore be quantified, but one thing remains clear; isolation is crucial, either directly or indirectly (respectively).

Polyploidy

One important class of sympatric speciation is through polyploidy, a condition comparatively common in plants and many invertebrates, but not in higher animals: it is unknown in mammals and birds, for instance (Grant and Grant 1989; Rosenzweig 1995). Polyploid species are those that have arisen by an increase in chromosome number. Two main forms of polyploid can be distinguished. If the new species has twice the chromosome complement of a single parent species it is termed an **autopolyploid**. If it has the chromosomes of both of two parent species (i.e. a crossing of lineages) it is called an **allopolyploid**. According to Rosenzweig (1995), autopolyploids are rare, but at a conservative estimate 25% or more of plant species may be allopolyploids. Barrett (1989) states that current estimates for the angiosperm subset are as high as 70–80% of species being of polyploid origin.

Rosenzweig (1995) presents an intriguing analysis of the proportionate importance of polyploids in different floras, showing that there is a general decrease in the relative contribution of polyploids in floras with decreasing latitude. The best trend was found for continental floras; for islands there was much more scatter in the relationship. Rosenzweig does not distinguish the nature of the islands in his analysis, i.e. whether they were predominantly oceanic or continental in origin, and this may be a confounding factor in his analysis. It is necessary to look a little further into the data to establish whether polyploidy has provided a significant contribution to island evolution and whether it has developed *in situ*. The following studies are cited in order of increasing proportion of polyploidy.

Stuessy *et al.* (1990) examined about one third of the 148 endemic plants of the Juan Fernández islands (33°S), which are volcanic, oceanic islands off the coast of Chile. Remarkably, none of the species examined was found to be polyploid. Borgen (1979) reports 25.5% polyploidy among 360 Macaronesian endemic plant species from the Canaries (28–29°N). This is a low proportion in comparison to most known floras, and is significantly lower than the 36.4% polyploids amongst the 151 non-endemic Canarian species for which data were available (Borgen 1979). It appears also that several of the polyploid events were fairly ancient ones, subsequent to which gradual speciation within certain polyploid lineages has occurred (an example being the genus *Isoplexis* in the Scrophulariaceae), but to a lesser degree than within diploid lineages (Humphries 1979). It may be concluded that polyploidy has not been a major evolutionary pathway in Macaronesian lineages. The New Zealand (*c*.35–47°S) flora consists of approximately 1977 species, 40% of which are known cytologically according to Hair (1966). Polyploidy is absent in the gymnosperms, but characterizes 417 (63%) of the 661 angiosperms investigated, a figure approaching the 70–81% polyploidy suggested for Oceanic–Subarctic and Arctic areas of high northern latitudes. New Zealand is an ancient archipelago, and its flora has been claimed to have been primarily of continental origin, relictual of Gondwanan break-up. However, this has been disputed by Pole (1994), who argues against the large-scale continuity of lineages in the New Zealand flora, claiming that palynological data support

derivation of the entire forest flora from long-distance dispersal (predominantly from Australia). This remains a hotly contested issue (see e.g. Heads 2004, Cook and Crisp 2005, McGlone 2005) making it difficult to put a time frame on the interpretation of the New Zealand data.

Few firm conclusions can be drawn from the data currently available. Polyploidy is more important in some island floras than others. The latitude and age of the islands in question may each be of relevance, although often, as shown, age and origin of the flora are imperfectly known. If the Macaronesian data can be taken as a good guide, it appears that the more spectacular radiations of island plant lineages do not tend to rely upon polyploidy.

8.4 Lineage structure

A number of different models and facets of island evolution have been introduced in this chapter. Too many frameworks may serve to confuse, but there is one more that appears particularly helpful in making sense of island speciation patterns, and although the terms may not be instantly memorable, the ideas seem useful. In some cases, evolutionary change takes the form essentially of the continuation of a single lineage, whereas in other cases the splitting of lineages is involved. Stuessy *et al.* (1990) provide three terms that codify what they see as the chief outcomes (Fig. 8.2).

- **Anagenesis** is when the progenitor species/form becomes extinct.
- **Anacladogenesis** is when the progenitor survives with little change alongside the derived species. What is meant by alongside is open to different interpretations—it could be within the same island, or less restrictively, within the same archipelago.
- Finally, **cladogenesis** is where the progenitor is partitioned into two lines and becomes extinct in its original form, a model suggestive of the classic examples of adaptive radiation. This framework sets aside cases where lineages converge or cross, via hybridization.

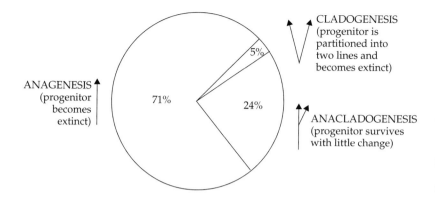

ANAGENESIS
(progenitor
becomes
extinct)

71%

5%

24%

CLADOGENESIS
(progenitor is
partitioned into
two lines and
becomes extinct)

ANACLADOGENESIS
(progenitor survives
with little change)

Figure 8.2 General patterns of phylogeny in the endemic vascular flora of the Juan Fernández islands. The figures give the percentage of each form of lineage development evident in 123 endemic plant species. (From data in Stuessy *et al.* 1990.)

8.5   Summary

In order to understand evolutionary change it is necessary to have a basic familiarity with the currency of evolution and knowledge of the problems associated with it. This chapter therefore begins by discussing some of the problems of the species concept, the arguably arbitrary distinction between species and varieties and the higher-order phylogenies of relationships that can be constructed for - particular lineages. Recent developments in phylogeny, including the use of cladistic and molecular biological techniques, have provided many new insights into long-standing island biogeographical issues, although such techniques should be viewed with proper caution as merely constructing realistic hypotheses of the unknown real genealogies.

A distinction is drawn between distributional, locational (and historical), mechanistic, and phylogenetic frameworks for describing island evolution. It is shown that divergence can take place both through isolation (allopatry) and in conditions of distributional overlap (sympatry), and indeed that these concepts can be applied on different scales: between island and mainland populations, within an archipelago, and within an island. Evolutionary change within a lineage may be accomplished by a variety of genetic mechanisms, including hybridization and polyploidy, and by more subtle alterations following from founding events, genetic drift, and natural and sexual selection in the novel biotic conditions of oceanic islands. Sympatric speciation may occur through more than one route. Perhaps the classic island examples are those that follow the competitive speciation model of niche expansion and break-up in to daughter populations exploiting different resource peaks. Polyploidy does not seem to have been a major route of speciation on islands, although it undoubtedly occurs. The final framework considered in the chapter is that of phylogenetic trees, which can be used to denote whether lineages develop without branching (anagenesis), with branching and extinction of the original form (cladogenesis), or by branching and survival of the original line (anacladogenesis).

CHAPTER 9

Emergent models of island evolution

Studies of island biotas are important because the relationships among distribution, speciation, and adaptation are easier to see and comprehend.

(Brown and Gibson 1983, p. 11, citing one of A. R. Wallace's general biogeographic principles)

Having considered the evidence for microevolutionary change on islands and the frameworks of speciation in the previous two chapters, we now examine how these building blocks fit together to produce neoendemics on islands. Although other categorizations are possible, we identify the following emergent patterns or models of evolutionary change on islands. The first is **anagenesis**, which refers to speciation without much radiation of lineages, and which may be most applicable to single isolated islands of moderate size and limited habitat heterogeneity. The second and third models, the **taxon cycle** and **adaptive radiation**, find their most dramatic illustrations within archipelagos and constitute multiagent evolutionary–ecological models: both stress adaptive changes. In contrast, the notion of **non-adaptive radiation** refers to the diversification of monophyletic lineages without apparent niche alterations, and emphasizes geographic isolation within or between islands.

Until comparatively recently, these scenarios relied heavily on distributional data (sometimes with supporting ecological data), for species classified using traditional taxonomic systematics. Over the last two decades, numerous studies have applied modern phylogenetic techniques to the description and analysis of island lineages, allowing the construction of scenarios for the historical patterns of evolutionary and distributional change on a lineage by lineage basis, at scales ranging from intraisland to across whole ocean basins. These scenarios are based around genetic relationships rather than adaptive characteristics. We consider some examples of these scenarios in the section on *island hopping radiations*. We also draw attention to the importance of changes to the platforms on which island evolution takes place and the implications of the life cycle of volcanic islands for patterns of phylogenesis. The final part of the chapter attempts to place the principal island evolutionary concepts and models into a simple framework of land area versus island or archipelago isolation.

9.1 Anagenesis: speciation with little or no radiation

The Juan Fernández archipelago contains two main islands, Masatierra (4 Ma) and Masafuera (1–2 Ma), located respectively some 670 km and 850 km west of mainland Chile. Of the native vascular plants, 18% of the genera and 67% of the 148 species are endemic to the archipelago. There is even one monotypic endemic family, the Lactoridaceae, represented by its only species *Lactoris fernandeziana* Stuessy *et al.* (1990) estimate that the presence of the 69 genera with endemic taxa on these islands may be explained by 73 original colonization events: in most cases only a single introduction per genus. On the basis of phylogeny and distribution of the endemic flora, Stuessy *et al.* (1990) have attributed 71% of endemic species to the **anagenesis** model, in which the progenitor form becomes extinct; 24% to the **anacladogenesis** model, where the progenitor survives little changed while a peripheral or isolated population diverges rapidly (perhaps while entering new habitats); and 5% to the **cladogenesis** model, in which the lineage divides into two lines and the original form fails to survive (Fig. 8.2).

The significance of anagenesis as a model of island evolutionary change may well have been overlooked given the obvious fascination provided by spectacularly branching lineages like the Hawaiian honeycreeper-finches. In the absence of systematic molecular data, a simple yet fairly conservative basis for estimating the proportion of speciation events fitting the anagenesis model is to count the number of cases where a genus is represented on an island/archipelago by a single endemic species. On this basis, Stuessy *et al.* (2006) estimate that for 2640 endemic angiosperm species from 13 island systems, about one quarter of speciation events match the anagenesis model.

Anagenesis describes an emergent phylogenetic pattern and does not imply a tightly specified explanatory model. For instance, a single endemic on an island might arise through a combination of a pronounced founder effect followed by genetic drift of the newly isolated island population. On the other hand, the colonist populations may be subject to strong directional selection pressure in the novel biotic and abiotic environment provided by a remote island, and show apparently clear adaptive features in response. There is insufficient genetic data to reveal whether species conforming to this model typically show unusual levels of genetic difference and thus whether chance or selection dominates. Where anagenesis is important on an island it indicates that intraisland vicariant events and competitive speciation have each been relatively infrequent, such that the motor for speciation is the further arrival of new colonists combined with environmental novelty and possibly environmental (e.g. climatic) change through time. It seems likely that anagenesis will be of greatest relative importance on islands of relatively limited environmental amplitude (Stuessy *et al.* 2006), and especially where such islands are themselves isolated (as the case of the Juan Fernández archipelago) so that opportunities for repeated phases of island hopping leading to multiple speciation events are limited.

9.2 The taxon cycle

Until recently it was generally held that islands represent a form of evolutionary blind alley, i.e. the flow of colonization events was considered essentially unidirectional from continents to islands, implying a generally lower fitness of island endemic forms. We now have evidence that sometimes island forms successfully colonize mainlands (e.g. Nicholson *et al.* 2005), but the general trend is undoubtedly mainland to island, and the taxon cycle helps explain why this is so.

The term 'taxon cycle' was coined by Wilson (1961) in his studies of Pacific ants. The model has been adapted to fit circumstances by different authors working on different taxa and regions, and hence it is not a discrete model allowing easy refutation. The shared features of different invocations appear to be that the evolution takes place within an island archipelago, in which immigrant species undergo niche shifts, which are in part driven by competitive interactions with later arrivals, such that the later arrivals ultimately may drive the earlier colonizing inhabitants towards extinction. The methodology employed in all the early studies was that patterns of geographic distribution and taxonomic differentiation provided the empirical basis from which were inferred cycles of expansion and contraction in the geographical distribution, habitat distribution, and population density of species in island groups (Ricklefs and Cox 1978).

Melanesian ants

As developed by Wilson (1959, 1961), the taxon cycle described 'the inferred cyclical evolution of species [of Melanesian ants], from the ability to live in marginal habitats and disperse widely, to preference for more central, species-rich habitats with an associated loss of dispersal ability, and back again' (MacArthur and Wilson 1967, glossary).

Wilson (1959, 1961) recognized differences in the ranges of the ponerine ants as a function of their habitat affinities (Fig. 9.1). Marginal habitats (littoral and savanna habitats) were found to contain both small absolute numbers of species and higher percentages of widespread species. These data were interpreted as a function of changes in ecology and dispersal capability as ants moved from the Oriental region, and particularly its rain forests, through the continental islands of Indonesia, through New Guinea, then out across the Bismarck and Solomon islands, on to Vanuatu, Fiji, and

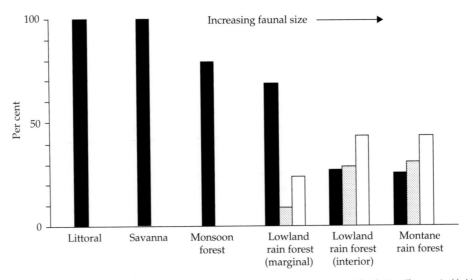

Figure 9.1 Proportions of ponerine ant species in different habitats, as a function of their geographical distribution. The marginal habitats, to the left, contain both smaller absolute numbers of species and higher percentages of widespread species. Dark columns, species widespread in Melanesia; stippled columns, species restricted to single archipelagos in Melanesia but belonging to groups centred in Asia or Australia; blank columns, species restricted to single archipelagos and belonging to Melanesia-centred species groups. (Redrawn from Wilson 1959.)

Samoa. Wilson bundled the species into three stages. Stage 1 species are those that dominate in the marginal habitats: open lowland forest, grassland, and littoral habitats (Fig. 9.2). They have greater ecological amplitudes than the other species, as they also occur in other habitats. They tend to be trail-making ants, nesting in the soil. Stage 1 species typically show a continuous distribution with no tendency to break into locally distinct races. Stage 2 of the process sees ants 'returning' to the dominant vegetation types, of interior and montane forest, within which they are more likely to be found nesting in logs or similar habitats. If they succeed in adapting to the inner rain forests they eventually differentiate to species level within Melanesia, forming superspecies or species groups. As they differentiate, they are liable to exhibit reduced gene flow across their populations. Stage 3 species occupy similar habitats to those of stage 2, but evolution has now proceeded to the point where the species group is centred on Melanesia and lacks close relatives in Asia. This may be, in part, a function of change in Melanesia, and in part of the contraction in the group remaining in Asia. Progress within the cycle is thus marked by a general pattern of restriction of

species to narrower ranges of environments within the island interiors. Meanwhile, these lineages have been replaced by a new wave of colonists occupying the beach and disturbed habitats.

Pivotal to this inferred evolutionary system is that the process is driven by the continuing (albeit infrequent) arrival of new colonist species. Their colonization initiates interactions that push earlier colonists from the open habitats. This is because the earlier colonists are likely to have become more generalized in the time since they themselves colonized, as they spread into additional habitats, thus losing competitive ability in their original habitats (Brown and Gibson 1983). This model is persuasive in that it provides answers to some seemingly puzzling phenomena. For example, it helps explain why species inhabiting interior forests of the oceanic islands are more closely related to species of disturbed habitats on New Guinea, than to those species that have similar niches and which occur in mature forest on New Guinea. Although largely a unidirectional model, it was suggested that some lineages did find a way out of their alleyway, perhaps temporarily. Occasionally, a stage 3 species may readapt to the marginal habitats, becoming

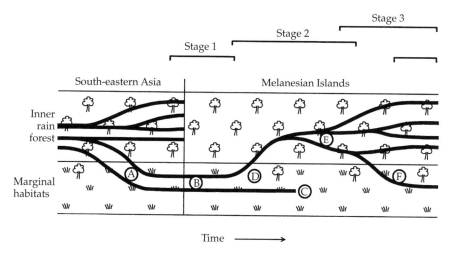

Figure 9.2 The inferred taxon cycle of ant species groups in Melanesia, in which is traced the hypothesized histories of groups derived originally from Asia. The following sequence was postulated: stage 1—species adapt to marginal habitats (A) in the mainland source region (SE Asia) and then cross the ocean to colonize these habitats (B) in New Guinea. These colonizing populations may become extinct in time (C). Alternatively, the cycle enters stage 2—the ants invade the inner rain forests of New Guinea and/or surrounding islands (D). If successful in re-adapting to inner rain forest habitats they in due course diverge (E) to species level. Stage 3—diversification progresses within Melanesia, while the group remaining in Asia may, in cases, be contracting its range, such that the lineage becomes centred in Melanesia. A few members of these lineages, especially those on New Guinea, may re-adapt to the marginal habitats (F), and expand secondarily. (Redrawn from Wilson 1959.)

a secondary stage 1 species, and expanding in its distribution (Fig. 9.2).

Caribbean birds

Ricklefs and Cox (1972) put forward a taxon cycle model to account for the biogeography of the avifauna of the Lesser Antilles. Their version distinguishes an additional stage (Table 9.1, Fig. 9.3), but the basics are shared with the ant study.

Stage I The first stage is the invasion of an island or archipelago by a mainland species. In the Caribbean, this means a species spreading across many islands. It is thus taken to be a dispersive form with interchange occurring between mainland and islands and among islands. The species is likely to be one of disturbed or coastal habitats, as such species typically have good dispersive ability. There is therefore little differentiation initially between mainland and island forms.

Stage II The colonist may then expand its niche, invading other habitats, and becoming more generalized in the use of resources. Species at this stage have more spotty distributions as selection

against mobility reduces gene flow. They gradually evolve local forms and become restricted to a subset of the islands. Species then become vulnerable to being out-competed in their original colonizing niche by further specialized colonists and may become restricted to interior forest habitats so that their niche breadth narrows again.

Stage III As they proceed through the stages of the cycle, species become highly differentiated endemics that ultimately become extinct and are replaced by new colonists from the mainland. Stage III species thus evidence a longer history of evolution in isolation, being found as scattered endemic forms.

Stage IV The fourth stage of the cycle is when a highly differentiated endemic species persists as a relict on a single island. From here the next step is extinction.

One reason for the success of colonists is that a recent arrival may have left behind predators, parasites, and competitors on colonizing an island, enabling it to flourish despite the existence of local species with a longer period of evolutionary adjustment to the conditions on that island. An important

Table 9.1 The taxon cycle as applied to the avifauna of the Caribbean by Ricklefs and Cox (1972, 1978) (from Ricklefs 1989)

(a) *Characteristics of distribution of birds in the four stages of the taxon cycle*

Stage of cycle	Distribution among islands	Differentiation between island populations
I	Expanding or widespread	Island populations similar to each other
II	Widespread over many neighbouring islands	Widespread differentiation of populations on different islands
III	Range fragmented due to extinction	Widespread differentiation
IV	Endemic to one island	N/A

(b) *Number of species of passerine bird on each of three islands*

	Jamaica	St Lucia	St Kitts
Stage I	5	8	6
Stage II	10	7	6
Stage III	8	9	2
Stage IV	12	2	0
Total	35	26	14
Area (km²)	11 526	603	168
Elevation (m)	2257	950	1315

part of the mechanism for the cycle as put forward by Ricklefs and Cox (1972) was the evolutionary reaction, or **counter-adaptation**, of the pre-existing island biota to the newly arrived species, such that over time they begin to exploit or compete with the immigrant more effectively, thus lowering its competitive ability. The subsequent arrival of further immigrant species will then tend to push the earlier colonists into progressively fewer habitats and reduce their population densities. By this reasoning, range of habitat use and population density should each be diminished in species at an advanced stage of the taxon cycle.

Ricklefs and Cox (1978) examined these propositions from their earlier paper on the basis of standardized frequency counts taken in nine major habitats on three islands: Jamaica, St Lucia, and St Kitts. Each species was first assigned to one of their four taxon cycle categories (Table 9.1a). Non-passerines did not show distinct trends in relation to stage of cycle, but the data for passerines were more interesting. The most common stage I species demonstrated habitat breadths and abundances rarely attained by mainland species, indicating that

colonization of islands must involve some degree of ecological release (cf. Cox and Ricklefs 1977), although the phenomena appeared to be confounded by variation in the proportion of species in each stage of the taxon cycle. Passerine species at 'later' stages of the cycle tend to have more restricted habitat distributions and reduced population densities. Early-stage species tend to occupy open, lowland habitats, whereas late-stage species tend to be restricted to montane or mature forest habitats. Similar findings were reported for birds on the Solomon Islands by Greenslade (1968). Ricklefs and Cox (1978) found these trends to be clear for Jamaica, with 35 passerines, but they are difficult to detect on St Kitts, with 13 species, none of which is endemic.

The Ricklefs and Cox model was subsequently criticized by Pregill and Olson (1981), who contested that the four stages are simply criteria that define a set of distributional patterns and that almost any species in any archipelago would fall into one of these categories without necessarily following the progression from stage I to IV in turn. To take a particular island, many of the endemics of Jamaica, especially those endemic at the generic level, are

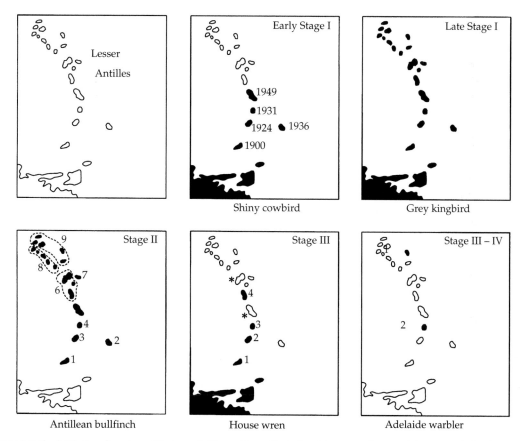

Figure 9.3 The distribution of members of the Lesser Antillean avifauna, illustrative of the proposed stages of the taxon cycle. Species are as follows: shiny cowbird, *Molothrus bonairiensis*; grey kingbird, *Tyrannus dominicensis*; Antillean bullfinch, *Loxigilla noctis*; house wren, *Troglodytes musculus*; and Adelaide warbler, *Dendroica adelaidae*. The small figures indicate differentiated populations (subspecies). The house wren became extinct on the islands marked* within the 20th century. Dates refer to first colonization. (Redrawn from Ricklefs and Cox 1972, Fig. 1.)

widespread and abundant, contrary to the original model. Ricklefs and Cox (1978) took this as illustrating that species on Jamaica have been able to diverge sufficiently in ecological terms for these generic-level endemics to avoid competition from more recent colonists to the island. But is this a case of special pleading? In addition, Pregill and Olson argue that not all endemic species belong to groups that were once widespread. Fossil data indicate that some colonized an island from the mainland and differentiated at the species level without having dispersed to other islands, an example being the Jamaican becard (*Platypsaris niger*). Similarly, they argued that very similar distributional patterns can arise from quite different evolutionary histories.

For example, the smooth-billed ani (*Crotophaga ani*) is absent from fossil deposits in the West Indies, whereas the white-crowned pigeon (*Columba leucocephala*), a species that is practically endemic to the West Indies but which, nonetheless, has a stage I distribution like that of the ani, occurs commonly in Pleistocene deposits.

Generally, analyses of faunal turnover in island studies are based on the unstated premise that environments are effectively stable, i.e. that biotic forcing factors, and not environmental factors, are responsible for turnover. In contrast, Pregill and Olson's (1981) analyses of the fossil record and of relict distributions, particularly amongst the xerophilic vertebrates of the Caribbean, indicated that conditions of climate and habitat were appreciably different during the last glacial period and indeed earlier in the Pleistocene (cf. Buskirk 1985). Lowered sea level combined with climatic changes to produce connections between islands presently separated and also produced a general increase in the extent of xeric habitats, such as arid savanna, grassland, and xeric scrub forest. They list several examples from the fossil record of species characteristic of dry, open country, which once had a much wider distribution than they have had in late Holocene times; these include the burrowing owl (*Athene cunucularia*), the Bahaman mockingbird (*Mimus gundlachii*), thick knees (*Burhinus*), falcons and caracaras (*Polyborus* and *Milvago*), curly-tailed lizards (*Leicophalus*), and rock iguanas (*Cyclura*). The significance of such losses of species on particular islands is exemplified by the figures for New Providence Island, where 50% of the late Pleistocene avifauna and 20% of the fossil herpetofauna no longer occur, many of these species being xerophilic.

Pregill and Olson (1981) provide a number of other examples of species distributions that they regard as most parsimoniously explained by reference to long-term environmental changes. For instance, they list a number of pairs of Hispaniolan bird species, such as the todies (*Todus subulatus* and *T. angustirostris*) and the palm-tanagers (*Phaenicophilus palmarum* and *P. poliocephalus*), which they regard as most probably having differentiated at times during the Quaternary when Hispaniola was divided into north and south islands. Hence, in opposition to the taxon cycle model, they offered the argument that the distributional patterns identified might be primarily an evolutionary outcome of climates becoming wetter (and warmer) since the end of the Pleistocene rather than being interpretable simply in terms of interspecific interactions.

At the time of this debate the distributional and fossil data were inadequate for a clear resolution of the argument. However, recent genetic analyses have provided a new angle on this controversy. Within the taxon cycle model it is assumed that a widespread species moves out in a single colonization phase from a mainland source, and spreads out according to a stepping-stone process of colonization through the archipelago, producing a

monophyletic group within which birds on islands close to one another should be more closely related than those distant from one another. Klein and Brown (1994) point out that the ornithologist James Bond (after whom the fictional character was named) proposed multiple colonizations from mainland sources for West Indian island populations of widespread species. Multiple colonization events would mean that the representatives of a species on different islands would not form a monophyletic group relative to samples from the various mainland source pools. Modern techniques allow Bond's idea to be tested.

Klein and Brown (1994) studied mitochondrial DNA (mtDNA) from specimens of yellow warbler (*Dendroica petechia*) collected from North, South, and Central American sites, as well as from the West Indies. The most parsimonious tree constructed from their phylogenetic analyses indicated colonization of some islands in the Lesser Antilles by Venezuelan birds, and colonization of the Greater Antilles by Central American birds. Furthermore, they found evidence of multiple colonizations not only of the West Indies as a whole, but also of individual islands. Such events can involve the introgression of characters from one lineage to another. Moreover, the phylogenetic data indicated that birds of adjacent islands are not always each other's closest relatives, against the assumptions of a stepping-stone model for colonization. The multiple colonizations of the West Indies by yellow warbler are at variance with the taxon cycle model as outlined above. At least some of the differences between populations are a consequence of what could be thought of as a series of founder effects, rather than exclusively because of *in situ* selective pressures and drift. Klein and Brown (1994) note that a number of recent studies of bats in the Caribbean provide similar lines of evidence for multiple colonizations of individual islands, and of one group of related haplotypes being widespread while another was confined to the Lesser Antilles.

Similar analyses based on mtDNA have now been conducted on many other bird lineages in the West Indies (papers reviewed in Ricklefs and Bermingham 2002), allowing a comprehensive re-evaluation of the debate over the taxon cycle model. Multiple colonization, as exemplified by the yellow warbler, occurred in a few other cases, but overall the results provide remarkably good support for the original model as a fair representation of the temporal sequence followed by many species. Some 20 lineages colonized the Lesser Antilles between 7.5 and 10 Ma, with individual island populations typically persisting for about 4 million years, and with single island endemics typically amongst the oldest species, as required by the model. Some species have undergone phases of secondary expansion. Phases of expansion and contraction do not appear to be correlated across species and contrary to Pregill and Olson (1981) many do not correlate with glacial climate transitions. Indeed, many of the cycles predate the most intense glaciation episodes of the last 2 million years. However, some 17 lineages appear to have colonized the Lesser Antilles between about 0.75 and 0.55 Ma, which suggests that although taxon cycles occur independently of climatic fluctuations, the onset of the most extreme Pleistocene climatic fluctuations may indeed have influenced the composition of the Lesser Antillean avifauna.

In conclusion, Ricklefs and Bermingham (2002) find evidence for expansion phases occurring at intervals on the order of 10^6 years, and that these episodes are not synchronous across independent lineages. They interpret these data as supporting the idea that taxon cycles are related to changes in specific aspects of population–environment relationships, such as host-parasite interactions: as demanded by the 'counter-adaptation' argument of Ricklefs and Cox (1972). Future analyses of the phylogenies of the parasites may allow a more rigorous test of this mechanism.

Caribbean anoles

The anoline lizards of the Caribbean comprise about 300 species, or between 5 and 10% of the world's lizards (Roughgarden and Pacala 1989). On Caribbean islands they are extremely abundant, replacing the ground-feeding insectivorous birds of continental habitats. The distribution of species/subspecies relates best not to the present-day islands but to the banks on which the islands stand, separated from other banks by deep water. The anoles have long provided one of the classic illustrations of **character displacement**, whereby island banks have either one intermediate-sized species or two species of smaller and larger size respectively. This has been taken to indicate that where two populations have established, interspecific competition between them has resulted in divergence of size of each away from the intermediate size range so that they occupy distinct niche space (Brown and Wilson 1956). This progresses to the point at which the benefit of further reduction in competition is balanced by the disadvantage of shifting further from the centre of the resource distribution (Schoener and Gorman 1968; Williams 1972). However, Roughgarden and Pacala (1989) contend that within the islands of the Antigua, St Kitts, and Anguilla banks in the northern Lesser Antilles, a series of independently derived facts contradict this model, and instead support a form of taxon cycle (see also Rummel and Roughgarden 1985).

The anolid taxon cycle of the northern Lesser Antilles begins with an island occupied by a medium-sized species, which is joined by a larger invader from the Guadeloupe archipelago. Rather than the two species moving away from the middle ground, as in the character displacement scenario, both species evolve a smaller size. The medium-sized species becomes smaller as it is displaced from the middle ground by its larger competitor. This opens up niche space in the centre of the resource axis, selecting for smaller size in the invader, which thus approaches the medium (presumably optimal) size originally occupied by the first species. The range of the original resident species therefore contracts, culminating in its extinction. At this point, the invader has taken up occupancy of the niche space of the first species at the outset of the cycle. The islands remain in the solitary-lizard state until (hypothetically) another large invader arrives.

The key points in the Roughgarden and Pacala (1989) analysis are as follows. Intensive scrutiny of the systematics of anoline lizards over a period of four decades resulted in a phylogenetic tree comprising five main groups. The northern Lesser Antilles are populated by the *bimaculatus* group,

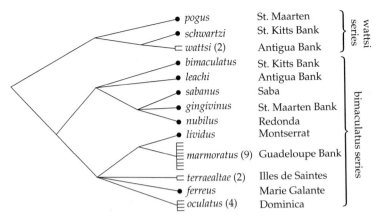

Figure 9.4 Phylogenetic tree for *Anolis* of the bimaculatus group (containing the wattsi series and the bimaculatus series) in the northern Lesser Antilles. Those end-points represented by dots are monotypic species, the others are geographic races or subspecies according to current nomenclature. This tree is merely one trunk of a larger phylogenetic tree for eastern Caribbean anolids, constructed on the basis of detailed systematic/phylogenetic data by Roughgarden and Pacala (1989). (From Roughgarden and Pacala 1989, Fig.1.)

which is subdivided into two series at the next level of the hierarchy (Fig. 9.4). The *wattsi series* comprise small brown lizards that perch within a few feet of the ground. The *bimaculatus series* comprise large or medium-sized green or grey-green lizards, which are relatively arboreal. Islands (for which read *island banks*) have either one species or two living in the natural habitat, i.e. excluding small enclaves of introduced anoles living near houses. Anguilla, Antigua, and St Kitts are the island banks having two species, in each case a larger *bimaculatus* and a smaller *wattsi*.

Earlier research by Williams (1972, cited by Roughgarden and Pacala 1989) had produced two empirical 'rules' concerning the sizes and ratios of the anoles:

1 Species from islands in which only one anole is present are intermediate in size (snout–vent length) between the body sizes seen on islands in which two anole species coexist. This rule is correct for 11 of 12 island banks having a solitary species (they range in length from 65 to 80 mm) and it applies throughout the eastern Caribbean.

2 Species on a two-species island differ in body length by a factor of 1.5–2.0, such that the larger species exceeds 100 mm and the smaller is less than or equal to 65 mm in length, the size of the smallest solitary species. This rule is correct for four of the

five two-species island banks, and again applies throughout the eastern Caribbean.

These patterns appeared to be consistent with the character displacement model. It was also thought that the original residents were members of the *bimaculatus* series, and that the invading species were members of the *wattsi* series, with both invading from Puerto Rico.

Although the character displacement mechanism appeared to explain much of the variation in anoline lizards through the Lesser Antilles, there were some exceptions and further research produced more problems in respect to the northern islands (Roughgarden and Pacala 1989):

● On St Maarten there are two anoles of nearly the same body size. By the character displacement hypothesis, one or both should be a new arrival. However, both are differentiated (visibly and biochemically) from other anoles in the region.

● Medium-sized lizards should become larger but Pleistocene cave deposits demonstrated unequivocally that larger anoles have become smaller, and no fossil data show a medium-sized species becoming larger.

● The two-species community supposedly has its origins in two colonists of similar size, which diverge in size. However, experimental studies of competition and accidental introductions of anoles serve to demonstrate that an invader very similar in size to an established resident does not easily establish on an island. A second species arriving on an occupied island needs to be larger than the occupant in order to become established.

Having dismissed the original model, Roughgarden and Pacala (1989) put forward 12 strands of evidence in further support of their taxon cycle model, from which the following points are taken. The only known historical extinction of an anole in the Lesser Antilles is of the smaller species from a two-species island. North of Guadeloupe, only the *wattsi* series has sufficient geographic variation within an island bank to have led to a subspecific nomenclature, the Antigua bank having *A. wattsi wattsi* on Antigua proper and *A. wattsi forresti* on Barbuda. This may be interpreted as indicating a longer presence of the smaller-sized lineage, the *wattsi* series, than the larger *bimaculatus* series on the northern Lesser Antilles. This is contrary to the original model but in accord with the taxon cycle model. The source for the *bimaculatus* series appears to be Guadeloupe, and this is geographically sensible given that the prevailing currents run towards the northern islands. In short, data on the ecology of colonization, the phylogenetic relationships among anoles, their biogeographical distribution, and the fossil record, refute the character displacement explanation for the northern Lesser Antilles, but are consistent with the taxon cycle model. Yet, in advocating this model, Roughgarden and Pacala (1989) caution that such a cycle does not appear to happen often, and there is no evidence for its operation other than in the northern Lesser Antilles.

Phylogenetic analyses shed further light on this, by showing that the Lesser Antilles are occupied by two distinct lineages of anoles. Dominica and islands to the north contain species related to those on Puerto Rico, whereas islands to the south harbour species with South American affinities (Losos 1990, 1996). The patterns found of character displacement in conditions of sympatry in the northern Lesser Antilles contrast with the evidence for change in body size in the southern Lesser Antilles.

In the latter case change appears to be unrelated to whether a species occurred in sympatry with congeners, and appears instead to result from a process of ecological sorting, such that only dissimilar-sized species can colonize and coexist (Losos 1996).

Evaluation

The taxon cycle has, until recently, been difficult to test. This is in part because of the fluidity of the theoretical framework, and in part because of the possible competing influences of environmental change, and of humans in 'messing up' the biogeographical signal. It has, for instance, been postulated that size reductions and extinctions of some larger lizards in the Caribbean were linked not to intrinsic biological processes so much as to human colonization of the islands (Pregill 1986)—an entirely believable proposition. However, the molecular phylogenies now available for Lesser Antillean avifauna and anoles have served to reaffirm the existence of taxon cycles. A history of taxon cycling has also been proposed for weevils of the genus *Galapaganus* on the Galápagos (Sequeira *et al.* 2000), suggesting that it might be worthwhile testing for taxon cycles in other island and indeed mainland settings.

9.3 Adaptive radiation

The word 'adaptation' gives an erroneous impression of prediction, forethought or, at the very least, design. Organisms are not designed for, or adapted to, the present or the future—they are consequences of, and therefore adapted by, their past.

(Begon *et al.* 1986, p. 6)

Island evolutionary forces reach their most spectacular embodiment in the patterns termed adaptive radiations. The Hawaiian honeycreepers (sometimes 'honeycreeper-finches') and drosophilids and the Galápagos finches are the most famous illustrations from the animal kingdom. Hawaii and the Canaries provide some of the better-known plant examples. The term **adaptive radiation** is taken here to refer to the evolutionary development of distinct species (or varieties) from a single ancestral form (i.e. the lineage is monophyletic), where the radiation is distinguished by niche differentiation amongst the members of the lineage. As we have seen, selection and adaptation does not necessarily lead to radiation. Similarly, as we will discuss later, radiation within monophyletic lineages does not have to be adaptive. The term **non-adaptive radiation** can be applied to cases where allopatric speciation via founder effect and genetic drift mechanisms has been dominant in lineage development, although it is best to think in terms of a gradient between entirely adaptive and non-adaptive end points, with most systems occupying a space somewhere in the middle of such a conceptual gradient. As with the taxon cycle, adaptive radiations are the result of a combination of driving forces and mechanisms; and they are drawn from a common tool kit. Adaptive radiation scenarios differ from the taxon cycle model in being concerned solely with the expansion phase of a monophyletic lineage, whereas taxon cycle theories attempt to describe the expansion and subsequent contraction of particular taxa within an entire evolutionary assemblage (e.g. ponerine ants, or land birds).

The availability of vacant niche space is a key feature in adaptive radiations, as it allows a form of ecological release, allowing the diversification that sometimes leads to speciation within a lineage. As suggested by its name, *Metrosideros polymorpha* is very diverse and multiform. It occupies habitats ranging from bare lowlands to high bogs, and occurs as a small shrub on young lava flows and as a good sized tree in the canopy of mature forest. It is the principal tree of the Hawaiian wet forests and, as an aside, at least some individuals can be found in flower at any time of the year, a pattern supported by enough of the nectar-producing flora to have allowed the evolution of the nectar-feeding lineages among the honeycreepers. Despite its radiation of forms, all members of this complex were originally allocated to *M. polymorpha*, i.e. *Meterosideros* has radiated a lot but has speciated to a lesser degree. Some highly distinct populations have now been recognized as segregate species, but active hybrid swarms also exist (Carlquist 1995), indicating continuing gene flow.

We now consider in a little more detail what we mean by the term adaptive. First, according to the

above quote from Begon *et al.* (1986), adjustment to the abiotic and biotic environment works by natural selection. This is effected by the successful forms contributing more progeny to the following generations, a form of filtering that tells us about the past and which may be of quite varying fit to contemporary circumstances (e.g. of altered climate, depleted native biotas, altered habitats and numerous exotic species). Gittenberger (1991) has argued that there should be no automatic labelling of radiations as adaptive, but that a degree of specialization of species in to different niches should be involved before the term is used.

Once these points are registered, there is little to the core theory of adaptive radiation that we have not already covered. Adaptive radiation invokes not so much a particular process of evolutionary change as the emergent pattern of the most spectacular cases, the crowning glories of island evolution. For this very reason the best cases are liable to involve a wide range of the evolutionary processes previously introduced.

The data in Table 9.2 provide recent estimates for the degree of endemism and an idea of the degree of radiation involved in the biota of Hawaii. These figures are not cast in stone, as they depend on the taxonomic resolution and assumptions involved in their calculation. For example, Paulay (1994) cites an estimate that there may be as many as 10 000 Hawaiian insect species, and that they may have evolved from 350–400 successful colonists. Notwithstanding such uncertainties, Hawaii clearly provides spectacular examples of radiations in taxa as diverse as flowering plants, insects, molluscs, and birds. It is estimated that the 980–1000 flowering plant species arose from between just 270 and 280 original colonists (Fosberg 1948, Wagner and Funk 1995), the pattern of radiation being indicated by the following statistics: there are 88 families and 211 genera, the 16 largest genera account for nearly 50% of the native species, about 91% of which are endemic (Sohmer and Gustafson 1993).

The biogeographical circumstances in which radiations take place are reasonably distinctive. Radiations are especially prevalent on large, high, and remote islands lying close to the edge of a group's dispersal range (Paulay 1994). MacArthur

Table 9.2 Number of presumed original colonists, derived native species, and endemic species for a selection of the Hawaiian biota (Sohmer and Gustafson 1993)

Animal or plant group	Estimated no. of colonists	Estimated no. of native species	% Endemic species
Marine algae	?	420	13
Pteridophytes	114	145	70
Mosses	225	233	46
Angiosperms	272	*c.* 1000	91
Terrestrial molluscs	24–34	*c.* 1000	99
Marine molluscs	?	*c.* 1000	30–45
Insects	230–255	5000	99
Mammals	2	2	100
Birds	*c.* 25	*c.* 135	81

and Wilson (1963, 1967) termed such peripheral areas the **radiation zone**. Here the low diversity of colonists, and the disharmony evident in the lack of a normal range of interacting taxa, facilitate *in situ* diversification (Diamond 1977). In illustration, ants dominate arthropod communities across most of the world, but in Hawaii and south-east Polynesia they are absent (or were, prior to human interference). In their place, there have been great radiations of carabid beetles and spiders, and even caterpillars have in a few cases evolved to occupy predatory niches (Paulay 1994).

For less dispersive taxa, their radiation zones may coincide with less remote archipelagos, which have a greater degree of representation of interacting taxa than the most remote archipelagos. Hence, the circumstances for radiation reach their synergistic peak on the most remote islands, the epitome being Hawaii (below). Examples that fit this idea of maximal radiation near the dispersal limit include birds on Hawaii and the Galápagos, frogs on the Seychelles, gekkonid lizards on New Caledonia, and ants on Fiji; exceptions—taxa that have not radiated much at remote outposts—include terrestrial mammals on the Solomons and snakes and lizards on Fiji (MacArthur and Wilson 1967). Equally, it is clear from even the most remote island archipelagos that not all lineages within a single taxon have radiated to the same degree. Nearly 50% of the *c.*1000 native flowering plant species of Hawaii are derived from fewer than 12% of the *c.*280 successful

original colonists (Davis *et al.* 1995). Most of the rest of the colonists are represented by single species. Such differences may reflect the length of time over which a lineage has been present and evolving within an archipelago, but it is clear that some early colonists have nonetheless failed to radiate. In short, although these geographical circumstances may be conducive to radiations, they are not the only factors of significance and they do not lead to radiations within all lineages.

Darwin's finches and the Hawaiian honeycreeper-finches

Although the Galápagos are renowned for other endemic groups, notably the tortoises (Arnold 1979) and plants (Porter 1979), the most famous group of endemics must be Darwin's finches (Emberizinae; genera *Geospiza*, *Certhidea*, *Platyspiza*, *Camarhynchus*, and *Cactospiza*). The context within which these birds and the other creatures have evolved is as follows. The Galápagos are in the east Pacific, 800–1100 km west of South America (Fig. 9.5a). Although equatorial, they are comparatively cool and average rainfall in the lowlands is less than 75 cm/year (Porter 1979). There are some 45 islands, islets and rocks, of which 9 are islands of area greater than 50 km². Isabela, at 4700 km², represents over half of the land area and is four times the size of the next largest island, Santa Cruz. Isabela and Fernandina have peaks of some

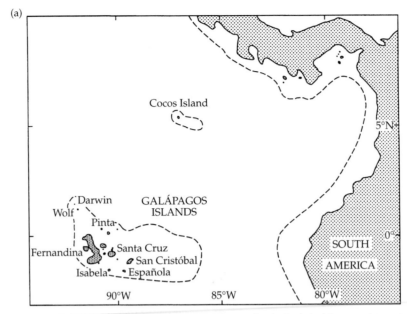

Figure 9.5 The Galápagos archipelago and Cocos Island and their finches. (a) Map, whereby the dashed line approximates the 1800 m depth contour. Five degrees on the equator represent approximately 560 km. (Redrawn from Williamson 1981, Fig. 9.1.). See (b) on the next page.

1500 m ASL, but most of the islands are relatively low. Volcanic in origin, they are true oceanic islands, never having been connected to mainland (Perry 1984). They remain volcanically active, many lava flows being recent and still unvegetated. Geological evidence suggests islands to have existed in the Galápagos for between 3.3 million and 5 million years, although the four westernmost islands (Pinta, Marchena, Isabela and Fernandina) are just 0.7 to 1.5 million years old (Simkin 1984).

Including the 13 finch species, 28 species of land birds breed on the Galápagos, of which 21 are endemic. There are four mockingbirds, which are recognized as separate species on the basis of morphological differences, although as they have allopatric distributions, it is not known if they could interbreed or not (Grant 1984). It is also unclear whether their genus (*Nesomimus*) is sufficiently distinct morphologically to warrant separation from the mainland genus (*Mimus*). Similar problems exist with the finches. As noted by Darwin (1845) in the quotation at the start of Chapter 8, such uncertainties do nothing to diminish their interest. Darwin's finches will be the focus of attention here as they have radiated to the greatest degree, 13 Galápagos

species being generally recognized (e.g. Sato *et al.* 1999; but see Zink 2002), with a fourteenth (*Pinaloroxias inornata*) occurring on nearby Cocos Island (Fig 9.5b; Chapter 8). In fact it is extremely difficult to identify all the Galápagos finches, as the largest members of some species are almost indistinguishable from the smallest members of others (Grant 1984). Collectively, they feed on a remarkable diversity of foods: insects, spiders, seeds, fruits, nectar, pollen, cambium, leaves, buds, the pulp of cactus pads, the blood of seabirds and of sealion placenta. It is principally through changes in beak structure and associated changes in feeding skills and feeding niches that the differentiation between the finches has come about (Lack 1947a; Grant 1994). In illustration, the woodpecker finch (*Camarhynchus pallidus*) uses a twig, cactus spine or leaf petiole as a tool, to pry insect larvae out of cavities. Small, medium and sharp-billed ground finches (*Geospiza fuliginosa*, *G. fortis*, and *G. difficilis*) remove ticks from tortoises and iguanas, and perhaps most bizarre of all, sharp-billed ground finches on the northern islands of Wolf and Darwin perch on boobies, peck around the base of the tail, and drink the blood from the wound they inflict.

(b)

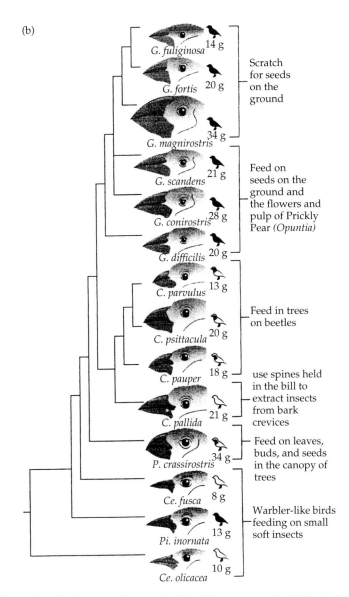

G. fuliginosa 14 g — Scratch for seeds on the ground

G. fortis 20 g

G. magnirostris 34 g

G. scandens 21 g — Feed on seeds on the ground and the flowers and pulp of Prickly Pear (Opuntia)

G. conirostris 28 g

G. difficilis 20 g

C. parvulus 13 g

C. psittacula 20 g — Feed in trees on beetles

C. pauper 18 g — use spines held in the bill to extract insects from bark crevices

C. pallida 21 g

P. crassirostris 34 g — Feed on leaves, buds, and seeds in the canopy of trees

Ce. fusca 8 g

Pi. inornata 13 g — Warbler-like birds feeding on small soft insects

Ce. olicacea 10 g

Figure 9.5 (b) Darwin's finches, the Emberizinae, are endemic to these islands. There are a number of different phylogenetic hypotheses of how these taxa relate to each other. The genetic distance between each species is shown by the length of the horizontal lines. The maximum amount of black colouring in the male plumage and the average body mass are shown for each species. This version is taken from Townsend *et al.* (2003) after Petren *et al.* (1999).

A comprehensive theory for this radiation was developed by Lack (1947a) and has been updated and summarized by Grant (1981, 1984; and see Vincek *et al.* 1997). The key points are as follows. One of the Galápagos islands was colonized from the mainland in a one-off founding event. Genetic analyses suggest that the effective population size of the founding flock was at least 30 individuals (Vincek *et al.* 1997). The founding population expanded quite rapidly, undergoing selective changes and/or drift. After some time, members of this population colonized another island in the archipelago, where conditions were slightly different. Further changes occurred through a combination of random genetic changes, drift, and selection (e.g. for different feeding niches). A degree of differentiation would then have been evident between the separate populations. At some point, individuals of one of the derived populations flew to an island already occupied by a slightly differentiated population. The result of this might sometimes have been an infusion of the newcomers into the established population, but where the two populations exhibited a greater degree of behavioural separation, hybridization would have been limited. Selection would have favoured members of the two groups that fed in different ways from each other and so did not compete too severely for the same resources.

Recent research has shown that female finches appear able to distinguish between acceptable and unacceptable mates on the basis of beak size and shape and also on the different patterns of songs (Grant 1984). This may have been the means by which females were able to select the 'right' mates, thus breeding true, and producing progeny that corresponded with the 'peaks' rather than the 'valleys' of the resource curves. Such characteristics would have a selective advantage in the populations over time, thus allowing two or more species to exist in sympatry.

To put these points together: the radiation of the lineage has taken place in the context of a remote archipelago, presenting extensive 'empty niche' space, in which the considerable (but not excessive) distances between the islands has led to phases of interisland exchange only occasionally. Differing environments have apparently selected for different feeding niches both between and within islands. Thereafter, behavioural differences between forms maintain sufficient genetic distance between sympatric populations to enable their persistence as (to varying degrees) distinctive lineages.

As Grant (1984) cautioned, this is not the only way in which speciation can occur. Theoretically a lineage may have split into two non-interbreeding populations on a single island, as suggested by the model of competitive (sympatric) speciation (Chapter 8), provided that the environment presented two or more sufficiently distinct resource peaks (i.e. was sufficiently heterogeneous). Further

progress has been made in understanding the interactions and exchanges that may occur between sympatric populations of Galápagos finches, highlighting the importance of sympatric episodes in lineage development (Grant and Grant 1996b), but before discussing this work we will outline the case of the Hawaiian equivalents to Darwin's finches.

The Hawaiian islands have formed as a narrow chain from a hotspot, which appears to have been operational for over 70 million years, although the oldest high island of the present group, Kauai, is only about 5.1 million years old (Table 2.3). It is believed that the hotspot has never been closer to North America than it is today and its position relative to Asia has probably also been stable. Molecular clock data suggest that few lineages exceed 10 Ma, i.e. the pre-Kauai signal in the present biota of the main islands is actually rather limited (Wagner and Funk 1995; Keast and Miller 1996).

At least 20 natural avian colonizations have been suggested (Tarr and Fleischer 1995), and endemics comprise approximately 81% of the native avifauna. There have been other avifaunal radiations on Hawaii. They include the elepaio (*Chasiempis sandwichensis*), a small active flycatcher endemic to Hawaii. Distinctive subspecies occur on Kauai and Oahu, and a further three occur within the youngest and largest island, Hawaii itself (Pratt *et al.* 1987). The native thrushes of Hawaii are placed in the same genus (*Myadestes*) as the solitaires of North and South America. Five species are recognized by Pratt *et al.* (1987), each occurring on its own island or island group. One was last seen in 1820, and three of the remaining species are severely endangered, with populations—if they survive—numbering fewer than 50 individuals (Ralph and Fancy 1994).

The Hawaiian honeycreepers (or honeycreeper-finches) have shown an even greater radiation than Darwin's finches. They are a monophyletic endemic group perhaps best considered a subfamily, the Drepanidinae (see Pratt 2005). For a long time it was believed that this radiation had resulted in 23 species in 11 genera. From a single type of ancestral seed-eating finch, the group had radiated to fill niches of seed-, insect-, and nectar-feeding species, with a great variety of specialized beaks and tongues; thus providing one of the most popular illustrations of evolutionary radiation (Fig. 9.6; Carlquist 1974; Williamson 1981). It is now known that there were actually many more species in the recent past. Estimates of the number of species known historically range from 29 to 33, with another 14 having recently been described from subfossil remains: most extinctions occurring between the colonization of the islands by the Polynesians and European contact (Olson and James 1982, 1991; James and Olson 1991; Tarr and Fleischer 1995). As noted, different genera are generally recognized within this subfamily, including *Psittirostra* (the Hawaiian name of which is ou) which have short, conical beaks, for seed eating (essentially they are still finches), and *Pseudonestor* (Maui parrotbill), which uses its powerful beak to

tear apart twigs in order to reach wood-boring beetles. The main evolutionary line, however, is quite different. Its members have longer, narrower bills, and they feed on insects (in the case of *Paroreomyza*, the Molokai creeper) and on nectar (*Loxops* (Akepa) and *Hemignathus* (e.g. Akialoa)).

Tarr and Fleischer (1995) observe that genetic differentiation between species of drepanidines is less than would be expected on the basis of their morphological divergence, indicating that they are the product of a relatively recent arrival (in the order of 3.5–8 million years), followed by rapid adaptive radiation. They concluded that evidence for an allopatric model of speciation was fairly convincing for some of the honeycreepers, but they were not able to rule out sympatric speciation (particularly in the light of recent fossil finds). *Himatione sanguinea*

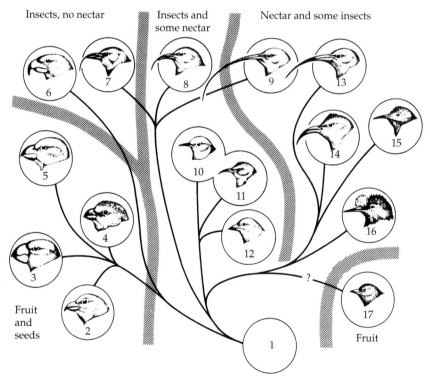

Figure 9.6 The inferred pattern of evolution of dietary adaptations as represented by 16 of the Hawaiian honeycreepers (family Fringillidae, Subfamily Drepanidinae). 1, Unknown finch-like colonist from Asia; 2, *Psittirostra psittacea* ; 3, *Chloridops kona* (*Psittacirostra kona*) (extinct); 4, *Loxioides bailleui* (*Psittacirostra bailleui*) ; 5, *Telespyza cantans* (*Psittacirostra cantans*) ; 6, *Pseudonestor xanthophyrs* ; 7, *Hemignathus munroi* (*H. wilsoni*) ; 8, *H. lucidus* ; 9, *H. obscurus* (*H. procerus*); 10, *H. parvus* (*Loxops parva*) ; 11, *H. virens* (*Loxops virens*) ; 12, *Loxops coccineus*; 13, *Drepanis pacifica* (extinct); 14, *Vestiaria coccinea* ; 15, *Himatione sanguinea* ; 16, *Palmeria dolei* ; 17, *Ciridops anna* (extinct). Source: Cox and Moore (1993, Fig. 6.11). The taxonomy preferred here follows Pratt *et al.* (1987), with that given in Cox and Moore (1993) in brackets where different.

sanguinea (the apapane) and *Vestiaria coccinea* (the i'iwi), two of the more widespread taxa, show an intriguing lack of differentiation, and it has been suggested that this may reflect relatively recent range expansion. This may have occurred in response to the arrival and spread of the tree *Metrosideros polymorpha*, or possibly because of the loss of competitors due to the extinction of other avian taxa. If these interpretations are correct, these two species may represent the early stage of a new cycle of dispersal and differentiation (Tarr and Fleischer 1995).

Having set out the basics of the Hawaiian picture, we can pick up the thread of the discussion of the interactions and exchanges between sympatric populations by reference to comparative studies between the Hawaiian and Galápagos lineages. Studies of morphological variation within populations of Darwin's finches have revealed that there is considerable variation in body size and beak traits within populations of ground finches (*Geospiza*), and that the amount of variation itself differs among populations (Grant 1994). In studies of six of these species, Grant and Grant (1994) found evidence for the limited occurrence of introgressive hybridization, allowing gene flow between populations of different species and contributing to the intraspecific variation observed.

Grant (1994) set out to establish from the study of 524 museum specimens of the 7 species of Hawaiian honeycreepers with finch-like bills (which might be termed honeycreeper-finches) whether the same hybridization process was at work in a lineage that has diversified to a much greater extent. The measurements demonstrated that variation within populations of Hawaiian honeycreeper-finches was *less* than among the *Geospiza*. The single Hawaiian species with both sympatric and allopatric populations did not show greater variation in the sympatric population, as would have been expected if hybridization was effective. It was concluded that there was no evidence of hybridization occurring within the past 100 years. Grant reported that G. C. Munro had suggested in 1944 that *Rhodacanthis flaviceps* might be a hybrid produced from *R. palmeri* and *R. psittacea*, and indeed the only morphological hint of hybridization among the Hawaiian species came from the close similarity found between *R. flaviceps* and *R. palmeri*. If

R. flaviceps is indeed of hybrid origin, then hybridization must have been going on for a long time, as the species is recognizable in fossil material.

Grant (1994) put forward two hypotheses which might explain why the hybridization evident in the Galápagos finches was not apparent in the Hawaiian equivalents.

• Geological and electrophoretic studies indicate that the honeycreeper-finches have been present in the Hawaiian archipelago for a much longer period, perhaps three times longer than have the finches of the Galápagos, and over this time have diversified further and have evolved prezygotic (behavioural) and/or postzygotic isolating mechanisms (Table 8.1).

• The Hawaiian honeycreeper-finches have evolved greater dietary specializations in the generally less seasonal and floristically richer Hawaiian islands. In an environment with more distinctive resource peaks, and following the line of reasoning outlined in the section on competitive speciation, stabilizing selection for specialist feeding may have provided strong selective pressure against an ecological niche intermediate between the two parental species.

Grant put forward a model to capture the important distinctions in a hypothetical phylogeny for the radiation of sympatric taxa (Fig. 9.7). An early phase of divergence is characterized by occasional genetic contact through interbreeding, followed by a genetically independent phase. The duration of the phase of introgression will depend on the ecological isolation attained, which will be related to environmental heterogeneity. From genetic distance measures of finch species known to hybridize from the Galápagos and from North America, Grant (1994) estimates the apparent duration of the phase of introgression, i.e. of hybridization, as up to 5 million years. He speculates that it may have been occurring over the whole of the estimated 2.8 million years of the radiation of Darwin's finches and quite possibly for much of the 7.5 million years estimated by some (but see Tarr and Fleischer 1995) for the diversification of the Hawaiian honeycreepers.

All such figures involve error margins and in the case of the Hawaiian species the evidence for such a model is very largely indirect. Hybridization has

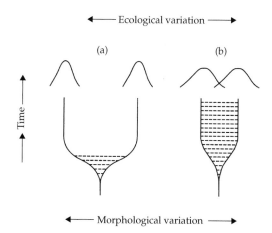

Figure 9.7 A model illustrating the relative significance of introgressive hybridization during divergence, as a function of resource space. In the early introgressive phase of divergence, newly formed species remain in genetic contact through occasional interbreeding, but this is followed by a genetically independent phase. The length of time over which introgression occurs may vary because of ecological factors and the heterogeneity of the resource base. Resource-use curves are shown at the top of the diagram. Species that have become resource specialists (model a) are less likely to hybridize than are generalists (model b). These hypothetical representations were offered by Grant (1994) for the Hawaiian honeycreepers (model a) and Galápagos finches (model b), respectively. (Redrawn from Grant 1994, Fig. 5.)

also been observed between the Galápagos land and marine iguanas which, although of monophyletic origin, are actually placed in separate genera. However, analysis of the DNA of iguana populations on Plaza Sur suggested that if gene exchange occurs it must be at a very much lower rate than in the finches (Rassmann *et al.* 1997).

The studies reviewed in this section allow us to conclude that both allopatric and sympatric episodes can be involved in radiations of island archipelago birds. Although events must vary from one lineage to another, Grant and Grant (1996*b*) suggest the following general scenario for sympatric congeners, based largely on their work in the Galápagos:

- First, following colonization of different islands, there is an initial phase of differentiation of allopatric populations.
- Then, interisland movements re-establish sympatry, following which further differentiation takes place between the two populations.

- Over a period of several million years, occasional introgressive hybridization occurs, often inhibiting speciation; but perhaps in cases the resulting recombinational variation plays a creative role, facilitating further divergence.
- Eventually, when the lineages have diverged far enough, this form of exchange effectively ceases to play a role.

Hawaiian crickets and drosophilids

The Hawaiian endemic crickets are thought to have derived from as few as four original colonizing species, each being flightless species arriving in the form of eggs carried by floating vegetation. The ancestral forms were a tree cricket and a sword-tail cricket from the Americas, and two ground crickets from the western Pacific region (Otte 1989). Three of the successful colonists have radiated extensively, and Hawaii now has at least twice as many cricket species as the continental United States. Much later, a further eight species have colonized, but these are considered to have been introduced by humans and will not be discussed further.

The tree crickets (Oecanthinae) have been calculated on phylogenetic grounds to have colonized Hawaii about 2.5 million years ago, radiating into 3 genera and 54 species (43% of the world's known species), with the greatest diversification seen within the older islands, which were occupied earliest (Otte 1989). They have radiated into habitats not occupied by their mainland relatives. Otte considers competitive displacement to have had a role in the within-island evolution of the Hawaiian crickets, and indeed distributional and cladistic data suggest that virtually all speciation takes place within islands. However, much of it likely occurs in locally isolated habitats, as the active geology and geomorphology of lava tubes, lava islands, incised valleys, and mega-landslips promotes repeated intraisland vicariance events (Carson *et al.* 1990; Carson 1992). This emphasis on within-island speciation concurs with data for the plant genus *Cyrtandra* and the land snail genus *Achatinella*, each of which is represented by over 100 species on Oahu alone.

Drosophila and several closely related genera in its subfamily include about 2000 known species, of which Hawaiian drosophilids (the closely related genera *Drosophila* and *Scaptomyza*) account for some 600–700 species (Brown and Gibson 1983, after Carson *et al.* 1970)—although, as noted earlier, the eventual figure could be as great as 1000 species (Wagner and Funk 1995). The ancestors of the Hawaiian drosophilids (a single species, or at most two) probably arrived on one of the older, now submerged islands and have radiated within and between islands. Carson and colleagues have studied the chromosome structure of members of the picture-wing group, using the karyotypic phylogeny as a means of reconstructing the radiation.

The sequence, as might be expected, appears to stem from the oldest island, Kauai (age 5.1 million years), and, in general, the older islands to the west contain species ancestral to those of the younger islands to the east, i.e. most, but not quite all, of the interisland colonizations have been from older to younger islands (this is termed the **progression rule**). In undertaking this analysis, the islands of the Maui Nui complex (Maui, Molokai, Lanai, and Kahoolawe) may be taken as a single unit, as they have fused and separated at least twice in their short history, as a result of sea-level changes, adding another element of complexity to the picture. Starting from Kauai, at least 22 interisland colonizations are required to account for the phylogeny of the picture-wing species found on the islands of Oahu, the Maui Nui complex, and the Big Island of Hawaii itself. The precise details of this sequence of events may vary according to the means and data used to construct the phylogeny, but the broad pattern is clearly established (Fig. 9.8). Furthermore, the trend of colonization from old to young islands found for these drosophilids is the most frequent found amongst the other lineages investigated to date (Wagner and Funk 1995). However, the majority of speciation events among the drosophilids have occurred within islands and, for example, Carson (1983) notes that of 103 picture-wing *Drosophila* species, all but 3 are endemic to single islands or island complexes. Brown and Gibson (1983, citing Carson *et al.* 1970) report that nearly 50 species of *Drosophila* have been reared from the leaves of *Cheirodendron* (in which they occur as leaf miners). If species can currently coexist in a *Cheirodendron* forest, and even be found in the same

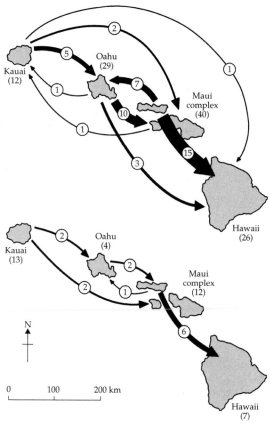

Figure 9.8 The dispersal events among the Hawaiian islands suggested by the inter-relationships of the species of picture-winged *Drosophila* flies (above) and of silverswords (tarweeds) (below). Arrow widths are proportional to the number of dispersal events, and the number of species in each island is shown in parentheses. (Redrawn from Cox and Moore 1993, Fig. 6.9, after an original in Carr, G.D. *et al.* (1989). Adaptive radiation of the Hawaiian silversword alliance (Compositae—Madiinae): a comparison with Hawaiian picture-winged *Drosophila*. In *Genetics, speciation and the founder principle*, (ed. L. Y. Giddings, K. Y. Kaneshiro, and W. W. Anderson), pp. 79–95. © Oxford University Press, New York).

leaf, it is difficult to rule out the possibility that in some cases populations may have diverged while in sympatry. However, as already indicated, in common with other large oceanic islands (Box 9.1) these islands have provided tremendous opportunities for within-island isolation, and it is generally considered that isolation of this form has been crucial to lineage development.

One form of within-island isolation prevalent on Hawaii is when lava flowing down the flanks

Box 9.1   Canarian spiders, beetles, and snails

The Canarian invertebrates provide an outstanding example of radiation on rather less isolated islands than Hawaii and the Galápagos. There may be as many as 6000 native Canarian invertebrate species, of which half are considered to be endemic (Oromí and Báez 2001). In general, the Canarian invertebrate fauna is characterized by the absence of certain groups (e.g. of scorpions, light-worms, Cicadae, Solifugae, and dung-beetles) but this has been compensated for through active speciation within colonist lineages, producing a high species–genus ratio. Additionally, 99 genera are considered exclusive to the archipelago, 58 of them being present in more than one island (multi-island genera) and 41 restricted to one island, of which 25 are found on the island of Tenerife (Oromí and Báez 2001). Not all lineages have radiated extensively; for example, as many as 57 of the Canarian endemic genera are monotypic, indicating a lengthy period of insular isolation and change without radiation.

Radiation is particularly evident in land snails, spiders and beetles, with at least 24 different genera producing 15 or more endemic species (see Table). The largest genera of beetles (*Laparocerus*), diplopods (*Dolichoiulus*), and spiders (*Dysdera*) are represented in all the major habitats, from the coast up to 3000 m, in arid and wet zones, in forested and open areas (even in lava tubes), all over the archipelago, a pattern strongly supportive of the label adaptive radiation. Several of the genera listed in the table are also present on other Macaronesian islands, although the degree to which they have radiated is variable between archipelagos (Oromí and Báez 2001). For instance, among diplopods, there are 2 species of *Dolichoiulus* on Madeira but 46 on the Canaries; in contrast, the figures for *Cylindroiulus* are 25 species on Madeira and 2 on the Canaries. This may reflect the timing of the colonization of each archipelago by different genera, but this explanation requires testing via molecular phylogenies.

As on other geologically young volcanic islands (e.g. Hawaii), there are fine examples of lava tubes on the Canaries. These features provide very similar habitats that are also very different from the environment above ground. Moreover, they are typically quite isolated within each island. Different tubes often feature their own sets of albino and blind *Dysdera* spiders and *Loboptera* cockroaches. Given the difficulty of such forms dispersing between lava tubes, they are most parsimoniously explained as examples of parallel evolution from above-ground relatives.

Number of species of endemic invertebrates of the Canary Islands, within Canarian endemic (bold) and non-endemic genera represented on the islands by at least 15 endemic species (source: Oromí and Báez 2001)

Group	Name	Number of species
Snails	***Hemicycla***	**76**
	Napaeus	**45**
	Obelus	21
	Plutonia	18
Spiders	*Dysdera*	43
	Oecobius	35
	Spermophorides	**22**
	Pholcus	16
Diplopods	*Dolichoiulus*	46
Hemiptera	*Cyphopterum*	24
	Asianidia	17
	Issus	15
Beetles	*Laparocerus*	68
	Attalus	51
	Cardiophorus	31
	Tarphius	30
	Acalles	27
	Calathus	24
	Hegeter	22
	Nesotes	20
	Oxypoda	16
	Pachydema	16
	Trechus	15
Isopods	*Porcellio*	18

leaves undamaged 'islands' of forest, termed *kipukas*, which can vary from a few square metres to many hectares in area. Whether *kipukas* maintain their isolation long enough to allow for speciation may vary between taxa, but in one particularly well-studied picture-wing, *Drosophila silvestris*, it has been found that populations in different isolates show a remarkable degree of genetic differentiation. Carson *et al.* (1990) and Carson (1992) have suggested that the many phases of development of these shield volcanoes provide conditions of continuing division and flux, acting as a general crucible for parallel active evolutionary change in the associated organisms.

The forces that created the Hawaiian islands and their long-ago foundered predecessors not only formed the firmaments upon which life could diversify but also may have played a heretofore unappreciated direct role in the acceleration of evolutionary processes as they operate in local populations (Carson *et al.* 1990, p. 7057).

Geological evidence suggests that the surface of Mauna Loa (the main volcano of the island of Hawaii), has been replaced at an average rate of about 40% per 1000 years during the Holocene, and accordingly the populations of *Drosophila* and other species of the volcano flanks must have been repeatedly 'on the move' (see also Wagner and Funk 1995). It is thus entirely reasonable to posit that species which are sympatric today were at least locally allopatric at the time their lineages diverged. At this sort of scale, however, the terms allopatry and sympatry are best regarded as hypothetical end points, potentially involving little or no interchange on the one hand, and free interchange on the other. Given the dynamism of these environments in evolutionary time, and the fluxes in population sizes and connectivity which are implied, the reality may lie somewhere between these absolutes, with each species existing in the form of metapopulations (Chapter 10), i.e. units experiencing a degree of interchange and of supply and re-supply as local isolates wax and wane (Carson 1992).

Adaptive radiation in plants

Figure 9.8 shows that the general trend of colonization from old to young Hawaiian islands in *Drosophila* (the progression rule), is matched by similar trends in the radiation of the silversword alliance (tarweeds), which is a group of 30 species in 3 genera (*Dubautia, Argyroxiphium,* and *Wilkesia*) in the Asteraceae. The alliance appears to have descended from a colonization event by a bird-dispersed herbaceous colonist (itself a hybrid from the genera *Madia* and *Raillardiopsis*) about 5 Ma, matching the age of the oldest island, Kauai (reviewed by Emerson 2002). They have diversified into a range of life forms, including monocarpic rosette plants, vines, mat-forming plants, and trees, and they occur in habitats ranging from desert-like to rain forest.

For Macaronesia, Humphries (1979) estimated a native vascular flora of 3200 species, of which about 680 (*c.*20%) are endemic. The Canaries are the largest in area, the highest, and the richest, with 570 endemics, about 40% of the native flora. Forty-four Macaronesian lineages are endemic at the generic level also, with exactly half of them restricted to the Canaries (Santos-Guerra 2001). Some lineages have radiated to a significant degree, with *Argyranthemum* (26 species; Asteraceae), *Monanthes* (17 species; Crassulaceae), and *Aichryson* (14 species; Crassulaceae) the most diverse endemic genera. Other non-endemic genera have radiated to even greater extents, notably *Aeonium* (31 species; Crassulaceae), *Sonchus* (29 species; Asteraceae), and *Echium* (28 species; Boraginaceae) (Table 9.3).

Support for the use of the term 'adaptive radiation' in several of the above Macaronesian genera comes from studies of features such as habit, leaf morphology, and habitat affinities (e.g. Table 9.4). Humphries (1979) argues that evidence of similar morphology in species of the same and of different genera (termed parallel or convergent morphology) occurring in similar habitats, is supportive of a model of selection having favoured particular adaptive outcomes.

Within the Macaronesian flora as a whole, most genera are represented by only one or two species, and most of those with over four occur in the Canaries. In general, radiation of lineages has been greatest on the larger islands with the greatest diversity of habitats. The island with the greatest number of endemic species (320 species) is Tenerife and it also has the largest numbers of *Aeonium* (12 species), *Sonchus* (11 species), and *Echium*

Table 9.3 Endemic (in bold) and non-endemic Macaronesian vascular plant genera represented by 10 or more endemic species. (Sources: Santos-Guerra 2001 for Macaronesia; Borges *et al.* 2005 for the Azores; Sziemer 2000 for Maderia; Izquierdo *et al.* 2001 for the Canaries; Arechavaleta *et al.* 2005 for the Cape Verde islands). When total species number does not match the sum, some species are shared among archipelagos

Plant genus	Azores	Madeira	Canaries	Cape Verde	Total
Aeonium	–	2	28	1	31
Echium	–	2	23	3	28
Lotus	1	5	19	5	28
Argyranthemum	–	4	**22**	–	**26**
Sideritis	–	1	23	–	24
Limonium	–	2	18	5	23
Sonchus	–	3	18	1	22
Monanthes	–	1	**16**	–	**17**
Cheirolophus	–	1	15	–	16
Micromeria	–	1	15	1	16
Pericallis	1	1	**13**	–	**15**
Euphorbia	2	2	10	1	14
Aichryson	1	3	**11**	–	**14**
Crambe	–	1	12	–	13
Teline	–	1	9	1	11
Convolvulus	–	1	10	–	11
Tolpis	2	2	7	1	11

Table 9.4 Adaptive radiation in the endemic genus *Argyranthemum* (Asteraceae) in the Macaronesian islands (data from Humphries 1979). Francisco-Ortega *et al.* (1996) recognize 26 monophyletic species in this genus

Environment	Adaptive features	Islands	Species
Broadleaved forests	Type A: shrubby, unreduced leaves	Tenerife and Gomera Madeira	*A. broussonetii* *A. adauctum* ssp. *jacobaeifolium* and *A. pinnatifidum*
South-facing, lowland arid	Type B/C: reduced lignification, habit, capitulum size, and leaf area	Tenerife	*A. frutescens* ssp. *gracilescens*, and *A. gracile*
		Gran Canaria	*A. filifolium*
Sub-alpine and high montane	Type D: dies back after flowering each year, leaves very dissected, hairy	Tenerife	*A. tenerifae* and *A. adauctum* ssp. *dugourii*
Exposed northern coastal areas	Type E: reduced habit, large capitula, increased succulence	Canaries	*A. coronopifolium*, *A. frutescens* ssp. *canariae*, ssp. *succulentum* and *A. maderense*

(9 species). Most endemic species have a restricted distribution, 48% occurring on a single island, and a further 15% on two. From this and other considerations, Humphries (1979) concluded that allopatric speciation has been crucial in the flowering plants.

Furthermore, from studies of a sample of 350 of the endemic species, he concluded that 90% had evolved by gradual divergence following the break-up of a population (including migration from one area to another), with the remainder accounted for

by abrupt speciation involving hybridization, polyploidy, and other forms of sudden chromosomal
change. However, it appears that those notably few
groups which have radiated adaptively into a wide
range of different conditions, have done so without
major chromosomal change (Humphries 1979).

In general, it appears that the 'species swarms'
represented by the species-rich Macaronesian
genera are essentially interfertile, i.e. they readily
hybridize, particularly in disturbed areas. However, stabilized hybrids are rare and so it seems that
the ecological differentiation between species is sufficient to maintain species integrity in most groups.
It follows that although gene flow through introgressive hybridization may occur between species,
hybridization has led to few speciation 'events'
(Humphries 1979; Silvertown *et al.* 2005). Thus,
while details differ, it appears that the essential features of evolution in the Macaronesian flora, in
cases involving spectacular radiations, are common
to other studies encountered in this section (see also
Kim *et al.* 1996). In terms of Rosenzweig's (1995)
three modes of speciation, the data support a role
for each of the geographical, competitive, and polyploid models in declining order of significance.

9.4   From valley isolates to island-hopping radiations

Both the taxon cycle and adaptive radiation scenarios involve a focus on evolutionary–ecological
mechanisms to account for the ways in which lineages develop within islands and archipelagos. In
this section we consider scenarios that are essentially biogeographical rather than mechanistic
models of island evolution. Here the focus is on
reconstructing the sequence of movements across
archipelagos that have been followed by particular
lineages, and establishing if in sum there are emergent tendencies within biotas which may be related
to the developmental history of the islands, or if the
patterns found could be accounted for by chance
dispersal and isolation events within and between
archipelagos. We start with an illustration of how
the lack of dispersal can lead to a form of non-adaptive (selectively neutral) radiation within a single

island and progress to tracing lineages that have
hopped back and forth across entire ocean basins.

Non-adaptive radiation

Gittenberger (1991) discusses the radiation of land
snails of the genus *Albinaria* within the island of
Crete. They have diversified into a species-rich
genus without much niche differentiation, to occupy
more or less the same or only a narrow range of habitats, yet only rarely do more than two *Albinaria*
species live in the same place. Gittenberger thus
posits non-adaptive radiation as a logical alternative
to adaptive radiation, while recognizing that the
process of radiation may involve a blend of the two
over time (cf. Lack 1947*a*). Cameron *et al.* (1996), in
their study of land snails from Porto Santo island
(Madeira), also concluded that the majority of the 47
endemic species recorded had arisen from radiation
of an essentially non-adaptive character by repeated
isolation within this highly dissected volcanic island.
Barrett (1996) suggests that the term non-adaptive
radiation may also be an appropriate way of describing diversification in particular plant lineages, citing
specifically Aegean island populations of *Erysimum*.
Macaronesian plant examples might include many
of the species of *Limonium* (23 species), *Cheirolophus*
(16 species), or *Helianthemum* (9 species) that exploit
similar habitats in different islands.

Non-adaptive radiation as described above is
attributable to stochastic mechanisms, such as genetic drift acting on small, isolated populations, producing an array of sibling species within an island
or archipelago. Within animals, another possibly
important mechanism is sexual selection, which has
been invoked in some studies of cichlid fishes as
contributing a selective yet non-adaptive component within a spectacular radiation (see Box 9.2).

Speciation within an archipelago

We may differentiate endemic species within an
archipelago on simple distributional grounds into
single- and multiple-island endemics. **Single-
island endemics** (SIEs) are those species restricted
to a single island. They may be the result of recent
in situ speciation or may represent the relictual

Box 9.2 The explosive radiation of cichlid fishes in the African Great Lakes

When considering freshwater organisms, lakes can be viewed as islands, and the African Great Lakes (Tanganyika, Malawi, Victoria, etc.), clustered in the Rift Valley, can be considered as an archipelago. They possess one of the most impressive examples of radiation in the form of about 1000 species of cichlid fishes (Cichlidae), with some 200 in Lake Tanganyika, 300 in Lake Victoria, and over 400 in Lake Malawi. Although Lake Tanganyika's cichlids very likely derive from multiple riverine ancestors, those in Lake Victoria apparently derive from just two separate ancestral species and those in Lake Malawi from a single founder population (Danley and Kocher 2001).

Lake Tanganyika developed as a shallow depression linked to the Congo River system about 20 Ma, and achieved its isolation about 9–12 Ma. Molecular clock estimates indicate that the Tanganyika lineages diverged from their ancestors at least 5 Ma. Lake Malawi is at most 1–2 million years old, and the cichlid radiation is estimated to date back only approximately 0.7 Ma. Finally, Lake Victoria is considered to date to between 0.2–0.7 Ma, but may have been completely dry as recently as 12 400 years ago, suggesting a truly explosive rate of radiation subsequently (Johnson *et al.* 1996).

Many models have been proposed in explanation of these spectacular cases of radiation. One of the more straightforward is the **three-stage radiation model** (Streelman and Danley 2003):

1 Major habitat diversification. In the first adaptive radiation stage, an initial divergence of the lineages takes place, resulting in the appearance of different clades related to the major available habitats: the rock-dwellers, the sandy-substrate dwellers, and the pelagic species.

2 Trophic diversification. This second adaptive stage is well studied in the rock-dwellers of Lake Malawi, and involves the rapid directional selection of the feeding apparatus, attaining resource partitioning between species. This process has yielded 10–12 recognized endemic genera, featuring morphological differences in features such as jaw shape and tooth structure.

3 Sexual selection. The most intriguing step of the model is the third phase, a non-adaptive radiation process, involving sexual selection through mate choice by females. This mechanism relies on colour differences in males and is thought to occur only in transparent waters, where the colours are evident. As cormorants are thought to be able to predate the colourful males more easily in transparent waters, it is assumed that bright colouration must come at a cost to fitness and thus can be considered essentially non-adaptive.

---

distribution of a formerly more widely distributed species, which may or may not have originated on the island to which it is currently restricted. For instance, *Saxicola dacotiae* (the Canary island chat) is today restricted to Fuerteventura, having been extirpated from the nearby islet of Alegranza a century ago. Although not endemic, another similar case is *Pyrrhocorax pyrrhocorax* (red-billed chough), now restricted within the Canaries to La Palma but known from fossil evidence to have occurred on Tenerife, La Gomera, and El Hierro (Martín and Lorenzo 2001). Such losses may be the product of natural causes as envisioned within the taxon cycle but in this case are attributable to humans.

In contrast, **multiple-island endemics** (MIEs) are attributable to three different scenarios.

● First, they may have originated on one island, most typically the oldest island, and have subsequently colonized a younger island or islands, where as yet there has been insufficient time (or too much interisland gene flow) for speciation to have occurred.

● Second, they may be palaeoendemics that have evolved little since colonizing an archipelago, but have been lost from their continental source areas.

● Third, in a small number of cases they may be examples of parallel microevolution, in which similar environmental pressures on separate islands

Table 9.5 Distribution patterns amongst the endemic plant species of the Canaries (Source: Izquierdo *et al.* 2001, reworked)

Islands occupied	1	2	3	4	5	6	7	Total
Species no.	289	75	44	30	45	9	19	511
%	56.6	14.7	8.6	5.9	8.8	1.8	3.7	100
More frequent distributions in order of importance	T 107	TC 18	PGT 9	PGTC 8	HPGTC 42	HPGTCF 5		
	C 75	PT 17	GTC 8	HPGT 7	GTCFL 2	PGTCFL 3		
	P 36	FL 17	HPT 5	HGTC 7				
	G 35	GT 11	HPG 5	TCFL 3				
	H 13	HP 4	CFL 5	HPTC 2				
	F 11	PG 3	HGT 4					
	L 12	HG 2						

T, Tenerife; C, Gran Canaria; P, La Palma; G, La Gomera; H, El Hierro; F, Fuerteventura; L, Lanzarote.

have favoured the same mutations, resulting in two related taxa of separate (i.e. paraphyletic) origin being incorrectly classified as one species.

The first scenario is considered by far the most frequent (e.g. Funk and Wagner 1995).

Examples of parallel microevolution have been documented within Canarian *Nesotes* beetles. On morphological criteria, *N. fusculus* was considered a MIE, occupying the xeric coastal zones of Gran Canaria, La Gomera and Tenerife. However, phylogenetic analysis showed this 'species' to be paraphyletic, with the '*fusculus*' phenotype having evolved independently on each of the three islands (Rees *et al.* 2001). A similar scenario applies to *N. conformis*, a laurel forest dweller, found on Gran Canaria, Tenerife, La Palma, and El Hierro. Phylogenetic analyses revealed that the laurel forest phenotype evolved twice within Gran Canaria and once more on the western Canaries (Emerson 2002). This last example appears to show that paraphylly can pertain even for an apparent SIE.

Just over half (57%) of the endemic Canarian flora (bryophytes +vascular plant species) are SIEs, with two thirds of these species being found either on Tenerife or on Gran Canaria (Table 9.5). Nevertheless, each island also harbours more MIEs than SIE species. For instance, Tenerife harbours 151 MIEs and 107 SIEs and Gran Canaria harbours

100 MIEs and 75 SIEs. Intriguingly, the number of endemic species found on two or more islands does not decline smoothly with increasing number of islands as might be expected, with some 45 species being shared by the five high islands of Gran Canaria, Tenerife, La Palma, La Gomera, and El Hierro (Table 9.5). This pattern is repeated for animals, of which 57.5% (1679 species) are SIEs and 42.5% (1239 species) are MIEs. Tenerife (644 species) and Gran Canaria (351 species) again account for the majority of SIEs (data from Izquierdo *et al.* 2001).

How proportions of SIEs and MIEs vary between taxa and archipelagos remains to be established. We provide a preliminary compilation of floristic data for three further volcanic archipelagos: the Azores (Borges *et al.* 2005), the Cape Verde islands (Arechavaleta *et al.* 2005) and the Galápagos (Lawesson *et al.* 1987) (Table 9.6). As in the Canaries, the SIEs of the Cape Verde and Galápagos are the largest single category, but they account for only 22% of endemics, in contrast to 57% SIEs in the Canarian endemic flora. In the Azorean flora, the commonest pattern is of MIEs found on all nine islands, again constituting 22% of endemics. These simple analyses serve to show that the proportions of SIEs and MIEs vary considerably between archipelagos. We can only speculate as to the reasons for these differences. It may be that the Canaries are in some respects more effectively isolated from one

Table 9.6 Endemic plant species distribution in the Azores, the Cape Verde and the Galápagos archipelagos (Azores and Cape Verde data include both bryophytes and vascular plants, whilst Galápagos data only include vascular plants). The data shown are the number and percentage of plant species in each distributional class. The largest class is highlighted in bold for each archipelago (Sources: Borges *et al.* 2005 for Azores; Arechavaleta *et al.* 2005 for Cape Verde; Lawesson *et al.* 1987 for Galápagos)

Archipelago	1 island	2 islands	3 islands	4 islands	5 islands	6 islands	7 islands	8 islands	9 or more islands	Total endemic species
Azores	4	5	2	5	8	8	6	13	**17**	77
(%)	5.2	6.5	2.6	6.5	10.4	10.4	7.8	17.0	**22.1**	(100)
Cape Verde	**16**	6	12	2	8	10	8	1	9	72
(%)	**22.2**	8.3	16.7	2.8	11.1	14.0	11.1	1.4	12.5	(100)
Galápagos	**32**	16	14	17	9	10	13	8	29	148
(%)	**21.6**	10.8	9.5	11.5	6.1	6.8	8.8	5.4	19.6	(100)

another than is the case for the other archipelagos, so that the spread of newly evolved endemics across the archipelago is hampered. For instance, several islands of the Galápagos archipelago (Santa Cruz, Santiago, Baltra, Pinzón, Rábida, Santa Fe, Isabela, and Fernandina) were merged into a single entity during the last glaciation event, as also happened in the Cape Verde islands. However, this was not the case for the archipelago showing the greatest tendency to multiple-island occupancy, the Azores, where only the Faial–Pico fusion occurred (F. García-Talavera, personal communication). Alternatively, and perhaps of greater relevance, the high proportion of SIEs on the Canaries may reflect greater opportunities for within-island speciation and radiation provided by the high topographic complexity and habitat diversity of the western islands of the group, especially within Tenerife, with its steeply dissected Tertiary age Anaga and Teno peninsulars, and the central Quaternary age massif, which is capped by the 3718 m peak of El Teide.

Island-hopping allopatric radiations: do clades respond to islands or to habitats?

Oceanic island archipelagos have been described as 'speciation machines' (Rosenzweig 1995). But, as we have seen, the proportions of single-and multiple-island endemics may be highly variable between archipelagos, indicating that these 'machines'

can function in rather different ways. As highlighted in Box 8.2, the application of the label 'archipelago speciation' does not tell us whether speciation is adaptive or non-adaptive. Advances in the application of molecular analyses in recent decades allow us to approach these phenomena in rather different ways, and to ask different questions (Emerson 2002; Silvertown *et al.* 2005), such as: (1) do clades respond to islands or to habitats? (2) do they move across oceans by routes determined by the shortest distances? (3) do they move back and forth rather than unidirectionally?

Within an archipelago we may posit two hypothetical extreme scenarios for the speciation machine that may be distinguishable in the phylogeographical structure of lineages (clades). First, a particular taxon may respond to islands more than to habitats, i.e. species inhabiting an island are genetically closer to other species on the same island than to species exploiting the same ecosystem on different islands. In this case, a single founder event per island is involved in the process, followed by an adaptive radiation driven by natural selection, which can be more or less complex according to the island's habitat diversity. Second, we can think of clades responding to habitats instead of to islands if the species in a certain habitat are taxonomically closer to other species occupying the same habitat all over the archipelago, than they are to their island congenerics. This scenario needs many more founder events, at least one for each occupied habitat on each island, and would likely involve colonization events back and forth within an archipelago. We have cast these extreme scenarios at the species level, but they might also be considered at the subspecies or population levels.

What do phylogeographical analyses show? Let us begin with evidence that the two extreme scenarios proposed are realistic. For the first scenario (responding to islands), an interesting animal example of intraisland radiation is provided by anolid lizards in the Greater Antilles, where as many as 140 different species occur. Cuba and Hispaniola are each inhabited by six different ecomorphs (species specialized to use a particular structural microhabitat: crown–giant, grass–bush, trunk, trunk–crown, trunk–ground, and twig), whereas five ecomorphs live on Puerto Rico and four on Jamaica. The presence of the same set of ecomorphs on each island might be accounted for in two ways: (1) each evolved once and then colonized other islands, or (2) each ecomorph evolved independently on each island (Emerson 2002). The phylogeny of anoline lizards depicted by Losos *et al.* (1998) for the Great Antilles suggests that, with two exceptions, the members of the same ecomorph class on different islands are not closely related. Rather, the species assemblages on each island seem to be the result of independent events of adaptive radiation. For the second scenario (responding to habitats) Francisco-Ortega *et al.* (1996) provide an analysis of the endemic Macaronesian plant genus *Argyranthemum*, which contains 26 monophyletic species. Between them, they have colonized almost all the habitats existing in the Canaries (see Table 9.4). The phylogeny supports the notion that *Argyranthemum* respond largely to habitats, which means that during their evolution within this island region they have often dispersed successfully between islands, but also that they have generally remained within the habitat type in which they originated. Similar vagility of species among the islands of the Canaries has been revealed in the phylogeography of the plant genera *Aeonium*, *Crambe*, *Sideritis*, and *Sonchus*, although these clades show more frequent evolutionary shifts between habitats than

have occurred in *Argyranthemum* (see e.g. Kim *et al.* 1996; Silvertown *et al.* 2005; Trusty *et al.* 2005).

Funk and Wagner (1995) analysed phylogeographical data from genetic analyses of more than 20 different Hawaiian taxa (among them crickets, *Drosophila*, spiders, honeycreepers, the silversword alliance, and *Clermontia* bellflowers). They inferred more than 100 speciation events associated with interisland dispersal, whereas intraisland radiation accounted for over 200 speciation events, thus yielding an approximate ratio of 1:2. This distinction is, however, oversimplistic as most of these events merely form one part of more complex evolutionary scenarios. They recognize nine of them: four basic patterns (progression rule, intraisland radiation, stochastic, and back-dispersal), four combined patterns (progressive clades and grades, terminal resolution, recent colonization, and extinction), and finally the possibility of an unresolved pattern; to which we have added two further scenarios derived from studies in the Canaries: repeated colonization, and fusion of palaeoislands (Table 9.7). The following exemplification is drawn from Funk and Wagner (1995) except where indicated.

• **Progression rule.** Taxa exemplifying this pattern show dispersion from the older to the younger islands of the archipelago, with or without *in situ* radiation. In the Hawaiian context, this implies initial colonization of Kauai, the oldest extant high island, when it was still young, either from the continent, or from an older now-submerged or eroded island (see Chapter 2). As each new volcanic island became available for colonization, dispersal occurred from the older to the younger island, with speciation following. This pattern is displayed by *Drosophila* flies, *Tetragnatha* spiders, and the plant genera *Hesperomannia*, *Remya* (Asteraceae), and *Kokia* (Malvaceae). In some archipelagos, island age correlates with distance from the principal source region thus potentially confounding interpretation. For example, the age of the Canarian islands decreases with distance from Africa (i.e. from east to west). This colonization path has been taken by many taxa, e.g. the wild olive tree *Olea*, *Gallotia* lizards, *Hegeter* beetles, and *Gonopteryx* butterflies (Fig. 9.9; Thorpe *et al.* 1994; Marrero and Francisco-Ortega 2001a,b).

Table 9.7 Biogeographical-evolutionary patterns exhibited by selected Hawaiian (Funk and Wagner 1995) and Macaronesian taxa (sources given in text). Some taxa provide examples of several patterns

Pattern typology	Hawaiian taxa	Macaronesian taxa
Progression rule	*Drosophila, Hesperomannia, Hibiscadelphus, Kokia, Remya, Tetragnatha*	*Olea, Gallotia, Hegeter, Gonopteryx*
Radiation	*Clermontia, Cyanea, Geranium, Prognathogryllus, Platydesma*	*Echium, Limonium, Pericallis*
Stochastic	Drepanidinae, *Scaveola,*	*Echium,* Crassulaceae
Back-dispersal	*Clermontia, Geranium, Kokia, Laupala, Sarona*	*Pimelia*
Progressive clades and grades	*Drosophila, Laupala, Dubautia*	
Terminal resolution	*Cyanea, Drosophila, Sarona*	
Recent colonization	*Tetramolopium, Clermontia*	*Parus caeruleus*
Extinction	*Clermontia, Geranium, Argyroxyphium*	
Repeated colonization	Unlikely to have occurred	*Calathus, Tarentola, Ilex, Lavatera, Hedera*
Fusion of palaeo-islands		*Gallotia, Steganecaurus, Pimelia, Eutrichopus, Dysdera*
Unresolved	*Clermontia, Sarona*	*Chalcides*

However, the Canaries also provide examples where phylogeographical analysis has shown a pattern of colonization not predictable by distance to the nearest continent and in opposition to the progression rule. For instance, finches of the genus *Fringilla* (chaffinches), originally from the Iberian peninsular, colonized the western Canaries via the Azores and Madeira (Marshall and Baker 1999), while *Pericallis* (Asteraceae) may have colonized the Canaries from America (Panero *et al.* 1999).

● **Intraisland radiation.** Within-island radiation following colonization by an ancestral species has already been discussed at length, and may involve both adaptive radiation and non-adaptive radiation. The latter may be related to periods of intraisland isolation and the complex geological and environmental history of an island. Hawaiian examples of intraisland radiation include *Geranium,* silverswords (*Argyroxiphium* and *Wilkesia*), and *Platydesma* (Rutaceae). This pattern is frequently found as part of a larger archipelagic radiation.

● **Stochastic.** Apparent stochasticity in distribution in relation to the developmental history of the archipelago may result from high vagility (as seems

to be the case for the Hawaiian honeycreepers (Drepanidinae) in so far as we can understand their now fragmented distribution), or from a recent establishment on the archipelago, colonizing adjacent islands in a random fashion (e.g. *Tetramolopium,* in the Asteraceae).

● **Back-dispersal.** This refers to lieages that generally may follow the progression rule, but in which one or more cases of dispersal from a younger to an older island occur, with subsequent speciation occurring from the back-dispersed colonist. This phenomenon can only be detected with certainty by molecular analysis. Examples include *Prognathogryllus* crickets, and *Kokia* (Malvaceae) on Hawaii (Funk and Wagner 1995) and *Pimelia* beetles in the Canaries (Juan *et al.* 2000). Recent studies have demonstrated that the back-dispersal pattern can even reach the continent from which the ancestral species of the lineage originated. Examples of plant lineages back-colonizing Africa from the Canaries include species in the genera *Sonchus* (*S. bourgaeui, S. pinnatifidus*), *Aeonium* (*A. arboreum, A. korneliuslemsii*) (Santos-Guerra 1999) and *Convolvulus* (*C. fernandesii*) (Carine *et al.* 2004).

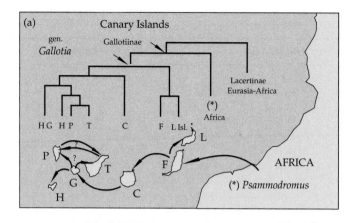

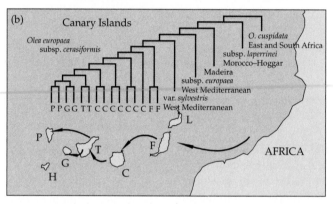

Figure 9.9 Phylogenies of (a) *Gallotia* lizards (based on mtDNA) and (b) *Olea europeaea* subspecies *cerasiformis* (based on internal transcribed spacer 1 (ITS-1) sequences, randomly amplified polymorphic DNAs (RAPD), and intersimple sequence repeats (ISSR) analyses) on the Canary Islands (after Marrero and Francisco-Ortega 2001 figures 15.5 and 15.6; in turn based on papers by González *et al.* 1996, Fu (2000) and Hess *et al.* (2000)). L = Lanzarote, L. Isl. = islets close to Lanzarote, F= Fuerteventura, C = Gran Canaria, T = Tenerife, G = La Gomera, H = El Hierro, P = La Palma.

• **Progressive clades and grades.** A clade is an entire group descended from one common ancestor (i.e. it is monophyletic). Grade refers instead to the level of adaptation, such that two organisms can be classed as of equivalent grade but not have a common ancestor (e.g. Old World and New World fruit bats). Progressive clades and grades are joined together because both result from radiations that differ only in which taxon of the group is involved in the dispersion to another island. This complex combined pattern presents a common feature in which species from a particular island are grouped together (either in a clade or in a grade) and the overall pattern in the area cladogram is consistent with the progression rule. Hawaiian examples are

shown by *Tetragnatha* spiders, *Prognathogryillus* and *Laupala* crickets, as well as *Hybiscadelphus* (Malvaceae) and *Dubautia* (Asteraceae) plant genera.

• **Terminal resolution.** Taxa showing this pattern have either an unresolved base or a base that has experienced a radiation on an older island, with subpatterns involving several well-defined clades. The latter can be either progressions or radiations, in each case indicating repeated introductions from an older, now largely or entirely submerged island. Hawaiian examples can be found in the plant genera *Schiedea* and *Alisinidendron* (Caryophyllaceae) and *Cyanea* (Campanulaceae), as well as in *Sarona* (Miridae) plant bugs.

● **Recent colonization.** Where a comparatively recent colonization event has taken place, landfall may have occurred on any of the extant islands, such that some taxa first colonize a young island, with subsequent dispersal from younger to older islands, while others may colonize an intermediate-aged island, subsequently dispersing to both younger and older islands. This pattern seems to be followed by *Tetramolopium* (Asteraceae) and *Geranium* (Geraniaceae) in the Hawaiian flora. The blue tit (*Parus caeruleus teneriffae* group) provides a Canarian example, with mitochondrial DNA data indicating first colonization on Tenerife and subsequent spread both to older and younger islands (Kvist *et al.* 2005).

● **Extinction.** If a taxon is specific to habitats such as high summits, found only on young islands, then the loss of such habitats from older islands may result in extinction from all but the youngest islands. This pattern can be distinguished from recent colonization in cases where the molecular clock age estimate indicates a much longer period of lineage development than could be provided by the islands on which the taxon currently occurs. In Hawaii this pattern is exemplified by *Clermontia* (Campanulaceae), *Geranium* (Geraniaceae), and *Argyroxiphium* (Asteraceae) plant genera.

● **Unresolved.** Some cladograms fail to provide a clear scenario either because of the inherent complexity of the evolutionary sequences involved or through lack of sufficient data.

● **Repeated colonization from outside the archipelago.** This pattern would not be expected in such a remote situation as Hawaii (although see Terminal resolution, above), but appears common in the Canaries (just 60 km from Africa during low sea-level stands). Repeat colonization produces paraphyly (separate clades) within a taxonomic group: for example, at least three different colonization events have been detected in *Calathus* beetles and *Tarentola* geckos, and two in the plant genera *Ilex* (Aquifoliaceae), *Hedera* (Araliaceae), and *Lavatera* (Malvaceae) (Juan *et al.* 2000, Emerson 2002). However, within the flora, the greatest degree of radiation occurs in genera that appear to be monophyletic (debated in Herben *et al.* 2005; Saunders and Gibson 2005; Silvertown *et al.* 2005).

● **Fusion of palaeoislands.** A repeated phylogeographic pattern found within Tenerife, is the occurrence of allopatric sister taxa in the Anaga (north-east) and Teno (north-west) peninsulas. These areas were separate islands until the construction of the Las Cañadas massif during the Pleistocene fused them together into a single, much larger island (Chapter 2). *Gallotia* lizards, *Chalcides* skinks, *Pimelia*, *Calathus*, and *Eutrichopus* beetles, *Steganecraus* mites, *Dysdera* spiders, and *Loboptera* cockroaches among animals (Juan *et al.* 2000) and *Aeonium*, *Cheirolophus*, *Limonium*, *Lotus*, *Sideritis*, and *Teline* among plant genera (Báez *et al.* 2001), have present-day distributions related to this historical sequence. For instance, *Pimelia* beetles show lineages associated with the west and the east of Tenerife, which have spread over the central regions during the Quaternary (Juan *et al.* 2000).

Island-hopping on the grand scale

As will already have become evident, molecular phylogenetics now allows biogeographers to reconstruct sequences of movement on the grand scale, across entire ocean basins. We have already given examples of how the Canaries have received colonists not only from Africa, the nearest continental landmass, but also from Europe via the Azores and Madeira, from the Cape Verde islands to the south, and even from America. In short, the Canaries have received colonists from all points of the compass, and have also supplied back-colonists to the African mainland.

The Indian Ocean also provides examples of complex scenarios of island-hopping. For example, coastal lizards of the genus *Cyrptoblepharus* appear to have colonized the western Indian Ocean islands by transoceanic dispersal from Australia or Indonesia. The genus appears then to have diversified in Madagascar from where it separately colonized the East African coast, the Comoro islands (twice) and Mauritius (Rocha *et al.* 2006). Galley and Linder (2006), in their review of the affinities of the southern African Cape flora, find that in addition to some ancient disjunctions across the Indian Ocean, there is also strong evidence for some very recent disjunctions, which can only be explained by dispersal events. Examples include *Pelargonium*, *Wurmbea*, and *Pentaschistis*, each of which has Australian members that are recently derived from Cape groups. For a further illustration, *Pentaschistis insularis* is one recently derived species, sister to an African species, which is present on Amsterdam Island, halfway between Cape Town and Australia. Galley and Linder (2006) therefore argue that the formation of biogeographic disjunctions across the Indian Ocean should be seen as ongoing and not just a relictual pattern from break-up of the Gondwanan supercontinent.

In biogeographical terms, New Zealand provides one of the most contested insular systems, with some panbiogeographers arguing that the vast bulk of New Zealand's biogeography is attributable to Gondwanan break-up and rafting of the fragments to their present-day positions. Here again, molecular phylogenies have made telling contributions to the story, showing that many New Zealand lineages have diversified only relatively recently, following the onset of late Tertiary mountain building (5–2 Ma) and Pleistocene climatic change (Winkworth *et al.* 2002). Moreover, as Winkworth *et al.* (2002) show in their review, molecular phylogenies are revealing complex patterns of transoceanic dispersal across the southern Pacific since the late Tertiary, in many cases in opposition to present day current systems. *Metrosideros* subgenus *Metrosideros*, one of the most widely distributed woody plant groups in the Pacific, provides a powerful illustration of this. Analyses based on nuclear ribosomal DNA indicate that one monophyletic clade underwent a dramatic range expansion across Polynesia (including Hawaii) from a Pleistocene dispersal event out of New Zealand. Wright *et al.* (2000) interpret this in terms of changes in wind-flow patterns in the southern hemisphere related to glacial periods.

Some have questioned whether the molecular clocks can be relied upon, arguing that they might in effect 'run fast' in explosive radiations. Bromham and Woolfit (2004) test this proposition using 19 independent cases of explosive radiations in island endemic taxa, but find no evidence of a consistent increase in rates compared to mainland relatives. The evidence of such recent island-hopping radiations does not rule out the existence of ancient historical elements in the biogeography of the southern oceans, but does conclusively demonstrate that long-distance dispersal has played a key role in shaping the contemporary biogeography of the region, especially for plants (Sanmartín and Ronquist 2004; Cook and Crisp 2005; McGlone 2005). Thus, island-hopping on the grand scale goes on, and has presumably done so for a long time and by diverse vectors. Not only has it been the means of colonization of all true oceanic islands, but it has also been important in shaping the biogeographical affinities of continental fragment islands. Not only that, but it is becoming increasingly apparent that island forms do successfully colonize mainlands (Santos-Guerra 2001; Carine *et al.* 2004; Nicholson *et al.* 2005; Rocha *et al.* 2006), providing explanations for at least a minority of the disjunctions in continental biotas on opposite sides of oceans (Galley and Linder 2006).

9.5 Observations on the forcing factors of island evolution

Darwin was right in 1845 to point to oceanic islands as being of great interest to the 'philosophical naturalist'. Evolutionary change on islands has been shown in this and previous chapters to come in many forms, from the microevolutionary alterations of morphology in near-shore mice populations to the macro-scale radiation of lineages. Many geographical, ecological, and genetic mechanisms have been invoked in explanation of the evolutionary changes recorded. These mechanisms provide the mode of delivery of change, but what can we say about the underlying forcing factors of evolutionary change on islands, such as competition, facilitation, impoverishment, disharmony and environmental change?

Competition has often been invoked in discussions of island evolutionary change, but as detailed in the island ecology section of this book such invocations are hard to bring to a definite conclusion. Even where micro-evolutionary changes appear attributable to competition, it is difficult to extend such reasoning to past evolutionary change with any degree of certainty as the extent of past competition between two currently sympatric species is unknowable (Connell's (1980) 'ghost of competition past'). In illustration, Silvertown (2004) has argued that genetic data for Canarian endemic plants indicate that niche pre-emption by earlier colonist lineages may have inhibited the success of later ones. Emerson and Kolm (2005*a*), on the other hand, argue for a sort of facilitation effect, claiming that diversification rate of a taxon is an increasing function of higher species richness of that taxon (diversity begets diversity), using Canarian plant data as one of four illustrative data sets. In practice, Emerson and Kolm's analysis is not based on a rate (i.e. per unit time) estimate but on the proportion of species that are endemic to particular islands, and all that they really show is that richer Canarian floras also have high proportions of endemic species. Our own unpublished analyses of the patterns of single-island endemism suggest that the opportunities for speciation show a roughly parabolic relationship to island age, with speciation rates and proportions of SIEs peaking on islands of young to intermediate age (cf. Silvertown 2004), and then declining, consistent with what we term the island immaturity–speciation pulse model (Box 9.3). It follows that the correlation between plant species richness and single-island endemism reported by Emerson and Kolm (2005*a*) fails to demonstrate that greater richness of a taxon leads to greater rates of (i.e. faster) speciation in that taxon. Both Silvertown's (2004) and Emerson and Kolm's (2005*a*) studies have been criticized for the logic involved in the interpretation (see Cadena *et al.* 2005; Saunders and Gibson 2005; and see replies by Emerson and Kolm 2005*b*; Silvertown *et al.* 2005), supporting our suggestion earlier in this book that the only certainty about claims regarding competitive effects is that they will be disputed.

Despite the problems in testing such ideas, ecological interactions including competition and interactions across trophic levels (mutualisms, parasitism, disease, predation, etc.) are clearly important within island ecology and evolution. This is demonstrated by the extent to which remote, species-poor, disharmonic islands have produced radiations of lineages now occupying ecological niche space that they occupy nowhere else, by the increased tendency towards generalist pollinator mutualisms on particular islands, by the

ground-nesting behaviour of many oceanic island birds (many now driven to extinction), and by many other phenomena reviewed in these pages. Alongside these biotic mechanisms and forcing factors, we must also consider how the role of abiotic forcing factors influences rates and patterns of evolution. This point is amply illustrated in studies reviewed already in this chapter.

How the two sets of forces (biotic and abiotic) interact is hard to determine with any precision. But we might set out the following hypothetical scenario (developed further in Box 9.3). In the earliest stages of the life cycle of a remote volcanic island, opportunities for evolution are severely limited by the barren and volatile nature of the system. As the island builds in area, elevation, and topographic complexity through time, it acquires colonists and complex food webs of interacting organisms. Opportunities for adaptive and non-adaptive radiation (and for mechanisms such as counter-adaptation to operate) increase accordingly. We might therefore expect the highest rate of evolution in this stage. As the island ages it becomes less active volcanically and erosion, mega-landslips, and subsidence combine to reduce its area, topographic complexity, and elevational range. Opportunities for speciation into vacant niche space and into isolated habitat pockets are reduced along with the overall carrying capacity of the island. Eventually the island slips back into the sea, in the tropics perhaps forming an atoll and so sharing an increasingly cosmopolitan strandline fauna and flora with other similar systems. The unique forms found on the island have by this point either colonized other islands or failed altogether.

Such a scenario might well apply to the Hawaiian islands, which have been likened to a conveyor belt, in which the islands undergo a pattern of birth, growth, maturity, and decline. The more recent lava flows within the youngest island are thus, today, the sites in which novel species and adaptations are most apparent. Kaneshiro *et al.* (1995, p. 71) in their analysis of species groups within the picture-wing *Drosophila* of Hawaii make the following observation:

Most of these species, like many other extant terrestrial endemic fauna, show a very strong but by no means exclusive tendency to single-island endemism. Most species thus appear to evolve on an island early in its history and thereafter remain confined to that island. Colonists arriving at newer emerging islands tend to form new species, a finding that has led to the serious consideration that speciation may be somehow related to founder events.

Studies from the Juan Fernández islands by Crawford *et al.* (1992) also support the idea of high *initial* rates of radiation. Electrophoretic data for the endemic genus *Robinsonia* (Asteraceae) suggest that the founding population arrived early in the 4.0 million year history of Masatierra island, radiating and speciating rapidly after colonization (Crawford *et al.* 1992).

Evolutionary change can be very slow paced but it can also occur with great speed, so the notion that speciation rates may vary during the life history of an island is not unreasonable. In illustration of rapid change, Losos *et al.* (1997) were able to demonstrate significant adaptive differentiation over a 10–14 year period in populations of *Anolis sagrei* introduced experimentally into small islands in response to varying vegetation of the recipient islands. Diamond's study of character displacement in the myzomelid honeyeaters is another example of the sort of context in which a fairly precise time frame, a maximum of *c.* 300 years, could be put to a given evolutionary change (Chapter 7; Diamond *et al.* 1989). The limitation of such studies is that they are concerned with microevolutionary changes rather than full speciation: for the latter we depend upon geological and biological dating techniques. Repeated speciation is demonstrated by the 70 species of *Mecyclothorax* beetles endemic to 1 million-year-old Tahiti, and the picture-wing *Drosophila*, in which 25 of the 26 species on the island of Hawaii are restricted to that approximately 0.6–0.7 million-year-old island (Paulay 1994), while the cichlid fishes of Lake Victoria provide the most spectacular example of explosive radiation (see Box 9.2).

Although such large islands as Hawaii were once assumed to be fairly well buffered from climatic variations, the large-scale global climatic fluctuations of the Pleistocene are now known to have affected them (Chapter 2; Nunn 1994) and must have some bearing on the biogeography of such

island systems (above). On ecological timescales, present-day weather conditions can, after all, be highly significant determinants of breeding success in bird species, including, it has been found, in one of the Hawaiian honeycreepers, the Laysan finch (Morin 1992). Some of the most interesting insights in this field come from the remarkable studies of hybridization in Darwin's finches by Grant and Grant (1996*a*), showing how influential the climatic anomaly of the 1982–83 El Niño was in affecting gene flow through hybridization. Thus, although speciation rates may indeed be higher at early stages in the life of an oceanic island, it remains to be established if and how such patterns have been influenced by global climate change during the Pleistocene (cf. Kim *et al.* 1996).

9.6   Variation in insular endemism between taxa

The majority of island lineages have not radiated spectacularly at the species level, but some types of organism have done so repeatedly, in widely separated archipelagos. Examples include genera within the Asteraceae in islands as remote from each other as Hawaii, Galápagos, and Macaronesia, and the finch lineages of Hawaii and the Galápagos. What traits might determine the extent to which particular taxa undergo speciation on islands?

The most obvious trait is dispersal ability. This trait varies hugely between different taxa, and largely determines which fail to colonize, which colonize very occasionally, and which may colonize repeatedly. The first group obviously cannot form island endemics, whilst very frequent gene flow between islands and source regions may also limit speciation. We can illustrate this by looking at endemism in ferns and seed plants on two archipelagos. As already mentioned, the Juan Fernández archipelago contains two main islands, about 150 km apart. Within the angiosperm genera containing endemic species, 31% have at least one species on both islands; the corresponding figure for pteridophytes (ferns) is 71% (Stuessy *et al.* 1990). Endemism within the Galápagos flora is distributed across major taxa as follows: 8 of the 107 pteridophyte species are endemic, 18 of the 85

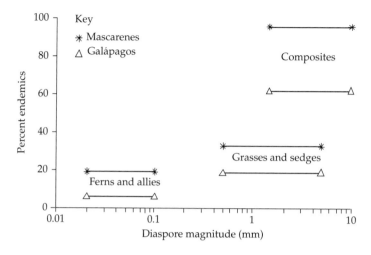

Figure 9.10 Percentage of endemism for selected plant taxa of the Galápagos and the Mascarenes, against diaspore magnitude (the rough length of the dispersal element). (Adapted from Adsersen 1995, Fig. 2.2.)

monocots are endemic, and 205 of the 351 dicots are endemic (Porter 1979, 1984). The low proportion of endemic ferns may be explained as a function of their better dispersability. This is indicated by the data in Fig. 9.10 which illustrate not only the tiny size of fern spores, but that seed sizes of grasses and sedges (monocots) are intermediate in size compared to the classic island endemics in the family Asteraceae. The proportion of endemism in the Hawaiian flora, is considerably higher than for the Galápagos flora, but again is greater in angiosperms (91% endemic) than in ferns and their allies (70% endemism) (Table 9.2).

Dispersal traits thus provide one key element of the evolutionary syndromes characterizing remote island biotas. This is again illustrated by the Hawaiian flora, for which Carlquist (1974) has attempted to determine the most likely means of colonization of each of the ancestral colonists of the current native flora (Table 9.8) (see also Keast and Miller 1996). Species dispersed by terrestrial mammals, for example, are not part of the game.

Porter (1984) has calculated that the Galápagos flora could be accounted for by 413 natural colonists. On this basis and assuming the age of the archipelago to lie between 3.3 and 5 million years (Simkin 1984), the present native vascular plant flora of the islands could be accounted for by one successful introduction every 7990–12 107 years. The error margins on such estimates may be large,

Table 9.8 Most probable means by which the original flowering plant colonists dispersed to Hawaii (data from Carlquist 1974)

Dispersal mode	Percentage of colonists
Birds	
Mechanically attached	12.8
Eaten and carried internally	38.9
Embedded in mud on feet	12.8
Attached to feathers by viscid substance	10.3
Oceanic drift	
Frequent (able to float for prolonged periods)	14.3
Rare (most likely to have arrived) by rafting	8.5
Air flotation	1.4

but they serve to indicate that successful dispersal events to such a remote location are very infrequent, judged on ecological timescales. Studies of less remote islands (e.g. Krakatau) find that the immigration rates vary through time, declining as the most dispersive species colonize and the richness of the flora builds up. It is hard to know if the same variation in rates occurs in very remote situations such as the Galápagos, but it seems likely (Box 9.3). In any event, the above data indicate that remote islands receive new colonists sufficiently infrequently as to provide long periods for those species that do colonize an archipelago early on to

Box 9.3 The island immaturity–speciation pulse model of island evolution

Emerson and Kolm (2005a) recently undertook analyses of the relationship between the proportion of single-island endemics (SIEs) and a number of other island properties for arthropods and plants of the Canaries and of Hawaii. They found that the strongest statistical explanation for the number of SIEs per island for each taxon was the species richness of that same taxon. This analysis is interesting, but inconclusive. It might tell us, as inferred by Emerson and Kolm (2005a,b), that high species richness creates the conditions for high rates of speciation (e.g. through competitive interactions), but it could

also be that the relationship is a by-product of circumstances whereby remote islands of high potential carrying capacity cannot be filled by immigration alone and so provide greater opportunities for speciation.

Here we set out the line of argument that rather than 'diversity begetting diversity', it is *opportunity* that is key to understanding variations in rates of speciation on remote islands, and that opportunity has a broadly predictable relationship to the life cycle of an oceanic island. We term the resulting model the **island immaturity–speciation pulse model**. In setting

Model of relationships between island history, species richness and key rates

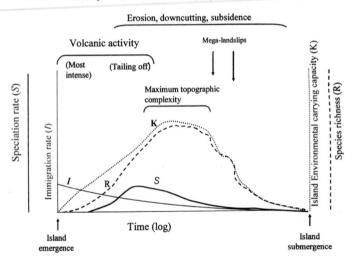

The island immaturity–speciation pulse model of island evolution, showing the form of variation expected in species richness (R), the intrinsic carrying capacity of the island (K), immigration rate (I) and speciation rate (S). Islands such as the Canaries experience a comparatively lengthy old-age of declining elevation, with speciation rate declining in tandem: hence we envisage time as some form of log function, as indicated on the figure. The greatest opportunities for speciation (i.e. where the gap between R and K is greatest) are when an intermediate number of the eventual colonist lineages have arrived, and where environmental complexity is great. This is likely to occur not in extreme youth, but during the youth of the island nonetheless, with opportunities for adaptive radiation greatest in this phase. At the very earliest stage of island colonization, increasing R is driven by I (mostly from older islands within the archipelago), but the rate of I declines (as per MacArthur and Wilson 1963, 1967) as R increases and the proportion of the archipelagic pool already on the island increases. Speciation takes time, and so the rate of speciation lags slightly behind the increased 'opportunity' (the latter is indicated by the gap between K and R), increasing to a peak early on in the islands' history and then slowly declining. The amplitude of I and S curves may vary as a function of effective isolation (compare with Heaney 2000, Fig. 5). Later speciation events, during the old age of an island, are more likely to be consistent with the anagenesis model than with anacladogenesis. For a discussion of extinction rate (E), see box text.

out this temporal model, we describe the observations on which it is based, and then examine what we expect in terms of critical rates, illustrating the model in a simple graphic, of similar format and terms to MacArthur and Wilson's (1967) EMIB graphic (Fig. 4.1).

Analyses of species richness variation reviewed in Chapter 4 provide broad support for the idea of an environmentally-determined carrying capacity *(K)* for richness. Yet, many oceanic islands fail to attain their hypothetical carrying capacity as they are too young or have had too little time since some past disturbance event(s) in which to rebuild their diversity (Heaney 2000).

Remote islands receive colonists so rarely that immigration and speciation occur on similar timescales, and thus increasing proportions of the biota of remote islands are made up of endemic species. Within archipelagos such as Hawaii and the Canaries, young islands should present the greatest opportunities in terms of the physical attributes of size and topographic range, and the biological attribute of 'vacant' niche space (see figure). This implies opportunities for radiation through sympatric, parapatric, and allopatric mechanisms (see text for details).

Following attainment of their maximum altitude, such islands are prone to suffer catastrophic losses due to mega-landslides, subsidence, and erosion (Chapter 2). In time, their eruptive activity slows, and although erosion then dominates, once their elevational range declines below about 1000 m these catastrophic losses in area become less likely (Hürlimann *et al.* 2004), so that the attrition of the island slows as the islands attain old age. Their eventual fate is to slip back into the sea, or, in tropical seas, to persist as atolls.

Opportunities for phylogenesis (maximal values of K−R), and especially for adaptive radiation, can thus be expected to be greatest at a fairly early stage in the process of biotic accumulation on a remote island, when there are sufficiently complete ecological systems to provide adaptive opportunities, but still plenty of 'unoccupied' niches. As opportunity allows, so speciation is initiated, and the rate of production of SIEs increases to an early peak. As the island ages, the complex eruptive and erosive history of the

stage may provide additional opportunities for non-adaptive radiation by within-island isolation (vicariance or peripheral isolation) (cf. Carson *et al.* 1990).

From this point in the island's history, we can consider two scenarios (using the notation in the figure, and with E being the extinction rate):

1. As R gradually approaches the hypothetical value of K, *S* might be expected to decline even as R continues to increase.

2 If the notion of a *fixed* K is considered an illusory concept (because increasing diversity allows more diversity to evolve, as MacArthur and Wilson 1967, Emerson and Kolm 2005*a*), then S must still *eventually* fall with the decline and submergence of the island itself. At this stage however fast evolution may be working, K is diminishing (e.g. see Stuessy *et al.* 1998), and so $E > (I+S)$ and the inevitable slide back into the ocean will necessarily draw R (and with it *S*) down to zero (see figure).

Note that in practice, the hypothetical carrying capacity of a volcanic island over the course of its existence can thus be predicted to describe something like a parabolic function, but skewed to present a peak capacity early on, followed by a far more extended period of gradual decline. This smooth curve will be interrupted by the occasional catastrophe (eruptions, mega-landslides; Chapter 2).

Within a single oceanic archipelago such as the Canaries or Hawaii, it is therefore to be expected that the islands with the greatest K and R values will tend to be those of greatest size, elevation and topographic complexity, and will thus also be relatively young. Hence, within the Canaries, as shown in this chapter, Tenerife, much of which is of ≤2 Ma and the oldest parts of which are ≤8 Ma, has greatest diversity. During 'late youth', islands have been in existence long enough to have sequentially built their substantial size, long enough to have accrued colonists, and long enough for speciation events to have occurred in a wide range of colonists. Hence, high proportions of SIEs are to be expected within such an island.

Hence, we expect to find that the relative contribution of *I* and *S* to the richness of a remote island will vary through time as shown in the figure. Further, although both S and I may each decline during the phase of island subsidence/destruction, the relative importance of S may be greater than in the very earliest stages (thousands of years) of the island's history, during which (as argued by Heaney 2000) most of the increase in richness comes from immigration.

To complete the model, we must consider the loss of SIEs. This can occur through two mechanisms: either colonizing another island, or simply going extinct.

• Colonizing another island. When islands exist for lengthy periods, opportunities exist for some of their new endemics to island hop to other, younger islands, thus becoming (at least for a while) an MIE and no longer an SIE. Many more lineages follow such a progression rule (see text) than back-colonize.

• Extinction. We have not attempted to sketch an extinction-rate curve in the figure, but we anti-cipate that in terms of *biotic* drivers of extinction, E should increase as R approaches K (see figure), although remaining very low in absolute terms. However, these biotic drivers of extinction are likely to be dwarfed by abiotic drivers; at early stages by mega-disturbance phenomena (damag-ing eruptions, mega-landslides), in the later stages of the island life cycle by the gradual attrition of topographic relief and area (Stuessy *et al.* 1998), and over the last 10 000 years, by 'relaxation' associated with the loss of area due to postglacial sea-level rise.

Over the life-cycle of an island as sketched in the figure, we would therefore anticipate that in extreme youth $I > S > E$, from youth towards island maturity $S > I > E$, interrupted by phases of high E(catastrophe), and that in advanced old age, $E > (S \equiv I)$. If I and S do come into approximate balance on an ageing island, then the attrition of SIE status within the biota through the progression rule should result in a decline in both the number and the *proportion* of SIEs on the oldest islands. This means that for archipelagos with a sufficiently complete range of island ages, we expect to see a humped relationship between speciation rate and island age, *and* a humped relationship between the *proportion* of SIEs and island age.

From our analyses (not illustrated) of Canarian arthropod and plant data taken from Izquierdo *et al.* (2004) we found that the proportion of SIEs on the islands is a humped function of island age. This is consistent with the logic that volcanic oceanic islands undergo a burst of speciation relatively early on in their evolution, reflecting the opportunities provided by the vacant niche space, and topographic complexity.

If we accept the logic that the inherent carrying capacity of volcanic islands describes a humped curve over time, it follows that older islands (which have longest to reach their K value) are likely to have greater competitive pressures operating over longer periods of time than younger islands. Hence, if the 'diversity begets diversity' logic holds, these islands (e.g. Lanzarote) should not have lower proportions of SIEs than islands of younger age (e.g. Tenerife), yet they show precisely this pattern. We therefore conclude that the 'diversity begets diversity' model (Emerson and Kolm 2005a,b) is at best an incomplete representation of the data, because it ignores the non-linearity of relationships between island life history and the key biological parameters and rates.

Unfortunately, the empirical evaluation of these ideas depends on simple exploratory regression models, based on several simplifying assumptions and tested on a necessarily small number of islands. This makes for a blunt instrument given the complexities of the environmental histories of archipelagos like the Canaries. The most profitable route to testing the model proposed herein would therefore seem to be via analyses of phylogenies of multiple taxa.

Acknowledgement. This box is a shortened version of an as yet unpublished manuscript, written with our colleagues R. J. Ladle, M. B. Araújo, J. D. Delgado and J. R. Arévalo.

exploit the opportunities available, through combinations of adapative, non-adaptive and archipelagic radiation.

Yet, still some taxa appear to radiate more than others, with some lineages persisting for millions of years on islands without radiating. This suggests that some taxa have a greater inherent degree of genetic plasticity than others (e.g. Werner and Sherry 1987) or that they possess traits related to their breeding systems that favour rapid evolutionary change on islands (see discussion in Chapter 7).

9.7 Biogeographical hierarchies and island evolutionary models

Williamson (1981) has suggested redefining island types as oceanic where evolution is 'faster' than immigration, and as continental if immigration is faster than evolution. He also offered the observation that it is useful for any particular island context to distinguish between those groups for which the island is continental, readily reached by dispersal, and those for which the island is oceanic, leading to little dispersal. For instance, the Azores could be regarded as oceanic for beetles, intermediate for birds, and almost continental for ferns. The gradual diminution in the range and numbers of taxa that have been able to cross ever-increasing breadths of ocean means that, in general, the more remote the island the more disharmonic the assemblage, and the greater the amount of vacant niche space. Thus, the greatest opportunities for adaptive radiation are towards the edge of the range of particular taxa, i.e. in the **radiation zone** of MacArthur and Wilson (1967), reaching their synergistic peak in the most remote archipelagos. Thus, in the absence of ants (amongst other things), there have been great radiations of carabid beetles and spiders in Hawaii and south-east Polynesia (Paulay 1994).

Although diversification in certain lineages has undoubtedly been enabled by the disharmony of very remote islands, the absence of particular interacting organisms (e.g. pollinators for particular plant species) may have restricted the colonization, spread, and evolution of other lineages. That is to say, hierarchical relationships within the island biota, partly predictable and partly the chance effects of landfall (e.g. the highly improbable arrival of the first drosophilids on Hawaii at a particular point in time), must have a strong influence on the patterns that have unfolded.

Once on an archipelago, the relative importance of interisland and intraisland speciation may vary (Paulay 1994). Only a handful of isolated archipelagos show large endemic radiations of birds or plants. The occurrence of a number of islands in proximity to one another appears to be particularly important to birds, suggesting that interisland effects must be important (cf. Grant and Grant 1996*b*). Intraisland speciation is restricted to islands large enough to allow effective segregation of populations within the island. This is taxon dependent, so that for land snails and flightless insects an island of a few square kilometres is sufficient.

As several authors emphasize, different island groups have their own special circumstances and histories, and few theories can span them all. A corollary of this is that each theory may have its own constituency. The solution, as elsewhere, must be to set up multiple working hypotheses and seek evidence to distinguish between them. Yet, frequently, a particular lineage reflects not the operation of a single process, but of several. Therefore it may not be possible to explain the overall biogeography of an island lineage by means of a single model, such as character displacement, a double invasion, or the taxon cycle. Moreover, alongside the evolutionary hypotheses we should also be on the look out for anthropogenic effects. The historical impact of humans in eliminating species from many islands, and thus selectively pruning phylogenies, turning multiple-island endemics into single-island endemics, and altering distributions of extant species (e.g. Steadman *et al.* 1999, 2002), requires careful consideration in evaluating island evolutionary models and phylogenies (e.g. compare Pregill and Olson 1981; Pregill 1986 with Ricklefs and Bermingham 2002).

Having established the relevance of factors such as the regional biogeographic setting, environmental change, dispersal differences between taxa, island area, island habitat diversity, and island isolation and configuration, we should consider how

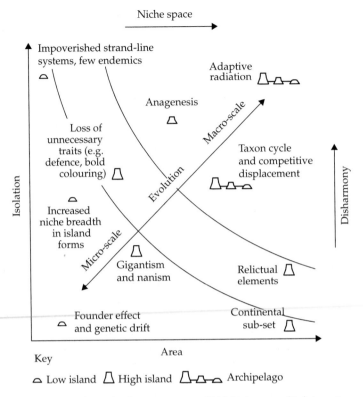

Figure 9.11 Conceptual model of island evolution. This diagram attempts to highlight the geographical circumstances in which particular evolutionary phenomena are most prominent. The scalars will vary between taxa.

they fit together (as, e.g. Fig. 7.7; Box 9.3). Figure 9.11 is one attempt to place island evolutionary ideas and models into a simple island area–isolation context. The aim is to highlight the geographical circumstances in which particular evolutionary phenomena are most prominent. This is not to imply that, for instance, founder events and drift are insignificant in large and isolated oceanic islands, as they do form elements of the macro-evolutionary models. Rather, it is that they emerge as the main evolutionary storyline principally for small, near-mainland islands.

Working up the isolation axis, small islands of limited topographic range and degree of dissection are liable to be poorly buffered from pronounced environmental fluctuations, such as are associated with El Niño events or cyclonic storms. Those species or varieties that become endemic on such islands are likely to have relatively broad niches, at

least in so far as tolerating disruptions of normal food supplies is concerned, as illustrated by the Samoan fruit bat, *Pteropus samoensis* (see Pierson *et al.* 1996).

Large, near-continent islands are likely to have very low levels of endemism. Mainland–island founder effects and drift are also unlikely to be prominent in their biota. Islands such as (mainland) Britain fit this category, having essentially a subset of the adjacent continental biota, most species having colonized before the sea-level rise in the early Holocene that returned the area to its island condition. Large islands, with a somewhat greater degree of isolation and considerable antiquity, are the sites where relictual elements are most frequently claimed in the biota (but see Pole 1994; Kim *et al.* 1996).

As frequently noted, the greatest degree of evolutionary change occurs on remote, high islands. Where these islands are found singly, or in very

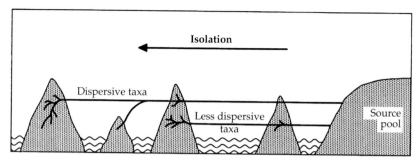

Figure 9.12 Dispersive taxa radiate best at or near to their effective range limits (the radiation zone), but only moderately, or not at all, on islands near to their mainland source pools. Less dispersive taxa show a similar pattern, but their range limits are reached on much less isolated islands. The increased disharmony of the most distant islands further enhances the likelihood of radiation for those taxa whose radiation zone happens to coincide with the availability of high island archipelagos.

widely spread archipelagos, speciation is frequent, but often without the greatest radiation of lineages, fitting the model of anagenesis. Non-adaptive radiation features principally in the literature on land snails, for high and remote islands, where the fine-scale space usage of snails and their propensity to become isolated *within* topographically complex islands allows for genetic drift of allopatric populations. The adaptive radiation of lineages is best seen on large, diverse, remote archipelagos in which inter-island movements within the archipelago allow a mix of allopatric and sympatric speciation processes as a lineage expands into an array of habitats.

Most cases proposed for the taxon cycle are from archipelagos which are strung out in stepping-stone fashion from a continental land mass (the Antilles) or equivalent (the smaller series of islands off the large island of New Guinea). Here the degree of disharmony is less than for systems such as Hawaii and the Galápagos, and, crucially, there are likely to be repeat colonization events by organisms quite closely related, taxonomically or ecologically, to the original colonists.

Of course, Fig. 9.11 is a simplification, and the placing of islands into this framework is likely to vary depending on the taxon under consideration, because of the different spatial scale at which different organisms (snails, beetles, birds, etc) interact with their environment, both in terms of space occupied by individuals, and dispersal abilities. Therefore, the scalars for area and isolation for Fig. 9.11 should vary between taxa, such that, for

instance, the radiation zones for different taxa may coincide with different archipelagos (Fig. 9.12). Such radiation is, however, best accomplished on higher islands, and archipelagos of such islands, than it is on low islands or single high islands.

These simple graphical portrayals of island evolutionary outcomes pay no attention to the temporal development of island biotas, an aspect of island biogeography that warrants further attention in relation to the life cycle of particular islands (Box 9.3), the role of long term environmental change, and in terms of how particular lineages develop. Questions that we might address include, for example: (1) how typical is the rapid expansion of *Metrosideros* across Polynesia documented by Wright *et al.* (2000)? (2) how well do such episodes match to periods of environmental change, altered current systems, or island configurations? and (3) to what degree did the colonization of this particular woody plant lineage on Hawaii transform the evolutionary theatre? Our ability to track the movements of such proficient island hoppers across whole ocean-basins should allow us to address many such questions in the near future. It is an exciting time to be working in or following island evolutionary biogeography.

9.8 Summary

This chapter outlines several emergent patterns by which island evolution may be understood. First, we note that some island speciation occurs with little or no radiation (anagenesis), but instead by change in the island form away from the colonizing phenotype along a singular pathway. Change may also occur in the mainland source region, such that these island forms are in part relictual. Next we consider the taxon cycle, wherein evolution takes place within the context of an island archipelago, driven by a series of colonization events of taxonomically and/or ecologically related forms. Immigrant species undergo niche shifts, driven by interactions with both existing island species and later arrivals. Through time, the earlier colonists lose mobility, decrease in distributional range, and ultimately may be driven to the brink of extinction. Evidence for the taxon cycle in the early literature was largely distributional in nature, but the existence of taxon cycles is now increasingly well supported by molecular phylogenetic data.

The most impressive evolutionary patterns, the radiation of lineages, are to be found on the most isolated of large, oceanic islands. Many such radiations involve clear alterations of niche and may be termed adaptive. Examples are drawn from plant, bird, and insect data from Hawaii, Galápagos, and Macaronesia. The most spectacular cases involve multiple islands and some even spread across different archipelagos. They typically involve a mix of sympatric, parapatric and allopatric phases and 'events', although distinguishing the imprint of these different forms of speciation within the environmentally dynamic situation of oceanic archipelagos is typically difficult. Some radiations have been postulated to be essentially non-adaptive, as a product of genetic drift in within-island isolates rather than clear niche changes. The clearest examples come from studies of land snails on topographically complex islands.

The recent explosion in phylogenetic analyses of island lineages has not only allowed the testing and refinement of the above evolutionary-ecological models, but has also allowed different sorts of models to be developed and evaluated, providing dynamic biogeographical interpretations of lineage development within archipelagos and even across ocean basins. Patterns documented include the progression rule, whereby early colonizing taxa disperse from old to young islands, speciating on each in turn; back-colonization, whereby a minority of movements occur from young to old islands, or even to the continent from which the lineage originated; and the repeated colonization pattern, whereby two or more colonization events are required to explain the island phylogeny.

It appears on theoretical and empirical grounds that the fastest rates of radiation occur among early colonists at relatively early stages of the existence of an island, declining as the island itself ages and declines: we term this the island immaturity–speciation pulse model. Radiation of lineages may also be enhanced by the physical dynamism and environmental variation of large oceanic islands, which can drive repeated phases of within-island isolation. The importance of environmental change, including Pleistocene climatic change, has been gradually gaining more attention, but it is often very difficult to separate abiotic from biotic forcing factors. Differential dispersal abilities, and evolutionary changes in dispersal powers, provide another key element to understanding the disharmony of island ecosystems and the context in which evolutionary change takes place in those lineages that do reach remote islands.

The final section of this chapter attempts to place the principal island evolutionary concepts and models into a simple framework of area versus island—or archipelago—isolation. It is suggested that different ideas come to prominence in different regions of this natural experimental factor space. Thus, for example, the taxon cycle may be an appropriate model to test in a not-too-remote archipelago, while adaptive radiation is more typical of the most isolated archipelagos of high islands. As effective dispersal range varies between (and within) taxa, different evolutionary patterns may emerge within the same archipelago by comparison of different types of organism. These geographical and taxonomic contexts provide a framework within which many of the ideas of island biogeography can be seen to be complementary rather than opposing theories.

PART IV

Islands and Conservation

This view across the cemetery of Old San Juan, Puerto Rico is illustrative of the density of settlement that can be found in the coastal zones of oceanic islands. Habitat loss is a key driver of biodiversity loss on many oceanic islands, and on Puerto Rico has been a key factor in the decline of the emblematic Puerto Rican parrot (*Amazona vittata*), only a few dozen of which persist in the wild (see text).

CHAPTER 10

Island theory and conservation

The classical paradigm in ecology, with its emphasis on the stable state, its suggestion of natural systems as closed and self-regulating, and its resonance with the nonscientific idea of the balance of nature, can no longer serve as an adequate foundation for conservation.

(Pickett *et al.* 1992, p. 84)

The fragmented world of biological communities in the future will be so different from that of the past, that we must reformulate preservation strategies to forms that go beyond thinking only of preserving microcosms of the original community types.

(Kellman 1996, p. 115)

10.1 Islands and conservation

Our species has had a profound influence on the ecology and biodiversity of the planet, altering the composition and functioning of ecosystems across the globe. Our actions in some respects represent merely the latest driver of change, concentrated largely within one interglacial of a period, the Quaternary (the past *c*.2 Ma), which has seen repeated dramatic fluctuations in climate. During this time, individual species have experienced alternating episodes of expansion and of contraction and fragmentation of ranges and have consequently been re-sorted into combinations that have often differed radically from those found today (e.g. Bush 1994). There is nothing new about range alterations. Now, however, it is humans who are driving range shifts and extinctions. We have done this for a long period of our history, and we are doing so in an accelerating fashion and on a global scale (Saunders *et al.* 1991; Bush 1996). Whether we see particular changes in ecosystem properties as beneficial or detrimental can depend as much on

our cultural perspective as on the nature of the changes themselves. However, there is no doubting the profound importance of the human agent in biogeographical terms: not only in extinguishing species but, as significantly, by bombarding areas with sets of species gathered up from far-distant regions of the globe, breaking bounds of isolation that have lasted for many millions of years, and in the case of oceanic islands for their entire existence.

The accelerated extinction of species on a planetary scale is something that concerns all societies and it should therefore be a high priority for natural scientists to study the processes involved, to document the changes involved accurately, to develop predictive models, and to provide scientific guidance for the conservation of biodiversity. This field of study is termed **conservation biology**, and it is an area of research that has expanded greatly in the last 20 years: yet, there remain many uncertainties concerning the big picture. In illustration, we are currently unable to say to within an order of magnitude how many species we share the planet with (Gaston 1991; Groombridge 1992). A second illustration is that although a large proportion of land plants and animals are believed to be rain forest species, we cannot be sure how large a proportion, nor how much tropical forest exists, nor with much precision, how fast these ecosystems are being degraded of their biodiversity on a global scale (Whitmore and Sayer 1992; Grainger 1993). We know there to be a crisis, but we cannot be sure of its precise magnitude.

Nature conservation as a social movement is not motivated solely by a concern to reduce global species extinction, but is also imbued with other values and goals. However, a full consideration of the theoretical and operational principles of nature conservation and the long history of the conservation movement are beyond our present remit. Our purpose in this final section of the book is to consider two broad themes. First, in the present chapter, we examine the application of ideas derived from island ecological biogeography to conservation problems. This research programme is focused largely on the loss of species resulting from the reduction in area and fragmentation of 'natural' habitats, and is based on the premise that island ecological rules can be applied to 'islands' of particular habitat types within mainland landmasses. This work forms part of the emerging field of conservation biogeography (Whittaker *et al.* 2005). Second, we return in the final two chapters to a focus on real islands, examining the particular conservation problems of oceanic islands and some of the management solutions that have been developed. The rationale for focusing on such islands in an island biogeography book is obvious, but this focus is also justified by the wider significance of islands as repositories of biodiversity and by the extent to which island endemics feature in the lists of recently extinct and currently endangered species.

10.2   Habitats as islands

... the abundance of publications on the theory of island biogeography and applications [of it] has misled many scientists and conservationists into believing we now have a ready-to-wear guide to natural area preservation ...

(Shafer 1990, p. 102)

It is generally accepted amongst conservation biologists, that the ongoing fragmentation and reduction in area of natural habitats is a key driver of terrestrial species extinctions at local, regional, and global levels (Whitmore and Sayer 1992). The remaining areas of more-or-less natural habitats are increasingly becoming mere pockets within a sea of altered habitat; in short, they are **habitat islands**. It was therefore to island biogeography that many scientists turned in the search for predictive models of the implications of fragmentation and for guidance as to how best to safeguard diversity.

The implications of increasing insularity for conservation were recognized in the seminal works of Preston (1962) and MacArthur and Wilson (1963, 1967), with Preston astutely observing that in the long run species would be lost from wildlife preserves, for the reason that they constitute reduced areas in isolation. MacArthur and Wilson's dynamic (EMIB) model provided the theoretical basis to develop this argument, allowing conservation scientists to explore the outcomes of differing configurations of protected areas, assuming them to act as virtual islands in a 'sea' of radically altered and thus essentially empty habitats. One early manifestation of this research programme was the so-called SLOSS debate, which posed the question 'given the opportunity to put a fixed percentage of land into conservation use is it better to opt for a Single Large Or Several Small reserves?' Here the focus is on the entire regional assemblage of species and how to maximize the carrying capacity of the protected area (i.e. habitat island) system. We will come back to this debate a little later, as we have organized this discussion to begin with the smallest units: populations of particular species.

The most basic question is: 'how many individuals are enough to ensure the survival of a single isolated (perhaps the last remaining) population? The resulting figure is termed the **minimum viable population**. We will consider this first. Many species of conservation concern are not, however, restricted to a single locality—or at least, not yet. Rather, their survival in a region might be dependent on a network of habitat islands. Most of this book has been concerned with real islands in the sea. Is it realistic to expect habitat islands within continents to behave according to the same principles as real islands? Instead of the barrier of salt water, a forest habitat island might be separated from another patch of forest by a landscape of meadows, hedgerows, and arable land. The implications for species movements between patches might be radically different. As we have seen, however, this is probably too simple a distinction and too simple a question. Scale of isolation is crucial, as well as the nature of the intervening landscape filter. Mountain habitat islands within continents, but surrounded by extensive arid areas, may be much more effectively isolated than real islands that happen to be located within a few hundred metres of a mainland. Many isolates are actually sufficiently close to one another that their populations are in effect linked: they form metapopulations, the second theoretical element that we will consider. When we have added these tools to our armoury, we can then consider the whole-system implications of fragmentation and how 'island' approaches have contributed to their understanding.

Over the past few decades there has been a shift in ecology from what Pickett *et al.* (1992) term the **balance of nature** paradigm, which sees the world as a self-regulating, balanced system, towards the **flux of nature** paradigm, which Pickett *et al.* (1992) believe provides a more realistic basis for conservation planning and management. This framework does not deny the possibility of equilibrium, but recognizes equilibrium as merely one special state. The flux of nature viewpoint stresses process and context, the role of episodic events, and the openness of ecological systems.

Conventionally, the 'island theory' input to conservation debates has been based on the general assumption that a 'natural' area is in a state of balance, or equilibrium. Humans then intervene to remove a large proportion of the area from its natural state, leaving behind newly isolated fragments, which are thus cast out of equilibrium. Subsequently, these fragments must shed species as the system moves to a new equilibrium, determined by the area and isolation of each fragment. This way of thinking is entirely consonant with the balance of nature paradigm.

By contrast, in the present volume we have stressed the importance of considering physical environmental factors in island biogeography, and have shown that ecological responses to environmental forcing factors are often played out over too long a time frame for a tightly specified dynamic equilibrium to be reached. Moreover, relatively few areas of the planet, if any, are pristine, untouched by human hand, so most areas that we may be fragmenting already contain the imprint of past land uses (e.g. cultivation and abandonment, hunting) (e.g. Willis *et al.* 2004). Hence, if the starting assumptions concerning the prior equilibrial state

of the system are flawed, some of the theoretical island ecological effects may, in practice, be fairly weak. These points are addressed in the present chapter in the applied, conservation setting. Hence we will consider the physical effects of fragmentation as well as the biotic, and we will examine what sorts of factors in practice cause species to be lost from fragments.

10.3 Minimum viable populations and minimum viable areas

How many individuals are needed?

What is the **minimum viable population** (MVP)? By this we mean the minimum size that will ensure the survival of that population unit, not just in the short term, but in the long term. It is often defined more formally in terms such as the population size that provides 95% probability of persistence for 100 or for 1000 years. Attempts to calculate the viability of single populations, i.e. whether the population exceeds the minimum requirement, are termed **population viability analyses (PVA)**. Important elements that have to be considered in PVA include demographic stochasticity, species traits, genetic erosion in small populations, and extreme events such as hurricanes or fires.

Demographic stochasticity of initially small populations can lead to losses from a series of isolates without a need to invoke any common factor such as predation or loss of fitness. However, where small populations persist for a reasonable length of time, they may also lose genetic variability as they pass through bottlenecks. They may then lack the genetic flexibility to cope with either the normal fluctuations of environment or an altered environment, and they may also accumulate deleterious genes. In short they lose **fitness** (Box 10.1). It is generally held as axiomatic that an increase in **inbreeding** in small populations reduces fitness in animals, although unambiguous demonstrations of the effect in natural populations are relatively scarce (Madsen *et al.* 1996). The basic rule of conservation genetics is that the maximum tolerable rate of inbreeding is 1% per generation. This translates into an effective population size of 50 to

Box 10.1 Fluctuating asymmetry in a fragmented forest system

Fluctuating asymmetry (FA) refers to differences between the right and left sides in animal anatomical characteristics that are usually bilaterally symmetrical, and is a form of developmental instability that may have genetic and/or environmental causes. Whatever the nature of the control, a high incidence of FA is regarded as a sign of a population under environmental stress. Recent work has suggested that small, isolated populations in fragmented habitats may exhibit a higher incidence of FA. Anciães and Marini (2000) test this hypothesis for wing and tarsus FA on 1236 mist-netted passerine birds of 100 species captured from 7 fragments and 7 control areas in the Atlantic rain forest area of Brazil. They recorded a higher mean level of FA in both features in the forest fragments than in the control areas, suggesting that passerine birds in these forests exhibit morphological alterations as a result of forest fragmentation and habitat alteration. Possible proximal causes might include increased pressures from predators, competitors, parasites and disease, and/or decreased food supplies, shelter or nest site availability.

In general, terrestrial and understorey insectivores showed greater relative FA changes than other foraging guilds, in line with the general sensitivity of this guild to habitat fragmentation. An example is *Dysithamnus mentalis*, which forages in the medium strata and may also follow army ants. It has previously been found to be an area-sensitive species, and indeed was observed in only two of the fragments in the study, where it showed higher levels of FA in the tarsus than in the control forests. By contrast, the most abundant frugivores sampled in the study (*Chiroxiphia caudate, Manacus manacus*, and *Ilicura militaris*) were commonly found in altered forest patches and seemed unaffected by fragmentation in their incidence and in terms of FA values.

This form of developmental instability may provide a form of early warning system, indicating stresses on vertebrate species in isolates prior to more profound changes, such as rapid population declines and species extinctions.

ensure short-term fitness according to the calculations of I. A. Franklin in 1980 (reported in Shafer 1990). However, there will still be a loss of genetic variation at such a population size, and Franklin recommended 500 individuals in order to balance the loss of variation by gains through mutation. These commonly cited values appear to derive respectively from data from animal breeders and bristle number in *Drosophila*, and so should be regarded as merely first approximations (Fiedler and Jain 1992). Lacy (1992) noted that crudely estimated MVPs for mammals based on data for inbreeding effects from captive populations varied over several orders of magnitude, and that it would be premature to generalize on MVPs in the wild on this basis.

Typically, only a proportion of the *adult* population participates in breeding; and it is these animals that form the **effective population size**, which is often substantially smaller than the total population size (Shafer 1990). A study of grizzly bears in the Yellowstone National Park showed that to prevent inbreeding rates exceeding 1% required an overall population size of at least 220 rather than 50 animals (Shafer 1990).

It is intuitively reasonable that the loss of a broad genetic base is likely to increase the probability of extinction. However, we learnt in an earlier chapter that putting a population through a bottleneck of just a few individuals may not be as damaging to the genetic base as one might anticipate, providing the population is allowed to expand again fairly quickly. In illustration, the great Indian rhino was reduced within the Chitwan National Park in Nepal (one of its two main havens) to an effective population of only 21–28 animals in the 1960s. In this case the bottleneck was short. Rhinos have a long generation time, and it has since recovered to about 400 individuals. Genetic variation within the population is still remarkably high (Tudge 1991). Nonetheless, the *Drosophila* studies reviewed in Chapter 7 showed that sustained or repeated bottlenecks do lead to loss of heterozygosity. Hence, although you may be able to rescue some, perhaps many, species from a bottleneck of short duration, you cannot assume that this will apply to a sustained or repeated bottleneck. Just because a suite of threatened species are hanging on in their reserves now, it doesn't mean that the same level of biodiversity can be sustained for a long time in a fragmented landscape.

A further complication in assessing genetic effects is that where a species is split into numerous separate populations in fragmented habitats, there may be multiple bottlenecks involved. This may result in reduced variation *within* each population but increased genetic differentiation *between* populations (see Leberg 1991). This may be of significance to lengthening persistence in metapopulation scenarios (below).

Pimm *et al.* (1988) have undertaken an analysis of island turnover that illustrates how some species traits may be important to determining minimum population sizes. They analysed 355 populations belonging to 100 species of British land birds on 16 islands. They found that the risk of extinction decreases sharply with increasing average population size. They examined extinction risk for large-bodied species, which tend to have long lifetimes but low rates of increase, and for small-bodied species. They found that at population sizes of seven pairs or below, smaller-bodied species are more liable to extinction than larger-bodied species, but that at larger population sizes, the reverse is true. This can be explained as follows. Imagine that both a large-bodied and a small-bodied species are represented on an island by a single individual. Both will die, but the larger-bodied species is liable to live longer and so, on average, such species will have lower extinction rates per unit time. On the other hand, if the starting population is large, but it is then subject to heavy losses, the small-bodied species may climb more rapidly back to higher numbers, whereas the larger-bodied species might remain longer at low population size, and thus be

vulnerable to a follow-up event. They also found that migrant species are at slightly greater risk of extinction than resident species. Numerous factors might be involved in migrant losses, such as events taking place during their migration or in their alternative seasons' range (e.g. Russell *et al.* 1994). Further studies will be necessary to establish the generality of this particular finding (see discussions between Pimm *et al.* 1988; Tracy and George 1992; Diamond and Pimm 1993; Haila and Hanski 1993; and Rosenzweig and Clark 1994).

Isolates subject to significant environmental change or disturbance may need to have much larger populations than is otherwise the case to ensure survival. A number of studies have recognized the *potential* of environmental catastrophes in this context but, typically, they have not attempted to evaluate the general significance of such catastrophes (see Pimm *et al.* 1988; Williamson 1989*b*; Menges 1992; Korn 1994). Of course, it is inherently difficult to incorporate extreme events into such analyses. As Bibby (1995) remarks 'how do you estimate the probability of a cat being landed on a particular small island in the next five years?' Yet, such an event may have a crucial impact on an endangered prey species.

Mangel and Tier (1994) argue that environmental catastrophes may often be more important in determining persistence times of small populations than any other factor usually considered, and therefore should be incorporated explicitly when formulating conservation measures. They go on to provide computational methods for modelling persistence times which do take account of catastrophes and which allow for quite complicated population dynamics. Unsurprisingly, they find that the resulting MVPs are larger than those derived from variants of the MacArthur and Wilson model, or indeed other analyses of population viability which ignore catastrophes (Fig. 10.1) (see also Ludwig 1996).

An empirical demonstration is provided by Hurricane Hugo, which in 1989 caused extensive damage to the forests of the Luquillo mountains, the home of the last remaining wild population of the Puerto Rican parrot (*Amazona vittata*). In 1975 the parrot population had reached a low of 13 wild

(a)

(b)

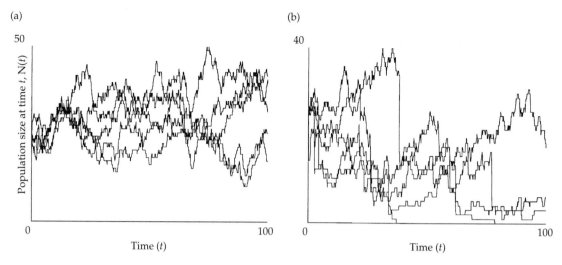

Figure 10.1 Simulation runs of population trajectories. (a) In the absence of catastrophes all five trajectories persist for the 100 time units shown. (b) In the presence of catastrophes two populations become extinct and two are at low population sizes after 100 time units. (From Mangel and Tier 1994, Fig. 5; original figure supplied by M. Mangel.)

birds, as a result of a combination of habitat destruction, capture for pets, and slaughter (for food and to protect crops). Given protection, the population built up from this low point, to pre-hurricane levels of 45–47, from which it was reduced by the hurricane to about 25 birds (Wilson *et al.* 1994). Intriguingly, post-hurricane conditions may have suited the parrots, as 6 pairs bred in 1991, the highest total since the 1950s, and 11 young fledged the following year. The post-breeding population thus swiftly climbed to between 34 and 37 parrots. However, the number of birds is clearly far too small for comfort, and the species could easily be pushed to extinction by repeated catastrophic hurricanes.

As endemic species tend not to occur on the smallest of islands, extinctions in the global sense caused primarily by natural disturbances are likely to be rare events on ecological timescales, unless the species concerned has already been reduced in range (Hilton *et al.* 2003). One possible example of a species finished off in this way, by hurricanes in 1899, is the St Kitts subspecies of bullfinch *Loxigilla portoricensis grandis*, although the last record of this bird was actually several years earlier, in 1880 (Williamson 1989*b*). A more recent example of a species reduced to the status 'critically threatened'

is the Montserrat oriole (*Icterus oberi*), which although confined to only 30 km² of forest within the island of Montserrat, was not considered globally threatened before the eruption of the Soufriere Hills volcano in 1995. The eruptive activity (as often the case) spanned several years (indeed it had not ceased as of the end of 2005), devastating much of the forest habitat of the oriole. The dramatic population decline extended even to the small areas of intact forest, suggesting knock-on consequences of the eruption such as reduced food supplies or increasing nest predation. It proved difficult to estimate the oriole population but it is thought that it had declined from several thousand individuals before the eruptions to perhaps 100–400 pairs in the year 2000 (Hilton *et al.* 2003). The populations are currently being closely monitored (<www.rspb.org.uk/science/diaries/montserrat/index.asp>).

In order to make MVAs more realistic, efforts have now been made to include data on age structure, catastrophes, demographic and environmental stochasticity, and inbreeding depression. Reed *et al.* (2003) use this approach to derive MVP estimates for 102 vertebrate species, using a working definition of MVP as 'one with a 99% probability of persistence for 40 generations'. Their chosen species included two amphibians, 28 birds, one fish, 53 mammals, and 18 reptiles. Across this data set, mean and median estimates of MVP were 7316 and 5816 *adults*, respectively. The values estimated did not differ systematically between major taxa, or with trophic level or latitude, but unsurprisingly were negatively correlated with population growth rate. Interestingly, they also found that MVPs based on shorter studies systematically underestimated extinction risk, which appeared to be due to the greater temporal variation in population size captured in studies of longer duration. Although Reed *et al.* (2003) stress that the values they obtained do not provide a 'magic number', they do serve to indicate the rough size of vertebrate populations that may be needed for successful conservation, i.e. about 7000 adults, which is higher than most previous estimates. In many cases, the area required for a large vertebrate population of this size to be contained in a single reserve will be unobtainable, requiring conservation strategies that either allow population exchange to occur between major reserves, or that are not dependent on reserve systems.

How big an area?

How does a figure for a MVP translate into a **minimum viable area** (MVA)? With some species and communities, e.g. butterfly populations, a fairly small area may suffice to maintain the requisite number of individuals. In general, the higher up the trophic chain, the larger the area needed. Simply considering the home range of a few vertebrates is illustrative. It has been calculated that a single pair of ivory-billed woodpeckers (*Campephilus principalis*) may require 6.5–7.6 km^2 of appropriate forest habitat; that the European goshawk (*Accipiter gentilis*) has a home range of about 30–50 km^2; and that male mountain lions (*Felis concolor*) in the western United States may have home ranges in excess of 400 km^2 (Wilcove *et al.* 1986). Even some plant and insect species may need surprisingly large areas, as illustrated by Mawdsley *et al.*'s (1998) analyses of strangler fig species, which indicate that the MVA for some species may be as large as 200 000 ha because of their low population densities. Thus, for many species, reserves must be really

rather large if their purpose is to maintain a MVP entirely within their bounds.

There is of course, one very large proviso attached to the calculation of a minimum critical area. The approach assumes that the area concerned acts rather like a large enclosed paddock in a zoo, with freedom of association within it, but no exchange with any other paddocks or zoos. Unless the reserve does indeed contain the only population of the species, or they are completely immobile creatures, there is always the possibility of immigration of other individuals from outside the reserve. Where a significant degree of exchange takes place, this may buffer populations from genetic erosion and extinction, such that the concept of a fixed critical minimum area loses its meaning. In practice, this must be assessed on a taxon by taxon, and probably species by species, basis.

Applications of incidence functions

Area requirements of particular species can be examined by means of incidence functions estimating the probability of a species occurring for habitat islands of given combinations of area and isolation (Box 5.2; Wilcove *et al.* 1986; Watson *et al.* 2005). The information returned from such incidence functions is not equivalent to estimating MVAs. This is because the incidence functions constitute the product of population exchange within the network of suitable habitats within a region. They therefore tell you the properties of 'islands' on which a target species currently occurs, not those on which it may persist in the long term, or in an altered ecological context.

Simberloff and Levin (1985) examined the incidence of indigenous forest-dwelling birds of a series of New Zealand islands, and of passerines of the Cyclades archipelago (Aegean Sea). In both systems they found that most species occur remarkably predictably, with each species occupying all those and only those islands larger than some species-specific minimum area. A minority of species in each avifauna did not conform to this pattern, possibly because of habitat differences among islands and because of anthropogenic extinctions. Although stochastic turnover might be

occurring, the structure in the data indicates that the habitat and human influences dominate. These findings are consistent with the position expressed by Lack (1976), that most absences of birds from islands within their geographic range can be explained by unfulfilled habitat requirements.

Hinsley *et al.* (1994) quantified the incidence functions of 31 bird species in 151 woods of 0.02–30 ha in a lowland arable landscape in eastern England over three consecutive years (Fig. 10.2). For many woodland species the probability of breeding was positively related to woodland area, although some breeding occurred even in the smallest of the woods. Only the marsh tit (*Parus palustris*), nightingale (*Luscinia megarhynchos*), and chiffchaff (*Phylloscopus collybita*) failed to breed in woods of area less than 0.5 ha in any year of the study, and each was uncommon in the area. Small woods appear to be poor habitat for specialist woodland species, but are preferred by others. Their work showed that harsh winters could

temporarily but significantly alter the incidence functions of particular species. Specialist woodland species were more likely to disappear from small woods after severe winter weather than were generalists, and could take more than a year to recolonize. Variables describing the landscape around the woods were important in relation to woodland use by both woodland and open country species. For instance, long-tailed tits (*Aegithalos caudatus*) were found to favour sites with lots of hedgerows around them, whereas yellowhammers (*Emberiza citrinella*) preferred more open habitats, small woods, and scrub.

To establish the broader reliability of incidence functions as indicators of area requirements, they need to be shown to be reproducible across the range of a species. The results of Watson *et al.* (2005) reported in Box 5.2 for birds in the Canberra area, Australia, are not encouraging. They calculated incidence functions for woodland habitat islands in three different landscapes, with similar ranges in

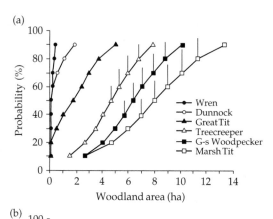

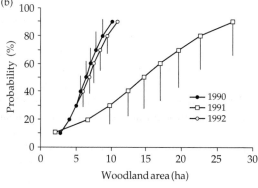

Figure 10.2 Incidence functions for the probability of breeding as a function of woodland area, based on 151 woods of 0.02–30 ha area, in a lowland arable landscape in eastern England. (a) Data representative of widespread and common woodland species (wren, dunnock, great tit) compared with that for more specialist woodland species (treecreeper, great spotted (G-s) woodpecker, marsh tit). All relationships are for 1990, except that for the marsh tit, which is for 1991. (b) Interannual variation in the incidence functions for the great spotted woodpecker, reflecting a period of severe weather in February 1991. Small woods were reoccupied in 1992. Vertical bars represent 1 SE; those below 10% are not shown. (Redrawn from Hinsley et al. 1994.)

woodland area and isolation. Overall, they found a mix of area-sensitivity, isolation-sensitivity, and compensatory area-isolation effects. But they also reported significant variation in the form of the incidence functions, which they interpreted as most likely being the consequence of differences between the three landscapes in terms of the habitat matrix in which the woodlands were embedded (cf. Ricketts 2001). Watson *et al.*'s study illustrates that area–incidence functions cannot be assumed to be consistent features within the range of a species.

In conclusion, incidence functions can be useful tools, particularly if repeat surveys are available to establish temporal variability in their form, but they do not of themselves provide sufficient data on which to base conservation policies. Notions of threshold population sizes and area requirements, as ever in island ecology, must be offset against dispersal efficacy, habitat requirements, and specific factors influencing mortality of particular species. Watson *et al.*'s (2005) findings also suggest that the 'island–sea' analogy is at fault, and that we should be paying greater attention to the ecology of the matrix (below).

10.4   Metapopulation dynamics

Most species are patchily distributed. Many occupy geographically separated patches that are interconnected by occasional movements of individuals and gametes, constituting a **metapopulation**. The first metapopulation models were constructed by Richard Levins in papers published in 1969 and 1970 (Gotelli 1991). The basic idea can be understood as follows. Imagine that you have a collection of populations, each existing on patches of suitable habitat. Each patch is separated from other nearby habitat patches by unsuitable terrain. Although these separate populations each have their own fairly independent dynamics, as soon as one crashes to a low level, or indeed disappears, that patch will provide relatively uncontested space for individuals from one of the nearby patches, which will soon colonize. Thus, according to Harrison *et al.* (1988), within a metapopulation, member populations may change in size independently but their probabilities of existing at a given

time are not independent of one another, being linked by mutual recolonization following periodic extinctions, on time scales of the order of 10–100 generations.

Studies of the checkerspot butterfly (*Euphydryas editha bayensis*) in the Jasper Ridge Preserve (USA) provide one of the better empirical illustrations of metapopulation dynamics (Harrison *et al.* 1988; Ehrlich and Hanski 2004). The checkerspot butterfly is dependent on food plants found in serpentinite grasslands. The study area is of 15×30 km and includes one large patch (2000 ha) and 60 small patches of suitable habitat. The large 'mainland' patch supports hundreds of thousands of adults and is effectively a permanent population. The smaller populations are subject to extinctions, principally due to fluctuations in weather, but patch occupancy may also be influenced by habitat quality. A severe drought in 1975–1977 is known to have caused extinctions from three of the patches, including the second largest patch, and so was assumed to have eliminated all but the largest population. Unfortunately, only partial survey data were available from the period and this assumption cannot be tested. By 1987, eight patches had been recolonized. Small patches over 4.5 km from the 'mainland' patch were found to be unoccupied. Harrison *et al.* (1988) showed that the distribution of populations described an apparent 'threshold' relationship both to habitat quality and distance. Patches had to be both good enough habitat and near enough to the 'mainland' in order to be inhabited at that time. A similar relationship, but from a separate study, is shown in Fig. 10.3 (and see Ehrlich and Hanski 2004).

From models of the dispersal behaviour of the populations, assuming a post-1977 recolonization of the patches, it was predicted that more patches would become occupied in time. How many depends on the assumptions of the model. They offered two differing scenarios. The first assumed that colonization continues until reversed by the next severe drought, an event with an approximately 50-year periodicity in the study area. The second assumed that some extinction events occur between such severe events, thereby requiring a continuous extinction model and producing an equilibrium

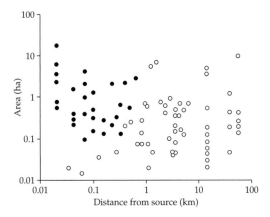

Figure 10.3 Occupancy of suitable habitat by the silver-studded blue butterfly (*Plebejus argus*) in North Wales in 1990. Most patches larger than 0.1 ha were occupied (filled circle), provided that they were within about 600 m of another occupied patch. Beyond this distance, no patches were occupied (open circle), regardless of patch size. (Redrawn from Thomas and Harrison 1992.)

number of populations, with some turnover. The authors concluded that there was evidence for both modes of extinction and also for variations in dispersal rates over time. We may note that the fit with equilibrium theory is thus uncertain. On the other hand, their data show a degree of independence of population dynamics between the isolates, combined with a demonstration of recolonization, thereby fulfilling requirements of metapopulation theory.

How does this view of extinction and recolonization differ from that of the EMIB? In the latter, the emphasis is on species number as a dynamic function of area and isolation. Whereas in the checkerspot butterfly story the focus is on the status of populations of a single species, with extinction seen principally as a function of (temporary) alterations in the carrying capacity of the system. It is consistent with the metapopulation approach—although not necessarily with all metapopulation *models*—that rates of movement and of population gains and crashes may vary greatly over time. As Haila (1990) suggested, within a metapopulation system, the dynamics of the islands are to some extent interdependent. If the patches were much more isolated, their population fluctuations would be entirely

independent, and one of the alternative theoretical models would be required. If, on the other hand, a more dispersive species was being considered, the entire system might constitute a single functional population, albeit one spread across a fragmented habitat. Thus, metapopulation models form a bridge between the study of population ecology and island theories of the EMIB form (Gotelli 1991).

Relaxing the single population assumption of the MVP model to allow for such ebb and flow between local populations will lengthen projected persistence times. Hanski *et al.* (1996) coin the term **minimum viable metapopulation** (MVM) for the notion that long-term persistence may require a minimum number of interacting local populations, which in turn depend on there being a minimum amount of suitable habitat available.

Conservationists have suggested the creation of metapopulation configurations for endangered species as a means of maintaining populations across areas of increasingly fragmented habitat. The most celebrated example of this strategy is that of the Interagency Spotted Owl Scientific Committee. They proposed the creation and maintenance of large areas of suitable forest habitat in proximity, so that losses from forest remnants will be infrequent enough and dispersal between them likely enough that numbers of the northern spotted owl (*Strix occidentalis caurina*) will be stable on the entire archipelago of forest fragments. Much of the work on metapopulations up to this point had been theoretical (Hess 1996), and Doak and Mills (1994) expressed some concern that despite intensive study of the owl, the metapopulation models have to assume a great deal about owl biology. The safety of the strategy cannot be demonstrated simply by running the models and so, in effect, the experiment is being conducted on the endangered species. Criticisms of the strategy led to a US district judge ruling against it on the grounds that the plan carried unacknowledged risks to the owl (Harrison 1994).

Similar concerns were expressed by Wilson *et al.* (1994) concerning plans to subdivide the small captive flock of the Puerto Rican parrot into three groups, following the argument that a metapopulation structure might bestow better prospects of survival. The captive flock, of about 58 birds, is maintained within the same forest area as the wild population. In this case, the suggestion was that on subdividing the captive flock, one of these populations should be transferred to a multispecies facility in the continental United States, with individuals subsequently exchanged between the populations. Wilson and colleagues pointed out the danger of the accidental spread of disease, picked up in this multispecies context, back to Puerto Rico. They also pointed out that mate selection is idiosyncratic in this, as in some other *K*-selected species, and that subdividing the population while it was still at a small size might be counter-productive for this reason also.

As with the other models considered in this chapter, it is important to understand that metapopulation models represent simplified abstractions, allowing 'what-if' scenarios to be explored (Harrison 1994). It is therefore dangerous to assign primacy to any single model in determining conservation policy.

The core–sink model variant

The classic metapopulation models assume equivalence in size and quality of patches, with all patches susceptible to extinction and all occupied patches able to act as re-supply centres (Fig. 10.4). Dawson (1994) noted two common scenarios that are inadequately dealt with in such simplified models. First, if one patch is big, and the others small, as at Jasper Ridge, then they may have a **core–satellite** or **source–sink relationship**, in which the core is effectively immortal and the smaller patches are essentially sinks (Table 10.1). In a study of orb spiders on Bahamian islands, larger populations were found to persist, whereas small populations repeatedly went extinct and re-immigrated from the larger areas and then become extinct again (Schoener and Spiller 1987), thus contributing little or nothing to the chances of persistence of the whole metapopulation (see also Kindvall and Ahlen 1992; Harrison 1994). Dawson (1994) concluded that it is usually survival on the large

(a) Classic metapopulation

(b) Source-sink

Figure 10.4 Metapopulation scenarios. (a) In classic models, habitat patches are viewed as similar in size and nature: the occupied patches (hatched) will re-supply patches that have lost their populations (unshaded), as indicated by the arrows, and at a later point in time will, in turn, lose their population and be re-supplied. (b) Much more commonly, large, effectively immortal patches (source or 'mainland' habitat islands) (dense shading) re-supply smaller satellite or sink-habitat patches after density-independent population crashes (e.g. related to adverse weather): close and larger patches may be anticipated to be swiftly re-occupied (hatched) but more distant patches take longer to be re-occupied, especially if small (unshaded).

habitat fragments that ensures species survival, not an equilibrium turnover on many smaller patches (see Thomas *et al.* 1996 for a counter-example). Secondly, even seemingly isolated populations may be saved from extinction through continued supplementation by individuals migrating from the core population. In other words, the 'rescue effect' may operate. Gotelli (1991) has shown that building this effect into metapopulation scenarios has a highly significant impact on the outcome.

Table 10.1 Problems that have been identified with classic metapopulation models

1. Occurrence of core–satellite relations
2. The rescue effect is not dealt with adequately
3. Distance effects and varying dispersal abilities
4. The occurrence of 'sink' habitats
5. Metapopulations are very difficult to replicate, tests are problematic
6. Much extinction is deterministic

It is easy to underestimate the power of natural dispersal to re-supply isolated sites because the process is so difficult to observe. We can illustrate this by stepping outside the metapopulation literature and considering the regular supply of migrant moth species to the remote Norfolk Island, 676 km from the nearest land mass (Holloway 1996). A survey over a 12 year period suggested that the resident Macrolepidoptera total 56 species, and migrants as many as 38 species. In addition, some of the resident species are also reinforced by the occasional arrival of extrinsic individuals. The rapid recolonization of the Krakatau islands, some 40 km from the mainland, by so many species of plants and animals, provides another example of the ability of many species to cross substantial distances. There are many such demonstrations to be found in the biogeographical literature, but the Norfolk Island data provide a powerful illustration because quantifying the extreme, the tail of the distribution, is generally difficult.

In contrast, there are many species of relatively limited powers of dispersal, which like the checkerspot butterfly may show distance limitation to patch recolonization on a scale of a few kilometres. For them to survive in increasingly fragmented habitats may require active intervention by people, to enable dispersal of viable propagules between isolates (Primack and Miao 1992; Whittaker and Jones 1994a,b). Most metapopulation simulation models assume no distance effects, and thus may perform poorly in simulating species that exhibit a strong gradient in arrival rates within a particular set of isolates (Dawson 1994; Fahrig and Merriam 1994).

The importance of continual immigration into so-called 'sink habitats' is such that in some populations the majority of individuals occur in sink habitats (Pulliam 1996). Furthermore, a habitat may contain a fairly high density of individuals but be incapable of sustaining the population in the absence of immigration. Density can, therefore, be a poor indicator of habitat quality. Thus, while some suitable habitats will be unoccupied because of their isolation from source populations, other 'unsuitable' habitats will be occupied because of their proximity to source populations. As pointed out earlier, if these effects operate, snapshot incidence functions may be misleading.

Deterministic extinction and colonization within metapopulations

Extinction models tend to be based on stochastic variation, in some cases emphasizing demographic stochasticity, sometimes taking genetic stochasticity and feedback into account, and more recently incorporating environmental stochasticity. However, Thomas (1994) has argued that, in practice, most extinctions are deterministic, and that they can be attributed directly to hunting by humans, introductions of species, and loss of habitat. In many cases documented in the literature, virtually the entire habitat was lost or modified, causing 100% mortality. Stochastic extinction from surviving habitat fragments is minor by comparison. According to this view, stochastic events are superimposed as decoration on an underlying deterministic trend. To put this in a metapopulation context, if the reason for patches 'winking out' is habitat changes rather than—for sake of argument—a period of unfavourable weather, then the local habitat is likely to remain unsuitable after extinction and so will be unavailable for recolonization. In cases, fairly subtle changes in habitat may produce a species extinction that might be mistakenly classified as 'stochastic'. Thomas cites examples from work on British butterflies, whereby even changes from short grass to slightly longer grass can result in changes in the butterfly species found in the habitat (cf. Harrison 1994).

Thomas (1994) usefully reminds us that there are many reasons for species colonization of a patch, and they are not all island effects. He suggests six circumstances in which natural colonizations by butterflies are observed:

- succession following disturbance
- where new 'permanent' habitat is created close to existing populations
- introductions of species outside former range margins
- regional increases in range
- turnover-prone peripheral patches (which act in essence as 'sinks' and are therefore unimportant to overall persistence)
- seasonal spread.

Thomas regards butterflies as persisting in regions where they are able to track the environment, and becoming extinct if they fail to keep up with the shifting habitat mosaic, or if the shift in habitat regionally is against them. Metapopulation models therefore need to be superimposed on an environmental mosaic, which in many cases will itself be spatially dynamic. 'An environmental mosaic perspective shifts the emphasis on to transient dynamics and away from the equilibrium (balance) concept of metapopulation dynamics, for which there is little evidence in nature' (Thomas 1994, p. 376). The possibility of incorporating patchy disturbance phenomena into metapopulation models has since been developed by Hess (1996).

Value of the metapopulation concept

As Gotelli (1991) concludes, the empirical support for classic metapopulation *models* is fairly thin on the ground and tests are liable to be difficult. Metapopulations are difficult to replicate, and the timescale of their dynamics (many generations of the organisms involved) may be of the order of decades. In addition, it is readily appreciated that populations are subdivided on many scales and so the delimitation of the local population is often subjective. Hence, to achieve greater realism, metapopulation simulations have to be extended from the original, simplistic models, to allow for the

differing degrees of population connectivity, and differing forms of interpatch relationship, to be found in real systems (Harrison 1994; Hanski 1996; Ehrlich and Hanski 2004). Fahrig and Merriam (1994) identify the following factors as potentially significant:

- differences among the patch populations in terms of habitat area and quality
- spatial relationships among landscape elements
- dispersal characteristics of the organism of interest
- temporal changes in the landscape structure.

These factors can be difficult to capture and may require rather different variables to be assessed for different species.

In modelling exercises there is typically just such a trade-off between greater realism and improved model fit on the one hand and generality on the other. In this respect metapopulation models are no better or worse than most other ecological models. The value of the metapopulation concept is that it has led conservation biologists to give greater attention to the spatial structure and temporal interdependency of networks of local populations, which is a clear step forward from considering them as essentially independent entities.

10.5 Reserve configuration—the 'Single Large or Several Small' (SLOSS) debate

Given a finite total area that can be set aside for conservation as a natural landscape is being converted to other uses, what configuration of reserves should conservationists advocate? At one extreme is the creation of a single large reserve; the alternative is to opt for several smaller reserves amounting to the same area, but scattered across the landscape. This question, reduced to the acronym **SLOSS**, was cast in terms of the assumptions and predictions of the EMIB (e.g. Diamond and May 1981). As one of the main aims of conservation is to maximize diversity within a fragmented landscape, might not island ecological theory, which focuses on species numbers, provide the answer as to the optimal configuration of fragments?

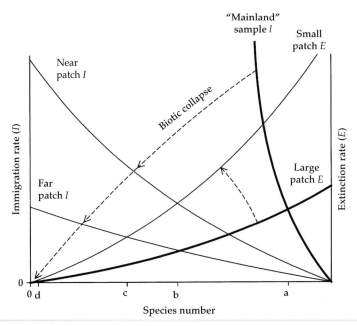

Figure 10.5 According to the assumptions of the EMIB, reducing area causes supersaturation as immigration rate declines and extinction rate rises. This causes loss of species, i.e. 'relaxation' to a lower equilibrium species number from a → b → c → d. In the extreme scenario of 'biotic collapse', the immigration rate is so low that the equilibrium species number is near zero. (Redrawn from Dawson 1994, Fig. 4.) In practice this abstract theory may lack predictive power as to either numbers lost or rate of loss, being founded on assumptions of the system passing from equilibrium (pre-fragmentation), to non-equilibrium (at fragmentation), to a new equilibrium condition (when *I* and *E* balance again): circumstances which may not apply in complex real-world situations. This is not to deny species losses, they do occur, but very often they are not a random selection, but instead are drawn from predictable subsets of the fauna or flora, structured in relation to: particular ecological characteristics of the species in question; habitat or successional changes in the isolate; and the nature and dynamics of the matrix around it. Equally, isolates may acquire increased numbers of some species, or gain new species, sometimes of an 'undesirable' nature judged in terms of their impact on the original species.

Following the island analogy, the first factors to be included are isolation and area. Increased isolation of reserves reduces migration into them from other reserves. If a reserve is created by clearance of surrounding habitat, then it follows that on initial isolation the immigration curve should be depressed. The contiguous area of habitat is also reduced and thus extinction rate should increase. At the point of creation, therefore, the habitat island contains too many species (it may even gain fugitive displaced populations), and the result is that it becomes **supersaturated**. It follows that it should in time undergo '**relaxation**' to a lower species number, a new equilibrium point (Fig. 10.5). Given knowledge of isolation and area of patches, it should be possible to estimate the number maintained at equilibrium in a variety of configurations of habitat patches. On these grounds, Diamond and

May (1981) favoured larger rather than smaller reserves, short rather than long interreserve distances, circular rather than elongated reserves (minimizing edge effects), and the use of corridors connecting larger reserves where possible (Fig. 10.6). These widely cited suggestions spawned a largely theoretical debate in which much appeared to hang on the validity and interpretation of the EMIB. Having considered the EMIB, MVP, and metapopulation ideas ahead of it, some of the limitations of the SLOSS approach will be immediately apparent and so can be dealt with quickly.

The simplest criticism of this approach follows the line that the EMIB is a flawed model. Hence it provides no firm foundation for the development of conservation policy. If policy-makers adopt such theories as though providing formal rules, as some did in this case, this criticism is justified (Shafer

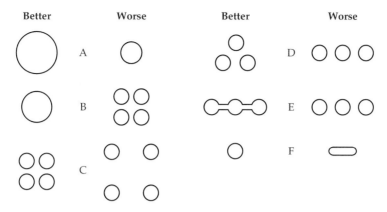

Figure 10.6 The suggested geometric principles for the design of nature reserves which were supposedly derived from island biogeographic studies, and which were at the centre of the so-called SLOSS, or Single Large Or Several Small debate (redrawn after Diamond 1975*b*). These 'principles' have been challenged on both theoretical and practical grounds (see text).

1990). The position adopted in this book, however, is that the EMIB has a place in a larger framework of island ecological theory. It is not that the theory is either true or false, but that the effects it models may be either strong or weak. If the effects are very weak for the habitat island system being considered, as often is the case, then the answer to the SLOSS question will not be supplied by the equilibrium theory.

The next obvious criticism of the attempts to solve SLOSS using the EMIB is that the theory is insufficiently precise on the amount of compositional overlap that will occur across a series of isolates of varying area. Deployment of the EMIB rationale often comes down simply to the use of species–area equations. Such an approach may give a rough idea of numbers of species on habitat islands, but not which habitats contribute most to richness, nor which species are most likely to be lost from the remnants (Saunders *et al.* 1991; Simberloff 1992; Worthen 1996). At its simplest, a strongly non-nested series of small reserves may be anticipated to hold more species than a single large reserve, but where the system is perfectly nested, the largest reserve will hold most species (discussed more fully below).

Interpretations of species–area relations based on equilibrium theory have it that the regression line provides the equilibrium number of species. The

statistical interpretation of the scatter is simply that the regression line represents the average number of species for a given area and a point above it represents a positive error (Boecklen and Gotelli 1984; Boecklen 1986). If a newly created habitat island lies above the line it doesn't necessarily follow that it is supersaturated and destined to lose species. It could just be that habitat quality or heterogeneity allows it to maintain more than the average number of species. In his analyses of a large US bird census data set, Boecklen (1986) found that habitat heterogeneity is a significant predictor of species number even after area has been factored out. The significance of habitat effects in the data set points to the possibility that in particular cases several small reserves can incorporate a wider (or better) array of habitat types and thus support more species than the single large option.

The discussion of SLOSS issues that follows will mostly be limited to biological and physical parameters. However, it should be recognized that there are often overriding practical considerations. For instance, the costs involved in establishing and maintaining conservation and game parks in developing countries depend on questions such as the following (Ayers *et al.* 1991). Is perimeter fencing required? Is it necessary to patrol to prevent poaching or encroachment into the reserve? How many wardens do you need? Do you have to equip and manage park headquarters? In many situations, fewer larger reserves make more sense from these practical, financial, and inevitably political perspectives. Such considerations often have overriding importance to decision makers, and many recent commentators regard the theoretical debates as having contributed little of direct practical value to conservation (e.g. Saunders *et al.* 1991). However, the *implications* of different reserve configurations still demand attention, irrespective of the fate of the theoretical frameworks or practical realities of land ownership and opportunity that spawned them.

Dealing with the leftovers

In most cases, natural scientists are unable to exert great influence on the basic ground-plan of fragmentation, and even where they do have a role to play, it is never a free hand. There is, moreover, a range of biogeographical, economic and political considerations which can be crucial to the strategy adopted. This has been recognized even by the advocates of the SLOSS 'principles'. For instance, the proposed reserve system for Irian Jaya, on which Jared Diamond acted as an adviser, rightly paid more attention to pre-existing reserve designations, distributions of major habitat types, and centres of endemism, and to economic and land-ownership issues, than it did to the SLOSS principles (see Diamond 1986).

Most often, conservationists today are in the position of dealing with and managing a system of habitat fragments, and perhaps modifying its configuration slightly. Fragmentation may have been carried out in a very selective fashion. Thus, on the whole, farmers take the best land and leave or abandon the least useful. In much of southern England, this has meant that most remaining native woodland is on very heavy, clay-rich soils, with the lighter soils in agricultural use. The message is that the species mix in your fragments may not be representative of what was in the original landscape. To give just one example, the short-leaved lime (*Tilia cordata*) does not favour the heavier soils, and it is now known as a species of hedgerows. In the past it was probably an important forest tree

(Godwin 1975). Equally, the juxtaposition of landscape units may be critical. For instance, some Amazonian forest frogs require both terrestrial and aquatic environments to be present within a reserve in order to provide for all stages of the life cycle (Zimmerman and Bierregaard 1986).

It must also be recognized that humans may interfere directly in events within reserve systems. Bodmer *et al.* (1997) point out that although much work has focused on extinction rates caused by deforestation, many of the recorded extinctions in the past few hundred years have been the result of over-hunting. They collected data on the relative abundance of large-bodied mammals in the northeastern Peruvian Amazon in areas with, and without, persistent hunting pressure. They found that species with long-lived individuals, low rates of increase, and long generation times are most vulnerable to extinction. In one reserve in northeastern Peru, it was found that nearly half the meat harvested by hunters in the buffer zones of the reserve originated from mammals that are categorized as vulnerable to over-hunting. Similarly, Peres (2001) uses game-harvest data to argue that there may be synergistic effects of hunting and habitat fragmentation, such that persistent over-hunting in fragmented landscapes drives larger vertebrate populations to local extinction, especially from smaller habitat patches. Such results can be used to help formulate appropriate policy towards such activities.

Trophic level, scale, and system extent

The optimal configuration of areas is liable to vary depending on the type of organism being considered, and on the spatial scale of the system. If forced to generalize with regard to trophic level, larger reserves are more appropriate for large animal species needing a large area per individual, pair, or breeding group, and/or requiring 'undisturbed' conditions, e.g. the Javan rhino (*Rhinoceros sondaicus*) (see Hommel 1990). Species requiring large areas are often the ones most threatened by us and in need of protection, e.g. top carnivores, primates, and rhinoceros. Whereas top carnivores, such as big cats, may require large territories, it might be possible to keep the highest diversity of butterfly species by means of a number of small reserves, each targeted to provide particular key habitats (Ewers and Didham 2005). They should not, however, be too small, lest they consist principally of 'sink' populations. However, trophic level and body size are not the only considerations. We have also to think about how species use their habitat space. For instance, birds of prey may have large territories incorporating both good and poor habitat. Conservation of large birds of prey in Scotland, such as the golden eagle (*Aquila chrysaetos*) and the osprey (*Pandion haliaetus*), has required a range of measures taken to discourage and prevent killing of the birds within a mixed-use landscape. It has also required intensive but highly localized efforts to protect nesting sites from egg collectors. The 'reserves' required are tiny compared to the ranges of the individuals.

Although, in general, most studies advocate relatively large reserves at the expense of larger numbers, in cases it may be better to have a set of several small reserves. First, they may incorporate more different habitats, i.e. they may capture beta diversity across a landscape. Secondly, competition may, in theory, lead to the exclusion of species of similar niches from any given reserve and so it may be good to have several reserves so that different sets of species may 'win' in different reserves. Thirdly, there is an epidemiological risk inherent in having 'all your eggs in one basket', and, fourthly, particular species may rely upon small islands. For instance, small estuarine islands in the Florida Keys area appear to provide important breeding sites for several species of waterbirds (Erwin *et al.* 1995).

In organizing our thinking about the implications of the SLOSS debate, we would do well to consider two aspects of spatial scale: study system grain and extent (cf. Whittaker *et al.* 2001). **Grain** refers to the size of the sampling unit in spatial analyses. In this context, grain size is variable: we refer in fact to the range of areas of the habitat islands. Log–log plots of species richness and area (e.g. Fig. 4.2) often indicate simple linear trends of converging richness between fragmented and unfragmented areas as sample area increases. However, recent theoretical discussion has reopened the question of whether

this is the right approach to fitting trend lines to such data (e.g. Lomolino 2000*c*; Tjørve 2003; Turner and Tjørve 2005), and in particular whether more complex forms of relationship should be tested for at the 'small island' end of the relationship. Empirically, it appears that superior model fits can be derived for many 'island' data sets using a form of break-point regression, consistent with the existence of a so-called small-island effect. In practice, this means that there is no systematic increase in richness until a critical area threshold is crossed, producing a two-phase relationship shifting from flat to increasing (Fig. 4.6; Lomolino and Weiser 2001; Triantis *et al.* 2006). The implication of this finding is that the slope and intercept of ISARS fitted using log–log models may be more sensitive to the range in area of isolates sampled than generally realized. This in turn may impact on our assessment of the relative richness gain from having larger-sized reserves: although it should again be noted that simply being able to predict the richness of each isolate does not reveal the overall richness of the system.

Turning to the study system extent, if we imagine two networks of a dozen habitat islands, each network identical in terms of the starting richness of each island, the conservation implications of increasing distance separating each of the habitat islands work in perhaps two key ways. First, the further apart they are, the lower the probability of their species populations mutually reinforcing one another, thereby preventing extinction from the network. Second, the further apart they are, the more likely they are to sample different habitats, and if they are really far apart they may even sample different biomes or different biogeographical source pools. Thus a widely scattered system of reserves may capture more differentiation diversity (*sensu* Whittaker 1977): but whether it can retain it in the long run may depend on the extent to which each reserve can sustain its initial species complement in the long term. Hence, the answer to whether it is better to have one large or several small reserves may be sensitive to the grain and extent of the system. Of course, the SLOSS 'principles' were advocated most strongly for circumstances in which the biogeography of an area is poorly known. When working at a fine scale of analysis, the 'island' assumption is at least a rational starting point for thinking about the SLOSS question. However, at coarse scales of analysis, the relevance of island theory is less clear, and so conservation biogeographers have developed other approaches, designed to make use of the improved availability of distributional data in digital form (at least for some taxa). For instance, computer algorithms have been developed to achieve goals such as finding the 10% of sites/grid cells that maximize representation of rare or endemic species (e.g. see Araújo *et al.* 2005a; O'Dea *et al.* 2006; and papers cited therein). Such analyses provide a far more direct approach to conservation planning than the original SLOSS guidelines and effectively supercede them at coarse scales of analysis.

To sum up, on theoretical grounds, the answer to the SLOSS debate is equivocal (e.g. Boecklen and Gotelli 1984; Shafer 1990; Worthen 1996). The answer depends, as we have argued here, on the ecology of the species (or species assemblage), on system grain and on extent. It also depends on a range of other system properties, which we will now explore.

10.6 Physical changes and the hyperdynamism of fragment systems

Large environmental changes can be involved in ecosystem fragmentation, particularly where forests are fragmented (Lovejoy *et al.* 1986; Saunders *et al.* 1991). The immediate and subsequent differences in fluxes of radiation, wind, water, and nutrients across the landscape can have significant consequences for the ecology of the remnants, and have been summed up by Laurance (2002, p. 595) in the idea of **hyperdynamism**, which he defines as 'an increase in the frequency and/or amplitude of population, community, and landscape dynamics in fragmented habitats'. The following examples are illustrative.

- **Radiation fluxes**. In south-western Australia, elevated temperatures in fragmented landscapes reduced the foraging time available to adult Carnaby's cockatoos (*Calyptorhyncus funereus latirostrus*) and contributed to their local extinction.

- **Wind**. When air flows from one vegetation type to another it is influenced by the change in height

and roughness. At the edge of a newly fragmented woodland patch, increased desiccation, wind-damage, and tree-throw can occur (e.g. Kapos 1989). Increased wind turbulence can affect the breeding success of birds by creating difficulties in landing due to wind shear and vigorous canopy movement. Wind-throw of dominant trees can result in changes in the vegetation structure, and allow recruitment of earlier successional species.

- **Water and nutrient fluxes**. Removal of native vegetation changes the rates of interception and evapotranspiration, and hence changes soil moisture levels (Kapos 1989). In parts of the wheat belt of Western Australia, the new agricultural systems cause rises in the water table, which can bring stored salts to the surface, and this secondary salinity has caused problems both to agriculture and the remnant patches. In the fenlands of eastern England, drainage for agriculture has led to peat shrinkage and a drop of 4 m in land level in 130 years. Remnant areas of 'natural' wetland now require pumping systems to maintain adequate water levels (see also Runhaar *et al.* 1996).

- **Fire regimes**. Modified landscapes can be a major source of surface fires, for example from burning of adjoining pastures. Penetration of such fires into fragment interiors can increase plant mortality, disturb the fragment boundaries and in time cause an 'implosion' of forest fragments (Laurance 2002).

The physical effects of edges thus have important knock-on effects for the biota, especially soon after fragmentation (Table 10.2). In time, the system adjusts to the new physical conditions, and the woodland edge fills and becomes more stable. Yet, just as there is a wide range of disturbance regimes across real islands, so must there be for continental habitats, and as we create new habitat islands, we inevitably alter disturbance regimes (Kapos 1989; Turner 1996; Ross *et al.* 2002). Furthermore, land-use of the matrix in which the habitat islands are embedded typically continues to change (Laurance 2002). It might therefore be anticipated that many habitat islands will be characteristic of the dynamic non-equilibrial position identified in Fig. 6.1 (cf. Hobbs and Huenneke 1992). Indeed, it has been suggested that smaller habitat islands may be subject to greater disturbance impacts than larger

Table 10.2 Classes of edge-related changes triggered by the process of forest fragmentation, as informed by the Minimum Critical Size of Ecosystem project. The first-order effects may lead to second-order and, in turn, third-order knock-on effects (modified from: Lovejoy *et al.* 1986)

Class	Description of change
Abiotic	Temperature
	Relative humidity
	Penetration of light
	Exposure to wind
Biological	
First-order	Elevated tree mortality (standing dead trees)
	Treefalls on windward margin
	Leaf-fall
	Increased plant growth near margins
	Depressed bird populations near margins
	Crowding effects on refugee birds
Second-order	Increased insect populations (e.g. light-loving butterflies)
Third-order	Disturbance of forest interior butterflies, but increased population of light-loving species
	Enhanced survival of insectivorous species at increased densities (e.g. tamarins)*

*Does not apply to birds (Stouffer and Bierregaard 1995).

islands, thereby making such sites less suitable for, or removing altogether, particular species (McGuinness 1984; Simberloff and Levin 1985; Dunstan and Fox 1996).

10.7 Relaxation and turnover—the evidence

If habitat islands behave according to the expectations of the EMIB, they should be supersaturated with species immediately after system fragmentation. Subsequently, species numbers in the isolates should 'relax' to a lower level (Fig. 10.5). Immigration and extinction should continue to occur, both during the relaxation period and subsequently, when the island has found its new, lower equilibrium richness level. The time taken for relaxation to occur is called the 'lag time' and the anticipated eventual species loss is termed the **extinction debt** (Ewers and Didham 2005). If these anticipated effects are strong, even high-priority species—the very species that you wish to conserve—may in time disappear from a reserve. As we have noted, however, the empirical evidence from a wide range of real and habitat islands is that turnover tends to

be heterogeneous (i.e. structured). Some species tend to be highly stable in their distribution across habitat islands. Most turnover involves 'ephemerals', species marginal to the habitat, or successional change (Chapters 4–6). The metapopulation studies examined above provide further insights into turnover. They suggest a tendency for large populations to persist, while smaller satellite populations may come and go, without jeopardizing the metapopulation. Nonetheless, species do disappear from particular habitat islands, and from entire landscapes, and in cases the loss is a global one. Earlier, we noted that many species losses are attributable to deterministic causes, meaning hunting, habitat changes, and the like. The relaxation effect, in contrast, is based on stochastic processes. How strong is this island relaxation effect?

One of the best-known examples of relaxation on ecological timescales is that of birds lost from Barro Colorado island, in Panama. It is not an ideal study in the present context in that, unlike most habitat islands, it is actually now surrounded by water. The island was formerly a hilltop in an area of continuous terrestrial habitat, but it became a 15.7 km^2 island of lowland forest when the central section of the Panama Canal Zone was flooded to make Lake

Gatun in 1914. Of about 208 bird species *estimated* to have been breeding on Barro Colorado island immediately following isolation in the 1920s and 1930s, 45 had gone by 1970 (Wilson and Willis 1975). Of these, some 13 have been attributed to 'relaxation'.

The other losses could be attributed to particular forms of ecological change. At the time of the analysis, much of the forest was less than a century old, following abandonment of farming activity before 1914. Many of the birds lost were typical of second growth or forest-edge, suggesting a probable successional mechanism as forest regeneration reduced the availability of these more open habitats. Some were ground nesters, which may have been eliminated by their terrestrial mammalian predators. The latter became abundant because of the disappearance of top carnivores (such as the puma) with large area requirements (Diamond 1984). This effect, of increasing numbers of smaller omnivores and predators in the absence of large ones, has been termed **mesopredator release** (Soulé *et al.* 1988), and it has now been documented in several other systems (Fig. 10.7; Crooks and Soulé 1999; Laurance 2002). Some of the birds lost were members of the guild of ant followers, which have been found to be vulnerable to fragmentation elsewhere (below). In a simple sense, Barro Colorado island provides evidence of relaxation, but the stochastic signal is seemingly much smaller than that produced by changing habitat and food-web relationships, and it might therefore be asked 'which is the signal and which the noise in the system?' (cf. Sauer 1969; Lynch and Johnson 1974; Simberloff 1976).

That species numbers and composition may change as a consequence of fragmentation is not in dispute. However, it is remarkably difficult to find good quantitative data for the island relaxation effect for systems of habitat islands (Simberloff and Levin 1985; and see review in Shafer 1990). Most studies, including that for Barro Colorado island, lack precise information about the species composition before fragmentation (e.g. Soulé *et al.* 1988). The Barro Colorado island study is also fairly typical in the losses recorded being mostly 'deterministic' in nature. The predictions of species loss through 'relaxation' derive from the EMIB, but theoretically relaxation may also affect metapopulations. Hanski *et al.* (1996) speculate that the pace of fragmentation in many regions has been so rapid that 'scores of rare and endangered species may already be "living dead", committed to extinction because extinction is the equilibrium toward which their metapopulations are moving in the present fragmented landscapes' (p. 527). However, their findings are heavily dependent on models of the data of a single species of butterfly, the Glanville fritillary (*Melitaea cinxia*). The effect identified appeared to be far from instantaneous, and might turn out in practice to be a relatively weak effect in relation to some of the other processes discussed in this chapter.

A classic illustration both of genuine reasons for concern and of the uncertainty involved in projections is provided by the tropical moist forests of the Atlantic seaboard of Brazil, which have been reduced over the past few centuries to only about 7% of their estimated former cover (Ribon *et al.* 2003). Based on the 'rules' derived from island theory, some 50% of species are expected to go extinct. To date, no extinctions have been documented with certainty (Whitmore and Sayer 1992;

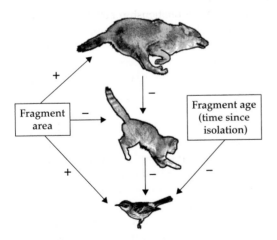

Figure 10.7 Model of the combined effects of trophic cascades and island biogeographical processes on top predators (for example, coyote), mesopredators (domestic cat) and prey (scrub-breeding birds) in a fragmented system. Direction of the interaction is indicated with a plus or minus. (From Crooks and Soulé 1999.)

Ribon *et al.* 2003; Grelle *et al.* 2005), although many species are Red-listed as vulnerable, endangered or critically endangered based on range, or population estimates (Ribon *et al.* 2003). Brooks and Balmford (1996) compare losses of birds projected using a species–area model with those listed by the IUCN as 'threatened' and find congruence, i.e. they argue that the predictions are basically correct, but that there is a substantial lag between the habitat loss/fragmentation process and global extinction of the species. There is, moreover, evidence of local extirpation within the existing range for bird species. Thus, within the Viçosa region (a 120 km² area in southeastern Brazil) over the last 70 years, it appears that at least 28 bird species have become locally extinct, with 43 'critically endangered', and 25 'vulnerable': in total 61% of the original avifauna has been significantly reduced in incidence (Ribon *et al.* 2003). Nectarivorous species were least at risk, next came omnivores and carnivores, with frugivores and insectivores in greatest jeopardy.

Assuming relaxation to be underway in these fragmented systems, the next problem is to estimate the lag time between fragmentation and eventual species losses. Brooks *et al.* (1999) attempt this for tropical forest birds in an area of fragmented upland forest in Kenya. They base their analyses on island theory, assuming z values (slope of the SAR) pre-isolation of 0.15, and post-isolation of 0.25 (i.e. for the new ISAR), and that the rate of decline in species is approximately exponential. This postulates that fragments will lose approximately half the species that will eventually be lost every x years: this they termed the 'half-life'. The difficulty is finding systems with 'before' and 'after' data. The system they chose for the test is the Kakamega forest, where forest fragmentation occurred over 60 years ago (but where the time frame is not perfectly specified). Their analyses of five fragments suggests a half-life varying in relation to fragment area from 23 (100 ha) to 80 (8600 ha) years, and of approximately 50 years for a 1000 ha fragment. As might be anticipated from island theory, relaxation times were suggested to be scale dependent in relation to both range in area ('grain size') and degree of isolation (cf. Box 10.2).

One important caveat to this study is, however, that the historical data for the Kakamega area were derived from general collecting/observations from the Kakamega area, with values assigned to the fragments on the basis of the assumption of $z = 0.15$ (slope of SAR) for their pre-isolation states. Hence, the figures derived for relaxation times should be seen as still of the nature of first approximations. Moreover, the relaxation times per fragment do not tell us overall species losses from the system. In practice, the data reviewed by Brooks *et al.* (1999) suggest that some loses have occurred, with 10 of an estimated 73 Kakamega species not having been recorded from these forests in recent years. Of those species counted as lost at least one is thought to have been trapped to extinction for the caged-bird trade; three others were at their eastern range limits within the study area, so their loss from this area might be related to other processes and may not be indicative of their risk of global extinction. In practice, a single study system, of five fragments, based on relatively poorly specified empirical data, is a rather limited basis for building predictive models, and there are surprisingly few other such studies with which to compare the values obtained.

Box 10.2 appraises the use of species–area models in predicting regional species extinctions as a function of regional habitat loss, arguing that although there is a crude relationship between habitat loss and species loss, there is no logical connection between such models and island theory. It should also be recognized that even where species–area analyses are linked phenomenologically to island theory (i.e. they are being applied to systems of fragments using parameters derived from other such systems, which also vary over similar ranges of area and isolation), they provide only a coarse, stochastic simulation of the impacts of fragmentation. The actual drivers of extinction may be structured, and in some instances avoidable by appropriate management. For instance, hunting and collection of birds and mammals may each be important drivers of the decline of particular species (Peres 2001; Ribon *et al.* 2003). With respect to the Atlantic forest system, conservation action has likely already been and will continue to be crucial to staving off the final extinction of highly

Box 10.2   Does island theory provide a basis for the use of species–area regressions to predict extinction threat?

In the text, we discuss a number of studies that have used species–area models to predict the eventual species losses following fragmentation (the so-called 'extinction debt'). A growing body of literature has applied this approach at varying, sometimes very coarse, scales (e.g. Brooks *et al.* 2002), to forecast species losses in the tropics as a function of forest loss. Recent work has attempted both to assess extinction threat and validate the application of species–area models by comparing at a regional scale the forecasts of species–area models to lists of threatened species. Where congruence is found, the analyses are regarded as mutually supporting: island effects are working, systems are 'relaxing' to lower species number, and given time (the 'lag effect'), will re-equilibrate to a lower species richness. The application of species–area models is generally taken to be supported theoretically by work on island species–area relationships, from which the slope parameter z is derived.

Returning to the Atlantic forest system discussed in the text, the study by Grelle *et al.* (2005) assesses the sensitivity of species–area based predictions of extinction to the z parameter (slope of the relationship) selected. Their paper provides an analysis of data for Rio de Janeiro State, in which present and historical coverage of forest are estimated at 19% and 100%, respectively. Following other recent studies, they first predicted extinctions using a species–area model with a z value of 0.25, and compared the predicted species losses with the number of species identified by an expert workshop as threatened with extinction. Their analysis suggested that the species–area model overestimated extinction threat for each of mammals, birds, reptiles and amphibians. When they repeated the analysis for endemic species only, they found that the species–area model successfully predicted threatened endemic mammals when using a z of 0.25, and threatened endemic birds when using a z ranging between 0.1–0.19.

We have detailed this study as it is one of several in the literature (e.g. Brooks and Balmford

1996; Brooks et al. 2002; Thomas et al. 2004) that use species–area regression as a means of forecasting species threatened by, or committed to, extinction. However, the way in which the species–area models are used in this and other studies is conceptually decoupled from the island theory from which it seemingly derives. In Chapter 4 we cite MacArthur and Wilson (1967) for the observation that ISAR z values may vary between 0.2 and 0.35, and note that the range in values may indeed be much greater (Williamson 1988). Therefore, first, a z of 0.25 is a largely arbitrary 'middle' value to take. Second, and more crucially, this z value has been derived from analyses of true isolates: it tells us approximately how many species are held in each of a series of isolates/islands of different size. Note that the total number of species held across the series of isolates is not derivable from the z parameter of the isolates: as the overall richness of the system depends on the degree of compositional overlap (nestedness) among the isolates. Yet, in recent studies such as these, the z value is applied not to separate fragments, but to the entire region. That is, the authors assume that the whole system (in this case Rio de Janeiro State) is acting as a single fragment, which has lost 81% of its area and which is therefore travelling along a trajectory towards its lower equilibrium point. However, in practice, we are talking about an area in which forest habitat persists in very large numbers of patches, scattered across that region in varying configurations. This loss of habitat is certainly bound to vastly reduce the number of individuals of forest species across that region, but whether that results in particular species falling below their minimum viable population in all isolates is not knowable simply from the regional figure for habitat loss (see e.g. Loehle and Li 1996).

There is thus no link of scientific logic that can carry us from empirical demonstrations of z values for islands and isolates, via MacArthur and Wilson's EMIB, and the related theoretical idea of species relaxation (Diamond 1975b), through to the assumption of a particular shape and slope of relationship between the percentage of forest loss on a regional basis and regional species losses. In short, the use of species–area regressions in this way appears to be currently without a theoretical justification, and thus represents merely an exercise in curve-fitting.

Taken as such, what do we learn from such studies? The Grelle *et al.* (2005) study illustrates that, in practice, there is no reason to favour a z value of 0.25 over other slope parameters in making what amounts to a back-of-the-envelope sketch of how regional habitat loss may relate to loss of species regionally. In practice, we would be better to use species-level assessments (whether underpinned by rigorous survey data or expert assessments) than to rely on such species–area fits. Such assessments enable

attention to be given to species judged most at risk, and necessarily involve an assessment of what factors threaten particular species. This sort of information would seem to be of inherent value to conservation agencies. Moreover, the argument that we should use species–area modelling in the absence of adequate data for other regions or other taxa, arguably breaks down with the realization that the success of the study in fitting species–area models to independent data on species threats varied between taxa, with extinction risks of reptiles and amphibians not successfully predicted in any of their models. This is one area within conservation biogeography where current theory appears inadequate and in urgent need of attention (as Whittaker *et al.* 2005).

threatened species (Birdlife EBA factsheet 75, <www.birdlife.org/datazone/>, visited July-2004). Given these varying forms of human intervention, it may be impossible to resolve the question of whether the Atlantic forests system provides support for, or a refutation of, the theory of species relaxation.

The assumption that fragmentation will *necessarily* lead to ongoing species losses through relaxation has been challenged by Kellman (1996). He argues that reference to past environmental change and species responses to it shows that many species are more flexible in their ecological requirements than generally recognized. Kellman contends that many plant communities are actually undersaturated and that systems of fragments may be able to sustain increased species densities for significant periods of time (although not necessarily indefinitely). He bases his case in large part on studies from Neotropical gallery forests, which constitute narrow peninsulars of forest extending through a matrix of savanna ecosystems. According to equilibrium theory, the tips of the peninsulas should be impoverished because of their relative isolation. This extension of island theory is termed the **peninsula effect**. It has been tested at a coarse scale for Florida and for Baja California, and has generally been found to be a fairly weak effect (Brown 1987;

Means and Simberloff 1987; Brown and Opler 1990). In the Neotropical gallery forests, Kellman (1996) reports that there is no evidence for floral impoverishment with distance from the peninsular base, possibly because the frugivores that disperse the seeds of the plants are capable of adjusting their ranging behaviour to accommodate the long, thin peninsular configuration, or indeed to move through the sparsely wooded savanna matrix surrounding the peninsulas. These peninsulas have a long history, and thus appear to provide evidence for adequate gene flow of plant populations to the peninsular tips over long periods, i.e. there has been plenty of time for 'lag effects' to unfold, but at least for plants, there is no evidence of species impoverishment. However, it is unclear how good an analogue these peninsulas provide for the *rapid* conversion of forested areas to fragmented systems.

There is a long-term fragmentation project which sheds some light on these questions, and in which data were collected before fragmentation: the Minimum Critical Size of Ecosystem project. It is located near Manaus, in the Brazilian Amazon, and sites were turned into habitat islands, isolated from intact forests by just 70 to 650 m of pasture, between 1980 and 1984. While some patches remained isolated, the land surrounding others was abandoned during the study and became infilled by forest regrowth dominated by *Cecropia*.

Stouffer and Bierregaard (1996) examined the use of these forest fragments by understorey hummingbirds. The three species found before fragmentation persisted at similar or higher numbers in the fragments through the 9 years of the study. Use of the fragments did not differ between 1 and 10 ha fragments. Furthermore, it mattered little whether the surrounding matrix included cattle pasture, abandoned pasture, or *Cecropia*-dominated second growth. This particular guild of birds thus appears to be able to persist in a matrix of fragments, secondary growth, and large forest patches.

By contrast, the guild of insectivores, which dominate the understorey bird community, was much more responsive to fragmentation: both abundance and richness declined dramatically (Stouffer and Bierregaard 1995). Three species of obligate army-ant followers disappeared within 2 years. Mixed-species flocks drawn from 13 species disaggregated over a similar timescale, although three of the species persisted in the fragments. The 10 ha fragments were less affected than the 1 ha fragments, and in time the species flocks reassembled in those 10 ha fragments that were surrounded by *Cecropia*-regrowth. The communities in these reconnecting fragments converged compositionally on pre-isolation communities, while communities in the smaller or more isolated fragments continued to diverge. Particular terrestrial insectivores (e.g. *Sclerurus* leafscrapers and various antbirds) failed to return to any of the fragments during the study. To sum up: (1) the impact of fragmentation on birds is strongly ecologically structured, (2) different 'guilds' respond in divergent fashion, and (3) the types of species most vulnerable to fragmentation are fairly predictable, with insectivores that participate in mixed-species flocks being the most vulnerable guild (e.g. Lovejoy *et al.* 1986; Stouffer and Bierregaard 1995, 1996; Christiansen and Pitter 1997; Ribon *et al.* 2003).

We may thus conclude that habitat fragmentation is associated with compositional change and species losses. Theory predicts significant losses where fragmentation is extreme, but over an uncertain time frame. However, it has yet to be satisfactorily demonstrated that the stochastic 'island' relaxation effect is generally a strong one, and it seems likely that there will be a degree of scale dependency in relation both to area and isolation: factors that warrant further theoretical and empirical attention. Recent work suggests that there may be taxon dependent thresholds of habitat loss, above which few species losses occur, but beyond which losses accelerate (Ewers and Didham 2005). Given the significance of the assumptions of relaxation to the debate on species extinction rates, it is surprising that more attention has not been given to demonstrations of these effects (see Simberloff and Levin 1985; Simberloff and Martin 1991; Whitmore and Sayer 1992; Brooks *et al.* 1999; Laurance 2002).

10.8 Succession in fragmented landscapes

The processes of land-use change which create fragmented systems typically initiate successional changes in the habitat island remnants. This was apparent in the Barro Colorado island study. Another example is provided by Weaver and Kellman's (1981) study of newly created woodlots in southern Ontario, Canada. The species losses that occurred did not take the form of stochastic turnover, with different species 'winning' in different woodlots. Instead, the 'relaxation' involved the successional loss of a particular subset of species. With appropriate management, the observed losses could be avoided. If occurring at all in the absence of disturbance and succession, turnover appeared to be very slow-paced among the vascular flora. In this system, the EMIB effects were clearly weak, and subordinate to other 'normal' ecological processes.

In practice, nature-reserve management often pays considerable attention to ecological succession, particularly in small reserves. Failure to do so often leads to the loss of desirable habitats. This applies to many of the lowland heath reserves of southern England, which have long been anthropogenically maintained. Without appropriate management, most areas suffer woodland encroachment. The concepts of **minimum dynamic areas** and **patch dynamics** (Pickett and Thompson 1978) are thus relevant to management plans. Simplified representations of woodland dynamics view a stand of trees as going through phases of youth, building, maturity, and senescence, each of which may support differing suites of interacting species. Reserves should therefore be large enough (or managed) to ensure that they contain enough habitat patches at different stages of patch life-cycles to support a full array of niches. A related and often contentious issue is how fire regimes are managed. In fire-prone regions there is often a cyclical pattern of post-burn succession and fuel accumulation, leading to an increased likelihood of fire, repeating the cycle. The maintenance of a particular fire regime and patch mosaic structure may be of crucial relevance to species diversity in a reserve system, but is often poorly understood (Short and Turner 1994; Milberg and Lamont 1995). Fire is also politically contentious because of the threat it poses to people and property.

Much active management of nature reserves is thus about keeping a mosaic of different successional stages, and not allowing the whole of the reserve to march through the same successional stage simultaneously. The relevance of such successional dynamics to species changes in reserve systems appears to have received much less attention in the theoretical conservation science literature than its significance warrants (but see Pickett *et al.* 1992). Moreover, continuing changes in the habitats of the matrix can also be extremely important to the fate of species populations in the reserves themselves (Stouffer and Bierregaard 1995, 1996; Gascon *et al.* 1999; Daily *et al.* 2003).

10.9   The implications of nestedness

As we established in Chapter 5, many island and habitat island data sets exhibit nestedness. That is, when organized into a series of increasing species richness, there is a significant tendency for the species present in species-poor (small) patches to be found also in successively richer (larger) patches. Although nestedness can be driven by selective colonization or by habitat nestedness, it appears that differential extinction plays a major role in producing nested structure in many habitat island data sets (Wright *et al.* 1998). A nestedness analysis can

contribute a simple answer to the SLOSS question, as a strong degree of nestedness implies that most species could be represented by conserving the richest (largest) patch. A low degree of nestedness, on the other hand, would mean that particular habitat patches sample distinct species sets and that an array of reserves of differing size and internal richness may be required to maximize regional diversity (cf. Kellman 1996). Knowledge of nested subset structure might therefore provide a basis for predicting the ultimate community composition of a fragmented landscape, particularly if it is possible to attribute patterns to particular causes (Worthen 1996, Fischer and Lindenmayer 2005).

Blake's (1991) study of bird communities in isolated woodlots in east-central Illinois shows how nestedness calculations can provide useful insights. He demonstrated a significant degree of nestedness, particularly among birds requiring forest interior habitat for breeding and among species wintering in the tropics. In contrast, species breeding in forest-edge habitat showed more variable distribution patterns. Blake's findings resonate with those reported from São Paulo by Patterson (1990), who found significant nestedness amongst sedentary bird species, but that the full data set (i.e. also including transient species) was non-nested. These results indicate that some species, often those of most conservation concern, will be lacking from any number of small patches, but can be found in the larger, richer patches.

Tellería and Santos (1995) have demonstrated the apparent importance of nestedness of habitat. They studied the winter use of 31 forest patches (0.1–350 ha) in central Spain by the guild of pariforms (*Parus* and *Aegithalos, Regulus, Sitta,* and *Certhia*—tits, goldcrests, nuthatches, and treecreepers). They found that birds with similar habitat preferences tend to disappear simultaneously with reduction in forest size, thereby producing a nested pattern of species distribution.

Simberloff and Martin (1991) have argued that establishing nestedness across a whole system is no longer particularly exciting, but that some benefit can come from examining discrepancies from nestedness. Cutler (1991) developed an index, *U*, which was designed to account for both unexpected presences and unexpected absences. He used it to identify what he termed holes and outliers. **Holes** are where widespread species are absent from otherwise rich faunas, and **outliers** are where uncommon species occur in depauperate faunas. He applied his index to the data for boreal mammals and birds of the North American Great Basin.He found that for mammals most of the departures from perfect nestedness were due to holes, whereas for birds most departures were due to outliers. Cutler speculates that this difference could be a function of the superior dispersability of birds, allowing them to generate greater numbers of outliers by recolonization events. 'Supertramp' species (Chapter 5) might also appear as outliers in analyses of this sort. The preponderance of holes in the mammal data supports the importance of extinction as a driving force in this system, with the holes representing departures from the generally predictable sequence of species losses that have occurred during the Holocene.

Such analyses might appear to provide important information for conservation. However, Simberloff and Martin (1991) rightly caution that, to a large degree, extinction across these mountain-top habitat islands has been inferred rather than demonstrated to have happened. The simple statistics can, moreover, mask more complex and important underlying details. In illustration, they show that the same nestedness score can be generated for species with widely differing underlying distributions. In the Maddalena archipelago (Sardinia), two species have a nestedness score of zero (by the Wilcoxon statistic). One, the rock dove (*Columbia livia*), is found across the entire range of island sizes, and is absent from only 2 of 16 islands. The second, the little ringed plover (*Charadrius dubius*), has been observed only once in the archipelago, breeding only in one year. As they point out, there may be different explanations for accordance or deviation from nestedness in particular species. The explanation of such patterns requires basic information on species' presence–absence, abundances, minimum area requirements, habitat use at different sites, and temporal regularity in occupying differing types of patch. In their view, given such data, it becomes possible to offer sensible input into conservation

decision-making—input in which the contribution of the nestedness statistics is fairly limited.

Moreover, there are many different indices for assessing nestedness, and much debate over which is the best to use for particular purposes (Rodríguez-Gironés and Santamaría 2006). Cook (1995) undertook a comparative study of 38 insular systems using 6 different indices of nestedness, ordered both by area and richness. The analyses ordered by richness returned higher nestedness scores than those ordered by area in between 35 and 37 cases (depending on the metric used). Cook's analyses show that the broad patterns of nestedness are robust, but that there is a danger of over-interpreting the details of a nestedness analysis.

Formal consideration of compositional structure was, for too long, sidelined in the SLOSS debate and related literature. The attention now being given to species compositional patterns across systems of habitat islands is thus welcome (Fischer and Lindenmayer 2005). A nestedness index can provide one compositional descriptor and can perhaps aid identification of risk-prone species; but should not be given primacy in conservation planning. The discovery of nestedness at a particular point in time does not necessarily provide clear insights as to the probability of maintaining the same sets of species (or any particular species) over time (Simberloff and Martin 1991). The isolates may be subject to turnover and/or species attrition in new ways dictated by the changing biogeographic circumstances of the landscape in which the fragments occur. As Worthen (1996, p. 419) put it, nestedness is not a 'magic bullet', ' . . . no single index should be expected to distil the informational content of an entire community, let alone predict how it will react to habitat reduction or fragmentation'.

10.10 Edge effects

The boundary zone, or **ecotone**, between two habitats, being occupied by a mix of the two sets of species, is often richer in species per unit area than either of the abutting core habitat types. In the reserve context, attention has focused on the reserve edge, typically where a woodland reserve is surrounded by a non-woodland matrix. For some species groups, possession of a large expanse of such edge habitat can increase the overall species number found on a reserve, an apparently beneficial effect (Kellman 1996). But, the argument goes, if the additional species supported by the edge habitat are those of the matrix, then they do not depend on the reserve and should be discounted in evaluating the merits of the reserve system.

In illustration, Humphreys and Kitchener (1982) classified vertebrate species into those dependent on reserves and those which persist in disturbed areas in the intervening agricultural landscape of the Western Australian wheat belt (Fig. 10.8). They found that the small habitat islands, which were mostly edge habitat, were disproportionately rich in ubiquitous species. Those species restricted to woodlots typically required a larger area (cf. Hinsley *et al.* 1994). This pattern was found separately for lizards, passerine birds, and mammals. By analysing all species within a taxon together, small reserves came out in a better light than they warranted. Similar results are reported by Watson *et al.* (2004*a*,*b*) in a study of birds in littoral forests and surrounding habitats in south-eastern Madagascar. Core forest locations were found to be richer than edge or matrix habitats, with some 68% of the forest dependent species found to be edge-sensitive. Frugivorous species and canopy insectivores (including six endemic vanga species) were generally edge-sensitive, while in contrast, sallying insectivores were edge-preferring. At least part of the edge-sensitivity recorded was attributable to changes in vegetation structure at the remnant edges. Consistent with this edge sensitivity, forest dependent species were generally lacking from fragments of less than 10 ha, thereby demonstrating a small-island effect (Chapter 4), masked in analyses of all bird species richness (Fig. 10.9). Hansson (1998) used a rather different approach in his studies of edge effects in ancient woodlands in Sweden, but again found differences in bird community composition that appeared to relate to habitat structure at the edges. He analysed the degree of nestedness, finding that whereas census data for birds from the centres of the patches were nested, the samples from the edge habitat were non-nested.

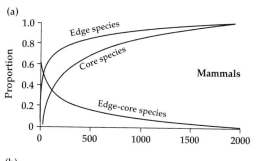

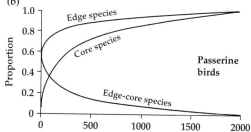

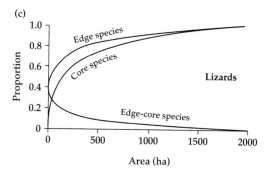

Figure 10.8 The proportion of the core species (those dependent on the reserves) and the edge species (those which can also survive outside the reserves in disturbed habitat) which were present in a series of 21 nature reserves in the Western Australian wheatbelt in relation to the number of each present in a reserve of 2000 ha. The curves with negative slopes represent the difference between the curves for edge and core species, and represent the relative excess of edge species over core species compared with their distribution at 2000 ha. Data for (a) mammals, (b) passerine birds, and (c) lizards. (Redrawn from Humphreys and Kitchener 1982, Fig. 1.)

In particular circumstances, ecotones may have negative implications for core woodland species (Wilcove *et al.* 1986), in that although they may support additional species or larger populations of non-woodland core species, these populations may have negative interactions with deep-woodland species. Studies in North America have suggested that the nesting success of songbirds is lower near

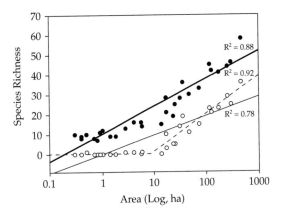

Figure 10.9 Bird species richness–area relationships for littoral forests, southeastern Madagascar. Closed circles (and bold, solid line): all bird species; open circles: forest dependent bird species, fitted by linear regression (solid line), and by break-point regression (broken line), demonstrating a small-island effect. (From Watson *et al.* 2004*b*.)

the forest edges than in the interior because of higher densities of nest predators—e.g. blue jay (*Cyanocitta cristata*), weasel (*Mustela erminea*), and racoon (*Procyon lotor*)—around forest edges. Wilcove *et al.* (1986) estimated that as a result, reserves of less than 100 ha will not support viable populations of forest songbirds. A few studies of birds and invertebrates have suggested that some species prefer edge habitats for breeding even though mortality rates at edges may be higher, a phenomenon termed an **ecological trap** (Ewers and Didham 2005). Paton (1994) has undertaken a critical review of studies of nest predation in edge habitats. He concluded that, in general, there does appear to be evidence for a depression of breeding success, due either to enhanced predation or through parasitism. However, these forms of edge effect usually operate only within about 50 m of an edge, a narrower belt than some have claimed (e.g. Terborgh 1992). Nonetheless, such an effect implies that the effective reserve area may be less than that described by the perimeter of the isolate.

Edge effects may be taxon or even system specific. For instance, Burkey (1993) undertook an experimental study of egg and seed predation with distance from the edge of a patch of Belizian rain forest. Egg predation rates were higher in a 100 m edge zone, but, conversely, seed predation rates

were found to be higher 500 m into the forest than 30 m and 100 m from the edge. In contrast to Burkey's egg predation data, Delgado *et al.* (2005) report higher predation pressure under closed canopy than within edge habitats in their study of artificial nest predation by ship rat (*Rattus rattus*) in laurel forest on Tenerife. This illustrates that the relationship between a reserve and its surrounding matrix is not subject to easy generalization. There are species that share both zones, and just as there are matrix species which may impact negatively upon core reserve species, there may also be reserve species (dependent on it for breeding and cover) which exploit resources in the matrix. The concepts of source and sink populations may again be useful in this context, as the maintenance of maximal diversity across a landscape depends on sufficient source or core habitat for species of each major habitat. The identification of edge effects provides a part of such an analysis.

10.11 Landscape effects, isolation, and corridors

The benefits of wildlife corridors

The premise of much of the literature discussed in this chapter is that habitat connectivity is beneficial to long-term survival, as it enables gene flow within populations and metapopulations. Habitat connectivity might be achieved by having stepping stones, or corridors, of suitable habitat linking larger reserves together. In practice, habitat corridors act as differential filters, enabling the movement of some species but being of little value to others. As Spellerberg and Gaywood (1993) point out, we are not merely concerned with forest peninsulas or hedgerows. All sorts of linear landscape features, such as rivers, roads, and railways, may act as conduits for the movement of particular species. Equally, they may represent barriers or hazards for others (Reijnen *et al.* 1996). Harris (1984) points out that landscapes with great topographic relief channel their kinetic energy via dendritic tributaries, main channels, and distributaries, and that these in turn influence the landscape template in characteristic ways. The stream-order concept of the fluvial geomorphologist may be a means of understanding species distributions within the landscape (Fig. 10.10), and of understanding movements of species between disjunct habitat patches (Fig. 10.11). Studies of features such as hedgerows have not always found them to be as effective in connecting up woodlands as was hoped, with some species moving as well through surrounding fields and others simply not moving well along the hedges. However, habitat corridors may be important in particular cases.

A useful illustration of how corridors can be beneficial comes from the study by Saunders and Hobbs (1989) of Carnaby's cockatoo (*Calyptorhyncus funereus latirostrus*) from the Western Australian wheat belt, an area of 140 000 km^2 in the south-west of the state, 90% of which has been cleared for agriculture. The Carnaby's cockatoo is one of Australia's largest parrots and was once the most widely distributed cockatoo in the region. It breeds in hollows in eucalypt trees and eats seeds and insect larvae from plants in the sand-plain heath. It congregates in flocks, both to nest and forage. Individuals may live 17 years and breeding status is not attained until at least 4 years of age, each pair rearing a single offspring at a time. The widespread clearance of the land has removed extensive areas of the vegetation type in which they feed, replacing it with annual crops that are useless to the birds. In recent developments, wide verges of native vegetation have been left uncleared along the roads. These act to channel the cockatoos to other areas where food is available.

Cockatoos have not survived in areas of earlier clearances, carried out without these connecting strips, as once they have run out of a patch of acceptable habitat it takes a long time for the flock to find another patch of native vegetation. The more patchy the vegetation, the less successful the birds are at supplying adequate food to rear their nestlings. Furthermore, the narrow road verges of these early clearances result in a higher incidence of road deaths. The example illustrates that the degree and nature of connectivity of different landscape elements—in this case of the breeding and feeding habitats, and of the vehicular hazards—are critical to the survival of this species in these newly

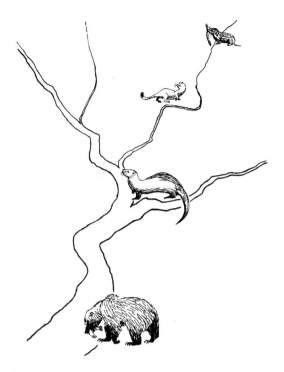

Figure 10.10 Association of different-sized carnivorous mammal species with stream order and typical food particle size in accordance with the stream-continuum concept. (Redrawn from Harris 1984, Fig. 9.8.).

fragmented landscapes. Given the relatively long life cycle of these birds, and their flocking behaviour, failure can have a sudden and wholesale expression in the bird's disappearance. The big reduction in area is fairly recent, and the cockatoo is not yet in equilibrium with the new regime. It is still on a downward path and is viewed as a species in danger.

The benefits of isolation

Although much of the literature is concerned with maintaining population movements, there are also contexts in which *isolation* of populations of a target species might be desirable (Simberloff *et al.* 1992), e.g. where there is a threat from a disease organism. Young (1994) evaluated 96 studies of natural die-offs of large mammal populations, defined as cases where numbers crashed by 25% of the individuals or more. He noted that populations

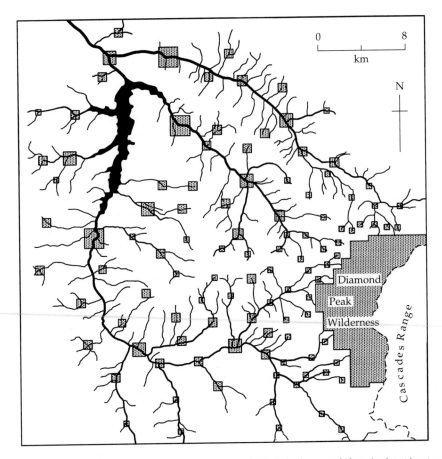

Figure 10.11 A possible spatial and size-frequency distribution of old-growth habitat islands arranged along riparian strips at progressively greater distances from a current wilderness area in the Willamette National Forest, USA, designed to provide the optimal degree of connectivity between patches in a system structured by the dendritic pattern of drainage. In this hypothetical system, the most distant islands are generally larger in order to counter presumed lower dispersal. (Redrawn from Harris 1984, Fig. 10.2.)

subject to large-scale phenomena such as drought and severe winters may not be protected from such die-offs by population subdivision. But, where the crashes are produced by disease epidemics, subdivision may be beneficial, and the creation of linking corridors, and translocation efforts, may be harmful. Young noted in his review that most herbivore die-offs were due to starvation, but carnivore die-offs were more often attributed to disease. An epidemic need not eliminate all the individuals in a population for it to have a crucial role in causing the eventual extinction of a population from a reserve. Combining these considerations with metapopulation modelling, Hess (1996) argues that there should theoretically be an optimal degree of

movement within a metapopulation to enable the movement of propagules of the target species but not of disease. But much will depend on the nature of the intervening landscape matrix and the extent to which it filters target and 'pest' species (Ricketts 2001; Ewers and Didham 2005).

The spread of exotic competitors and predators and of fire into reserves can also be problematic (Spellerberg and Gaywood 1993; Lockwood and Moulton 1994). It has been suggested that alien plant invasion can be reduced by isolating reserves and surrounding them with simplified cropland oceans (Brothers and Spingarn 1992). Short and Turner (1994) found that the decline and extinction of certain species of native Australian mammals were most parsimoniously explained by reference to the roles of introduced predators (foxes and cats) and herbivores (rabbits and stock). This explained the continued persistence of the native species on offshore islands from which the exotic species were absent. In these cases, isolation is beneficial (as Janzen 1983) and many reserves set in 'seas of altered habitat' within mainland Australia have proved to be insufficiently isolated to prevent their invasion by exotic pests and pathogens.

Corridors or isolation?

Is it possible to offer guidance about the situations when corridors may be beneficial and when not? First, as Dawson (1994) notes, all-purpose corridors do not exist, as each species has its own requirements for habitat, its own ability to move, and its own behaviour. Corridors, like other landscape components, therefore act as filters. He points out that many rare and threatened species are unlikely to benefit from corridors, because the corridor would have to contain their presumably rare habitat, i.e. rare species may require odd corridors.

Some corridors are essential in providing links between preferred habitats for animals that undertake regular seasonal migrations, e.g. some fish, amphibians, reptiles, and for large mammals in both the seasonal tropics and the Arctic. On the other hand, for some populations, corridors may act as 'sinks', drawing out individuals from the main habitat area but not returning individuals to supplement it, in which case they may do more harm than good. In other cases, they may be fairly neutral in their ecological cost–benefit, but perhaps be quite expensive to purchase and set up if not already existing in a landscape.

Simberloff *et al.* (1992) point out that numerous corridor projects have been planned in the USA, potentially costing millions of dollars, despite the lack of data on which species might use the corridors and to what effect. Once again, the cost-- benefit equation for corridors may depend on properties of the system involved, including scale considerations and assessment of the major drivers of species loss. In the longer term, it has been argued that climate change is likely to drive sub- stantial shifts in the distribution of species, and that the resulting species migrations will be impeded by the human sequestration of land to agriculture and other purposes. Hence, on these grounds, it would seem to be prudent to plan more or less continuous habitat corridors that straddle major climatic/ elevational gradients where this is feasible (e.g. Bush 1996, 2002).

Reserve systems in the landscape

As we have seen, the debate about habitat corridors has broadened from fairly simple theoretical beginnings to a consideration of how a whole range of differing natural, semi-natural, and artificial features are configured within a landscape. For instance, countless thousands of birds are killed each year through collision with motor vehicles and with overhead power cables (Bevanger 1996; Reijnen *et al.* 1996). Research has shown that deaths through collisions with cables can be greatly reduced by consideration of important flight paths during construction, by design features of the gantries, and by attaching a variety of objects to the cables to enable birds to sight them (Alonso *et al.* 1994). Paved roads and tracks can play an important role as corridors favouring the introduction of alien plants (Arévalo *et al.* 2005) or animals (Delgado *et al.* 2001) into otherwise well-preserved ecosystems, but also creating corridors of suitable habitat that can in cases allow endemic species (e.g. lizards) to disperse through inhospitable forest matrix, enabling genetic exchange between otherwise isolated populations (Delgado *et al.* in press). On the other hand, wide roads and, especially, fenced highways, may act to impede the movements of large terrestrial animals, thus fragmenting populations and interfering with migration. Such information needs to be integrated into improved management of whole landscapes (Spellerberg and Gaywood 1993). In short, a landscape ecological framework is needed.

Landscape components include both habitat patches and the matrix in which the patches are embedded. Modelling exercises typically favour clumped configurations of reserves, treating the matrix as a uniform 'sea'. In practice, the optimal configuration must take account of the details of the landscape involved (e.g. Fahrig and Merriam 1994). For instance, there may be greater opportunity for dispersal between two distant reserves linked by a river and its adjacent riparian corridor, than between two similar but closer reserves separated by a mountain barrier of differing habitat type (Fig. 10.11). For butterflies, differences in land use as subtle as switches from one woodland type to another can significantly influence the passage of butterflies from one favoured habitat patch to another (Ricketts 2001). An early attempt to place island theories into realistic landscape contexts was provided by Harris (1984). Recognizing that there were limits to the amount of land that would be put over to reserve use, Harris advocated a series of reserves placed within the landscape as dictated by geographical features such as river and mountains, to encourage population flows between reserves. Most reserves would remain in forestry management. Within such reserves, there should be an undisturbed core of forest that is never cut, providing habitat for species requiring undisturbed, old-growth conditions, around which commercial operations might continue, following a pattern of rotational partial felling. This would provide a mosaic of patches of differing successional stage, maximizing habitat diversity, while maintaining income (Fig. 10.12). His strategy aimed to satisfy the requirements of island theories, patch dynamic models, and economic realities. The implementation of such a policy requires concerted action from a wide variety of agencies and is thus easier to sketch out than to bring to a realization.

Species that don't stay put

In the Northern Territory of Australia, as in many other systems across the globe, mobility and massive population fluctuations in response to a highly variable climate are features of the wildlife. Woinarski *et al.* (1992) cite as a particular example the magpie goose (*Anseranas semipalmata*). They note that a conventional reserve network consisting of discrete national parks (such as they have developed for plants in the region) will not cater for the

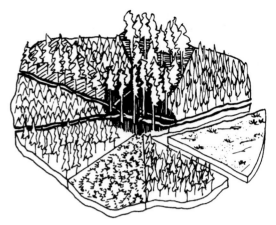

Figure 10.12 Harris's (1984) proposed system for the management of forest habitat islands on a long-rotation system, such that different sectors are cut in a programmed sequence, but the core old-growth patch is left uncut. This recognizes the importance of patch dynamics to habitat diversity and how such management may serve both economic and conservation goals. (Redrawn from Harris 1984, Fig. 9.5.).

need of the magpie geese to relocate in response to patchiness of rainfall. They identify four possible strategies for conservation in the Northern Territory:

• The status quo, with improvements to ensure all vegetation types are included in the network. This would not solve the problem, as it is not just vegetation types but the year-to-year variation in carrying capacity across the region that matters.

• Seeking inclusion of known habitat of important species into the reserve system. For the magpie geese, this might mean inclusion of all wetlands. However, it would be difficult to decide which species to concentrate on and it would be impractical to gazette all the land in question because of competing claims for use.

• Developing very large reserves which span extensive slices of the environmental gradient, just as does the present Kakadu National Park, a 20 000 km² reserve. However, this would require vast reserves in the semi-arid and arid region, in which, at the present time, the largest reserve is of 1325 km².

• Supplementing the representative reserve network with measures that protect wildlife habitat on large areas of unreserved land. They see this as the only viable answer in the long term, requiring a broad educational and political strategy to get there (Woinarski *et al.* 1992; Price *et al.* 1995).

So, wildlife conservation must in part be concerned with the practicalities of conserving outside reserves. The Northern Territory example may seem an extreme case when viewed from many parts of the world, yet the argument surely applies to some degree everywhere.

10.12 Does conservation biology need island theory?

Zimmerman and Bierregaard (1986) undertook an analysis of Amazonian frogs which demonstrated that knowledge of the **autecological requirements** of the species would be critical to planning a reserve system (not that a system would necessarily be designed just for frogs). They undertook a prediction by a species–area approach and showed that it lacked relevance in the light of the available autecological data. Frog species differ in the habitat required for breeding: whereas some are strictly streamside breeders, others require landlocked permanently or periodically flooded *terra firma* pools. The authors concluded that if such data are available for enough species in the system being considered, then there need be no recourse to abstract principles. If they are not available, it is still preferable to conduct a rapid survey and collect biogeographic and ecological data. This is just one of many studies that have championed empirical over theoretical 'island' approaches. Does conservation biology need island theories?

A non-equilibrium world?

Worthen (1996) bemoans the fact that the bias towards species–area relationships over compositional–area relationships continues to be a feature of the literature, including some of the most widely used introductory ecology texts. This bias continues despite the demonstration, in numerous studies, of nestedness, community similarity of habitat fragments, the relevance of successional patterns, and a variety of other forms of compositional structure. These non-random patterns in how communities assemble, or alternatively how they change in response to fragmentation, demand that we reappraise the contribution of 'island ecology' within conservation biology.

Much of the earlier applied island literature seemed to assume that matrix contexts were relatively insignificant and that stochastic decay towards equilibrium would restore a new, depauperate balance. These were arguably useful simplifying assumptions, but the behaviour of habitat islands is perhaps more realistically captured within the notions of *flux of nature* described by Pickett *et al.* (1992). This framework stresses physical as well as biotic processes (e.g. succession), the role of episodic events (e.g. fires or hurricanes), and the differing degrees of connectivity of populations across 'filtering' landscapes. We have also seen that the nature of the matrix can be crucial to what goes on in the habitat islands within the landscape. How much of this is 'island biogeography' is a moot point, but insularity remains a key property and issue within these other topics.

The application of equilibrium (EMIB) assumptions to reserve systems has been neatly critiqued by Saunders *et al.* (1991, p. 23):

It is commonly assumed that at some stage the remnant will re-equilibrate with the surrounding landscape. It is however, questionable whether a new stable equilibrium will be reached since the equilibration process is liable to be disrupted by changing fluxes from the surrounding matrix, disturbances, and influx of new invasive species. The final equilibrium can be likened to an idealized end point that is never likely to be reached, in much the same fashion as the climatic climax is now conceptualized in succession theory. Management of remnant areas will thus be an adaptive process directed at minimizing potential future species losses.

The general premise of much of the applied island biogeography literature is that (1) the alarming fragmentation of many 'natural' systems will continue and therefore (2) if much wildlife is to survive, it will have to do so in reserve systems, and that (3) we have very little time in which to act. These concerns explain the initial attractiveness of deriving general principles for reserve systems from island theory. Increasingly, the resulting lines of reasoning have been attacked on the grounds: first, that many valid generalizations are self-obvious platitudes; secondly, that models are often both too simplistic and effectively untestable; and, thirdly, that many theoretical generalizations have achieved the status of dogma (Doak and Mills 1994). Yet, the general principles derived from island theory have influenced conservation policy and they remain pervasive. Because of this, it is important to understand both the strengths and, particularly, the weakness of the 'island' theories discussed in this and the previous chapters.

If our focus is on global losses of species diversity, then clearly the loss of—and fragmentation of—species rich habitats is a key concern. But, the stochastic consequences of reduced population sizes are not necessarily simple to understand and model. Some species can persist well in fragmented landscapes providing that the areas of remaining habitat are not too small and isolated. Much may depend on the previous history of the region (extent of initial departure from equilibrium assumptions), the taxa considered, and the scale sensitivity of the systems, and how the habitat matrix (the land outside the fragments) is managed. Simple species–area models, particularly as applied on regional scales, may not in the end provide reliable estimates of how many species really are 'committed to extinction', i.e. still around but heading for an inevitable nemesis. Some would regard this assessment as complacent and over-optimistic. For instance, Pimm and Askins (1995) argue that concerns about the precise causal mechanisms of extinctions, as expressed here, miss the point and that, in practice, simple predictions of species losses based on species–area relationships are actually fairly reliable. From their analyses of North American forest birds they suggest that if such estimates err, they tend to be conservative, i.e. predicting fewer extinctions than actually occur. We are not convinced that their claim can be generalized, as so much seems to depend on issues such as system configuration and the actual drivers of species threat and extinction (see discussion of the Atlantic forest analyses, above).

The problem of biodiversity loss (including population-level decline, genetic erosion and species losses) is a depressingly real one and the pace of species extinctions is undoubtedly accelerating (Whitmore and Sayer 1992; Turner 1996). Yet, as Simberloff *et al.* (1992) observe, ideas derived from island theory, and the so-called SLOSS principles (specifically the benefits of corridors), became embedded in policies of bodies such as the IUCN and World Wildlife Fund without having been validated. Thus, policy-makers and legislators have in cases acted upon such uncertain scientific 'principles'. Once released from the bounds of scientific journals, scientific information is often poorly understood. It is therefore important not to offer up general principles and rules where the evidence for them is equivocal and, in cases, contradictory (Shrader-Frechette and McCoy 1993).

Nonetheless, we should not simply discard the theoretical frameworks of island theory, metapopulation theory, etc. They have their relevance to particular systems and a wider and important role in helping us to understand the sorts of effects that might be operative in real-world contexts, and that therefore should be tested for or monitored. As Doak and Mills (1994) have put it, we should face up to the problems rather than use them as an excuse for ignoring the theoretical debates: 'it is incumbent on us to teach such complexity to managers and nonbiologists, rather than attempting to snow them with undefendable overgeneralizations'.

The notion that species disassembly (as well as assembly) is fairly structured and predictable, at least in terms of functional ecological types or guilds, is one that has reappeared in several guises in this chapter (e.g. from studies of relaxation, experimental fragmentation, incidence functions, nestedness, and edge effects). In the case of the Manaus habitat fragmentation study, similar structuring was evident both in the initial losses (disassembly) and, in particular circumstances, in the recovery of bird communities that followed. This reminds us, first, that habitat change does not *have* to be a one-way process, and, secondly, that it might be possible to target conservation efforts fairly directly on to particular sets of vulnerable species. In this sense, 'island' approaches have produced some dividends. It is now a question of learning what they are, and differentiating between those effects that are weak and those that are strong (the world is neither wholly non-equilibrial nor wholly equilibrial). It seems that we can work out which sorts of species are most threatened and some of the major reasons why they are threatened, and it should therefore be possible to offer some management solutions. Some of these insights have indeed been derived from 'island' studies.

For many species, reserves are essential. Very often this is because of specific threats posed by people (or our commensals) and the protection that can be afforded within the reserve. Simberloff (1992) comes down in general on the side of large, continuous blocks of habitat. He argues that no existing theory adequately describes the joint effects of loss of area and fragmentation, but that there are empirical observations, such as those to do with predation discussed earlier, to suggest that for many species the combined impact is indeed strongly deleterious: 'Probably any species that has evolved in large, relatively continuous habitat has traits that are maladaptive in small, isolated fragments' (p. 85). Beyond this generalization, the means, rate, and extent to which extinction is delivered, are likely to be idiosyncratic. The theoretical basis for calculating likely regional and global extinctions as a result of the rapid phase of habitat loss currently under way remains highly uncertain. For a careful review of the evidence and basis for calculations see Whitmore and Sayer (1992).

Ecological hierarchies and fragmented landscapes

As with other branches of island theory, approaches to the debates on reserve configuration that take no account of hierarchical interactions between organisms of differing trophic levels are incomplete (Terborgh 1992). For many animal groups, plants (collectively or individually) determine the ability of particular animal species to occupy a habitat island. Successional changes in plant communities may often explain turnover in bird or butterfly communities (Chapter 5; Thomas 1994). Equally

important may be the activities of animals as dispersers and pollinators of plants (and as predators), both in contexts of forest maintenance and of recolonization of disturbed areas. Different types of animals disperse different, albeit overlapping, suites of plant species. Studies of cleared areas in the Neotropics have suggested that bats may be more significant in seeding open habitat than are day-flying birds. The efficacy of birds, in particular, may also be positively influenced by the availability of some woody cover in an otherwise open area (Estrada *et al.* 1993; Gorchov *et al.* 1993; Guevara and Laborde 1993). Thus, for instance, cheap plantings of small clumps of plants may encourage speedy return of forest around the inocula.

The Krakatau recolonization data demonstrate the significance of birds and bats in transporting seeds between sites, and thus also supports their significance for population interchange between forest patches in the tropics. Maintenance and restoration of forest might thus necessitate protection for vertebrate species, if not in their own right, for their interactions with plants. It is not 'merely' biodiversity which is at stake, as in cases there may be huge economic costs to the loss of pollinator species on which economically useful plants depend (Pannell 1989; Fujita and Tuttle 1991; Cox *et al.* 1992; Ricketts 2004). At the same time, the Krakatau study demonstrated that particular groups of plants, those dispersed by non-volant frugivores and the large-seeded, wind-spread species, are likely to be greatly hampered by patch isolation (Whittaker and Jones 1994*a,b*; Whittaker *et al.* 1997). These topics have recently seen a resurgence of research interest (Ewers and Didham 2005), with studies in both Africa and the Neotropics demonstrating the negative impacts on frugivore-mediated plant dispersal of forest fragmentation (e.g. Cordeiro and Howe 2001; Martínez-Garza and Howe 2003).

It seems likely that some area effects involve threshold responses and that these operate via hierarchical links within ecosystems (Soulé *et al.* 1988; Terborgh 1992). For instance, below a certain size, an isolated habitat island may be incapable of supporting populations of large predatory vertebrates, with the consequence that higher densities of their prey species may be maintained, with potentially significant consequences further down the food chain. Thus, the densities of medium-sized terrestrial mammals are between 8 and 20 times greater on Barro Colorado Island, which lacks top predators, than in a comparable 'mainland' site in which they occur. Terborgh (1992) suggests that this may be having important selective impacts on plant regeneration. On some smaller islets nearby in Lake Gatun, not only are the (true) predators absent, but so too are those smaller mammals that act as seed-predators, the squirrels, peccaries, agoutis, and pacas which have such high densities on Barro Colorado island. These small islets became dominated in the 70 or so years since their isolation by large-seeded tree species. This is plausibly because these plants no longer suffer high rates of attrition at the point of germination and establishment, and have therefore been able to out-compete the smaller-seeded species to a greater extent than usual.

Predators undoubtedly have a crucial role to play in fragmented landscapes, not just in terms of predation within the habitat island, but also within the matrix. Simberloff (1992) notes that in the Manaus fragmentation study the bat falcon (*Falco rufigularis*) has repeatedly been seen to chase birds flying over cleared areas from fragment to fragment. It may therefore be unsurprising that most understorey birds will not willingly cross gaps of even 80 m. Similarly, in the USA, the northern spotted owl (*Strix occidentalis caurina*), of metapopulation theory fame, previously was a species which rarely left closed forest. In the newly fragmented landscapes, as many as 80% of yearling males die, apparently from predation by great horned owls (*Bubo virginianus*) and goshawks (*Accipiter gentilis*) as they disperse over cleared areas. This was one consideration that persuaded the interagency federal committee responsible for the management plan to shift their strategy towards larger, more continuous blocks. This brief consideration of predation helps us recall the point that the particular mechanisms responsible for failures to sustain populations or to recolonize habitat islands may thus be poorly predicted by 'species–area' type-approaches, or for that matter by metapopulation models.

There may also be important dynamic features, such as the spatial and temporal oscillations that characterize elephant populations in the Tsavo National Park in Kenya, on a cycle of roughly 50 years. Even with very large reserves, migratory animals may ignore park boundaries and, especially in lean years, may move into areas outside, where they are most unwelcome and perhaps also unsafe. These patterns can provide big management problems, even when a very large area is enclosed as a reserve (Woinarski *et al.* 1992). The elephants illustrate another idea: that of the **keystone species**, meaning a species that is so critical to the functional character of an ecosystem that its removal would cause a chain of alterations (e.g. Soulé 1986; Pimm 1991). Elephants have a key role as grazers. Without them successional changes occur in the vegetation, altering the carrying capacity of the system for other species. High densities of elephant, on the other hand, can lead to overgrazing and widespread destruction of trees. In such cases, conservation management naturally focuses around managing the keystone species.

Climate change and reserve systems

Range shifts driven by significant climatic change have been a feature of the Quaternary period. There is no reason to consider these climatic fluctuations to be at an end, and there is also mounting concern that humans have initiated a phase of rapid climatic warming. Researchers are now focusing on the abilities of plant and animal species to track their required climatic envelope across the now fragmented landscapes (Huntley *et al.* 1995). Key factors are the speed of climatic change, the dispersal abilities of the target species, and the characteristics of the landscapes across which species may have to move, which may range from the optimal to the benign to the hostile. Grappling with such scenarios requires a variety of research tools (Bush 1996, 2002; Pearson and Dawson 2003; Araújo *et al.* 2005*b*), amongst which are the tools of island biogeography reviewed in these pages (Boggs and Murphy 1997).

10.13 Concluding remarks: from island biogeography to countryside biogeography?

As made clear at the outset of this chapter, this review was not intended as a summary of all important branches of wildlife conservation. However, we have stressed along the way the importance of autecological, distributional, and other forms of data. The student of conservation biology will also need to have a knowledge of such themes and issues as *ex situ* conservation efforts in zoos, and other dedicated breeding facilities; germplasm (seed) banks; species translocation schemes; reserve management practices; CITES and other wildlife legislation; poaching and trade in endangered species; strategic conservation policy; the philosophy, politics, and economics of conservation and other fields besides (e.g. Primack 1993; Shrader-Frechette and McCoy 1993; Whittaker *et al.* 2005).

It must also be recognized that there can be many aims and purposes for shaping conservation management, ranging from the aesthetic, through the scientific, to the economic. We may wish to conserve systems or species which are 'representative', 'typical', rare, speciose, nice to look at, of recreational value, or provide economic return (Ratcliffe 1977; Shrader-Frechette and McCoy 1993). Such multiplicities of purpose require requisite tools; the island theories have their place in the tool kit, but they should not always be the first to be reached for.

Allied to the habitat remnant/reserve strategy, wherever possible, conservationists should argue for greater priority to *extensive* rather than *intensive* conservation, i.e. for environmental management policies to encourage the survival (or passage) of many species outside of the more closely protected reserves systems (cf. Rosenzweig 2003; Daily *et al.* 2005). In short, to work to prevent habitat islands and reserves becoming more and more like real islands—except in those biogeographical contexts where insularization is actually beneficial to survival prospects. We have seen in this chapter that issues such as size, shape, and configuration within a landscape are important to reserve success, not just in terms of how many species will be held within a reserve, but also which sets of species. The number of species held in a reserve (or reserve system) is actually less important than to conserve those species which cannot survive outside the remnants (e.g. Newmark 1991).

Some recent efforts have been made to move beyond an exclusive focus on (forest) fragments towards understanding the role of such habitat islands within mixed-use landscapes. This switch in emphasis comes under varying headers. For example, Watson *et al.* (2005) show that the incidence functions of woodland bird species in three different landscapes in the Canberra area (Australia) differ significantly, seemingly as a function of differences in properties of the landscape matrix within which the woodlands are embedded. Hence, Watson *et al.* (2005; and see Box 5.2) join others (e.g. Gascon *et al.* 1999; Cook *et al.* 2002, Ewers and Didham 2005) in calling for greater attention to 'matrix effects'. Hughes *et al.* (2002) adopted a slightly different approach to their study in southern Costa Rica, focusing on the extent to which native forest species make use of the surrounding countryside: they found that some 46% of bird species foraged often kilometres away from extensive areas of native forest. Although they stress that not all species can be so readily accommodated outside large tracts of native forests, their work supports the importance of developing 'countryside' landscapes that are biodiversity-friendly and penetrable by native fauna (as Harris 1994). Daily and colleagues (e.g. Daily *et al.* 2001, 2005; Hughes *et al.* 2002) coin the term 'countryside biogeography' for this switch in attention from remnants *per se*, to the way in which remnants function within whole landscapes. This switch in emphasis is similar to that promoted by Rosenzweig (2003) under the heading 'reconciliation ecology'. Whether we label this shift in emphasis 'matrix effects', 'countryside biogeography', or 'reconciliation ecology', the common element is a realization that effective conservation must include consideration of what happens outside reserves, as the way we shape the countryside, whether we farm intensively or extensively, whether we retain hedgerows and trees within mixed landscapes, can have profound implications for regional diversity and for abundances of wildlife (e.g. see Gascon *et al.* 1999; Gates and Donald 2000).

Conservation requires pragmatic decision-making. As we continue to fragment landscapes, island effects may inform such decision-making, but should not be oversimplified. There is no single message, and no single island effect; indeed, insularity may, in at least a minority of cases, bring positive as well as negative effects (Lockwood and Moulton 1994). Island effects may be weak or strong. The implications of insularity vary, depending on such factors as the type(s) of organism involved, the type of landscapes involved, the nature of the environmental dynamics, the biogeographical setting, and the nature of human use and involvement in the system being fragmented. It is unfortunate that the term 'island biogeography theory' has become largely synonymous in a conservation setting with a limited conception of island ecology, stressing the inevitability of stochastically driven trends to equilibrium. Both pure and applied island biogeography are richer than this. This richness of ideas and information needs to be understood and integrated into teaching for both pure and applied purposes. Moreover, there is also a need for a renewed research effort within this area of conservation biogeography, in the search for improved scientific guidance.

10.14   Summary

The island analogy can be extended to patches of habitats within continents. The conversion of more-or-less wild habitats to other uses is fragmenting, isolating, and reducing 'wild' areas across much of the globe. The implications of this insularization are examined in this chapter from the scale of individual populations up to whole landscapes.

The minimum viable population (MVP) is the smallest number of individuals required to ensure long-term population persistence. We do not know how big MVPs should be. We do know that MVPs vary from species to species, and that the effective population size is generally smaller than the actual population size, i.e. not all individuals, or all adults, are involved in breeding. Population loss may be due to stochastic demographic and/or genetic

effects, or to environmental disturbance and change. Stochastic demographic effects may have been over-estimated. Genetic effects are potentially complex, and vary between taxa. The significance of environmental change or catastrophe is probably crucial to population viability, but can be difficult to model. MVPs therefore need to be established separately for different types of species, and management regimes must be responsive to changing circumstances.

The area required by a MVP is termed the minimum viable area, and may be estimated from knowledge of home range size, although this approach assumes no population exchange with other isolates. Another approach to area (or habitat, or isolation) requirements is to use incidence functions, although the patterns revealed may be confounded by multicausality, and may fail to predict changes that follow within fragmented landscapes. Moreover, recent work suggests that area requirements for species presence in fragments may vary significantly even within the same region.

Where geographically separated groups are interconnected by patterns of extinction and recolonization they constitute a population of populations, or a metapopulation. Here, the idea is that particular patches have their own internal population dynamics, but when they crash to local extinction they are repopulated from another patch within the metapopulation. Such a scenario alters the projections of population viability considerably. In many cases, however, habitat patches appear to describe a source–sink rather than a mutual support system. In such circumstances, a large patch acts as an effectively permanent population, with smaller sink habitats around it being re-supplied from the source population. The sink habitats may have little relevance to overall persistence-time of the metapopulation.

A heated debate developed around the SLOSS 'principles', which represented an attempt to answer the question of whether it was better to advocate Single Large or Several Small reserves of the same overall area, based on the assumptions of the EMIB. Unfortunately, island theory provides no resolution to this debate. In practice, the optimal reserve configuration for one type of organism, landscape, or scale of study system, may not be so for another. Physical environmental change in both fragment and 'matrix' habitats has not generally been given enough attention in the theoretical literature. Successional change and altered disturbance regimes may have long-term implications for species persistence in altered landscapes. The evidence for 'relaxation', i.e. stochastically driven decline in species number, is surprisingly thin. Species composition and richness are liable to change in fragmented systems: species are indeed lost. However, in relation both to SLOSS and metapopulation scenarios, it is apparent that much turnover may well be deterministic in nature and explicable in relation to specific ecological processes such as succession or meso-predator release. Various lines of data suggest strong structure to the disassembly as well as to the assembly of systems. For instance, many systems and species are strongly nested in their distributions, and this can, it seems, be produced in either colonization- or extinction-dominated systems.

It appears possible to predict the types of species that will be in greatest difficulty in fragmented landscapes, and it is these species that require primacy in relation to reserve configuration, population monitoring, and management planning. The consequences of fragmentation can be severe, but they are poorly predicted by classical island theories, which assume stochastic demographic effects to dominate in a system of fragments searching for a new equilibrium in a sea of altered habitats. The nature of the connectivity of habitat patches may be crucial. In some cases, corridors and short dispersal distances may be beneficial, whereas in other circumstances isolated reserves may actually provide the best chance of survival for a threatened species. Some species thrive in edge habitats, others are disadvantaged by increased edge effects. The 'filtering' properties of the matrix can be crucial and require careful assessment in conservation planning. Landscape effects may be complex, with the same linear feature acting as corridor, barrier, and lethal hazard for different organisms. Moreover, some species are mobile, seasonally or aseasonally, and their protection may

require a more extensive approach to conservation than provided for by any reserve system on its own.

These considerations demand a re-evaluation of the place of island biogeography within conservation biology. The equilibrium island theory turns out to provide rather equivocal guidance, and appears to have often been invoked as a basis for predictions of species loss where there is a logical disconnection between theory and prediction. Some of the effects traditionally associated with island approaches may actually be fairly weak in general, but other features of insularity may provide powerful and fairly immediate influences on species persistence and system change. The role of hierarchical ecological relationships, e.g. between plants and their dispersers, between vegetation change and altering habitat space for butterflies, between understorey insectivores and their predators, and so forth, may be crucial to the consequences of fragmentation. Such successional and food-web controls may be the most powerful forms of expression of islandness in the altered landscapes, and they typically involve interactions and exchanges between the fragments and the surrounding matrix. The attention being paid to the effects of the matrix on persistence in remnants, and to the use of matrix habitats by native species, signals an on-going switch in emphasis from island thinking to what some term 'countryside biogeography', i.e. to the importance of incorporating conservation concerns in managing whole landscapes, not just isolated reserves.

A line must be drawn somewhere in any text, and in this chapter discussion stops short on alternative approaches to conservation problems, and on themes such as the implication of large-scale climate change under global warming scenarios. Neither is any attempt made to draw out guiding principles for conservation from island studies. The grand hopes for simple unifying principles have proven to be as illusory as the end of the rainbow. It is nonetheless important to appreciate how these ideas have influenced conservation theory, just as it is important to tackle the ecological complexities of increasing insularity within real landscapes.

CHAPTER 11

Anthropogenic losses and threats to island ecosystems

We cannot discuss the ecology of islands without making a few disparaging comments on goats. These creatures must be the true embodiment of the devil for a plant lover.

(Koopowitz and Kaye 1990, p. 72)

The biodiversity crisis is nowhere more apparent and in need of urgent attention than on islands. Approximately 90% of all bird extinctions during historic times have occurred on islands . . .

(Paulay 1994, p. 139)

11.1 Current extinctions in context

Throughout our discussions of island ecology and evolution, we have mostly set to one side the significance of humans as shapers of island biogeography. As this chapter will show, this is a questionable practice, as human societies have had a profound influence on island biodiversity in both recent historical and prehistorical time (Morgan and Woods 1986; Johnson and Stattersfield 1990; Groombridge 1992; Pimm *et al.* 1995). This influence has been manifested in many ways, but the most profound must surely be the extinction of numerous island races and species.

Of all the species that have ever lived on the Earth, only a small fraction are alive today. Natural catastrophes, such as volcanic eruptions, meteorite impacts, climate changes, marine transgressions and regressions, in combination with biotic forcing, have been key agents of species extinction in the past. Extinction rates have varied through time, with five phases recognized as particularly intense and widespread events in the fossil record, the so

called mass-extinctions events. The most recent of these was the end-Cretaceous event, which resulted from a bolide impact approximately 65 Ma (Groombridge and Jenkins 2002). Whether the comparison is entirely apposite is a moot point, but many argue that a sixth mass extinction event is now under way, which uniquely is being driven by the activities of a single species.

Primack and Ros (2002) suggest that more than 99% of recent species extinctions can be attributed to human activities. Extinctions of vertebrate species over the last 400 years average 20–25 per century, a rate between 20 to 200 times the background extinction rate (i.e. the natural rate excluding mass extinction events) (Groombridge and Jenkins 2002). This anthropogenic episode began some 100 000 years ago with the expansion of modern humans from Africa throughout the world, and has been particularly intense in Australia and North and South America. However, many of the less recent continental extinctions occurred concurrent with episodes of pronounced climate change, and the extent to which humans caused many of the past losses has been much debated (a well-balanced evaluation is provided by Lomolino *et al.* 2005). During this anthropogenic period, some of the most dramatic and convincing illustrations of extinctions due to human action come from islands worldwide (Diamond 2005).

Estimating the extent of the biodiversity crisis is a hard task, because we still have a remarkably poor knowledge of the total diversity of the planet (the so-called Linnean shortfall), and of how that diversity is distributed (the Wallacean shortfall).

Although approximately 1.7 million species have been described, typical estimates of the total diversity of the planet range from 5 million to 30 million, with figures as high as 100 million occasionally being cited. Many of the supposed undescribed species are thought to occur amongst invertebrate taxa in tropical forests, and ocean floors, although huge uncertainty remains as to how diverse these systems really are (see, e.g. Lambshead and Boucher 2003).

Islands, too, are part of this uncertainty, undoubtedly holding many species currently scientifically unknown that are waiting to be discovered and described. For instance, the cataloguing of the Canarian biota has been under way since the French priest Louis Feuillee came to the archipelago in 1724. Despite almost three centuries of scientific attention, species new to science continue to be discovered, in recent decades at a rate approximating one species every 6 days (Fig. 11.1), including two large lizards (*Gallotia intermedia and G. gomerana*) and two trees (*Myrica rivas-martinezii and Dracaena tamaranae*) (Izquierdo *et al.* 2004). Many new finds have escaped detection until recently

because they persist only in small populations. Hence, no sooner are they discovered than many are categorized as in danger of extinction.

11.2 Stochastic versus deterministic extinctions

Two main types of extinction events operating on islands can be differentiated (Table 11.1): largely stochastic extinctions, inherent to the natural dynamics of island environments, and deterministic extinctions, related directly or indirectly to human activity. In the first group we include natural disasters, such as volcanic eruptions, but also the consequences of the taxon cycle dynamics, which involve species moving down an evolutionary cul-de-sac in becoming highly restricted local endemics, increasingly prone to natural disasters. The second group includes habitat-loss, degradation or fragmentation, the introduction of alien species and predation by humans, whether for the pot or for other purposes.

When one species goes extinct, any evolutionary interactions involving that species must also cease.

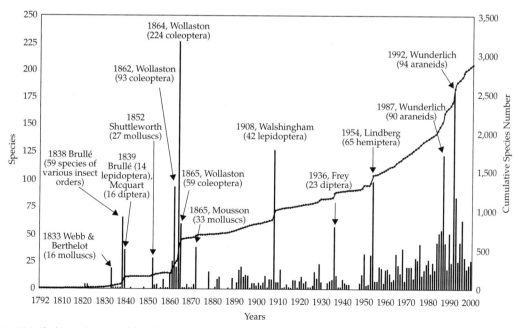

Figure 11.1 The historical pattern of description of endemic invertebrates of the Canaries. The bars represent the number of species described each year, and the line the cumulative total. (After Izquierdo *et al.* 2001.)

Table 11.1 Largely stochastic, natural causes of extinction, versus deterministic causes arising from human activities on islands

Stochastic causes (inherent to islands)	Deterministic causes (related to human activities)
Natural disasters (volcanic eruptions, landslides, hurricanes, tsunamis)	Habitat loss, degradation, or fragmentation
Taxon cycle dynamics (including demographic collapse and inbreeding depression)	Alien species introduction (including competitors, predators, hybridizing congeners, parasites, disease vectors, or diseases)
Pleistocene climatic fluctuations (including sea-level rise)	Direct predation (including hunting, fishing, or specimen collection)
—— Trophic cascade ——	

If this involves very tight mutualisms (e.g. as can be involved in pollination services), then the surviving co-evolutionary partner, or 'widow' species (*sensu* Olesen and Valido 2004) may in time follow the same fate, unless it is able to find similar services from another locally available species. Such losses may then multiply, as other mutualisms are interrupted, in a fashion analogous to the meso-predator release mechanism discussed in the previous chapter. Such trophic cascades may follow from both deterministic and stochastic extinctions. This term implies a cascading effect across trophic boundaries as an ecosystem unravels. There are some spectacular examples driven by 'ecosystem transformer' species (e.g. Rodda *et al.* 1999; O'Dowd *et al.* 2003), but it is by no means clear how often and how far such cascading effects operate. Many ecosystems in practice have a degree of ecological redundancy (Olesen and Jordano 2002), such that a cascading 'ecological meltdown' may be relatively uncommon.

11.3 The scale of island losses globally

Species populations go extinct locally all the time. Of itself, a population extinction event is not necessarily of conservation concern. It becomes a matter of concern when it marks a pattern of range collapse, and especially when it represents the loss of an endemic subspecies or species. In discussing island species losses, we can distinguish different levels of extinction by the terms local species extinction or island extirpation, as distinct from the final extinction of the last population of a species, i.e. when an island endemic form is entirely eliminated. We focus on the latter unless otherwise denoted.

Those collating records of species extinctions commonly take the historic period of island exploration and scientific record to have begun *c.*AD 1600. In the period so defined, significantly more species of plants and animals are known to have become extinct from islands than from continents: for the taxa listed in Table 11.2, about 60% have been island species. However, we know so little about the losses of invertebrates that it would probably be safer to discount them from the analysis. For the groups with the best data, the proportion of extinctions from islands varies from 60% for mammals (which are generally lacking from remote islands in the first place), to 79% for molluscs, 81% for birds, and as high as 95% for reptiles. Moreover, of 121 continental species losses cited by Groombridge (1992), about 66% can be classified as aquatic species, many of them inhabiting lakes, which can be regarded as 'negative islands' (Box 11.1).

The values in Table 11.2 reflect a highly cautious approach to counting extinctions, and these numbers are predicted to change radically over the coming decades as the various 'extinction debts' (Chapter 10) are called in the world over. To date, therefore, relatively small numbers of recent terrestrial animal extinctions have been recorded from mainland tropical forest regions such as Amazonia and the Atlantic forests of Brazil, although of course these are the areas from which large numbers of extinctions are estimated to be occurring or about to occur. Yet, when we place the figures for island extinctions and threat into the context of how many species originate from islands, it is clear that they have suffered disproportionate losses in recent time and also that they remain priority sites for conservation. In the case of birds, and allowing for the

---

Box 11.1 The decline of Lake Victoria's cichlid fishes

Lakes are 'negative islands', that is, they are more or less isolated freshwater areas surrounded by a hostile land matrix, usually connected to other lakes through rivers. Lakes thus behave as islands in many biogeographical and ecological respects. Long-lasting lakes (e.g. Lake Baikal in Siberia, the Great Lakes of the African Rift Valley, and Lake Titicaca in South America), have provided similar evolutionary opportunities to oceanic islands, and in the case of the African Rift Valley lakes (Tanganyika, Victoria, and Malawi) outstanding illustrations of radiations of fish species (Box 9.2).

Lake Victoria, shared by Kenya, Tanzania, and Uganda, is the largest tropical lake and the second largest freshwater lake in the world, comprising some 3000 km³ of water. It is the unique habitat of c. 300 endemic cichlid fishes, belonging to the genus *Haplochromis*, which are thought to have diversified since the lake recharged after 12 400 BP (Box 9.2). Unfortunately, much of this endemic richness has

been depleted following the introduction of the Nile perch (*Lates niloticus*) in 1954 in order to increase the size of the fishing catch (Mackay 2002). Today it is considered that at least half of the lake's native fish species have gone extinct or are so severely depleted that too few individuals exist for the species to be harvested or recorded by scientists (Groombridge and Jenkins 2002).

Although predation by the Nile perch is believed to be the major cause of decline, important additional factors may include increasing pollution and sediment load, excess fishing pressure, and possible competition from introduced tilapiine cichlids (Groombridge and Jenkins 2002). The lake itself has now become depleted of oxygen, as a consequence of the massive alterations in trophic organization within the lake. A shrimp tolerant to oxygen-poor waters today provides the major source food for the Nile perch, which itself forms the basis of a fishery increasingly focused on the export trade rather than providing a local protein source.

Table 11.2 Summaries of known extinctions from islands, continents, and oceans since *c.* AD 1600 (data from Groombridge 1992; Steadman 1997a; Primack and Ros 2002). The values for several taxa differ from those given in the first edition of this book. Figures given by different authorities vary depending on the criteria adopted, so the values given here should be taken as of largely indicative value

Group	Islands	Continents	Oceans	Total	% insular
Mammals	51	30	4	85	60
Birds	92	21	0	113	81
Reptiles	20	1	0	21	95
Molluscs	151	40	0	191	79
Insects	51	10	0	61	84
Plants	139	245	0	384	36
Total	**504**	**347**	**4**	**855**	**59**

numbers of island versus continental species, island forms have had about a 40 times greater probability of extinction over the period since AD 1600 than have continental species (Johnson and Stattersfield 1990).

Recent palaeontological and archaeological studies have made it abundantly clear that island

endemics have been extinguished in significant numbers from islands and archipelagos all over the world, before the explorations of modern collectors brought their biotas to the attention of Western science (Table 11.3). In the case of island birds, at least, the established prehistoric losses greatly

Table 11.3 Estimates of actual and threatened losses of island endemic birds (data from Johnson and Stattersfield 1990; Steadman 1997a). The total of 97 extinctions since AD 1600 is slightly at variance with the estimate of 92 species losses given in Table 11.2, and contrasts with the value of 21 species given in that table for the number lost from continental areas over this period

Region	Number of extinctions			Endemism	
	Prehistoric[a]	AD 1600–1899	AD 1900–94	Approximate number of endemics	% endemics threatened
Pacific Ocean	90	28	23	290	38
Indonesia and Borneo	–	0	2	390	22
Indian Ocean	11	30	1	200	33
Philippines	–	0	1	180	19
Caribbean Sea	34	2	1	140	22
New Guinea and Melanesia	10	2	3	500	10
Atlantic Ocean	3	3	1	50	50
Mediterranean Sea	10	0	0	–	–
Total	**158**	**65**	**32**	**1750**	**23**

[a] Data are lacking for prehistoric losses from Indonesia and the Philippines. The prehistory category consists of species from prehistoric cultural contexts, and includes only species already described. Sources of possible overestimation include: (1) the so-called Lazarus effect, i.e. where species thought to be extinct are eventually rediscovered, (2) identification of species that did not exist, e.g. there were once thought to have been up to 11 species of flightless giant Moa on New Zealand, but recent genetic analyses of bones indicates that there were only 2 species which, however, exhibited exceptional morphological dimorphism (Bunce et al. 2003). However, overall, there are doubtless many more extinct species yet to be discovered by the archaeologists; we can thus regard these estimates as minimum figures.

outnumber the historical losses. Table 11.3 provides a figure of 158 extinctions of island endemic birds in prehistoric times, but a slightly higher figure, of just over 200 species (recorded as subfossils, and whose disappearance is probably due to prehistoric man), is given by Milberg and Tyrberg (1993). There is no doubt that further archaeological and taxonomic work will add to the number of extinctions for the prehistoric period. It should therefore be understood that the year AD 1600 did not mark a key point in island history globally, and that its use by those compiling the statistics is essentially a matter of convenience. The primary place at which we ought to divide the analysis is the point at which islands have been colonized by humans.

Steadman (1997a) estimates that rates of natural, pre-human extinction in island birds may be at least two orders of magnitude lower than post-human rates. The world's avifauna would have been about 20% richer in species had the islands of the Pacific remained unoccupied by humans. As

many as 2000 bird species may have become extinct following human colonization but before European contact; the list being dominated by flightless rails, but including moas, petrels, prions, pelicans, ibises, herons, swans, geese, ducks, hawks, eagles, megapodes, kagus, aptornithids, sandpipers, gulls, pigeons, doves, parrots, owls, owlet-nightjars, and many types of passerines. The emphasis on rails stems from observations that virtually every Pacific island that has been examined palaeontologically has been found to have had one or more rails of such reduced flight capabilities that they must have been endemic species. By extrapolation, it is conceivable that more than 800 Pacific islands were once graced with one or more rail species in one of the most dramatic examples of island radiation to have persisted into the Holocene (cf. Trewick 1997). Only a handful of these species remain today.

Island species continue to be disproportionately at risk. For instance, approximately one in six plant species grows on oceanic islands, but one in three of all known threatened plant species are island endemics. The degree to which the remaining island endemic birds are believed to be threatened is quantified in Table 11.3. Although IUCN data show that the threat of extinction is switching significantly towards the continents (Groombridge 1992), statistics such as these underline the continuing importance of focusing conservation efforts on islands.

11.4   The agencies of destruction

There are four major reasons why island species are reduced by human action: (1) direct predation; (2) the introduction of non-native species; (3) the spread of disease; and (4) habitat degradation or loss. In a global context, the loss of habitat is commonly seen as the greatest problem for biodiversity (Lawton and May 1995), but on particular islands other factors from this list may be more pressing. For instance, on the Galápagos, introduced animals and plants have been described as: ' . . . absolutely the most insidious, silent, and dangerous scourge that exists in the archipelago' (*Galápagos Newsletter* 1996). Commonly, native endemic species are caught in a pincer action, reduced by habitat loss to a rump, attacked by disease or predation, and facing competition with exotic species. The synergisms between these forces result in the loss of species which conceivably could cope with and adjust to one force alone. Similarly, there may subsequent trophic cascade effects (above), such that the loss of prey species might prejudice the survival of native predators. For example, the very large extinct New Zealand eagle, *Harpagornis moorei*, was probably dependent on moas and other large and now extinct bird species (Milberg and Tyrberg 1993).

Predation by humans

Humans hunt island species to eat and also for other reasons, such as the societal values attached to brightly coloured feathers. In the case of the extinct Hawaiian mamo (*Drepanis pacifica*), it has been estimated that the famous royal cloak worn by Kamehameha I would have required some 80 000

birds to be killed (Johnson and Stattersfield 1990). Trade in such biological resources has a long history, and was an important feature of the period in which the Polynesians spread across and held sway over the Pacific. Present-day trade in fruit bats (flying foxes) provides an example of how hunting, not primarily for local consumption, but for export, can reduce oceanic island populations to crisis point. The critical role that these bats play as pollinators and as dispersers of plants may mean that their loss will have significant knock-on impacts, i.e. they act in effect as 'keystone' species (Rainey *et al.* 1995; Vitousek *et al.* 1995). Hunting for collections continues to be problematic. Today, the trade is driven mostly by individual collectors, but in the past hunting for museum collections was common and is, for example, believed to have dealt the final blow to the spectacular New Zealand bird, the huia (*Heteralocha acutirostris*) (Johnson and Stattersfield 1990).

Introduced species

As humans have spread across the islands of the world we have taken with us, either purposefully or inadvertently, a remarkable array of plant and animal species (see Box 11.2). These anthropogenic introductions are variously called exotic, alien or non-native species. Some species persist simply as domesticated (animal) or cultivated (plant) species, while others become naturalized, i.e. they form self-sustaining populations within modified habitats (Henderson *et al.* 2006). Of these, some become feral or invasive, expanding into intact or semi-intact habitats. Of this subset, a few species cause serious ecological impacts. These species are termed ecosystem transformers (Henderson *et al.* 2006). Based on a review of numerous case studies, Williamson (1996) has suggested that only about 10% of introduced species become established, with in turn about 10% of these species achieving pest status.

The impact of non-native species can take numerous forms. These include predation and grazing impacts that are often strikingly apparent, while the changes can also be more cryptic, but nonetheless profound, such as alterations to

Box 11.2 Some of the worst invasive alien species on islands

The examples given here demonstrate that problematic aliens that have invaded islands across the world are drawn from many higher taxa, and originate from all parts of the globe. Although most of these problematic invasives are continental in origin, some are drawn from other islands.

Sources: Cronk and Fuller (1995); IUCN/SSC Invasive Species Specialist Group (2004)

Invasive alien species on islands

Common name	Scientific name	Origin	Islands invaded
Insects			
Black twig borer	*Xylosandrus compactus*	Asia	Madagascar, Mauritius, Seychelles, Java, Sumatra, Fiji
Crazy ant	*Anoplolepis gracilipes*	West Africa	Hawaii, Seychelles, Zanzibar, Christmas Island
Common malaria mosquito	*Anopheles quadrimaculatus*	unknown	Hawaii
Formosan subterranean termite	*Coptotermes formosanus*	China	Hawaii
Tropical fire ant	*Solenopsis geminata*	Central America	Hawaii
Argentinian ant	*Linephitema humilis*	South America	Worldwide
Little fire ant	*Wasmannia auropunctata*	Tropical America	Galápagos, New Caledonia, Solomon Islands, Hawaii, Vanuatu
Worms			
Triclad flatworm	*Platydemus manokwari*	New Guinea	Hawaii, Guam, Marianas
Snails			
Rosy wolfsnail	*Euglandina rosea*	SW United States	French Polynesia, Hawaii, Bermuda, Bahamas
Giant Africa snail	*Achatina fulica*	NE Africa	Hawaii, Samoa, Philippines, Sri Lanka, Tahiti
Golden apple snail	*Pomacea canaliculata*	South America	Japan, Philippines, Borneo, New Guinea
Amphibians			
Giant toad	*Bufo marinus*	Central America	Hawaii, Fiji, Samoa, Guam, Marianas, Caroline and Solomon Is.
Caribbean tree frog	*Eleutherodactylus coqui*	Puerto Rico	Hawaii, Virgin Islands
Reptiles			
Brown tree snake	*Boiga irregularis*	Australasia	Guam
Birds			
Indian myna bird	*Acridotheres trisitis*	India	New Zealand, New Caledonia, Fiji, Samoa, Hawaii, Cook
Red-vented bulbul	*Pycnonotus cafer*	South Asia	Fiji, Samoa, Tonga, Hawaii
Mammals			
European hedgehog	*Erinaceus europaeus*	Western Europe	New Zealand
Brushtail possum	*Trichosurus vulpecula*	Australia	New Zealand
Grey squirrel	*Sciurus carolinensis*	North America	Great Britain, Ireland
Stoat	*Mustela erminea*	Holarctic	New Zealand
Long-tailed macaque	*Macaca fascicularis*	South East Asia	Mauritius
House mouse	*Mus musculus*	India	Worldwide
Feral pig	*Sus scrofa*	unknown	New Zealand, Hawaii, Galápagos
Small Indian mongoose	*Herpestes javanicus*	South Asia	Mauritius, Fiji, West Indies, Hawaii
Domestic cat	*Felis catus*	Unknown?	Worldwide
Feral goat	*Capra hircus*	Asia	Worldwide
Maori rat	*Rattus exulans*	South Asia	Marianas, New Zealand, Hawaii, Polinesia

(continued)

Common name	Scientific name	Origin	Islands invaded
Brown rat	*Rattus norvegicus*	NE China	Worldwide
Ship rat	*Rattus rattus*	India	Worldwide
Rabbit	*Oryctolagus cuniculus*	Iberian Peninsula	Worldwide
Plant climbers			
Old man's beard	*Clematis vitalba*	Europe	New Zealand
Hiptage	*Hiptage benghalensis*	India	Réunion, Mauritius
Merremia	*Merremia peltata*	Africa, South Asia, Australasia	Fiji, French Polynesia, Micronesia, Guam, Samoa, Solomon Is., Tonga, Vanuatu
Mile-a-minute weed	*Mikania micrantha*	Central America	Philippines, Solomon, Sri Lanka, Mauritius, Rarotonga
Banana passion fruit	*Passiflora mollisima*	South America	Hawaii
Herbs			
Wakeke	*Abelmoschus moschatus*	South Asia	Samoa, Tonga, Micronesia, Marianas, Hawaii, Polinesia
Crofton weed	*Ageratina adenophora*	Mexico	New Zealand, Hawaii, Canaries
Broom sedge	*Andropogon virginicus*	SE USA	Hawaii
Sweet vernal grass	*Anthoxanthum odoratum*	Eurasia	Hawaii
Arundo grass	*Arundo donax*	India	Micronesia , Guam, Palau, Fiji, Hawaii, Nauru, New Caledonia, Norfolk, Christmas Island
Kahili ginger	*Hedychium gardnerianum*	Himalaya	Micronesia, French Polynesia, Hawaii, New Zealand, Réunion, Jamaica, Azores, Madeira
Indian balsam	*Impatiens glandulifera*	Himalaya	Great Britain
Curly water weed	*Lagarosiphon major*	South Africa	New Zealand, Mascarenes
Elephant grass	*Pennisetum* sp.	Africa	Hawaii, New Zealand, Sri Lanka, Guam, Galápagos, Canaries
Water fern	*Salvinia molesta*	Brazil	Sri Lanka, New Guinea
Wandering jew	*Tradescantia fluminensis*	Brazil	New Zealand, Canaries
Shrubs			
Babul	*Acacia nilotica*	South Asia	Antigua, Barbuda, Anguilla
Red cinchona	*Cinchona pubescens*	Tropical America	Tahiti, Galápagos, Hawaii
Koster's curse	*Clidemia hirta*	Tropical America	Hawaii, Fiji, Vanuatu, Mascarenes, Comores, Seychelles
Broom	*Cytisus scoparius*	South Europe	New Zealand
El marabu	*Dichrostachys cinerea*	Africa	Cuba, Hispaniola, Guadeloupe, Marie-Galante, Martinique
Fuchsia	*Fuchsia boliviana, F. magellanica*	South America	Réunion
Cuban hemp	*Furcraea cubensis*	Greater Antilles	Galápagos
Mauritius hemp	*Furcraea foetida*	South America	Cape Verde
White sage	*Lantana camara*	Tropical America	Hawaii, Cape Verde, Galápagos, New Caledonia, New Zealand
Wild tamarind	*Leucaena leucocephala*	Central America	Hawaii, Mascarenes
Privet	*Ligustrum robustus*	South Asia	Réunion, Mauritius
Miconia	*Miconia calvescens*	Tropical America	Tahiti, Moorea, Raiatea, Marquesas
Wild tobacco	*Nicotiana glauca*	South America	St. Helena, Canaries
Strawberry guava	*Psidium cattleianum, P. guajava*	Tropical America	Hawaii, Norfolk, Mauritius, Galápagos, New Zealand
Rhododendron	*Rhododendron ponticum*	Mediterranean	Great Britain, Ireland
Blackberry	*Rubus argutus*	North America	Hawaii
Blackberry	*Rubus moluccanus*	Australasia	Réunion, Mauritius
Crack willow	*Salix fragilis*	Eurasia	New Zealand

(continued)

Common name	Scientific name	Origin	Islands invaded
Brazilian pepper tree	*Schinus terebinthifolia*	South America	Hawaii, Norfolk, Mauritius, St. Helena
Chilean guava	*Ugni molinae*	Chile	Juan Fernández, Chatham
Gorse	*Ulex europaeus*	Western Europe	Hawaii
Succulents			
Cook feet	*Carpobrotus* sp.	South Africa	Balearics
Coirama	*Kalanchoe pinnata*	Madagascar	Hawaii, Galápagos, Kermadec
Coastal pricklypear	*Opuntia dillenii*	Central America	Canaries
Common pricklypear	*Opuntia ficus-indica*	Mexico	Hawaii, Canaries
Trees			
Black wattle	*Acacia mearnsii*	Australasia	New Zealand, Hawaii
Sycamore	*Acer pseudoplatanus*	Europe	New Zealand, Great Britain, Ireland, Madeira
Shoebutton ardisia	*Ardisia eliptica*	South Asia	Hawaii, Okinawa, Jamaica
Australian pine	*Casuarina equisetifolia*	Australasia	Hawaii, Réunion
Embauba	*Cecropia peltata*	Tropical America	Hawaii, Tahiti, Raiatea
Red quinine tree	*Cinchona succirubra*	Ecuador	Galápagos, St. Helena
Cinnamon	*Cinnamomun zeylanicum*	East Indies	Seychelles
Paper bark tree	*Melaleuca quinquenervia*	Australasia	Hawaii
Fire tree	*Myrica faya*	Macaronesia	Hawaii
Cluster pine	*Pinus pinaster*	Mediterranean	Hawaii, New Zealand
Sweet pittosporum	*Pittosporum undulatum*	Australia	Norfolk, Lord Howe, Jamaica, Azores
African tulip tree	*Spathodea campanulata*	Africa	Hawaii, Fiji, French Polynesia, Samoa
Honduras mahogany	*Swietenia macrophylla*	Central America	Sri Lanka
Diseases			
Avian malaria	*Plasmodium relictum*	unknown?	Hawaii

pollination and dispersal networks, or hybridization. We illustrate both these forms of alteration in this section, reserving mention of other equally important forms of change, such as the development of extensive thickets of non-native plant species (Meyer and Florence 1996; Henderson *et al.* 2006), to later sections.

Predators and browsers

As often noted, it is the introduction of vertebrate predators, such as cats, rats, dogs, and mongooses, which have caused some of the worst problems (Cronk 1989; Groombridge 1992; Benton and Spencer 1995; Keast and Miller 1996). Feral cats are a feature of innumerable islands. They are opportunistic predators, eating what is most easily caught. Cats were introduced to the sub-Antarctic Marion Island in 1949 and at one stage it was

estimated that there were about 2000 of them, each taking about 213 birds a year, and thus in total killing well over 400 000 birds per annum, mostly ground-nesting petrels. Given these loses, only by eradicating the cat from the island could the long-term survival of the native bird species be ensured (Leader-Williams and Walton 1989). Cats have been present on another sub-Antarctic island, Macquarie, for a lot longer, and in that case have been responsible for the loss of two endemic bird species. The combination of cats and mongooses on the two largest Fijian islands has resulted in the local extinction of two species of ground-foraging skinks of the genus *Emoia*: they survive only on mongoose-free islands. In the Lesser Antilles, three reptiles have been extirpated from St Lucia within historical time, coincident with the introduction of the mongoose. Domestic dogs can also be devastating, and feral populations have been responsible for the local extinction of land iguanas in the Galápagos.

Even invertebrates have led to the extinction of island endemics, notably the losses of all endemic species of *Partula* land snails following the misguided introduction of the carnivorous snail *Euglandina rosea* to the island of Moorea (Williamson 1996). Similarly, exotic ants introduced to the Galápagos and Hawaii (from which they are naturally absent) predate endemic invertebrates and are an increasing cause for concern in both archipelagos. Browsing animals such as goats and rabbits have also led to losses (Cronk 1989; Johnson and Stattersfield 1990). Examples blamed on the rabbit include the Laysan rail (*Porzanula palmeri*) and Laysan millerbird (*Acrocephalus familiaris*), both of which disappeared during the early twentieth century because rabbits denuded their island.

Changes to pollination and dispersal networks

Species dispersing spontaneously or being introduced by humans to remote islands leave behind their usual web of interacting species, which may include pollinators, dispersers, competitors, food plants, prey, predators, parasites, and pathogens (Janzen 1985, Olesen and Valido 2004). If they cannot withstand the loss of evolutionary 'partners' they may fail to establish on the island, but at least a fraction of introduced species find a new web of interacting species they can interact with successfully. Indeed, as discussed in Chapter 7, there is some evidence that pollinator networks tend to be more generalist on remote islands, involving supergeneralist endemic species that readily include non-native species into their networks, thus aiding their establishment (Olesen *et al.* 2002).

The introduction of exotic pollinators such as the honey bee (*Apis mellifera*), continental bumblebees (*Bombus* spp.), or wasps (*Vespula* spp.) on to an island, will have consequences for both native plant and pollinator communities, and these consequences may be positive or negative for the island species. If the exotic pollinator is more effective in transporting pollen than the native pollinators, then native (but also exotic) plants will increase their fruit and seed production in relation to native plants not pollinated by the exotic species.

Conversely, if it is less effective in pollinating a particular plant than a displaced native pollinator, then this may reduce seed set. For instance, introduced honey bees have been found to compete with the endemic Canarian bumblebee (*Bombus canariensis*), resulting in diminished pollen transfer success for several Canarian endemic plants (such as *Echium wildpretii*) (Valido *et al.* 2002).

The introduction of predators or disease organisms that knock out native pollinator networks can be expected to have significant negative consequences for native plant species. For instance, the introduction in Hawaii of predatory insects such as *Vespula pensylvanica* and of diseases, like avian malaria, have extinguished entire groups of pollinators, such as 52 endemic species of *Nesoprosopis* bees (Vitousek *et al.* 1987*a*), and many endemic bird species, respectively. The disappearance of exclusive pollinators is considered responsible for the extinction of many plants pollinated by them, including some 30 endemic Hawaiian Campanulaceae species (Cox and Elmqvist 2000).

The introduction of exotic dispersers can have positive consequences on native (but also on exotic) plant species. For example, the Balearic endemic plant *Cneorum tricoccon*, was originally dispersed by *Podarcis lilfordi* lizards, which occupy elevations of less than 500 m ASL. Following the introduction to Majorca of an alternative disperser, the European pine marten (*Martes martes*), *Cneorum* begin to colonize altitudes up to 1000 m ASL, driven by the wider altitude range exploited by the marten (Traveset and Santamaría 2004). Conversely, the introduction of less-effective dispersers can result in negative consequences for native species. For example, two introduced mammal species, rabbit (*Oryctolagus cuniculus*) and Barbary ground squirrel (*Atlantoxerus getulus*), now compete for the fruits of the endemic shrub *Rubia fruticosa* in Fuerteventura (Canaries) with the indigenous dispersers (native birds and lizards). The seeds are dispersed less effectively by the mammals and have a lower viability. In this case there are thus negative consequences both for the native plant (poorer dispersal services) and for the native frugivores (competition for a food resource) (Nogales *et al.* 2005).

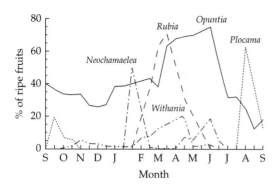

Figure 11.2 Availability of ripe fruits (% of the crop size) in a Canarian subdesert coastal scrub, Teno Bajo, Tenerife, for September 1993 through August 1994. The presence of the exotic invasive cactus *Opuntia* has altered the pattern of fruit availability for the local frugivore community, and this in turn may impact on the dispersal services provided to the native plant species. (From Valido *et al.* 2003)

Similarly, the introduction of an exotic plant can alter the insular networks in varying ways. The new plant can offer the native dispersers new resources that will produce positive effects for both of them, but simultaneously have negative consequences for the native plants that have to compete for resources with the exotics. For instance, the introduction of the prickly pears (genus *Opuntia*) on the Canaries has provided native dispersers, including ravens and endemic lizards, a new fleshy fruit to feed on, with a much larger year-around availability than the native resources (Fig. 11.2). The spread of *Opuntia* has thus been aided by native lizards and ravens, so that *Opuntia* species have successfully invaded large areas of Canarian coastal scrub rich in endemic *Euphorbia* and other plant genera (Nogales *et al.* 1999, 2005; Valido *et al.* 2003). Finally, it is often the case that interactions between two or more exotic species are crucial to the spread of problematic invasives. For example, within the Galápagos, introduced mammals are entirely responsible for the spread of the introduced guava (*Psidium guajava*), one of a number of ecosystem-transforming non-native plant species (e.g. Henderson *et al.* 2006).

Hybridization with native species

The introduction to an island of an allopatric congener, which is not fully reproductively isolated from the native taxon, can lead to a process of introgressive hybridization and the genetic dilution and eventual disappearance of the native form. This introgression process is currently affecting the Canarian endemic palm *Phoenix canariensis*, the plant symbol of the Canaries, as a result of the introduction of its closest relative, the date palm *Phoenix dactylifera*. *P. canariensis* grows wild in the lowlands of each of the main islands of the archipelago, whereas *P. dactylifera* is a fruit palm, of unknown origin, that has been cultivated for at least 5000 years in the Arabian peninsula and North Africa. Although it was introduced before the Spanish settlement of the Canaries, the problem of introgression is essentially a recent one, following the importation of massive numbers of date palms to decorate the new tourist resorts and paved roads across the archipelago (Morici 2004). Hybrid swarms are now found all over the archipelago, especially in the tourist honeypot areas on the eastern islands. In addition to hybridization, the endemic palm is also threatened by a lethal pest, the weevil *Rhynchophorous ferrugineus*, introduced with date palms imported from the Middle East. As a result of these pressures, this particular plant symbol of the Canaries faces an uncertain future. Genetically pure stands of the Canarian palm populations remain in the interiors of Gran Canaria and La Gomera, far away from the resorts (González-Pérez *et al.* 2004), providing potential seed supply for future conservation programmes. A similar process of introgression is under way on the Cape Verde islands, between the endemic palm *Phoenix atlantica* and the date palm (Morici 2004).

Disease

The problem of disease is closely associated with exposure to exotic competitors and, indeed, the separation of exotic microbes as a category from other introduced organisms is largely arbitrary. The most striking exemplification of the impact of disease on insular populations has, without doubt, been among the native peoples of islands, particularly on first contact with Europeans. This factor, in combination with more purposeful persecution, led to the decimation of many island peoples, such that in some cases little or no trace of them remains (Kunkel 1976; Watts 1987; Diamond 2005).

The most commonly cited example of exotic disease afflicting native island birds is from Hawaii. Hypotheses to explain why so many species of birds became extinct in the period since Cook's landfall in AD 1778 include habitat destruction, hunting, competition with introduced birds, and introduced predators. Undoubtedly, each has had its role in the degradation of Hawaii's ecosystems. Early work by Warner (1968) identified avian malaria (and avian pox), carried by the introduced mosquito *Culex quinquefasciatus*, as an important factor. Warner proposed an 'imaginary line' at about 600 m ASL, above which there were no mosquitoes and below which native birds were not found because they had succumbed to the disease. However, van Riper *et al.* (1986) suggest that avian malaria probably took some time to build up a large enough reservoir and that it did not have a major impact upon the numbers of Hawaiian birds until the 1920s, some time after the extinction pulse of the late nineteenth century. Van Riper *et al.* (1986) also report that the malarial parasite *Plasmodium* is found up to treeline elevations, i.e. well above Warner's 600 m threshold. The parasite is nonetheless concentrated in the mid-elevational ranges, where introduced bird species and native birds have the greatest distributional overlap. The introduced species have been found to be less susceptible to malaria and thus to act as vectors for the disease. Although a number of native bird species have developed immunogenic and behavioural responses reducing the impact of the parasite, van Riper and colleagues concluded that avian malaria is currently a major limiting factor, restricting both abundance and distribution of the endemic avifauna.

The omao (*Myadestes obscurus*) is one of four surviving species of thrushes on the Hawaiian islands. It occupies no more than 30% of its former range, and although locally abundant in rain forests in particular parts of the Big Island of Hawaii, it is absent from other areas in which it was once common. Its disappearance from the Kona and Kohala districts during the early part of the twentieth century remains an enigma. A plausible hypothesis is that it failed to develop resistance swiftly in these parts of its range to avian malaria, but in other parts of its range, such as the Puna district, it did develop resistance, thereby allowing its present pattern of coexistence with both the mosquitoes and the malarial parasites in that area (Ralph and Fancy 1994).

Given the possibilities of local variations in the development of resistance to disease, it is clearly going to be difficult to be certain about the role of disease in the demise of Hawaii's birds. Added to which, it needs to be remembered that most species were lost before European contact, and that a host of other changes have occurred over the past 200 years or so. The pattern of post-European contact losses is bimodal, being concentrated between 1885 and 1900, and 1915 and 1935. The former phase of losses included many species historically confined to higher altitudes (> 600 m ASL) and therefore contradicting Warner's (1968) original malarial scenario. The explanation for this phase of losses may relate to another introduced disease, avian pox, or to the impact of habitat modifications and introduced predators. However, the second period does appear to relate to the arrival and spread of malaria. Birds which succumbed during this period were found in the mid-elevational forests in which the highest prevalence of avian malaria was found during studies around 1977–80 (van Riper *et al.* 1986). As the introduced tropical mosquito *Culex quinquefasciatus* slowly adapted to cooler elevations, the die-off of native bird species slowly advanced up the mountains, reaching 600 m by the 1950s and 1500 m by the 1970s, sweeping away several mid-elevation species in the process, and restricting the survivors to ever higher refugia. This process is ominously described by Pratt (2005) as 'the third and ongoing wave.'

Two things emerge. First, disease has played and continues to play a part in the problem, causing range reductions and precipitating extinctions. Secondly, we are dealing with a complex system involving opportunities for the spread of exotics due to habitat alteration and the interactions of several introduced organisms. The exotic birds and mosquitoes act as vectors, and the exotic disease provides the proximal cause of mortality. The fact that introduced birds act as vectors illustrates that important interactions between organisms at the same trophic level are not all obviously 'competitive' in nature.

Habitat degradation and loss

Probably the most renowned example of insular habitat degradation is the case of Easter Island, a depressing experiment in unsustainable exploitation of natural resources, which some have seen as a warning for the rest of us (Bahn and Flenley 1992). Environmental degradation has actually been widespread across the Pacific, and has been linked to both Polynesian and European expansion (Nunn 1990, 1994; Diamond 1991b). Habitat changes are hard to separate from other negative influences, but there is no doubt that the reduction in area of particular habitats (notably forest) and the disturbance of that which remains, have been potent forces in the reduction and loss of numerous island endemics (Steadman 1997a,b). A specific example is the loss of the four-coloured flowerpecker (*Dicaeum quadricolor*) following the almost complete deforestation of the island of Cebú in the Philippines (Johnson and Stattersfield 1990).

A popular image of pre-industrial hunter-gatherers and neolithic agriculturalists is that such peoples lived in harmony with their environment, practising a conservation ethic and avoiding the exploitative destruction ('development') typical of later, industrial societies. This idea is associated with notions of the 'noble savage' and a mythical 'golden age' (Milberg and Tyrberg 1993). Sadly, the evidence for widespread degradation of island ecosystems across the world before they had any contact with modern European societies is now overwhelming (e.g. Nunn 1990; Diamond 1991b; Milberg and Tyrberg 1993; Schüle 1993; Diamond 2005).

Prehistoric humans on islands, although dependent on limited animal resources, regularly failed to exploit these in a sustainable way. Several cases where human populations disappeared altogether from Pacific islands were due to overexploitation of natural resources, including animals, vegetation and soils (Diamond 2005). It thus seems likely that the Polynesian taboo system, which prevents overexploitation, was started as a result of the extinctions and misuse of other resources, such that the 'ecological balance' only applied to biota that could not be profitably overexploited with the available technology and that could survive habitat destruction and the introduction of rats, dogs, and pigs (Rudge 1989; Bahn and Flenley 1992; Milberg and Tyrberg 1993; Benton and Spencer 1995). Indeed, Pimm *et al.* (1995) estimate that the first colonists typically wiped out about half of the native avifauna of an average Pacific island. Yet, there is also a danger that some extinctions may be falsely attributed to the first colonists, because intensive collection often began half a century or so after the damage initiated by European discovery. Furthermore, while in many systems it appears that the first colonists had eventually reached some sort of balance with the native faunas and floras, ongoing processes of globalization continue to imperil many oceanic island species.

11.5 Trends in the causes of decline

It is easiest to identify trends in the nature of the threat with reference to island birds, for which the most systematic assessments have been carried out. In prehistoric times most losses can be attributed to hunting, followed by the introduction of commensal species. Over the last 400 years, it appears that introduced species, followed by hunting, have been the key forces (Table 11.4). In the contemporary era, Johnson and Stattersfield (1990) cite habitat loss in respect to over half the threatened island bird species, with exotic species the principal threat in a fifth of the cases. Many of the potential introductions of predators to once predator-free islands have already occurred and in at least some cases the impacts of exotics are being mediated by protection measures. However, it remains the case that many of the most severely threatened bird species are endangered by exotics (e.g. Box 11.3). Alien species also continue to pose a severe threat to other island

Table 11.4 Principal causes of rarity, decline, and extinction of island endemic bird species (from data in Johnson and Stattersfield 1990)

Threat	Number of species	
	Extinct since *c.* AD **1600**	Rare or declining
Habitat destruction	19	206
Limited range	–	165
Introduced species	34	76
Hunting	25	35
Trade	–	16
Human disturbance	1	10
Natural causes	1 or 2	11
Fisheries	–	2
Unknown	41	64

For many species in both 'extinct' and 'rare or declining' categories, more than one cause has been assigned and the values given above reflect this; the overall totals in the data set were 97 'extinct' and 402 'rare or declining'.

taxa, and for example are regarded as the principal threat to over 90% of 282 endangered plant species on Hawaii (Cox and Enquist 2000).

In terms of habitat change, the loss of forest and woodland is the biggest concern (Fig. 11.3). Johnson and Stattersfield (1990) estimate that about 402 threatened species of birds are restricted to islands (a further 67 threatened species have both island and continental distributions). Of these 402 species, 77% are forest dwellers, divided between 200 rain forest (lowland and montane) species and 113 seasonal/ temperate-forest species. A recent assessment of threatened birds in insular South-East Asia found a close relationship between the numbers of bird species considered threatened, and the numbers of extinctions predicted from deforestation using a species–area approach (Brooks *et al.* 1997). Although certain caveats must be attached to such crudely derived estimates (cf. Chapter 10; Bodmer *et al.* 1997), there is no doubt that habitat loss is a key driving force no matter which means of assessment is used.

Box 11.3 The brown tree snake in Guam

The most notorious recent example of the impact that the accidental introduction of a predator can have on an oceanic island is provided by the brown tree snake (*Boiga irregularis*). A native of the Australasian region, it was introduced accidentally to Guam, possibly as a stowaway in military cargo, soon after the end of the Second World War. Guam is the largest (541 km²) and most populated (100 000 inhabitants) island of the Marianas archipelago, in Micronesia. The snake was first sighted in Guam in the 1950s, and by 1968 it had spread all over the island. Its nocturnal activity, its ability to live in close proximity to humans and, especially, the wide range of prey it consumes (lizards, rats, fruit bats, birds- including native, introduced and domestic species), enabled a demographic explosion, with densities of 12 000 to 15 000 snakes per square mile being reported (Patrick 2001).

Before the arrival of the brown tree snake, the only snake on the island was a small, blind snake, which lives in the soil and feeds on ants and

termites. The vertebrate fauna of Guam, having evolved for millennia without such native predators, lacked the usual defensive behaviour mechanisms of continental forms, and this is thought to have been an important contributory factor to their demise (Rodda *et al.* 1999). Brown tree snakes are considered responsible for the extirpation from Guam of all its breeding population of seabirds, as well as 10 of the 13 species of native forest birds: Guam rail (*Rallus owstoni*), white-browed rail (*Poliolimnas cinereus*), white-throated ground-dove (*Gallicolumba xanthonura*), Mariana fruit-dove (*Ptilinopus roseicapilla*), Micronesian kingfisher (*Halcyon cinnamomina*), nightingale reed-warbler (*Acrocephalus luscinia*), Guam flycatcher (*Myiagra freycineti*), rufous fantail (*Riphidura rufifrons*), cardinal honeyeater (*Myzomela cardinalis*), and bridled white-eye (*Zosterops conspicillatus*) (Rodda *et al.* 1999). Two of the extinct endemic species, the Guam rail and the Micronesian kingfisher, are being bred in zoos in the hope that they can eventually be released back into the wild, once the snake is controlled or eradicated.

Furthermore, two out of the three species of native mammals: the Pacific sheath-tailed bat (*Embollonura semicaudata*) and the little Marianas fruit bat (*Pteropus tokudae*), as well as 5 of the 10 species of native lizards: the snake-eyed skink (*Cryptoblepharus poecilopleurus*), the azure-tailed skink (*Emoia cyanura*), Slevin's skink (*Emoia slevini*), the island gecko (*Gehyra oceanica*), and the Micronesian gecko (*Perochirus ateles*) have also been extirpated. It is not possible to be certain that all these losses are solely attributable to the snake, but careful assessments of the evidence suggest that it has had a major role (Rodda *et al*. 1999), especially for birds and lizards, as shown in the figure.

The ecological (and socio-economic) consequences go beyond this immediate biodiversity loss. A trophic cascade can be considered to be under way, due to the loss of birds and bats that acted as pollinators and dispersers of trees, and there are concerns that there may be consequent outbreaks of insect populations that are no longer subject to predation by insectivorous birds and lizards (Rodda *et al*. 1999). In recent years, large populations of mosquitoes have propagated dengue fever among the people of Guam (Fritts and Leasman-Tanner 2001). Economic damage due to the brown tree snake includes electrical outages every third day, including island-wide blackouts, that have been costed at $1–4 million a year. Finally, the central role of Guam in the Pacific transportation network has facilitated the further extension of the brown tree snake range to other oceanic islands, such as several Mariana islands (Saipan, Rota, Tinian), Okinawa, Pohnpei, Oahu, and even Diego García in the Indian Ocean. Considerable efforts are under way to alert people to the danger of introducing this species to other islands.

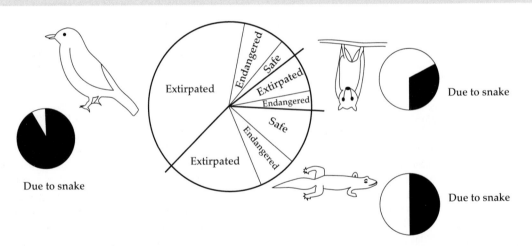

Status of Guam's native forest vertebrates (as present in 1950), with estimates of the degree to which their decline may be attributed to the introduction of the brown tree snake (black segments). The central pie graph represents the vertebrates, and the small pie graphs represent the degree of responsibility for the declines suffered by mammals (upper right), lizards (lower right), and birds (left); heavy lines delineate those major taxa in the central graph. (Redrawn from Fig 2.13 of Rodda *et al*. 1999.)

The species–area relationship overestimates species extinction threat in the Lesser Sundas, and this was suggested by Brooks *et al*. (1997) to be because the natural vegetation of these islands is deciduous monsoon rather than moist tropical forest. Most endemic birds are found in this deciduous monsoon habitat and they speculate that these birds may adapt better to secondary growth and scrub than do the moist forest species. In contrast, in the Philippines, 78% of the endemic birds are confined to the accessible lowland

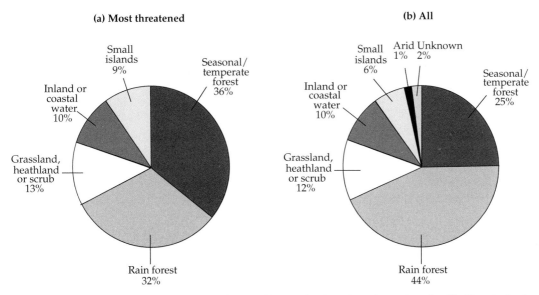

(a) Most threatened

Small islands 9%

Seasonal/temperate forest 36%

Inland or coastal water 10%

Grassland, heathland or scrub 13%

Rain forest 32%

(b) All

Small islands 6%

Arid 1%

Unknown 2%

Seasonal/temperate forest 25%

Inland or coastal water 10%

Grassland, heathland or scrub 12%

Rain forest 44%

Figure 11.3 Habitats of threatened island endemic bird species: (a) species listed as either endangered or vulnerable, (b) species listed as endangered, vulnerable, rare, indeterminate, or insufficiently known. Some species have been assigned to more than one habitat in this analysis, hence, for example, the relative importance of the combined forest habitat in part (b) (69%) differs slightly from the estimate for forest species (77%) given in the text. Data from Johnson and Stattersfield (1990), according to the then IUCN *Red Data Book* categories: *endangered*—those in danger of extinction and whose survival is unlikely if the causal factors continue operating; *vulnerable*—those likely to move into the endangered category in the near future; *rare*—taxa with small world populations; *indeterminate*—taxa known to belong to one of the foregoing categories, but with insufficient data to say which; *insufficiently known*—taxa suspected but not definitely known to belong to one of the foregoing categories. (The data given in this figure exclude the category *extinct*—taxa no longer known to exist in the wild after repeated searches of their type localities.)

forests in which deforestation has been concentrated, hence over half of the Philippines' 184 endemic birds are considered threatened.

How these various assessments of risk translate into future patterns of loss will depend on a variety of factors; not least, the protection afforded to particular island habitats and species, and conflicting economic pressures on the resource base of particular islands. It is one thing to identify causes of species extinction, quite another to provide remedies. Solutions to the problems demand political will, legislation, finance, education, and broad public support (Chapter 12). Nonetheless, the first step is to assess and articulate the problem.

11.6 A record of passage—patterns of loss across island taxa

To set the following sections in context, it is worth noting variations in the timing of human settlement

(Table 11.5). When Europeans began their major phase of expanding around the world from the sixteenth century onwards, they encountered islands with varying antiquity of human occupation. We might arbitrarily distinguish three groups: (1) palaeoinhabited islands, which were populated several millennia or tens of millennia before European expansion (e.g. New Guinea, the Solomons, Tasmania, the Antilles); (2) neoinhabited islands, which were populated just one or two millennia before European contact (e.g. the Canaries, Madagascar, Marquesas, Hawaii, New Zealand), and finally, (3) previously uninhabited islands, still pristine when the European expansion began (e.g. the Azores, Madeira, Cape Verde, St Helena, Tristan, Mascarenes, Galápagos, Juan Fernández).

The record of species extinctions from the islands that were colonized before modern European contact is largely dependent on proxy sources. Important techniques include the analysis of pollen

Table 11.5 Approximate date of first human colonization in relation to European contact dates for several archipelagoes around the world (from several sources)

Island or archipelagos	Pre–European establishment	European contact (AD)
Palaeo–inhabited		
New Guinea	40–50 000 BC	1545
Tasmania	20–30 000 BC	1642
Cyprus	c. 8000 BC	–
Corsica–Sardinia	8600–6700 BC	–
Great Antilles	c. 7000 BC	1511
Crete	6250 BC	–
Neo–inhabited		
Majorca	c. 2000 BC	–
Fiji, Samoa	1500 BC	
Canaries	c. 500 BC	c. 1400
Marquesas	200 BC	1595
Easter	AD 400–900	1722
Hawaii	AD 400	1778
Madagascar	AD 800–900	1500
New Zealand	AD 900	1643
Previously uninhabited		
Azores	–	1432
Cape Verde	–	1456
St Helena	–	1502
Galápagos	–	1535
Juan Fernández	–	1574
Mascarenes	–	1598
Tristan da Cunha	–	1812

from lake sediments and the analysis of subfossil bones recovered from middens. Such studies have led to important insights into the past vegetation and animal communities of many oceanic islands. Examples of extinct island birds known to science only from bones include New Zealand's extinct moa species (*Dinornis novaezealandiae* in the North Island and *D. robustus* in the South Island), giant flightless ratite birds, hunted to extinction by the Maori by about 500 BP (Bunce *et al.* 2003), and the elephant bird (*Aepyornis maximus*), extinguished by the Malagasy people (Groombridge and Jenkins 2002). Similarly, two giant rats (*Canariomys tamarani* and *C. bravoi*) endemic to the Canaries, were very likely wiped out by the first inhabitants (the guanches) or by the animals introduced by them (Cabrera 2001). Considering the diversity of their endemic biota, some island groups colonized in the distant past, such as the Canaries or Madagascar, appear to have suffered remarkably few extinctions after AD 1600 (Table 11.6), probably because the biota that were prone to extinction (conspicuous, highly specialized species with low rates of recovery or small population sizes) had already been extinguished by the first colonists.

By contrast, the pristine biota of the previously uninhabited group of islands was quite often studied by ships' naturalists in the early voyages of discovery, providing a mix of descriptions, drawings, and specimens, famously including the flightless dodo. Unfortunately, these earlier voyagers also introduced exotic species, such as goats, and also predated native species, including giant tortoise and again the dodo (although it was unpleasant to eat), beginning the slide of many island species towards extinction (below).

Table 11.6 Islands or archipelagoes that have lost at least five terrestrial vertebrate species after AD 1600 as a result of anthropogenic activities (Source: Groombridge and Jenkins 2002)

Island or archipelago	Mammals	Birds	Reptiles	Total
Hawaii	0	21	0	21
Mauritius	1	10	5	16
New Zealand	1	13	1	15
Hispaniola	11	1	0	12
Rodrigues	0	8	3	11
Réunion	1	7	2	10
Cuba	8	1	0	9
St Helena	0	7	0	7
Society Islands	0	6	0	6
Cayman	5	1	0	6
Jamaica	2	2	1	5
Madagascar	4	1	0	5
Martinique	1	2	2	5
Total	**34**	**80**	**14**	**128**

Extinctions of island vertebrates since AD 1600 have affected all the major oceans, but have been particularly concentrated in three nuclei: the Mascarenes in the Indian Ocean, the Caribbean arch (the Antilles), in the Atlantic Ocean, and Polynesia (Table 11.6). The lack of extinctions of mammals from some remote islands (e.g. Hawaii, St Helena) is simply a reflection of the failure of terrestrial mammals to colonize them in the first place.

Pacific Ocean birds and the Easter Island enigma

Birds are typically the most species-rich vertebrates on remote islands. They leave good fossil or sub-fossil evidence, and have been relatively well studied, often from sites associated with human settlement (i.e. middens). The picture that has emerged of their demise in prehistoric times can thus be more closely related to human activities than is the case for any other taxon—and the oceanic region where the most dramatic and convincing evidence of human involvement has been found is the Pacific. Not only has the Pacific lost large numbers of bird species in the past, but today it holds more threatened bird species than any other oceanic region: approximately 110 species, of

which 31 are classified as endangered and 29 vulnerable (the two highest *Red Data Book* categories; Johnson and Stattersfield 1990).

It is believed that the Australian and New Guinea land masses on the Sahul shelf were colonized by people at least 50 000 and 40 000 BP respectively. Voyagers reached New Ireland and the Solomons by 29 000 BP, but then paused for over 25 000 years—islands beyond this zone quite possibly were undetectable to these people (Keast and Miller 1996). Then the ancestors of the Polynesians became the first wave of human colonists to spread across the Pacific, and they did so very rapidly, between about 1200 BC and AD 1200 (Diamond 2005). So extensive were their explorations that nearly all islands in Oceania (Melanesia, Micronesia, and Polynesia) were inhabited by humans within the prehistoric period (Fig. 11.4; Pimm *et al.* 1995; Steadman 1997*a*). The human colonists cleared forests, cultivated crops, and raised domesticated animals. Birds provided sources of fat, protein, bones, and feathers for people and so were hunted as well as being indirectly affected by the changes that followed human arrival. European exploration began in the sixteenth century, but their phase of colonization started much later, such that, for instance, the first missionaries arrived in Hawaii in AD 1779 and in Tahiti in AD 1795 (Pimm *et al.* 1995).

In the Hawaiian islands, at least 62 endemic species of birds have gone extinct since Polynesian colonization began some 1600 years BP, and most of these were lost before European contact began (Steadman 1997*a*). The processes involved are manifold, including habitat alteration, hunting, diseases, and exotic species. The commensals introduced by the Polynesians, especially the Maori rat (*Rattus exulans*) and the feral pig (*Sus scofra*), have been supplemented since European arrival by many other exotic species, which have had a powerful impact on the native biota, including black rats, feral dogs, cats, sheep, horses, cattle, goats, mongooses, and an estimated 3200 arthropod species (Primack 1993). The extinctions have included flightless geese, ducks, ibises, and rails, and the major part of the radiation of Hawaiian honeycreepers (Olson and James 1982, 1991; James

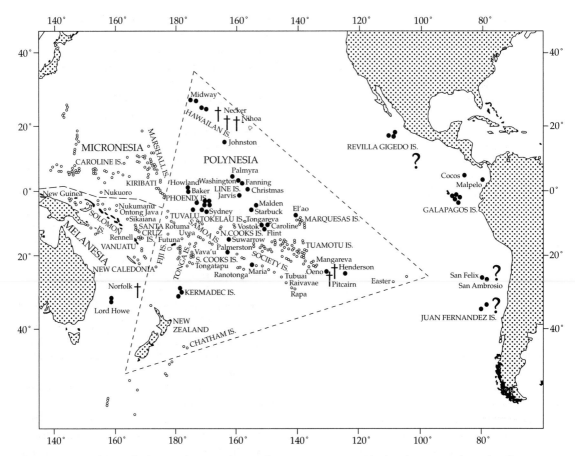

Figure 11.4 Map of the Pacific showing Polynesia. At the time of European contact, many islands in the eastern and central Pacific were uninhabited (large dots). Crosses mark islands colonized by the Polynesians prior to AD 1500 but then abandoned, usually after the extinction of many native species. Question marks indicate those islands where pre-European colonization was noted by Brown and Lomolino (1998) as suspected but not clearly documented. (Source: Brown and Lomolino 1998, after Terrell 1986.)

and Olson 1991). Pimm *et al.* (1995) believe that roughly 125–145 bird species once inhabited the main islands, of which 90–110 are extinct, with another 10 considered imminently threatened with extinction. The original avifauna contained only three non-endemic land birds, and all the passerines were endemic, yet in consequence of the enormous losses and of large numbers of species introductions by humans, only about a third of the passerines found on Hawaii today are endemic. Nearly all the birds seen now below 1000 m ASL are aliens. Today, some 48% of the land birds of Hawaii are exotics introduced and released in order to augment the fauna with game birds and birds of

beautiful plumage or song. Of 94 species known to have been introduced before 1940, 53 species became established at least locally and only 41 failed completely.

The Hawaiian islands illustrate another feature of prehistoric extinctions, namely, how they altered the structure of vertebrate—specifically bird—feeding guilds. All but one species of native predator became extinct (James 1995). Such was the fate of the terrestrial herbivores, with the exception only of the Hawaiian goose (*Branta sandvicensis*), which was rescued from the brink of extinction only by an *ex situ* breeding programme. Losses included the large flightless anseriforms (moa-nalos), which were browsers on understorey vegetation. The terrestrial omnivores all but disappeared too, along with the Hawaiian land crabs. Prehistoric extinctions among insectivores (37%) and arboreal frugivores/omnivores (40%) have been proportional to the losses among passerines generally, but granivores have suffered much worse and nectarivores much better in comparison. With predators disappearing, habitats shifting, and species richness falling during the shake-up following Polynesian settlement, it is likely that some species, even guilds, became both relatively and absolutely more widespread and abundant than under conditions before human contact. James (1995) argues that the differential effects of extinction on vertebrate feeding guilds in prehistoric times may still be affecting plant communities, i.e. that the losses must collectively have implications for the functional characteristics of the remaining ecosystems, but such a legacy is difficult to separate from the other driving forces for change.

The pre-human avifauna of New Zealand was probably about twice the size of the native avifauna today. The most famous losses were of the giant, flightless moas, the largest of which weighed around 250 kg (Rudge 1989; Holdaway 1990; Bunce *et al.* 2003). The moa exhibited exceptional variation in body size in relation to sex and habitat, such that until very recently there were thought to have been as many as 11 species. Genetic analyses of bones recovered from sites around New Zealand reveal there to have been only two distinct genetic lineages, *D. novaezealandiae* in North Island and *D. robustus* in South Island (Bunce *et al.* 2003).

An interesting story is attached to the loss of one of the smaller New Zealand birds, the Stephens Island wren, named *Traversia lyalli* after the lighthouse keeper (Lyall) who discovered it. Stephens Island is a tiny island in Cook Strait, off South Island. The wren was unknown to science until in 1894 the lighthouse keeper's cat brought in a few specimens it had killed. The exotic predator met the island endemic: end of story. At least, this was the conclusion reached at the time, prompting a correspondent to the *Canterbury Press* to suggest that in future the Marine Department should see to it that lighthouse keepers sent to such postings should be

prohibited from taking any cats with them, 'even if mouse-traps have to be furnished at the cost of the State'. Although cats have been claimed to have by far the worst record in exterminating New Zealand birds (King 1984), we should not place all the blame on the lighthouse keeper and his cat. Before the arrival of the Polynesians, the wren was actually a widespread mainland resident (Holdaway 1989; Rosenzweig 1995). Its loss from mainland New Zealand was probably due to a combination of habitat changes and the introduction of rats by the Polynesians; the loss from Stephens Island was but the final step to extinction. Interestingly, it is just such small offshore islands that provide the enclaves that may yet save some of New Zealand's remaining endemics from extinction as, cleared of exotic pests, translocated native populations are able to establish and persist. Here, in microcosm, can be found some of the essential ingredients of the diversity disaster of oceanic islands, and a lifeline of hope for the future.

The Polynesians discovered New Zealand around AD 950, having voyaged from the area containing the Marquesas, the Society Islands, and the South Cooks, more than 3000 km to the east. They brought with them six species of plants and two species of mammals. Polynesian settlement led to the destruction of half of the lowland montane forests, widespread soil erosion, and the loss or reduction of much of the vertebrate fauna. During the phase of their expansion, New Zealand lost its sea lions and sea elephants, and at least 25 bird species, including the 2 species of moa which, in the absence of terrestrial mammals, had evolved to occupy the browser/grazer role normally occupied by animals such as kudu, bushbuck, or deer. The most severe modification occurred between 750 and 500 years ago, when a rapidly increasing human population overexploited animal populations and used fire to clear the land. By about AD 1400, the major New Zealand grazing systems had ceased to exist, their place being taken by unbrowsed systems: the removal of the moas thus had significant effects on the structure and species composition of the vegetation.

Holdaway's (1990) analyses of subfossil remains indicate that the pre-human avifauna was dominated by forest species. The high rate of extinction in the avifauna prior to European settlement must have been related to the characteristics of the forest areas removed by the Polynesians: these being shown by palaeoecological data to have been mainly the drier, eastern forests, or those inland areas where drought or severe climate restricted regeneration after clearance. In pre-human times, it was the drier, more structurally diverse forests on more fertile soils which supported the greatest diversities of New Zealand's birds.

Pacific seabird populations have also suffered, as Steadman (1997a) points out; petrels and shearwaters have seen the greatest loss of species, and many other seabird species have had their distribution and population sizes diminished, including albatrosses, storm-petrels, frigatebirds, and boobies. Abbott's booby (*Papasula abbotti*) is a classic example: once widespread in the South Pacific, it now breeds only on Christmas Island in the eastern Indian Ocean.

Easter Island (27°9'S, 109°26'W) was, until the colonization of Tristan da Cunha, just two centuries ago, the most isolated piece of inhabited land in the world. When first contacted by Europeans on Easter Day AD 1722, it presented a fascinating enigma: how could this impoverished treeless island, with its impoverished people, have supported the construction of the remarkable giant statues (*moai*), and how and why had it all gone wrong? The flora of Easter Island consists today of over 200 vascular species, of which only 46 are native. However, formerly the flora was richer, containing several native tree and shrub species. Evidence for a forest cover is provided not just by pollen records, and by the discovery of palm nuts, but also by the discovery of extinct endemic achatinellid land snails, and by the identification to species or genus of over 2000 fragments of charcoal (Diamond 2005). The forests are now known to have contained a giant palm tree (*Paschalococos disperta*) and a number of other trees (e.g. *Sophora toromiro*), some reaching over 30 m in height. Palynological studies by Flenley and colleagues demonstrate that the forests of Easter Island persisted for at least 33 000 years (as far back as the record goes), during the major climatic shifts of the

late Pleistocene and early Holocene. We can thus be clear that deforestation by humans was responsible for the treeless state of the island (Flenley *et al.* 1991; Bahn and Flenley 1992; Diamond 2005).

Dated archaeological evidence indicates that Easter Island was settled by the Polynesians from the Marquesas or Gambier archipelagos (Oliver 2003) some time between AD 300 and AD 900, with the more recent date preferred by Diamond (2005) as the most reliable earliest date of occupation. Their effect on the forest was soon detectable in the pollen record (Bahn and Flenley 1992) and it was entirely eliminated some time between the early 1400s and the 1600s AD (Diamond 2005). The human population reached its peak of about 10 000 (or possibly as high as 15 000) around AD 1600, but subsequently collapsed along with the megalithic culture that had sustained the quarrying, sculpting, transport, and erection of the *moai*. Figure 11.5 sets out a model of the sequence of events which may have led to this collapse, within which a pattern of overpopulation and continuing overexploitation and thus resource depletion was central. This depletion involved not just the loss of a forest cover, but through it the loss of the raw materials (large trees in particular) needed to make ocean-going canoes, further diminishing resource availability. The pattern of decline may have also been influenced by the introduction of rats by the Polynesians, which could have had a key role in limiting tree recruitment (Fig 11.5). The loss of trees and other plant species is matched by a more complete loss of native birds than on any comparable island in Oceania (Steadman 1997a). Examination of bird bones associated with artefacts 600–900 years old has revealed that Easter Island once had at least 25 species of nesting seabirds, only 7 of which now occur on one or two offshore islets, and just one of which still nests on the main island. Native land birds are known to have included a heron, two rails, two parrots, and an owl. None survive.

The Easter Island story holds some interesting twists and turns that remain the subject of some debate amongst archaeologists. When the Dutch captain Jacob Roggeveen and his three vessels arrived in AD 1772, the overthrow of the statue-building culture and its chiefs and priests had

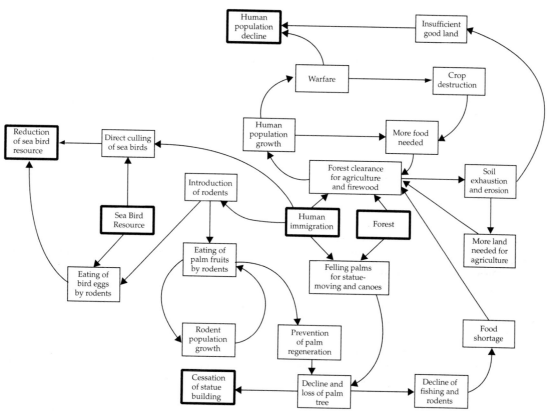

Figure 11.5 A hypothetical model to indicate the possible course of events on Easter Island as deduced largely from palaeo-ecological data. (Modified from Bahn and Flenley 1992, Fig. 191.)

already occurred, although the repercussions of this event were ongoing, and in fact, some statues were still standing, with the last European mention of an erect statue being in AD 1838 (Diamond 2005). The coup happened around AD 1680, led by military leaders called *matatoa*. After this event, a new culture and some degree of order had developed around the Birdman cult, which appeared to provide a mechanism for the distribution of resources amongst the people of Easter Island. However, it was a much altered society, in which human strife and cannibalism were now prevalent features of a society that had once been so stable and well provided for that it was able to devote surplus labour to the quarrying, carving, transportation and erection of statues weighing between 10 and 90 tons. Possibly hundreds of people would have been involved in some of these operations, and they

would have required large timbers for the sleds, ladders, levers, and ropes needed for the movement of the statues. At the time of European contact, the human population was estimated at about 3000, a figure at which it may or may not have stabilized (Erickson and Gowdy 2000). Unfortunately, the interaction of this society with the outside world generated a further collapse, this time fuelled by introduced diseases (e.g. smallpox epidemics, Diamond 2005). In 1862, slave traders forcibly removed about 1500 people, approximately one third of the population at the time. They were transported to Peru and sold at auction to work in guano mines and other menial jobs. Most soon died, but under international pressure, Peru repatriated a dozen survivors in the following year. The survivors brought with them another smallpox epidemic. By 1872, the population of Easter Island stood at only 111 people. Just 16 years later the Chilean government annexed the island, and it became in effect a sheep ranch, with the remaining islanders confined to a single village as forced labour.

The islands of the Pitcairn group lie between 23.9 and 24.7°S and 124.7 and 130.7°W in the South Pacific Ocean, and consist of volcanic Pitcairn Island, the raised coralline Henderson Island, and the small coral islands of Ducie and Oeno atolls. Pitcairn is of special historical interest, owing to its occupation by the HMS *Bounty* mutineers, in AD 1790, but before that it was occupied and then abandoned by the Polynesians. Although the precise timing of the settlement of Pitcairn by the Polynesians remains problematic, it now appears that Henderson Island was settled between AD 800 and 1050 and deserted some time after AD 1450 (Benton and Spencer 1995). During the occupation, marine molluscs, turtles, and birds were heavily predated. Settlement was accompanied by the introduction of cultigens and tree species, and by burning for crop cultivation. A number of endemic bird species and at least 6 out of 22 land-snail species became extinct during this period. It seems likely that sustained occupation was only possible through interaction with Pitcairn and, 400 km to the west, with the island of Mangareva.

Weisler (1995) has outlined a prehistory of Mangareva, which suggests a similar pattern of resource exploitation and environmental depletion as Easter Island. Mangareva experienced high human population levels, depleting the resource base to the point where they could no longer support long-distance voyaging, dependent as it was on large ocean-going vessels, skilled crews, food supplies, and exchange commodities. As on Easter Island, the requirement for large trees for canoe manufacture does not appear to have been matched with sustainable tree husbandry! Thus, according to Weisler's reconstructions, the overexploitation of resources by these linked communities resulted in the severing of voyaging linkages and the eventual abandonment of the most marginal locations, including both Henderson and Pitcairn. With its poor soils, sporadic rainfall, lack of permanent sources of potable water, karstic topography, and

low diversity of marine and terrestrial life, it is pretty remarkable that Henderson was settled by the Polynesians in the first place (see illuminating discussion in Diamond 2005).

The Henderson petrel (*Pterodroma atrata*), only recently described as a distinct species, breeds exclusively on Henderson Island. A study in 1991/92 showed that its breeding success was greatly reduced by predation by the Pacific rat (*Rattus exulans*) and the species was judged to be threatened (Benton and Spencer 1995). It has been calculated that this petrel has been undergoing a slow decline since the Polynesians introduced the rat some 700 years ago, and it is possible that this may continue until the petrel becomes extinct, although this would seem a remarkably long lag time for a new 'equilibrium' to be reached. Henderson Island has four surviving endemic land birds, including one fruit dove, *Ptilinopus insularis*. In the past, this dove shared the island with at least two other Columbiformes: *Gallicolumba*, a ground dove, and a *Ducula* pigeon. The dove became extinct during Polynesian occupation about the late thirteenth century. Indeed, five of Henderson's nine endemic land birds became extinct as a result of this occupation, as did most of the small ground-nesting seabirds (Wragg 1995). Human predation, as well as predation by (and competition with) the commensal Pacific rat, and habitat alteration by people, each had a role in these changes in the avifauna. The process of change is ongoing, and today a number of the endemic land snails of Pitcairn appear to be restricted to remnants of native vegetation of only about 1 ha or less. These stands are currently under great threat from the spread of invasive exotic plants, which cast a deep shade and create an understorey inimical to the snails (Benton and Spencer 1995).

Indian Ocean birds

As shown in Table 11.3, the islands of the Indian Ocean have also suffered significant losses of birds in prehistoric and historic times. Most notably, on Madagascar, 6–12 species of ratites became extinct, almost certainly as a result of the arrival of the Malagasy people. These losses included the largest bird ever recorded, the giant elephantbird (*Aepyornis maximus*). Other losses include 2 giant tortoises, pygmy hippo, and at least 14 species of lemur, most of them larger than any surviving species (Diamond 1991*b*; Groombridge 1992).

The islands of Amsterdam and St Paul, some 2500 km east of Madagascar in the southern Indian Ocean, and 80 km from each other, are known to have had their own avifauna, but nearly all of the native species were exterminated through fires and the introduction of domesticated and commensal mammals, notably feral cattle (Olson and Jouventin 1996). The extinct avifauna included a species of rail, and a different species of duck on each island. The ducks can be assumed to have been different species because remains of the Amsterdam form, *Anas marecula*, reveal it to have been small and flightless: whatever was on the other island (known only from an historical account) must perforce have been a distinct form.

Without doubt the most celebrated of all extinct island birds, the dodo (*Raphus cucullatus*) was a large, flightless pigeon (Columbiformes) endemic to Mauritius (Livezey 1993). A related species, the solitaire (*Pezophaps solitaria*), occurred on Rodrigues, also in the Mascarene group. The existence of a third species on Réunion is indicated by historical evidence, but has not been backed up by a specimen. The dodo and its relatives were wiped out by the early eighteenth century by a combination of hunting, habitat destruction, and the negative impacts of introduced vertebrates. Predators implicated include pigs, rats, cats (especially on Rodrigues), and monkeys (Mauritius only). Introduced herbivores responsible for significant destruction of native vegetation were cattle, goats, and deer (Mauritius only). Although they proved to be vulnerable to the evolutionarily unpredictable intervention of humankind, these giant, ground-dwelling pigeons were formerly dominant consumers within their respective island ecosystems.

The native forests and woodlands of Mauritius have been reduced in extent by 95% to just 5% of the land area (Safford 1997). This deforestation, combined with the introductions of many exotic animals and plants, has led to the extinction of at least eight other endemic bird species in addition to

the dodo and the reduction to critical levels of eight others. Five of the six extant forest-living native passerines are essentially restricted to native vegetation. Moreover, only a small fraction of this habitat is of good quality because of nest predation, reduced food supply, disease, and, in the past, the use of organochlorine pesticides. In 1993, each of these five species was estimated to have populations of the order of only 100–300 pairs. Safford (1997) recommends management of exotic browsers and predators, and weed removal, as part of a broader strategy for conservation on Mauritius. He also suggests that, given the impracticality of eliminating predators on the mainland, some of the offshore islets be considered for ecological rehabilitation and translocation programmes. However, he notes that bird populations on small offshore islets would be vulnerable to catastrophes, especially cyclones.

Reptiles

Case *et al.* (1992) review the data for reptilian extinctions worldwide over the past 10 000 years, noting that the great majority have occurred on islands (as Table 11.2) and as a consequence of human impacts. On small islands, humans have raised extinction rates by about an order of magnitude, but rates of loss have been much lower on very large islands. Introduced predators, principally mongooses, rats, cats, and dogs, have been the main agents of human-related extinctions, whereas competition with introduced reptiles appears to have had relatively little impact on native island reptiles. The reptiles most prone to extinction have been those with relatively large body size and endemics with a long history of island isolation.

Case *et al.* (1992) subdivide islands into three categories:

• First, they discuss islands colonized in prehistory by aboriginal people and then colonized later by Europeans. On such islands, many reptiles are known only as subfossils, having become extinct during the aboriginal period. In New Zealand, for example, three species of lizard known from the

Holocene are extinct, and a further nine reptile species are today found only on the off-lying islands, among them the 'living fossil' tuatara (*Sphenodon punctatus* and *S. guntheri*), the single representatives of the order Sphenodontia (the rest of the lineage died out by 60 Ma). These offshore island populations are relictual distributions following extinctions from the mainland consequent upon Polynesian settlement. A similar pattern is found in the Caribbean, where several species became extinct during aboriginal occupation on islands such as Hispaniola and Puerto Rico; others have become extinct within the past century. The large herbivorous iguanine *Cyclura* has become extinct on a number of islands in the recent past, and the giant *Cyclura pinguis*, lost from Puerto Rico during the Holocene, survives on the small offshore island of Anegada. Among the lost reptiles of the Caribbean were giant tortoises (*Geochelone* spp.) from the Bahamas, Mona island, and Curaçao. Giant tortoises also once occurred on Sicily, Malta, and the Balearic islands in the Mediterranean, as well as in the Canaries, and on the Mascarene islands. Their disappearance from the latter followed shortly after human contact, but in the Mediterranean the role of humans in the losses is not yet so clearly established (Schüle 1993). The Canaries, in the eastern Atlantic, lost a number of lizard species following their colonization by the aboriginal Guanches between 2500 and 2000 BP. In the Canaries, fossil lizards are found in association with human artefacts, but lizards were probably not a major item of the diet, and the introduction of commensals, such as rats, dogs, goats, and pigs, was probably of greater significance in the demise of the largest species. Interestingly, populations of two Canarian giant lizards (*Gallotia simonyi* and *G. gomerana*) thought to have been extinct, have been rediscovered on El Hierro and La Gomera respectively in recent years: examples of so-called Lazarus species.

• The second category identified by Case *et al.* (1992) was of islands with a colonial period but no aboriginal history, on which island endemics survived to be described as living species, very often to meet their demise shortly after. Two spectacular large geckos, *Phelsuma* spp., co-occurred briefly with Europeans on Rodrigues island in the Indian Ocean in the late seventeenth century, until hunted to extinction by rats and cats. A similar fate befell many endemic reptiles on Mauritius. None of the endemic reptile species of the Galápagos has become extinct yet, but population densities have declined and local extirpations, for instance, of giant tortoises, have occurred. Land iguanas have also been lost from Baltra and James islands, probably as a result of predation by feral dogs.

• Case *et al.*'s (1992) third category was of islands with no permanent human settlement to date. Such islands are generally too bleak or small to support human settlement. For this reason they also tend to support few endemic species. Losses from this category of islands have thus mostly been local rather than global extinctions, and of interest principally for the insights they offer into natural turnover.

Caribbean land mammals

Extinctions have dramatically altered the biogeography of West Indian land mammals over the past 20 000 years (Morgan and Woods 1986). Of 76 non-volant mammals, as many as 67 species (88%) have become extinct since the late Pleistocene, including all known primates and Edentata (Table 11.7). Late Pleistocene extinctions are attributable to climatic change and the postglacial rise in sea level, but most late Holocene losses are due to humans. Thirty-seven of the losses post-date the arrival of people in the West Indies, which Morgan and Woods assumed occurred some 4500 years ago (although it may have been rather earlier than this: Table 11.5). Based on the 4500 years time line, species have been lost at the average rate of 1 per 122 years since human colonization, the figure for the preceding 15 500 years being about 1 per 517 years. In contrast to the data for non-volant mammals, bats appear to have fared relatively well, only 8 of the 59 species having disappeared over the past 20 000 years.

Many species of extinct West Indian mammals, especially rodents, are common in archaeological sites, indicating them to have been an important part of the Amerindian diet. In common with other studies cited here, humans brought about extinctions in the West Indies through predation, habitat destruction, and (especially in post-Columbian

Table 11.7 Native West Indian land mammals known from living or fossil records from the past 20 000 years (source: Morgan and Woods 1986)

Order	Total number of species	Living species	Extinct species	% extinction
Non-volant groups				
Rodentia	45	7	38	84
Insectivora	12	2	10	83
Edentata	16	0	3	100
Primates	3	0	3	100
Volant group				
Chiroptera	59	51	8	14
Total	**135**	**60**	**75**	**56**

times) the introduction of exotic species. Knowledge of the precise timings of species disappearances is fragmentary, but it does appear that a number of species disappeared during the Amerindian period, including at least one species of ground sloth, whereas other species undoubtedly survived to post-Columbian times, including several species of rodent.

Island snails

Snail extinctions from islands appear to be largely attributable to two factors—the loss of habitat, and the introduction of exotic species, mostly predatory snails or ants. The endodontoid snails are tiny tropical land snails, only a few millimetres in diameter. Over 600 species have been described from the Pacific, but one in six of them is thought to have gone extinct during the twentieth century (Groombridge 1992). They are mainly ground dwellers in primary forest, and they are threatened by habitat loss and by introduced ants that prey on their eggs. Other important island families are entirely or largely arboreal, e.g. the Partulidae, a restrictively Pacific family of some 120 species, most of which are considered to be threatened. On Hawaii, populations of achatinelline snails have been lost because of their low fecundity combined with over-collecting and habitat modification. In the Caribbean and New Caledonia it is reported that land snails most at risk are those in dry lowland forests, which may be lost more rapidly than upland forest to

causes such as cattle grazing and other forms of development. In New Zealand, there are many very localized endemic snails that are entirely dependent on native plant associations and which are rapidly being lost. On Madeira, in the eastern Atlantic, it is reported that of 34 species of land snails present in samples of varying ages, 9 have become extinct since human settlement in AD 1419, principally, it seems, as a result of habitat loss (Goodfriend et al. 1994). Finally, there is evidence of at least 19 extinct snail species from the Canaries, although there is some uncertainty as to whether these losses are attributable to humans or to past climate change (Izquierdo et al. 2004).

Griffiths et al. (1993) have studied the diet of the introduced snail Euglandina rosea in Mauritius, where it was released in the hope that it would control introduced giant African snails, Achatina spp. As in some other cases of failed biological control schemes, the introduced predator ignored the target species and instead gorged on native snail species. Some 30% of Mauritian snails are now extinct. As most losses pre-dated the introduction of Euglandina, they have been attributed principally to habitat destruction, but the exotic predator now represents one of the major threats to the remaining endemic snail species. On Hawaii, the introduction of Euglandina has been implicated in the decline of 44 species of endemic Achatinella; and it has been a major factor in the extinction of Partula species in French Polynesia, 9 species being lost from Moorea alone. Thus far, on Mauritius, primary forests do not appear to have been penetrated by the exotic

Table 11.8 Islands and archipelagoes that account for more than five extinct or possibly extinct vascular plant species since AD 1600, attributable to anthropogenic activities. Sources: Walter and Gillet (1998); Primack and Ros (2002); Groombridge and Jenkins (2002). As noted in the text, the 2004 IUCN assessment is less pessimistic on both extinctions and threatened species. These figures should therefore be treated with caution

Islands	No. of extinct species	No. of possibly extinct species	No. of extinct + possibly extinct sp.	No. of threatened species
Hawaii	3	99	102	623
Mauritius	38	4	42	256
Cuba	23	0	23	811
Madagascar	0	19	19	306
French Polynesia	12	0	12	157
Rodrigues	9	2	11	51
New Caledonia	5	5	10	481
St Helena	10	0	10	41
Réunion	6	1	7	95
Tasmania	7	0	7	142
New Zealand	7	0	7	236
Fiji	0	7	7	72
Total	**120**	**137**	**257**	**3271**

species, and so habitat conservation remains key to the conservation of the native species.

Plants in peril

With all the attention given to animals in these pages, it is important to note, for the sake of balance, that the problems also extend to plants. According to the IUCN 1997 Red List assessment (Walter and Gillet 1998), the total number of confirmed plant extinctions globally since AD 1600 is 396, with a similar number of unconfirmed losses, and perhaps as many as 33 500 threatened species. Table 11.2 provides a figure of 384 extinctions. The equivalent figures from the 2004 IUCN Red List (<www.redlist.org/>;visited March 2006) are 100 extinct, and 8321 threatened, which amounts to a rather more conservative appraisal. The variation between these assessments highlights the uncertainty associated with such efforts, and the need for continued efforts to improve the resolution of conservation assessments (cf. Whittaker *et al.* 2005). With plants, given the lack of adequate species-level fossil data, it is particularly difficult to assess extinctions from the prehistoric period on islands: there undoubtedly have been numerous anthropogenic extinctions that will remain long undetected (for an example, see Flenley 1993). Once again, the figures given in Table 11.8 should thus be taken as of largely indicative value. The table summarizes the 1997 Red List data on plant extinction and threat in islands or archipelagoes that have lost more than 5 species. According to this assessment, anthropogenic extinctions of plants from the 12 most affected islands/archipelagos total 120 species (30% of all the extinct species), whereas 137 more were considered possibly extinct. In global terms, these figures indicate a disproportionate rate of loss and threat concentrated on islands. Geographically, the pattern of vascular plant species loss is rather similar to that of vertebrate animals (Table 11.6), with fairly high numbers of possible or confirmed extinctions in Hawaii, the Mascarenes, the Caribbean, and St Helena.

St Helena provides a classic example of the damage to a remote oceanic island brought about by incorporation into European trading routes. It is an isolated island of 122 km², located in the South Atlantic. Since discovery by the Portuguese in AD 1502, the vegetation has been completely transformed by browsing, grazing, erosion, cutting for timber and fuel, the introduction of alien plants, and clearance for cultivation, plantations, and pasture (Cronk 1989). No native mammals occurred on the island, as is typical of remote oceanic locations. The Portuguese introduced goats, pigs, and cattle, principally to supply homeward-bound carracks from India, and also brought with them rabbits, horses, donkeys, rats, and mice. Within 75 years of their introduction, goats had formed vast herds, devastating the vegetation.

The introduction of exotic plants began in earnest in the nineteenth century by way of British colonial botanic gardens, in part to provide a new plant resource to replace the exhausted native cover. Large numbers of herb and shrub species were introduced, along with timber trees such as *Pinus pinaster* and *Acacia melanoxylon*. By the time the first reliable botanical records were made in the nineteenth century most of the destruction had already occurred, but some documentary evidence, and the relict occurrence of endemic plants in isolated places such as cliffs, provide some indications of the former cover. Of the 46 endemic plants *known* from the island, 7 are considered extinct, the rest threatened (Groombridge 1992; compare with Table 11.8). Some species are restricted to only a few individuals, and in the early 1980s *Nesiota elliptica* and *Commidendrum rotundifolium* were reduced to just single plants, the latter becoming extinct in the wild in 1986 (Cronk 1989).

Goats have been particularly important in preventing woodland regeneration by grazing and barking saplings. So, as old trees were cut for firewood, areas became converted to a mosaic of anthropogenic formations, notably grassland and *Opuntia* scrub. Erosion has been very rapid, and on decadal timescales, isolated thunderstorms cause sheet and gully erosion where the vegetation cover has been removed by herbivores. After such erosion little vegetation cover can re-establish. Goats were nearly shot out in the early 1970s and all domestic goats are kept penned by law; however, sheep were still allowed to roam freely. Cronk (1989) considers that a process of rehabilitation is possible, based on trial plantings of native species, but that initially, at least, plantings require protection by fencing.

Turning to one of the showcases of plant evolution, Wiles *et al.* (1996) provide estimates for Hawaii of about 100 extinct taxa, 200 endangered or threatened, and a further 150 recommended for listing; however, 282 species appear on the list of federally endangered plants (Cox and Enquist 2000), and yet another estimate is given in Table 11.8. Although different criteria generate different numbers, it is clear that a disproportionate number of Hawaii's plant species are in trouble: for instance, the figure of 282 contrasts with 486 on an equivalent list for the mainland USA (Cox and Enquist 2000). Numerous introduced plant species have become naturalized on Hawaii, to the point at which the exotic flora outnumbers the endemic flora. Several have become serious pest species (Primack 1993).

In the Marianas (Micronesia) the forces involved in the demise of native vegetation include loss of forest to agriculture; war and development; the introduction of alien animals, plants, and plant pathogens; alterations to the fire regime; and declines in animal pollinators and seed dispersers. The impacts on several native plant species have been severe. *Serianthes nelsonii* (Leguminosae) is one of the largest native trees, and is endemic to Rota and Guam. It formerly occurred on volcanic soils but is now restricted to limestone substrates. Wiles *et al.* (1996) reported that just 121 adult trees survived on Rota, and 1 on Guam. The populations were senescent, with little or no regeneration occurring, and the main threats were from browsing by introduced deer and pigs, plus infestations of herbivorous insects (Wiles *et al.* 1996). On Guam, several plant species were found to be showing poor regeneration for these same reasons, and in one case at least, germination problems limit re-establishment. It appears that the seeds of *Elaeocarpus joga* need to pass through the gut of a bird, yet native frugivores have been wiped out by the introduced brown tree snake (*Boiga irregularis*) (Box 11.3). Another species, *Pisonia grandis*, is rarely seen with fruit, perhaps because of the loss of native pollinators such as fruit bats or birds. In addition to impacting on seed set and germination, the loss of frugivorous birds and the reduction of the one extant pteropid fruit bat species to a remnant population of a few hundred animals have had a predictably marked effect on seed dispersal on Guam (Rainey *et al.* 1995). Thus, while habitat removal is again a key driving force of plant reductions and extinctions, a broad array of factors and their synergistic interactions is at work.

In illustration of how differing factors have primacy in different situations, we can turn to Tahiti, where the overriding threat to endemic plants currently takes the form of another plant species. Tahiti is in the Society Islands (Polynesia), and is a large (1045 km²), high (2241 m) island, of diverse habitats, with a native flora of 467 species, 45% of which are endemic (Meyer and Florence 1996). Some 70% of the endemic plants occur in the montane cloud and subalpine forests, which are some of the largest, finest, and most intact to be found throughout the Pacific. The Polynesians colonized Tahiti perhaps 2000 years ago, bringing with them about 30 domestic and 50 other accidental plant introductions. European arrival in the AD 1760s led to massive resource overuse, and habitat loss, as well as a great flood of exotic plant species—in excess of 1500 now being known for the Society Islands. Many of these have become naturalized, and some have become problem plants. The lowlands and coastal zones were totally transformed by people, and such native forests as remain on Tahiti are endangered by pressures such as housing, agriculture, forestry (e.g. based on exotic *Pinus caribbea*), feral animals, and hydroelectricity schemes.

Yet, the highlands of Tahiti remained essentially intact until the arrival of what has become the worst of all exotic ecosystem transformers, *Miconia calvescens*, a small tree species introduced to the ornamental gardens in 1937. In less than 50 years it had spread to all the mesic habitats between 10 and 1300 m ASL, including the montane cloud forest, often forming pure stands. It now covers over two-thirds of the island of Tahiti, and has spread to the nearby islands of Moorea and Raiatea. It is a potential problem for other islands in the region, and is already recognized as one of the 86 plant pests of Hawaii, having been introduced there by horticulturalists as recently as the early 1970s. It is an effective invader thanks to traits such as: tolerance to a wide range of germination conditions, tolerance of low light levels, fast growth, early reproductive

maturity, prolific year-round seed production, and efficient dispersal by small introduced passerines such as the silvereye (*Zosterops lateralis*). In Tahiti alone, 40–50 endemic plant species are considered to be threatened directly, including 8 species of *Cyrtandra*, 6 *Psychotria*, and 2 *Fitchia*.

What can be done to halt this decorative invader? One approach is to dig it out. In 3 years (1992, 1993, and 1995), nearly half a million plants were uprooted by hand, and *Miconia* was cleared out of its bridgehead sites on the island of Raiatea. In addition to such plain labour, Meyer and Florence (1996) promote the need for a systematic assessment and inventory of alien and endangered native species, for research, monitoring, and control programmes, and for *ex situ* conservation in the botanic gardens—which still remains only an ornamental garden. Part and parcel of such an integrated approach to conservation, as commonly advocated in other similar studies, is a sustained educational programme.

The problem posed by invasive exotic plants may be first and foremost for native plant species but, if on sufficient scale, can also present a problem for native fauna (Cronk and Fuller 1995). For instance, the now pantropical shrub *Lantana camara* forms dense impenetrable stands in parts of the Galápagos, threatening both some endemic plant species and the breeding habitat of an endangered bird, the dark-rumped petrel (*Pterodroma phaeopygia*) (Trillmich 1992). Invasive plants may alter the environment in other, more subtle ways (Cronk and Fuller 1995). The shrub *Myrica faya*, native to Macaronesia, was introduced to Hawaii in the late nineteenth century, and has since been spread within the National Park, especially by an exotic passerine, the white-eye (*Zosterops japonica*), and by feral pigs. It has a highly effective nitrogen-fixing symbiont, which in part accounts for its ability to out-compete the native flora. By increasing nitrogen inputs relative to native species by 400%, it alters the nutritional balance of the Hawaiian soils. It poses considerable management problems and is predicted to have additional long-term consequences because the altered soil nutrient status is likely to facilitate the spread of other alien species (Vitousek *et al.* 1987b; Cuddihy and Stone 1990; Vitousek 1990). For further details see Box 11.4.

Box 11.4 The invasion of *Myrica faya* in the Hawaiian islands

Of the more than 13 000 plant species that have been introduced to Hawaii, c.900 have became established in the wild, and over 100 of them have became serious pests (Cox 1999). Among the more problematic tree species are guava (*Psidium cattleianum*), also a serious pest in the Galápagos, miconia (*Miconia calvescens*), tropical ash (*Fraxinus uhdei*) and, especially, the fire tree (*Myrica faya*, also known as *Morella faya*). *Myrica faya* is an interesting example of an oceanic island endemic species being introduced to other oceanic islands on the other side of the globe with a similar environment. A valuable constituent and early pioneer tree within protected forests in the moist zones within Macaronesia, it has become a pernicious forest weed in Hawaii (Cronk and Fuller 1995).

In the Canaries, *Myrica faya* is the only nitrogen-fixing tree species of the community, in symbiotic association with the actinomycete *Franckia*, and it exhibits a persistent pioneer strategy (Fernández-Palacios *et al.* 2004*b*). It is a dioecious species some 20 m in height, possessing medium-size fleshy fruits dispersed by indigenous birds, such as the blackbird (*Turdus merula*). *Myrica* maintains a local soil seedbank, and indeed its seeds germinate only after the appearance of a large canopy gap. Thus, *Myrica* plays an important successional role in the return of forest to previously cleared areas, but unlike some other Canarian woody pioneers (*Erica* spp.) also persists after the gap is filled through its ability to re-sprout via suckers from the base (Fernández-Palacios and Arévalo 1998).

Myrica faya was introduced to Hawaii as early as AD 1880 and it was used extensively in reforestation, especially during the 1920s and 1930s. In Hawaii its fruits are consumed and seeds dispersed by a large number of both native (*Phaeornis obscurus*—Hawaiian thrush) and introduced birds (e.g. *Zosterops japonica*—the Japanese white-eye, and *Carpodacus mexicanus*—house finch), as well as feral pigs. Between the 1960s and 1980s, aided by its ability to fix nitrogen, *Myrica* rapidly invaded native forests of *Metrosideros polimorpha*, which it out-competed in reaching the forest canopy and forming pure stands (Vitousek and Walker 1989). *Myrica* became invasive within the Hawaiian Volcanoes National Park in 1961 and has now spread over more than 35 000 ha. Today the new *Myrica–Metrosideros* forest extends across a belt of 400–1200 m ASL on the southern slopes of Hawaii (the Big Island), particularly on recent (< 1000 years) lava flows (Mueller-Dombois and Fosberg 1998), and it is a well-established and problematic invasive in all of the main Hawaiian islands.

Cronk and Fuller (1995) listed the following ecological characteristics of the fire tree as possible reasons for its success as an invader: prolific seed production, long-distance dispersal by birds, nitrogen fixation, possible production of allelopathic substances that may inhibit potential native competitors (specifically *Metrosideros polymorpha*) and the formation of a dense canopy under which native species are unable to regenerate. The impacts of *Myrica* on soil properties are particularly important, with knock-on consequences for ecosystem form and function. Vitousek *et al.* (1987*b*) calculate that the nitrogen availability in an unaltered Hawaiian forest is about 5.5 kg/ha per year (mainly through the fixation ability of the native *Acacia koa*). In stands invaded by *Myrica* this may be increased up to 23.5 kg/ha per year. These substantially altered edaphic conditions, in combination with the activities of feral pigs, have combined to enable the invasion of many continental plant species.

Options for the control of *Myrica* infestation have been the subject of considerable research efforts, particularly in the search for a safe biological control agent. However, the most effective means of control at present appears to be cutting and the use of foliar applications of herbicide, both of which are only really practical for dealing with newly establishing populations. The species continues to spread throughout the archipelago.

Exotic plants may alter other ecosystem properties, such as fire regimes or hydrological conditions. Examples of each can be drawn from Hawaii. Bunchgrass (*Schizachyrium* spp.) increases the flammability of the vegetation, and has prevented regeneration of native species that are not adapted to burning. *Andropogon virginicus* is another exotic grass, considered to be one of the most threatening of aliens. It not only carries fire, but also alters hydrological properties. It occurs in disturbed grassland and scrub on the island of Oahu in areas on red clay soils where native forest vegetation has been replaced by introduced woody and herbaceous plants. Its seasonal pattern of growth and tendency to form dense mats of dead matter are resulting in increasing runoff and accelerated erosion (Cuddihy and Stone 1990).

11.7   How fragile and invasible are island ecosystems?

It has become axiomatic to represent oceanic islands as fragile ecosystems, which, because of their evolution in isolation from continental biotas, are particularly susceptible to the introduction of alien species. Are island systems really particularly fragile or have humans just hit them especially hard? D'Antonio and Dudley (1995) contend that the data support two propositions. First, island ecosystems do typically have a higher representation of alien species in their biota than do mainland systems. Secondly, the severity of the impacts of invasions appears in general to increase with isolation.

Cronk and Fuller (1995) suggest six reasons why oceanic islands may be more susceptible to invasion:

- **Species poverty**. This may mean that there is more vacant niche space and less competition from native species. This is plausible, particularly where humans have disturbed habitats, but it is only one part of the picture.

- **Evolution in isolation**. It is often assumed that island species are competitively inferior to continental species as a resut of their long evolutionary isolation. The case of *Myrica faya*, transported from

one set of oceanic islands, in Macaronesia, to another, Hawaii, illustrates that the 'continentality' of the land mass from which a species hails may not be of great moment: simply to be from a different biogeographical pool may be sufficient. However, the evolution of flora and fauna in isolation, often without adaptation to grazing, trampling, or predation by land mammals has led to the loss of defensive traits in many oceanic island endemics, expressed in plants as the absence of typical grazing adaptations such as spines, thorns, and pungent leaf chemicals. In the presence of browsers these plants are at a competitive disadvantage and tend to disappear, as was the fate of many endemic plant species on St Helena following the introduction of goats shortly after the island's discovery in AD 1502. Similarly, many island endemic animals lack a fear of predators, and some birds have reduced flight or are flightless, and build their nests on the ground. These are traits that may have no competitive disadvantage when faced with a species of the same trophic level, but which have proven disastrous when faced with terrestrial vertebrate predators.

- **Exaggeration of ecological release**. Alien species generally arrive on islands without their natural array of pests and diseases, and this provides them with an advantage over native species. A similar ecological release mechanism is proposed as part of the taxon cycle in Chapter 9, and indeed also applies to many introductions to continents.

- **Early colonization**. Particularly in the Atlantic, Caribbean, and Indian Oceans, islands were the first landfalls and first colonies of the Europeans, hence they have had a long history of disturbance and of introductions. Although this is true, this line of reasoning fails to account for the severity of impact of particular introductions such as the brown tree snake on Guam or *Miconia* on Tahiti: the outcome has been a particular property of the species involved rather than the length of time over which disturbance and introductions have occurred.

- **Small scale**. The small geographical size of islands means that their history is concentrated in a small area, within which there are few physical features large enough to prevent exploitation, disturbance, and introduction on an island-wide basis. They also often sustain surprisingly high human populations (Watts 1987); for example, Cronk and Fuller (1995) note that Mauritius, with a population of 530 people/km^2, may be compared culturally and historically with India, which has only about 240 people/km^2.

- **Crossroads of intercontinental trade**. Particularly in the days of sailing ships, islands have often been used as watering points, re-supply posts, and staging posts for trade and for plant transport. As they have often changed hands between different colonial powers, they have, moreover, had opportunities to receive introductions from differing networks of nations and biogeographical regions.

The deluge of disturbance and of alien species which has swept across so many oceanic islands goes some way towards explaining the apparent susceptibility of islands to invasion. At least one tropical biologist, who has worked for many years on Hawaii, argues that this is the case:

Contrary to common opinion, many endemic island species are strong competitors. They would not be eliminated from their niche in the island ecosystem if it were not for the new stress factors introduced by man. The island species evolved with stress factors associated with volcanism, fire in seasonally dry habitats, and occasional hurricanes. The effects of volcanism resulted in superior adaptation of many native species to extreme edaphic conditions existing on volcanic rockland, where soil-water regimes fluctuate almost instantly in direct relation to rainfall. (Mueller-Dombois 1975, p. 364)

This general position is supported by D'Antonio and Dudley (1995) who conclude that, with exceptions, island habitats are not inherently more easily invaded than are continental habitats, but that, nonetheless, exotics have had a greater impact on island systems. Some more formal analysis provides a measure of support for Mueller-Dombois's views. It has been suggested that as a rough rule (the tens rule), about 1 in 10 introductions into exotic territories becomes established (Williamson 1996). The data for Hawaiian birds indicate a far higher success rate, of over 50% of introduced species establishing. However, Williamson (1996)

notes that the majority of introductions have colonized disturbed lowland habitats, and that the figure for species establishing into more or less intact native habitats (some lowland forest, together with upland habitat) falls at between 11 and 17%, which is not significantly out of line with the tens rule. Studies elsewhere within very disturbed islands, containing large numbers of naturalized exotic species, have also reported that those habitats that have remained largely undisturbed quite often contain few exotics (e.g. Watts 1970, on Barbados; Corlett 1992, on Singapore). These examples suggest that the susceptibility of island ecosystems to invasion is above all a reflection of the profound alteration of habitats. However, other case studies cited in this chapter demonstrate that some invaders have not required such assistance.

One of the curious features of particular invasions is that invading species sometimes undergo a pattern of increase to a peak of density, followed by a crash, sometimes to extinction. This boom-and-bust pattern appears to have occurred fairly often on islands (Williamson 1996). It has, for instance, been recorded for a number of the passerines introduced to the Hawaiian islands, with their time to extinction varying between 1 and 40 years. There may be a number of reasons for these boom-and-bust patterns, but Williamson favours those relating to resource deficiencies as providing the most general explanation.

From the point of view of island biogeography theory it is noteworthy that the introduction of a single exotic species, such as the brown tree snake (*Boiga irregularis*) on Guam and the shrub *Miconia calvescens* on Tahiti, may cause the local extinction of numerous native island species. Moreover, in the case of *Miconia*, the effect operates within the same taxon, i.e. it is a single species of plant which is implicated in the loss of numerous native plants. This provides an instance in which an equilibrium model, involving turnover on a replacement basis subject only to relaxation effects from area reduction (as discussed in Chapter 10), would be a wholly inappropriate device for predicting species losses.

11.8 Summary

As a result of human action, island birds have been 40 times more likely to become extinct within the past 400 years than continental birds. In the same period, the proportion of documented animal extinctions from islands varies from about 60% for mammals to 95% for reptiles. Moreover, the losses caused by humans in prehistoric times exceed those of the past 400 years. This is most clearly established for birds, but applies to other animal taxa and to islands across the world. Island plants have also been subject to a disproportionate rate of loss. The scale of environmental degradation on certain Pacific islands, exemplified by Easter Island and Henderson Island, has been shown to have been so great that it led to complete cultural collapse and, in the latter case, human abandonment of the island. The available evidence indicates that the key turning point in the transformation of island ecologies is typically the first colonization by humans, but that successive waves of contact with the outside world, and of innovation in resource use, can continue to ramp up the pressure on island biodiversity, sometimes generating multiple phases of extinction.

The major causes of local and global extinction from islands have been: (1) predation by humans, (2) the introduction of exotics, (3) the spread of disease, and (4) habitat loss, of which habitat loss and the introduction of alien species currently give greatest overall cause for concern. Predation remains a critical driver of extinction threat for a subset of threatened species and islands. The introduction of exotic species, in some cases a single exotic species, can cause the local extinction of numerous native island species. The shrub *Miconia calvescens* on Tahiti and the brown tree snake (*Boiga irregularis*) on Guam provide two classic examples. However, case details provided in this chapter from the Pacific, Indian Ocean, Caribbean, and Atlantic illustrate that while some losses can indeed be attributed to a sole major cause, many species losses have resulted from synergisms between several causes, such as habitat alteration allowing the invasion of exotic plants, which in turn are spread by exotic birds, which in turn carry exotic diseases. The term **trophic cascade** is used to describe the situation where the loss of one or a small number of species triggers the disruption of ecosystem processes more generally. Such knock-on effects can work through predatory relationships and, as just indicated, through other forms of ecological networks, such as pollination or dispersal networks.

While the big problem for continental organisms is the increasing insularization of their habitats, the problem for oceanic island biotas in recent history and today is the increasing breakdown of the insularity of theirs. As a result, many oceanic islands now have a high representation of exotic species. It does not necessarily follow that island habitats, if undisturbed, are always more invasible than continental habitats. However, exotic invaders have had a significant impact on many island ecosystems, and this susceptibility to exotic impact can be related both to peculiarities of human use of islands and to evolutionary features of the island biotas, especially to their evolution in the absence of terrestrial mammalian browsers and predators.

CHAPTER 12

Island remedies: the conservation of island ecosystems

Although generally distinct from continental environments, and therefore of interest in their own right, island environments have been generally misunderstood, misinterpreted, and mismanaged. The main reason is that they have been interpreted for management purposes largely by continent-trained observers . . . islands are not simply miniature continents and . . . continental solutions do not simply need to be scaled down in order to be successful.

(Nunn 2004, pp. 311 and 319)

In this final short chapter, we first illustrate some of the special problems facing contemporary island societies, the pressures on the environments and biotas of oceanic islands, and constraints within which those interested in promoting conservation must operate. Conservation in island archipelagos like the Canaries, the Maldives, the Lesser Antilles, and the Galápagos present distinctive challenges, which arise from the characteristics of the island environments and biota and from the socio-economic and political situation of the island societies (e.g. Nunn 2004). Therefore, secondly, we look at some of the conservation solutions that have been promoted in the context of the special circumstances of remote islands. Solutions 'parachuted in' from the outside without regard to this context are unlikely to produce the desired outcomes.

12.1 Contemporary problems on islands

The previous chapter illustrated some of the common elements of the problems afflicting island biodiversity: habitat loss/degradation and the role

of introduced species foremost amongst them. We start this section, by contrast, with a short selection of case studies drawn from different regions of the world, which illustrate some of the profoundly different circumstances and issues that may be of concern in different island groups. Although our focus throughout this final part of the book has been on biodiversity, we felt it instructive to include at the start of this chapter a couple of examples of islands where the loss of endemic species is not the central concern. As will become clear, however, whether species extinctions are at issue or not, the loss of ecosystem goods and services is, in the end, always of concern to island societies.

Maldives: in peril because of climatic change

The Maldives form an extended equatorial archipelago of about 1200 small, low coral atolls, extending for 800 km in a line across the Indian Ocean. Being small, low islands they have a low rate of endemism, with just five endemic plant species (all in the genus *Pandanus*) within a native flora of 277 species (Davis *et al*. 1995). They have a total land area of about 300 km² and a population of about 350 000 inhabitants. The foreign exchange economy is based almost exclusively on the naturalness and beauty of these islands, which have become a destination for so-called 'quality' tourism.

Together with the Lakshadweep Islands to the north and the Chagos Islands to the south, the Maldives form part of a vast submarine mountain range, on the crest of which coral reefs have grown. Actually, the term 'atoll' is derived from the native name *atholhu*, applied to these islands, which are examples of this structure. It is this mode of origin that explains their low elevation.

Projected twenty-first-century sea-level rise resulting from global climatic warming could see the complete inundation of Malé Atoll, the capital and most heavily populated island, if not of the entire archipelago, which has a maximum elevation of just 2 m above sea level. Even a lesser rise could be sufficient to jeopardize the continued existence of this society and its unique culture, along with numerous other similarly distinctive societies living on low-lying islands around the world (Nunn 2004). With few mitigation options open to them, the Maldivian government, with Japanese aid, has undertaken the biggest project ever seen in the archipelago, the construction of sea walls around the island of Malé.

Okino-Tori-Shima: the strategic economic importance of a rocky outcrop

As islands disappear under the sea by erosion, subsidence, or sea-level rise, so do their Exclusive Economic Zones (EEZ), the 200 nautical mile extension around a land mass, whereby a nation retains use rights to undersea resources, primarily fishing and seabed mining. This aspect of international law accounts for the interest countries have in claiming and holding on to desolate chunks of rock in the ocean.

A prime example of this is shown in the actions of Japan with regard to Okino-Tori-Shima (also known as Parece Vela), Japan's southernmost island, located some 1750 km south of Tokyo. The island is a largely submerged atoll, which at high tide consists of just two rocky outcrops jutting slightly more than 60 cm above the waters of the Pacific (Chang 2000). Typhoons rip through this region annually, and their destructive impact, alongside regular wave action, threatened to submerge the atoll altogether. Because this atoll is the basis for an EEZ of 400 000 km² (larger than Japan itself), of very rich fishing waters (alongside mineral rights), Japan could not allow this island to disappear, and in 1988 millions of dollars were spent in constructing iron and concrete breakwaters to prevent the loss of the atoll—so far successfully.

By contrast, two uninhabited islands, Tebua Tarawa and Abanuea (ironically meaning 'the beach which is long lasting'), part of the island nation of Kiribati, were lost in 1999. If sea-level continues to rise over the twenty-first century, as forecast, then the further loss of territory in this way will require the recalculation of the EEZ of Kiribati, which currently constitutes over 3.5 million square kilometres, almost 5000 times the land area, and equivalent to over a third of the land area of the USA (Chang 2000).

These examples serve to illustrate the importance of the loss of land area to island nations, even when the area lost is not particularly important in terms of human habitation.

Nauru: the destruction of an island

The island of Nauru (once known as Pleasant Island) is located in the Pacific Ocean, about 500 km northeast of New Guinea. It has a total land area of 21 km² and a population of about 13 000 as of 2006. The coast is fringed by a ring of sand, surrounded by a protective reef, but the interior is formed of guano deposits that have mixed and solidified with a coral limestone base. The mining of these deposits has devastated the island environmentally, and despite the initial economic returns, has created severe financial, legal, and cultural problems for the Nauruans.

Guano is a rich source of phosphate, which is naturally scarce and of high value as a key constituent of fertilizer. Phosphate mining began in 1908 when the Germans, then in control of Nauru, began to mine the island's substantial deposits. Between World War II and Nauru's independence in 1968, Australia became the Administrator of the island. The mining operation in this period produced 2 million tons yearly, which was mainly exported to Australia and New Zealand. Phosphate mining has been concentrated in the central plateau, where the mining has left behind deep pits and tall pillars, creating a moon-like scene of barren wasteland, which now comprises over 80% of the island area, with the residents living on a narrow strip along the coast. As a result of the mining, the vast majority of soil and vegetation has been stripped away, making agriculture impossible.

With independence from Australia, the Nauruans began to experience the financial benefits of phosphate mining for the first time. They became rich virtually overnight, creating one of the world's highest per capita incomes. The new Republic filed suit with the International Court of Justice against Australia in 1989, asking for compensation for the environmental damage sustained prior to independence. Five years later, a settlement was reached whereby Australia agreed to pay more than 100 million Australian dollars (Pukrop 1997). The money earned may have been considerable, but the phosphate extraction begun by the colonial powers and continued by the Nauruans has destroyed the traditional and cultural way of life. They no longer practice agriculture or fishing, relying instead on pre-packaged imported fatty foods, which have decreased the life expectancy, in the case of men to just 55 years (Economist 2001). Phosphate production peaked in the 1980s, and this, combined with a collapse in the market price, has undermined the main income source for the island. Moreover, the financial bonanza was badly mishandled, and the money was squandered. Offshore banking was the next big idea, but this shady trade with next to no internal regulation did nothing for the international reputation of Nauru, which soon found itself in trouble with international regulators.

With the end of the mining operations in sight, the attention of the island government turned to the future rehabilitation of the island. With an expanding population crowded into the coastal strip, the Nauruans need more space for new buildings, including a hospital and schools. This development can only occur in the central plateau, which, however, remains a barren wasteland of limestone and coral. One solution is to crush the pillars and to import topsoil, humus, and other nutrients, thus beginning a long process of rebuilding the ecosystem. The cost is estimated to be double the compensation received from the Australian government, and it could take more than 30 years to complete the scheme. Given the worsening financial situation that the island now finds itself in, such ideas seem unlikely to come to fruition. As The Economist (2001) puts it 'It is a melancholy sign of the islanders'

desperation that the idea of simply buying another island and starting afresh is once again under discussion. But who in his right mind would let the Nauruans get their hands on another island?' This may seem a rather unsympathetic remark, but is fair comment on the quality of leadership since independence, as this once isolated island society has attempted to come to terms with the disastrous results stemming from its interactions with the global economy.

The Canaries: unsustainable development in a natural paradise

The Canarian archipelago is one of the most biodiverse areas within Europe (or at least within the European Union, given that the islands lie closer to Africa than to Europe). The islands possess more than 12 500 terrestrial and 5 500 marine species in, or around, a land area of only 7500 km^2. Within this total of 18 000 or more species, about 3800 species and 113 genera are endemic (Izquierdo et al. 2004). Among them are many examples of spectacular radiations of animals (e.g. more than 100 species of *Laparocerus* weevils) and plants (e.g. 70 species of succulent rosette-forming members of the Crassulaceae). Furthemore, an average rate of description of one species new to science every 6 days over the last 20 years, suggests that the catalogue of Canarian species is still far from complete (Fig 11.1).

The Canaries were first settled by the Guanche people, believed to be of Berber stock, around the first century BC. The archipelago has been under European influence since the fifteenth century, with extensive deforestation of the mesic zones, goat grazing of the less productive land, and intensive agricultural use of whatever land could be put into production. Between 1960 and 1970, as a result of cheap air travel, the pattern of exploitation of natural resources began perhaps its third major phase, shifting from an agricultural society to a mass tourism model.

This tourist boom has given rise to very abrupt socio-economical and cultural changes (Table 12.1), with profound impacts for the natural and semi-natural ecosystems of the archipelago. Some 12-million tourists now visit the Canaries each year,

Table 12.1 The shift of the economic development model (1960–2000) on the Canaries (Source: Fernández-Palacios *et al.* 2004*a*).

Property	1960	1970	1980	1990	2000
Population (M)	0.94	1.17	1.44	1.64	1.78
Number of tourists (M)	0.07	0.79	2.23	5.46	12.0
Population density (inhabitants/km^2)	130	155	189	206	231
Cultivated area (K ha)	95	68	60	49	46
Oil consumption (K oil equiv. ton.)	–	827	1442	2473	3155
Electric energy consumption (GW)	–	890	1680	3423	6292
Concrete consumption (M ton.)	–	0.76	1.22	1.57	2.65
Number of cars (M)	0.02	0.08	0.28	0.5	1.08
Active population in agriculture (%)	54	28	17	7	6
Active population in services (%)	27	46	55	62	70
Unemployment (%)	2	1	18	26	13
Female life expectancy (y)	65	75	77	80	82
Literacy (%)	36.2	–	91.7	95.7	96.4
Per capita income (K dollars)	4.3	8.8	11.4	15.4	17.2

and current population growth, fuelled largely by immigration, is occurring at the rate of about 50 000 a year. The upshot is human population densities of about 550/km^2 in Gran Canaria and 450/km^2 in Tenerife. The increased resident and transient population increasingly demands more space for houses, hotels, roads, and infrastructure, with increasing demands for food, goods, water, and other resources, while simultaneously producing more waste (e.g. García-Falcón and Medina-Muñoz 1999). It can be estimated that on a daily basis the average Canarian resident contributes more than 20 kg CO_2 to the atmosphere and produces 5 kg waste (Fernández-Palacios *et al.* 2004*a*).

In terms of the indigenous resource base, this pattern of development is clearly unsustainable (Fernández-Palacios *et al.* 2004*a*). The arid regions are increasingly being built over, as land is gobbled up by the development instigated by the tourist sector. Coastal ecosystems have been fragmented by buildings, harbours, roads, and golf courses. The water table has been depleted as the water resources have increasingly been mined, and pressures have also increased on fisheries in the inshore

waters. As a by-product of tourist-driven urbanization, about half the agricultural area (50 000 ha) has been abandoned. Agriculture in the Canaries is of necessity a labour-intensive activity, with cultivation focused in small units not amenable to large-scale mechanization: easier and seemingly more attractive livelihoods are to be found in the urban areas.

The upshot of these changes has been a wholesale shift in the nature and geography of the human impact on Canarian landscapes, alongside a shift from the archipelago acting as a net exporter to a net importer of food. On Tenerife, the upper regions, which were once devastated by overgrazing and cutting, have now largely been handed over to replanting (especially in the endemic Canary Island pine belt) and to conservation. The laurel forest belt, although reduced to perhaps 10% of its original area, is stable in area and has protected status. Many areas once in cultivation in the less arid parts of the lowlands are gradually being recolonized by a mix of native and exotic plants (Box 12.1), or else built on. As current mass tourism favours the sunniest environments, it is in the dry lowlands that the pressure is now greatest, with

---

Box 12.1 Invasive plants in the Canaries: the incorporation of prickly pear (*Opuntia*) into the landscape

During the nineteenth century, several prickly pear species (*Opuntia*, Cactaceae) of Mexican origin were introduced into the Canaries, principally for the natural red-purple colorant (carminic acid) produced by the females of the parasitic insect *Dactylopius coccus* (Coccoidea, Homoptera), also introduced from Mexico, which feed on their succulent leaf-shaped stems. True cacti are not naturally found outside the New World. *Opuntia* rapidly naturalized and became invasive within the natural and semi-natural ecosystems of the Canarian lowlands (Otto *et al*. 2001, 2006), where they found similar arid, warm conditions to their original distribution. One of the keys to the success of prickly pears as invaders on the Canaries is the year-round availability of their large, fleshy, sweet fruits, which have proved attractive to both native and exotic dispersers (Chapter 11). They also regenerate and spread vegetatively, from falling leaves taking root: behaviour not found in the native Canarian *Euphorbia*, which are the indigenous dominants of this scrub vegetation. Finally, another clue to their success is that whereas all but one species of *Euphorbia* lose their leaves in summer drought conditions, *Opuntia* are still able to photosynthesize, using the crassulacean acid metabolism (CAM) photosynthetic pathway, which again is not possessed by the local *Euphorbia* species.

Although several prickly pear species grow wild today on the Canaries (*O. ficus-barbarica*, *O. dillenii*, *O. tomentosa*, *O. tuna*, *O. robusta*, and *O. vulgaris*) (Izquierdo *et al*. 2004) only two of them (*O. ficus-barbarica* and *O. dillenii*) are widely distributed. *O. dillenii* form part of the subdesert coastal scrub of the islands, whereas *O. ficus-barbarica* thrives in the disturbed zones at mid altitudes, where it forms dense, almost monospecific patches. Their succulent leaves were eaten by goats, providing them with a valuable water intake, whereas their fleshy fruits are not only eaten by humans, but also by endemic

Gallotia lizards (Valido and Nogales 1994) and several native birds, especially the raven (*Corvus corax*). Nogales *et al*. (1999) found *O. ficus-barbarica* to be the most common seed in the pellets regurgitated by ravens, and both *O. ficus-barbarica* and *O. dillenii* showed improved germination after passing through raven guts. Furthermore, *Sturnus vulgaris* has been observed feeding on the parasitic insect *Dactylopius coccus* growing on the prickly pear (Martín and Lorenzo 2001). *Opuntia* thus provides an illustration of how exotic plant species can compete successfully with native and endemic species, whilst in this case simultaneously contributing to the food resources of both native and exotic animal populations.

Besides *Opuntia*, other invasive plants competing successfully in otherwise well-preserved Canarian ecosystems include *Pennisetum setaceum*, *Nicotiana glauca*, and *Agave americana*, in the subdesert coast scrub; *Cytisus scoparius*, *Ulex europaeus*, *Tradescantia fluminensis*, *Ageratina adenophora*, *Ailanthus altissima*, *Zantedeschia aethiopica*, *Eschscholtzia californica* (Californian poppy) in the native laurel and pine forest ecosystem and several *Acacia* species at different altitudes. Altogether, alongside approximately 1300 native plant species in the Canaries (see Table 3.2), Izquierdo *et al*. (2004) list some 673 plant species as being introduced (or probably introduced), of which 79 are currently considered invasive. This amounts to 11.7% of the total introduced set, in line with Williamson's (1996) so-called 'tens rule' (Chapter 11).

Today, after several centuries of presence in the Canaries, prickly pears are considered part of the Canarian landscape and no attempt has been made to eradicate or control them. In contrast, a lot of money has been invested in ineffective efforts to control some other exotics, such as the grass *Pennisetum setaceum*.

some of the biggest tourist developments severely impacting on the most arid areas, previously only sparsely populated.

Contemporary problems in the Galápagos: a threatened evolutionary showcase

On the Galápagos islands, the main terrestrial conservation threats include introduced mammals (goats, pigs, cats, rats, dogs, cattle and donkeys), aggressive alien plants (e.g. *Lantana camara*, *Cedrela odorata*, *Cinchona succirubra*, *Rubus* sp., and *Psidium guajava*), habitat fragmentation, agricultural encroachment, over-exploitation and replacement of native woody species, and fires (Jackson 1995; Desender *et al.* 1999; Cruz *et al.* 2005). The possibility of the introduction of exotic disease, as has afflicted Hawaii's honeycreepers (Chapter 11), also constitutes a latent threat. The problems of the Galápagos are not solely land-based. Overfishing (e.g. for sea cucumber and lobster) and illegal fishing (especially for shark) by the artisanal fishing fleet in response to growing overseas markets (e.g. Merlen 1995) have led to the virtual collapse of the most lucrative legal fisheries and have resulted in violent and prolonged disputes between fishermen and conservationists.

As has happened in other island archipelagos, the growth of the tourist sector has brought mixed benefits to the Galápagos. Although many tourists would count themselves environmentalists, their interest in the islands has stimulated the local economy and boosted immigration. This population growth has, in turn, brought additional pressures to bear on the environment and biota, further fuelling conflicts between pro-sustainability and pro-development and extraction interests (Trillmich 1992; Davis *et al.* 1995).

It has been estimated that tourism in the Galápagos generates in the order of US $350 million per year but that over 90% of it is retained in the national economy rather than locally. The number of visitors to the islands has increased from about 41 000 in 1990 to about 110 000 in 2005, most of whom are non-Ecuadorians. Each foreign visitor pays an entrance fee of $100 (as of 2006), and the

additional revenue is generated through tours, boats, hotels, and other local services (Scott Henderson, personal communication). Given the totemic significance of the Galápagos in the development of evolutionary theory, it might be anticipated that there would be a matching conservation focus and effort. Indeed, the Galápagos National Park covers 97% of the archipelago and the island have also been designated a World Heritage Site, a Biosphere Reserve, and a priority in schemes such as the 2004 CI hotspots programme.

Research in the park is coordinated by the Charles Darwin Research Station and the Park Service. The management of the park aims to preserve the biological uniqueness and intactness of the islands and surrounding seas, through regulating resource use and research activities. A comprehensive zoning system is a key tool in both terrestrial and marine realms. Introduced animals remain the greatest problem, and although feral goats have been eliminated completely from a number of smaller islands—with notable recovery of vegetation—the destruction brought about by pigs, goats, and donkeys on the large island of Santiago has been so great that it is doubted whether a full recovery is now possible (Davis *et al.* 1995). Pigs eat plants, invertebrates, the eggs and hatchlings of endemic tortoises, lava lizards, and Galápagos petrels, amongst other species. As a result of one of the most ambitious conservation programmes, the last pig was eliminated from the large island of Santiago in 2000, at the end of a 30 year campaign. Santiago is now believed to be free of goats too (Campbell and Donlan 2005) and none has been spotted recently on Isabela, following a campaign in which over 100 000 were eradicated. Over the years, other ambitious conservation initiatives have been launched, the goals of which have included fencing against feral mammals, inventory of the worst alien plants, development of control and eradication methods, a programme to develop sustainable timber use, and an *ex situ* breeding programme for threatened plants to complement *in situ* efforts.

In May 1989, seven feral goats were observed on the Alcedo volcano on Isabela: the first confirmed report of goats in the habitat of the most intact race of Galápagos tortoise. By 1997, the goat population was close to 100 000, and the habitat of the tortoises and other fauna on Alcedo was reported to be collapsing. It is indicative of the problem facing many conservationists that appeals for donations to finance the goat control programme had to be issued, for example, in the journal *Conservation Biology*, as the Charles Darwin Foundation at that time lacked the core funding needed (Herrero 1997). One former director of the Charles Darwin Research Station, Chantal M. Blanton, noted with dismay that the national financial commitment to conservation research in the Galápagos slipped to zero in 1995, while tourist concessions doubled between 1992 and 1996 (*Galápagos Newsletter* 1996). When asked to name the most important problems facing the islands, she listed: first, introduced animals and the lack of a functioning quarantine system; secondly, inadequate political and financial support from government; and thirdly, the politicizing of the National Park Service and of posts requiring technical expertise. Although recent successes in reducing and in some cases eradicating goats and pigs from islands in the Galápagos archipelago are greatly encouraging—it seems that methods now exist for efficient eradication programmes (Box 12.2)—funding remains a key limiting factor in island goat eradications generally (Campbell and Donlan 2005).

The basic conservation problem of the Galápagos is the ever-accelerating breakdown of the islands' former isolation as a result of of dramatic changes in human population size, activity patterns, and mobility (Trillmich 1992). The real impact of the tourist business occurs not through the direct disturbance to animals by tourists armed with cameras but indirectly via the socio-economic changes on the four inhabited islands. The resident human population has grown rapidly due to the tourist and fishing industries, from about 10 000 in 1990 to 28 000 by 2005. This growth has come largely from immigration from mainland Ecuador, with migrants generating increased pressure on infrastructure and an ongoing influx of alien species. Indeed, in a 5 year period, about 100 new plant species were detected,

and although this sudden rise reflects new efforts by botanists to record non-native and invasive species (Tye 2006), there is no doubt that the increased flow of people from the mainland has contributed greatly to the non-native flora, as people have brought in ornamental plants in the urban areas, and have tested new agricultural plants in the upland areas, from where several have already spread to become ecologically and economically expensive invasives (Scott Henderson, personal communication).

The growth of the human population has meant more pets. A survey in the main settlement on Santa Cruz established that the dog population was about 300 at the start of the 1990s, with unwanted pups being released to become feral. Feral dogs prey on tortoise hatchlings and eggs, and in the late 1970s wild dogs decimated a large colony of land iguanas on Santa Cruz, killing over 500 animals (Jackson 1995). Remnants of endemic birds, reptiles, and even insects in cat faeces bear witness to the impacts these animals have in both feral and domesticated conditions. Although the National Parks Service is working to keep the feral population low, it will be difficult to reduce the number of domestic dogs as long as robberies in the villages continue to increase, as this creates an incentive for settlers to keep dogs (Trillmich 1992). An intensive community-based programme is underway to increase owner responsibility through education programmes, designed to demonstrate the long-term benefits of maintaining the archipelago's unique biological diversity and heritage (Scott Henderson, personal communication). The case articulated by numerous conservationists over more than 25 years for a quarantine system is overwhelming. But, in common with feral animal control and other conservation measures, this will only ultimately be successful with sustained political and societal support.

12.2 Some conservation responses

Adsersen (1995) notes the prevalence of the same invasive plants in archipelagos as diverse and remote from one another as the Mascarenes, the

Box 12.2 The arrival and eventual removal of ecosystem transformers: feral goats and Judas goats

Goats in the Canaries
Introduced from North Africa in the first millennium BC by the Guanches, goats (*Capra hircus*) enjoyed a feral status on all of the Canary Islands until just a few decades ago. Today herds of semi-wild goats roam freely only on Fuerteventura. With the exception of the now extinct giant tortoises (*Geochelone burchardii* and *G. tamaranae*), the Canarian flora evolved without large grazers for several million years, and thus generally lacked the defensive spines and toxic compounds typical of continental flora. The long history of goat grazing has thus undoubtedly been hugely significant in transforming the vegetation cover of the islands, and most likely in reducing native endemic species to extinction. It is difficult to be certain how many species populations were lost in the absence of fossil data, but it is clear that goats and other introduced herbivores have been directly responsible for reducing some persisting endemics to the very brink of final extinction (e.g. Marrero-Gómez *et al.* 2003), just as also occurred for example on St Helena (Cronk 1989).

Following the declaration of the summit region of Tenerife as a National Park in 1954, goats were removed from the uplands. After the shift of the economic development model from agriculture to mass tourism during the 1960s, the goats slowly began to disappear from the rest of the Canarian landscape. Finally, in 1980, the last wild goats were successfully extirpated from the uninhabited 10 km^2 island of Alegranza, marking the end (except on Fuerteventura) of an influence that lasted for more than two millennia. In response to the removal of the goats, in those areas escaping development, a trend of vegetation recovery can be seen. In the extreme (arid) climates of the high uplands this has not been a particularly rapid process, but nonetheless, several rare plant species have since greatly increased in abundance (e.g. the endemic *Pterocephalus lasiospermus*). Conservation efforts have focused on assessing the key life-history phases and autoecological (e.g. germination) requirements of highly threatened plant species. Breeding programmes

have been established and efforts are being made to reintroduce populations into the wild. This approach requires a good and sustained level of funding, alongside the availability of suitably trained scientists, appropriate infrastructure, and finely tuned management plans for protected areas (e.g. see Marrero-Gómez *et al.* 2003).

Removal of feral mammals
Goats have been introduced on numerous islands around the world, sometimes to provide a food resource on uninhabited islands for visiting sailors, other times by settlers. They are highly effective ecosystem transformers, and present huge conservation problems in many islands. However, they have now been eradicated successfully from over 120 islands across the world (Campbell and Donlan 2005). Whereas in the past, such campaigns typically took many years to complete and often struggled to eradicate the population altogether, modern technology has now been harnessed to enable swift and effective eradication of populations of goats numbering many thousands.

An illustration is the use of so-called Judas goats. Animals are trapped and radio-collared before being re-released, upon which (being social animals) they typically find and remain with other goats. They are then tracked and the other animals accompanying the Judas goat are shot with high-powered rifles. The Judas goat is then released again, and the process repeated. This approach allows the last members of the feral population to be tracked down and eliminated. Similarly, feral pigs, which were causing huge ecological impacts on Santiago island in the Galápagos, were finally eliminated from the island at the end of a 30 year campaign, during which some 18 000 animals were killed (Cruz *et al.* 2005), with many lessons learnt towards increasing eradication efficiency in future efforts.

Although some environmentalists raise principled objections to such culling programmes, there is no doubting their central importance in protecting the native biota of many oceanic islands (Desender *et al.* 1999, Campbell and Donlan 2005, Cruz *et al.* 2005).

Galápagos, the Canaries, and the Bahamas. Similarly, many other problems, such as feral animals, habitat loss, and burning, are common to numerous islands (e.g. Cronk 1989; Vitousek *et al.* 1995; Rodríguez-Estrella *et al.* 1996). Thus, although solutions have to be tailored to the particular circumstances of each island system, field managers and conservationists often face the same problems. There are thus still many benefits to be gained simply from improved information exchange between those involved in conservation management on islands (Adsersen 1995, Campbell and Donlan 2005).

The large catalogue of extinctions of native island species is bound to increase. What can be done to stem the tide? If we cannot save the remaining evolutionary wonders of the showcase islands such as Galápagos and Hawaii—and the signs are not hugely encouraging—then the prospect for many other less totemic islands and island endemics must be dismal indeed. Yet, it is not a hopeless task. Species can be saved by means of rigorous habitat protection, pest or predator control, and translocation of endangered island species (Franklin and Steadman 1991, Marrero-Gómez *et al.* 2003), given adequate political and financial support.

Biological control—a dangerous weapon?

Biological control is the term given to pest control not by means of chemical agents but by the introduction of a species that targets the pest organism without seriously affecting non-target species. Cronk and Fuller (1995) state that it is the long-term goal of conservation managers in Hawaii to achieve the biological control of most of the introduced weeds, but as yet only 21 are controlled biologically. Although in many cases biological control is the only practical approach, there are associated risks (above; Williamson 1996). A classic example in earlier times was the introduction of the mongoose as a pest-control agent. The mongoose has shown limited success in controlling rats but has been extremely effective in devastating many native island bird and reptile species, particularly ground-foraging skinks and snakes (Case *et al.* 1992). So, it is vital that biological control programmes begin

with rigorous screening and testing to ensure that the control agent's effects are highly specific to the target species (e.g. Causton 2005).

Translocation and release programmes

Translocation and repatriation from *ex situ* breeding and rearing programmes have been used as measures to rescue highly threatened species on islands, with some success. Translocation is used to remove a species from an overwhelming local threat or to re-establish a population on an offshore island where it may be safe from exotic predators. In general, translocated wild animals have been found to establish more successfully than captive-bred animals. Griffith *et al.* (1989) surveyed translocation programmes undertaken in Australia, Canada, Hawaii, mainland USA, and New Zealand over the period 1973–86. They found that only about 46% of release programmes of threatened/endangered species were successful, a far lower percentage than for translocations of native game species, for which the success rate was 86%. They attempted to identify the factors influencing success for 198 bird and mammal translocations from within their survey. The single most important factor was the number of animals released. They also found that herbivores were significantly more likely to be successfully translocated than carnivores or omnivores, that it was better to put an animal back into the centre of its historical range than the periphery, and that wild-caught animals fared better than captive-reared animals.

In New Zealand, a number of threatened bird species have been translocated into small offshore islands that either lacked or have been cleared of introduced predators, such as the ship rat (*Rattus rattus*), which are prevalent on the mainland (Lovegrove 1996). By this means the extinction of several species, including the South Island saddleback (*Philesturnus carunculatus*), have been averted. The kakapo (*Strigops habroptilus*), a flightless, nocturnal parrot, was once considered to be effectively extinct, until the discovery of a small population on Stewart Island (Clout and Craig 1995). They continued to decline following rediscovery, to fewer than 50 individuals, as a result principally of predation by feral cats. All known kakapo (i.e. the entire species) were therefore transferred to three predator-free island refuges, where, supported by supplementary feeding, the kakapo appeared to have better prospects. However, recovery was reported to have been initially disappointing due to the bias of survivors to older birds, a delay in resuming breeding after translocation, and periodic invasions of stoats on one of the islands.

Assessing the success of 45 release programmes, Lovegrove (1996) concluded that failures occurred where either (1) predators were still present, or arrived after release, or (2) too few birds, with an unbalanced sex ratio, were released. All releases involving more than 15 birds on islands lacking predators were successful, at least in the short term. Other New Zealand studies reportedly show that releases of as few as 5 birds on predator-free islands are likely to succeed, and, more generally, the review of Griffith et al. (1989) indicates that releases of 40 or more birds into good habitat are generally successful.

Clearly, the elimination of exotic predatory mammals is essential. Moreover, as on occasion even individual animals can deal a heavy blow to small populations of threatened species, it may be necessary to monitor islands closely to ensure that predators do not return. Much effort has been channelled into developing effective methods of predator control. Conservationists have been developing a wide range of techniques, often at considerable expense, including the introduction of new rodenticides, and the spreading of bait from helicopters, in order to target particular pest species to greatest effect (e.g. Clout and Craig 1995; Micol and Jouventin 1995). Such schemes are often unaffordable (Herrero 1997).

It is obvious that it is important to translocate populations into suitable habitat, but what this may be is not so obvious, as for many species populations the reasons for their original demise are not fully understood (e.g. Hambler 1994). Given the role of humans in the demise of island species, it is important that translocations take place into areas in which harmful human influences are appropriately managed, or mitigated. Because the prehistoric human impact was so intense in Polynesia,

Franklin and Steadman (1991) argue that the lack of many taxa from uninhabited, forested islands may be due to events that occurred centuries ago, and that with human abandonment and forest regeneration, such islands may once again be able to support the extirpated species. They therefore advocate the translocation of species into such islands, following careful assessment of the resources and problems involved. From their preliminary assessments they argue that low makatea islands, such as the Cooks, hold limited potential, but that there may be greater potential among the 20 or so Tongan islands that are both uninhabited and greater than 2 km² in area. Moreover, the fossil record demonstrates the former presence on one such island, 'Eua, of species that still survive in Fiji, Samoa, and other Tongan islands, including the parrots *Vini australis* and *Phigys solitarius*. Establishing multiple populations will greatly improve the prospects of a species surviving events such as the loss of a local population in a hurricane. Such initiatives require a sound biogeographical and ecological baseline, together with the political and legal frameworks to support the efforts.

The only wild giant tortoises in the Indian Ocean, *Geochelone gigantea*, survive on Aldabra atoll. Here, the culling of goats has enabled their population to persist in reasonable numbers. Between 1978 and 1982, 250 tortoises were translocated to Curieuse in the granitic Seychelles (Hambler 1994), but in a detailed survey in 1990 it was found that only 117 animals remained. Poaching probably accounted for most of the losses, although some of the surviving adults were diseased and low growth and reproductive rates suggested possible saturation of resources and nutrient limitation (particularly calcium deficiencies). Both feral cats and feral rats are abundant on the island, and probably account for the very low rate of recruitment. To ensure the persistence of the new colony, it became necessary to establish a facility to rear hatchlings until they are over 5 years old and thus relatively safe from predation (if not from poaching). In contrast, a translocation to another of the granitic Seychelles, Frégate, has been comparatively successful. The reasons for the difference in outcome on the two islands are as yet uncertain, but it is likely that both habitat differences and differences in predation and poaching are involved. This demonstrates that the success of translocation is not simply a function of initial size of the translocated population. Eradication of the feral predators, and reduction in poaching might be the answer on Curieuse, but it is also possible that the island is basically unsuitable for the long-term persistence of a viable population.

Programmes based on *ex situ* breeding and translocation back into the wild thus have their place in the conservation of island endemics. However, they have to be undertaken as part of a highly organized, managed programme. The haphazard removal of animals from the wild supposedly for captive breeding in zoos may otherwise be counter-productive. Christian (1993) notes that although several parrots have been taken out of St Vincent and the Grenadines for captive breeding, there have been only three known successes, and to date there is no evidence of any captive-bred parrots being returned to St Vincent for re-release. Thus, not only is it necessary to 'fix' the local environmental problems, but it is also necessary to control the *ex situ* part of the process.

Protected area and species protection systems: the Canarian example

The combination of high biodiversity value and severe pressure on natural environments and biota within the Canaries has been recognized by governmental and non-governmental organizations at all levels. For example, Birdlife International's Endemic Bird Areas scheme rates the islands a 'high priority', and Conservation International has recently included the islands within an expanded Mediterranean hotspot in its 2005 revision of the 'hotspots' scheme (CI 2005; see Chapter 3). Under UNESCO designation systems, the Canaries contain four Biosphere Reserves (the islands of Lanzarote, La Palma, El Hierro, and the south-western part of Gran Canaria), a World Heritage site (Garajonay National Park, in La Gomera), and one Ramsar Convention wetland (El Matorral, Fuerteventura). The archipelago also has four National Parks (IUCN management category II): Timanfaya in Lanzarote, Teide in Tenerife, Garajonay in La

Gomera, and Taburiente in La Palma, out of a Spanish network of just 14 National Parks, although the archipelago constitutes just 1.5% of Spain's land area. There are also two important conservation networks: the Canarian ENP network (Red Canaria de Espacios Naturales Protegidos) and the European Union Natura 2000 network of sites. Today, the archipelago has about 150 protected areas (Table 12.2), covering 45% of the Canarian land area. These are impressive figures, especially for such a densely populated and heavily visited archipelago.

The Canarian ENP system of protected area designation (Table 12.2) is a slightly modified implementation of the IUCN protected areas model. It includes both strictly protected sites (e.g. 'integral natural reserves' and 'natural monuments'), and 'rural parks' in which conservation and sustainable development go hand in hand (Martín-Esquivel *et al.* 1995). In the latter, the emphasis is on sustaining the rural populations and their culture while improving livelihoods and standards of living. Thus, the goals of species and habitat conservation within these parks are to be achieved within 'cultural landscapes' of traditional resources use.

Although the Canarian National Parks have hitherto been co-managed by the state and regional administrations, the rest of the protected area system has been managed by each island government. In total, the Canarian ENP network provides some measure of protected area designation for about 40% of the land area of the archipelago, with percentages for each island ranging from about 30% for Fuerteventura up to 60% for El Hierro. Together, the large islands of Gran Canaria and Tenerife (Fig. 12.1) contribute more than the half of the whole protected area in the archipelago. This Canarian network does not include protected marine areas.

The recent establishment of the European Union's Natura 2000 network on the Canaries has been used by the regional administration as an opportunity to complement the Canarian ENP network in those terrestrial habitats not previously well-represented (mainly thermophyllous woodlands). Otherwise, almost all the terrestrial Natura

Table 12.2 Designations within the Canarian Protected Areas system (sources: Martín-Esquivel *et al.* 1995; Fernández-Palacios *et al.* 2004*a*).

Name	Characteristics	No.
National Park (Spanish and Canarian networks)	Large areas relatively untransformed by human activity, with high importance due to the singularity of their biota, geology or geomorphology and representing the main natural Spanish ecosystems	4
Natural Park (Canarian network)	Large areas with similar characteristics as the National Parks, being representatives of the Canarian natural heritage	12
Rural Park (Canarian network)	Large areas where agricultural and livestock activities coexist with zones of a great natural and ecological interest	7
Integral Natural Reserve (Canarian network)	Small natural areas protecting populations, communities, ecosystems or geological elements deserving a special value for their rareness or fragility. Only scientific activities allowed	11
Special Natural Reserve (Canarian network)	Similar to the former, but allowing educational and recreational activities together with scientific ones	15
Site of Scientific Interest (Canarian network)	Small isolated sites comprising populations of threatened species	21
Natural Monument (Canarian network)	Small areas characterized by unusual geological or palaeontological elements	52
Protected Landscape (Canarian network)	Areas with outstanding aesthetic or cultural values	27
Special Area of Conservation (EU Natura 2000 network)	Areas which contribute in a valuable manner in maintaining or restoring natural habitat types or species in a favourable conservation status. The 151 terrestrial SACs largely overlap with the Canarian network of protected areas, adding 300 km² more, but the 23 marine SACs contributed an additional 1765 km² of protected marine ecosystems	174
Special Protected Area (EU Natura 2000 network)	Areas which contribute to preserve, maintain or restore the diversity and extension of the proper habitats for the 44 Canarian bird species included in the European directives	27
Biosphere Reserves (UNESCO network)	Areas protecting spaces where human activity constitutes an integral component of the territory, and where the management should focus on the sustainable development of the resources	4
World Heritage Site (UNESCO network)	Outstanding natural areas with unique features on a worldwide scale	1
Ramsar Convention Protected Wetlands (UNESCO network)	Wetlands offering important ecological services, such as the regulation of water regimes as well as important sources of biodiversity	1

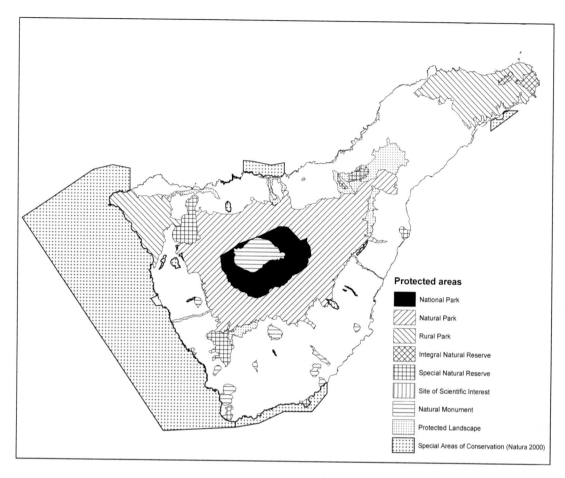

Figure 12.1 The Canarian ENP network (Red Canaria de Espacios Naturales Protegidos) and the Natura 2000 networks of Protected Areas on and around Tenerife. Some zones correspond to two different protection categories, for example, the highest peak of the island (Teide) is protected as a Natural Monument within a National Park, similarly, some zones of the Anaga massif (in the northeast corner of the island) are protected as Natural Reserves within a Rural Park. Also, the Natura 2000 terrestrial sites largely overlap with the ENP network. The Natura 2000 marine protected areas can be easily distinguished from the coastline of the island by the simplicity of their form. See Table 12.2 for explanation of the categories. (Map drawn by Ángel Vera, based on a figure in Martín-Esquivel et al. 1995.)

2000 protected areas were already covered by the Canarian ENP network. The additional land designated through this network is significant only in Gran Canaria, La Gomera, and La Palma, together contributing a little over 300 km² land area. This figure has increased the protected land area within the archipelago from 40.4 to 44.5% (Table 12.3). However, its main additional benefit is in incorpo-

rating 23 marine areas of high natural value, totalling 1765 km², into the protected area estate.

In addition to these 'area-based' schemes, there are two different protected species catalogues, one designated at national (Spanish state) level, seeking to protect 173 Canarian species and the other at the Canarian (autonomous region) level, seeking to protect 450 endemic species or Canarian populations

Table 12.3 Protected terrestrial areas designated by the Canarian and European Union Natura 2000 networks on the Canary Islands (sources: Martín Esquivel *et al.* 1995, Fernández-Palacios *et al.* 2004*a*).

Item	Lanzarote	Fuerteventura	Gran Canaria	Tenerife	La Palma	La Gomera	El Hierro	Canaries
Island area (km²)	846	1660	1560	2034	708	370	269	**7447**
Number of Canarian network protected areas	13	13	32	43	20	17	7	**145**
Canarian network protected area (km²)	350	477	666	989	250	123	156	**3011**
Additional protected area contributed by the Natura 2000 network (km²)	2.9	1.5	103	10.7	116	63.4	1.3	**298.8**
Island area protected (km²)	352.9	478.5	769	999.7	366	186.4	157.3	**3309.8**
Island area protected (%)	41.7	28.8	49.3	49.1	51.7	50.4	58.5	**44.5**
Island contribution to Canarian protected area system (%)	10.7	14.5	23.2	30.2	11.1	5.6	4.8	**100**

of non-endemic species with a high emblematic value or global conservation, like whales, dolphins, or birds.

Recently, a governmental proposal for the withdrawal of about 56% of the species included in the Canarian catalogue has been published (Martín-Esquivel *et al.* 2005). This action was based on a 4 year monitoring and assessment project of the conservation status of species listed in the Catalogue. The authors concluded that in about 200 cases, while the species concerned were restricted to only a few populations, these populations were healthy and stable. In addition, some 40 species previously listed as endangered were removed from that list on the basis of changed assessments of whether they were endemic, taxonomic clarification of status, etc.

Although some species populations may be healthier than was once thought, and intensive conservation efforts may be pulling some endangered species back from the brink of extinction, many others remain acutely at risk (e.g. Marrero-Gómez *et al.* 2003). The political imperatives within the archipelago remain focused on short-term economic interests and a model of increased tourism, development, and urbanization. This economic model casts a dense shadow of uncertainty over the future of the natural resource base and biodiversity of the archipelago, particularly of the warm, dry climate belt so popular with European tourists. Despite the attempts to provide legal protection for biodiversity and to invest in environmental conservation at varying political levels, the pressures on the natural resources of the Canaries continue to increase (García-Falcón and Medina-Muñoz 1999). These pressures include efforts to reduce the protection afforded to particular protected areas that have potential commercial value for development. Nonetheless, the various networks of protected area status and the other conservation measures discussed in this section show how it is possible to tailor protected area models to an insular context: without these legal instruments, the future of many Canarian endemic species and ecosystems would indeed be bleak.

12.3 Sustainable development on islands: constraints and remedies

The elements involved in the many case studies discussed in this and the previous chapter are repeated in varying combinations and with varying emphasis for numerous other islands for which conservationists have made management assess-ments. The remedies identified are often the same: control exotics, halt or reverse habitat loss, prevent hunting, assist breeding of endangered species, enlist political and legislative support, educate and enthuse the people to improve compliance and cooperation in meeting management goals. It is important, finally, to consider whether there are any special considerations on the human side of the equation which must be included in conservation thinking.

In many respects, oceanic islands are peculiar environments, both biotically (high proportion of endemics, disharmonic, etc.) and abiotically (expe-riencing wave action from all sides, tending to have small hydrological catchments, peculiar geologies, etc.) (Nunn 2004). Insular peoples have commonly evolved distinctive cultures and retain a strong allegiance to both home and culture (Beller *et al.* 1990). It may be worth noting also that insularity may provide essentially ecological problems for humans. Examples include the extermination of entire cultures, such as the Caribs of the Caribbean, killed off by disease introduced by Europeans; the introduction of malaria into Mauritius in the 1860s, which led to the death of 20% of the capital's popu-lation; and a decline of perhaps 50% in the popula-tion of Hawaii in the 25 years following the arrival in 1778 of Europeans, who brought diseases such as syphilis, tuberculosis, and influenza (Cuddihy and Stone 1990). Less directly, an epidemic of swine fever hit São Tomé and Principe in 1978, wiping out their entire pig population, and causing an 11-fold increase in meat imports in 3 years. Likewise, one of the most precious resources with respect to attracting outside finance is the possession of eco-logical interest, to attract international support, or tourist dollars. So, in many respects the problems of the plants and animals of islands are faced also by people on islands.

Yet, as islands include a vast array of climatic, geographic, economic, social, political, and cultural conditions, it is extremely difficult to offer meaning-ful generalizations as to the problems facing island peoples and how they influence and constrain the options from a conservation management perspec-tive. One approach might be to focus on those islands holding most threatened species. For

instance, over 90% of threatened island species are endemic to single geopolitical units and just 11 such units are home to over half the threatened island birds (Groombridge 1992, p. 245). However, even this does not allow much in the way of generaliza-tion, as the 11 units concerned are Cuba, Hawaii, Indonesia, the Marquesas, Mauritius, New Zealand, Papua New Guinea, the Philippines, São Tomé and Príncipe, the Seychelles, and the Solomons. This is a pretty mixed bag of islands and societies.

For the human societies and economies of smaller islands (up to about 10 000 km^2) the prob-lems include the following (Beller *et al.* 1990):

- **Lack of water.** This provides a particular problem for very small, low, narrow atolls, where salinity conditions the entire land area, but can also affect substantial, high islands such as Tenerife, where the tourist and horticultural industries have been sus-tained only on the back of the unsustainable mining of geological reserves of water (see e.g. Ecker 1976).

- **Susceptibility to rising sea level.** For example, the Bimini islanders have formally requested advice from the UN on how to plan for their entire land area being below the projected sea level according to greenhouse-warming predictions. It is not simply the loss of area that may be critical, but for small states like the Bahamas, the potential loss of much of their low-lying reservoirs of fresh water (Stoddart and Walsh 1992) and/or of their EEZs.

- **Vulnerability to storm or tectonic damage.** Hurricanes and cyclones can badly damage the economy of a small island (where population and key elements of infrastructure are often concen-trated in just a few coastal localities), forcing the human populations to exploit biological resources themselves damaged by such events. Certain types of islands are particularly prone to volcanic and tec-tonic hazards and to related phenomena such as tsunamis (Chapter 2). Such phenomena can deal a savage blow to the sustainable human use of an island, as evident from the devastation inflicted on the island of Montserrat by eruptions in 1997.

- **Climatic variation.** The agricultural base of a small island may be hard hit by atypical weather conditions. The El Niño event of 1983 caused immense disruption to marine and terrestrial habitats, both proximal and distal to the tropics where the unusually high water temperatures were recorded. Central Pacific seabird communities had a dramatic reproduction failure and briefly abandoned atolls such as Christmas Island. Island life is clearly vulnerable to short-term weather events as well as long-term climatic change (Stoddart and Walsh 1992). Because of the limited options provided by the discrete area of a small island, agricultural/fishing economies are similarly constrained.

- **Narrow agricultural base.** The constraints of the often narrow agricultural base of island societies is clearly demonstrated by the history of Barbados, which has been dominated until modern times by the boom-and-bust fortunes of the sugar industry since colonization of the islands by Europeans in 1492 (Watts 1987). When sugar was first established as the export crop, the island's ecology was devastated in just 20 years, at the end of which time 80% of the land area was cultivated for sugar cane. Many indigenous species were lost in this first, dramatic land-use conversion. Subsequently, changes in the global market for sugar, and occasional pest problems, have resulted in very significant fluctuations in the area of cane, and in the security and wealth of the island. The narrow environmental range provided by such small size, and the small land area, means that it is difficult for the local economy to diversify, and hard to produce exportable quantities of more than a limited range of agricultural crops. The boom-and-bust cycle of periodic export booms, subsequent deflation and resource exhaustion, and chronic emigration appears to be a feature common to many small island histories (Beller *et al.* 1990).

- **Lack of suitable land for agriculture.** Very much tied in to the previous point, is that there may be few crops suitable for the island soils and climate. The Bahamian island of Andros, for instance, is characterized by a viciously dissected coral rag substrate. Mechanized agriculture is not feasible, and the land essentially does not provide a living.

- **Heavy dependence on the seas.** Dependence on the seas is all very well when the fish are plentiful.

Problems arise not only from anomalies in climate or ocean currents, but also from the activities of larger, more efficient foreign fishing fleets, over which local peoples may have limited sway (cf. Merlen 1995).

- **Small or fragmented states.** Tied in to all of the foregoing is the political geography of small or fragmented states. In the case of countries such as the Bahamas, or Jamaica, with a shallow natural resource base, a fluctuation in the environment (e.g. a hurricane) can deal a significant blow to the resources of the state.

In his review of resource development on Pacific islands, Hamnett (1990) identifies five major trends within the region: (1) increasing pressure on the land and freshwater resources; (2) intensification of agriculture; (3) loss of native forest resources; (4) river and stream siltation and the loss of aquatic resources; and (5) increasing use of coastal areas and the degradation of harbours and lagoon environments. We might wish to add others to this list, notably the increasing interest from tourists, seeking an idyllic desert island on which to holiday, but expecting the facilities of a modern society—the hotels, roads, fresh water, and other amenities—as well, perhaps, as the exotic wildlife. This sixth trend is powerfully evident in the Caribbean, which, in the period after the Second World War, has witnessed rapid urbanization, a significant increase in tourism, and marine degradation from the construction of transport (air and sea) and petroleum facilities and hotels (Beller *et al.* 1990). Many of the islands of the Mediterranean, and certain Atlantic islands, notably the Canaries, have experienced similar patterns of expansion in the tourist industries. Tourist pressures are also extremely significant on Hawaii and the Galápagos (Cuddihy and Stone 1990; above) and indeed on other tropical and subtropical islands. This is just one way in which, in common with the problems facing native plants and animals, exogenously driven forces (social, economic, and technological) are affecting and disrupting many island cultures, and eroding traditional skills.

The task of confronting these pressures is not a simple one. Some of the themes involved in sustainable development planning for small islands are listed in Table 12.4. In many ways the ecological management goals are the easiest part of the equation to specify, while managing the societal, economic, and political issues is very difficult (Beller *et al.* 1990). As an illustration, concepts of ownership among island nations can be both extremely varied and complex. In Papua New Guinea, the Solomon Islands, and Vanuatu, ownership is defined by oral traditions, land is typically owned by family groups rather than individuals, and there may be secondary and tertiary levels of 'ownership' providing rights of use but not implying actual ownership of the land (Keast and Miller 1996). Success in such circumstances will depend on how well insular resource planning and management can provide solutions tailored to local circumstances and cultures (Nunn 2004), and whether they carry with them the support of the island communities. The involvement of major users, farmers, fishermen, charter operators, etc., is one element in this. The mobilization of non-governmental organizations (NGOs) in island problems is another element, whereby energies and expertise exist that could be harnessed provided interest can be raised. Perhaps above all, it is necessary to foster a general public awareness of the environmental and ecological constraints and problems, and the significance to the island societies themselves of the ecological 'goods' at stake (Christian 1993).

The importance of public awareness and support can be illustrated by reference to parrot conservation in the West Indies. Since the arrival of Columbus at the end of the fifteenth century, 14 species and two genera of parrots have been lost. The remaining nine endemic members of the genus *Amazona* are each confined to individual islands. The case of the Puerto Rican parrot, *Amazona vittata*, and its conservation was outlined in Chapter 10. Within the Lesser Antilles, seven endemic parrots have become extinct in historical times, with just four species remaining. They are now threatened by deforestation, predation, illegal hunting and collecting, competition with exotics, and natural disasters (Christian 1993; Christian *et al.* 1996). Impacts on the parrots resulting from habitat alteration include the loss of nesting cavities, food, and shelter. Given these pressures, management intervention is essential for their survival.

Measures for parrot conservation in the Lesser Antilles consist of environmental education, habitat protection, enforcement of appropriate legislation, and enhancement of wild breeding and captive breeding programmes. Expensive programmes, such as those adopted for the conservation of the Puerto Rican parrot, are not feasible in the Lesser Antilles because of shortage of funds. Despite this, effective targeting of conservation resources has delivered some degree of success. Nonetheless, more needs to be done; first, to protect habitats;

Table 12.4 Sustainable development options for small islands should aim at increasing self-reliance, and can be categorized under six basic headings (after Hess 1990).

Categories	Examples
Resource preservation	Conservation zones, multiple-use options, control of hunting
Resource restoration	Replanting, re-introduction, alien herbivore removal
Resource enhancement	Freshwater re-use, seawater desalination
Sustainable resource development	Small-scale diversified, closely managed forms of resource-based enterprises (agriculture, fisheries, tourism)
Provision of human services	Alternative energy generation sources and distribution systems, waste disposal
Non-resource-dependent development options	Financial services (tax havens), light industries processing imported materials, rental of fishing rights

secondly, to discontinue the export (legal or illegal) of live parrots; and thirdly, to involve local people as much as possible in both the rationale of and the employ of the management programmes.

Christian (1993) has described the efforts made by the government of St Vincent and the Grenadines to ensure the long-term survival of the Vincentian parrot, *Amazona guildingii*, which has been reduced to about 450 birds. Aided by financial support from bodies such as the World Wildlife Fund, the RARE Centre for Tropical Bird Conservation, and notably the local St Vincent Brewery, the Forestry Department has been able to bring the conservation message to the local population and enlist their support. A significant amount of the island's limited resources have been channelled into activities and programmes that support and complement parrot conservation. Thus this small island nation has already made a big contribution and shown its commitment to biodiversity conservation. Christian stresses the importance of continuing to demonstrate the links between habitat protection for the parrot, and the opportunities for ecotourism, extraction of minor forest products, and soil and watershed protection. He cautions that failure to demonstrate such direct linkages would risk the loss of public cooperation and support so vital for conservation.

This final section has served merely to introduce a few of the topics of sustainable development of islands and to present, briefly, the case that islands have many problems different in kind from those of larger land masses (for a fuller account see Beller *et al.* 1990). There is in this a parallel with the underlying ecological or nature conservation problems of oceanic islands. In short, small or remote oceanic islands are ecologically special: they constitute evolutionary treasure-houses, threatened by contact with other regions and especially the continents. The nature of the ecological problems involved is now well understood and the pattern of continuing degradation and species loss is all too obvious and, in the general sense, predictable. Dealing with these problems requires a sensitivity to, and respect for, the island condition of the human societies that occupy them (Nunn 2004). There are remedies, but they require solutions tailored to the islands and not simply exported from the continents.

12.4 Summary

Island societies face many problems, some common to those of societies everywhere, and others that are specific to the island context. We illustrate a few of these special problems by reference to the implications of the loss of very small islands to rights of access to marine resources, and by reference to the potential loss of entire island societies through (projected) future sea-level rise.

In an increasingly globalized world, the effective isolation of many oceanic islands has been broken in the last few hundred years and especially in the most recent decades. The devastation of Nauru through mining of phosphate-rich guano deposits, and the huge socio-economic changes (and consequent environmental pressures) brought by mass tourism within the Canaries, provide evidence of these processes, which can be particularly difficult for small island societies and governments (like those of Nauru) to cope with. The societal dimension of these processes and phenomena is a crucial part of any serious effort to plan for biodiversity conservation in island settings.

In considering some of the conservation responses that have been attempted on remote islands, we briefly review the use of biological control, translocation, captive breeding and re-release programmes, and of 'high-tech' approaches to the removal of feral animals from islands: just a small selection of the wildlife conservation tools that may be crucial to conservation efforts. Integrated programmes involving the re-insularization of populations, for instance by transferring them to small offshore islands from which predators have been eradicated, offer some hope for the persistence of particular threatened island endemics. We also illustrate through the Canarian example, how for larger oceanic islands, it is possible—and arguably crucial—to adapt the legal instruments of protected area planning and species protection, for application in an insular context.

Many practical problems continue to face island conservationists, and they are illustrated here by reference to the Galápagos, wherein the political, financial, and socio-economic problems are argued to be the key drivers of conservation problems. The strategy required for effective conservation of such island systems must be tailored to meet the particular features of island environments and their human societies, and the particular factors that threaten their environments and their biodiversity (sea-level rise, exotic predators, exotic grazers, mining, over-fishing, over-hunting, habitat degradation, etc.). To be successful, conservation actions must ultimately be part of broader management for the sustainable development of island ecosystems. It is stressed that it is crucial to enlist political and legislative support, and to build educational programmes to engender the interest and support of island peoples.

Glossary

actinomorphic Having flowers with radial symmetry.

adaptive radiation The evolutionary process by which an ancestral species gives rise to an array of descendant species exhibiting great ecological, morphological or behavioural diversity.

alien species A foreign (exotic, or non-native) species moved by humans to a region outside its natural geographical range or environment.

alleles Two or more forms of a gene occupying the same locus on a chromosome.

allelopathy A form of interference competition by means of compounds produced by one species of plant that reduce the germination, establishment, growth, survival, or fecundity of another plant species.

allochthonous Having originated outside the area where it now occurs.

allogamy Fertilization involving pollen and ovules from: (1) different flowers (whether on the same plant or not); (2) genetically distinct individuals of the same species.

allopatric speciation The differentiation of geographically isolated populations into distinct species.

allozyme One of several forms of an enzyme coded for by different alleles at a locus.

anacladogenesis Evolutionary change within a lineage where the progenitor species survives with little change alongside the derived species.

anagenesis Evolutionary change within a lineage where the progenitor species becomes extinct.

anemochory Dispersal by wind.

anemophily Pollination by wind.

area cladogram A cladogram in which the terminal taxa have been replaced by their respective geographical distributions.

assemblage The set of plant or animal species found in a specific geographical unit or area.

assembly rules Identified structure in the composition of a series of islands (or other operational units), typically determined within ecological guilds. Assembly rules include: (1) compatibility rules, (2) incidence rules, and (3) combination rules.

assortative equilibrium In the context of the equilibrium theory of island biogeography, the species number having initially equilibrated, may increase over time to a new equilibrium value reflecting the ecological interactions within the biota.

atoll Coral reefs built around subsiding volcanoes.

auto-compatibility Plant species that can self-fertilize (also termed simply 'compatible').

autogamy Self-fertilization.

auto-succession The assemblage of a community with no obvious environmental change or transition.

bauplan A particular body architectural plan.

benthic Living at, in, or associated with, structures on the bottom of a body of water.

biodiversity The variability of life from all sources, including within species, between species, and of ecosystems.

biodiversity hotspot May mean a geographic area with high levels of species richness and/or endemism. Also used to refer to areas that combine high biodiversity with high threat to that biodiversity (e.g. due to habitat loss).

biogeographical regions, kingdoms, and provinces Major biogeographical subdivisions of the world based largely on shared composition at the family level

biomass The total weight of living tissue of organisms accumulated over time, usually expressed as dry weight of organic matter per unit area.

biosphere reserve   Areas of terrestrial or coastal ecosystems that are internationally recognized within the framework of the UNESCO's Man and the Biosphere Programme (MAB).

biota   All species of plants, animals, and microbes inhabiting a specified region.

bottleneck   A severe reduction in population number that is often associated with reduced genetic diversity and heterozygosity and reduced adaptability of that population.

caldera   A large crater (typically 5–20 km diameter) formed by the collapse of a volcano following withdrawal of magma from an underlying storage chamber.

Cenozoic   An era of geological time comprising the Tertiary and the Quaternary periods, i.e. the last 65 million years.

character displacement   The divergence of a feature of two similar species where their ranges overlap so that each uses different resources.

chequerboard distribution   Pattern shown by two or more species that have mutually exclusive but interdigitating distributions across a series of isolates, such that each island (or habitat) supports only one species.

clade   Any evolutionary branch in a phylogeny, especially one that is based in genealogical relationships.

cladogram (or phylogenetic tree)   A line diagram derived from a cladistic analysis showing the hypothesized branching sequence of a monophyletic taxon and using shared derived character states to determine when each branch diverged.

cladogenesis   Evolutionary change within a lineage where the progenitor species is partitioned into two or more lineages and becomes extinct.

coenocline   Vegetation continuum along a gradient.

colonization   The relatively lengthy persistence of an immigrant species on an island, especially where breeding and population increase are accomplished.

competition   Negative, detrimental interaction between organisms caused by their need for a common resource. Competition may occur between individuals of the same species (intraspecific competition) or of different species (interspecific competition).

competitive exclusion   The principle that when two species with similar resource requirements co-occur, one eventually outcompetes and causes the extinction of the other.

connectance   Fraction of all possible pairs of species within a community that interact directly as feeder and food. In other words, number of actual connections in a food (or pollination) web divided by the total number of possible connections.

conservation   Human intervention with the goal of maintaining valued biodiversity (genetic variation, species, ecosystems, landscapes, natural resources).

continental fragments or micro-continents   Mainland fragments separated by tectonic drift from their continents millions of years ago, with the species they carried.

continental or land bridge islands   Emergent fragments of the continental shelf, separated from the continents by narrow, shallow waters. This separation is often recent, as a result of the postglacial rise of sea level, which isolated the species on the island from conspecific mainland populations.

continental shelf   The area of shallow sea floor (<200 m depth) adjacent to continents, underlain by continental crust and effectively a submerged extension of the continents.

convergent evolution   Evolution of similar features independently in unrelated taxa, usually from different antecedent features or by different developmental pathways.

corridor   A dispersal route of favourable habitat type, which permits the direct spread of species from one region or habitat fragment to another.

counter-adaptation   The evolutionary reaction of an island's native species to a newly arrived species, such that over time they begin to exploit or compete with the immigrant more effectively, thus lowering its competitive ability.

crassulacean acid metabolism (CAM) plants   Usually succulent xerophytes in which carbon dioxide taken up during the night is stored as malic acid until fixed using the C_3 pathway the following day.

cryptoturnover   Turnover of species (extinction followed by immigration) not detected because occurring between two surveys.

density compensation Higher than normal densities of a species on an island: typically explained as an outcome of the lower overall richness of the island assemblage.

density overcompensation A situation where density compensation occurs to an apparently excessive degree in one or more species.

density stasis A situation where the overall population of the community on the island is less than that of the reference mainland system, such that population sizes per species are the same as the mainland.

diffuse competition Community-wide interspecific competition among a group of species, subjecting each species to a range of competitive pressures exerted by other species.

dioecious species Having the sexes in separate individuals.

diplochory Plant species with two different dispersal mechanisms.

disassembly The process of a community/habitat patch losing species as, for instance, a result of habitat loss and increased isolation due to habitat fragmentation.

disharmony A characteristic of island biotas, such that they are biased in their representation of higher taxa compared to nearby mainland source areas.

disjunction A distribution pattern exhibited by taxa that have populations that are geographically separated.

dispersal The movement of organisms away from their point of origin.

disturbance Any relatively discrete event in time that removes organisms and opens up space which can be colonized by individuals of the same or different species.

dwarfism The tendency for some isolated populations to undergo significant decreases in body size in comparison to their conspecifics on the mainland (also termed **nanism**).

ecological release A colonizing species encounters an environment in which particular competitors or other interacting organisms, such as predators, are absent, and being released from these constraining forces responds, e.g. by expanding its feeding niche, or losing defensive traits.

ecological trap The preference of some species for breeding in edge habitats even though mortality is higher there than in the fragment.

ecoregion Major ecosystem type prevalent across a region, resulting from large-scale predictable patterns of solar radiation and moisture, which in turn affect the kinds of local ecosystems and animals and plants found there.

ecotone The transition zone between two adjacent communities/habitats, which typically shows a mix of the properties of each.

ecosystem transformer An introduced species that becomes naturalized, and invasive, and which significantly alters ecosystem properties, thus impacting on native species.

ecotourism Tourism based on an interest in observing nature, preferably with minimal ecological impact.

edge effect In fragmented landscapes, the edge of a habitat patch typically has altered physical and biotic properties, which extend some way towards the core of the habitat patch.

effective population size The proportion of the adult population that actually participates in breeding.

El Niño–Southern Oscillation (ENSO) Climatic anomalies yielding significant changes in rainfall, temperature, humidity, and storm patterns over much of the tropics and subtropics, due to the invasion of warm surface waters from the western part of the equatorial Pacific basin to the eastern part, disrupting the upwelling along the western coast of South America.

empty niche In the island context, refers to the lack of representation of a particular mainland ecological guild or niche, providing evolutionary opportunities for colonizing taxa to exploit the vacancy.

en echelon **lines** Short lines of crustal weakness, which are sub-parallel to each other.

endemic A taxon restricted to a given area, whether a mountain top, island, archipelago, continent, or zoogeographic region.

entomochory Dispersal by insects.

entomophily   Pollination by insects.

equilibrium   A condition of balance between opposing forces, such as between rates of immigration and extinction.

establishment   The successful start or founding of a population.

eustatic sea level changes   Sea-level changes originated by the changing volume of water in the sea.

explosive radiation   Evolutionary process resulting in an exceptional number of new monophyletic species, e.g. the Hawaiian drosophilid flies or Lake Malawi's cichlid fishes.

extinction   The loss of all populations of a species from an island, or the global loss of a species.

extinction debt   The anticipated eventual species loss from an area of fragmented habitat following fragmentation.

extirpation   The loss of a local (insular) population, not implying global loss of the species.

faunal drift   The random sampling of species reaching an 'empty' habitat, and thus providing a novel (disharmonic) biotic environment in which selection drives divergence into a variety of new niches.

fitness   The number of offspring an individual produces over its lifetime.

founder effect   Genetic loss that occurs when a newly isolated population is founded, by a small number of colonists.

founder event   The colonization of an island by the first representatives of a species.

frugivory   Action of those animals feeding on fruits.

geitonogamy   Pollen transfer occurring among different flowers of the same individual.

gene flow   The movement of alleles within a population or between populations caused by the dispersal of gametes or offspring.

generalist species   A species that uses a relatively large proportion, or in extreme cases all, of the available resource types.

genetic drift   Changes in gene frequency within a population caused solely by chance without any influence of natural selection.

Ghyben–Herzberg lens   Rainwater accumulation that floats on the denser salt or brackish water that permeates the base of an island.

gigantism   The tendency for some isolated populations to undergo significant increases in body size in comparison to their conspecifics on the mainland.

Gondwanaland   The southern super-continent, which split apart from Pangaea (the last unique land mass) some 160 Ma. It was composed of the areas known today as South America, Africa, Antarctica, Australia, India, Madagascar, New Zealand, and New Caledonia.

guild   A group of species exploiting the same class of resources in a similar way.

guyot   A type of **seamount** with a flat top, formed by accretions of carbonate sediments on the summit of a subsiding volcano.

habitat fragmentation   Anthropogenic process by which a large, continuous habitat is diminished in area and divided into numerous isolated habitat fragments.

habitat island   Discrete patch of a particular habitat type surrounded by a matrix of strongly contrasting habitat(s).

halophytes   Salt-tolerant plant species, commonly found in salt marsh, drylands, and saline lake environments.

hermaphrodites   Individuals having both male and female reproductive structures.

herpetofauna   Amphibian and reptile fauna.

heterozygote   An individual organism that possesses different alleles at a locus.

hybridization   The production of offspring by parents of two different species, populations, or genotypes.

hyperdynamism   An increase in the frequency and/or amplitude of population, community, and landscape dynamics in fragmented habitats.

hydrochory   Dispersal by water.

Holocene   The latest epoch, commencing about 11 500 years ago, at the end of the Pleistocene.

homozygote   An individual organism that has the same allele at each of its copies of a gene locus.

immigration   In island ecology, refers to the process of arrival of a propagule on an island not occupied by that species (may be distinguished from 'supplementary immigration'—subsequent reinforcements).

impoverishment The characteristic by which islands have fewer species per unit area than the mainland in comparable ecosystems, a distinction that is more marked the smaller the island.

inbreeding depression A reduction of the fitness in a normally outbreeding population as a consequence of increased homozygosity due to inbreeding.

incidence functions Biogeographical tool for describing how the probability of occurrence of a species varies with selected characteristics of islands (species richness, area, isolation, etc.).

introduction The voluntary or accidental human-assisted movement of individual species into a site that lies outside their natural, historical range.

introgressive hybridization (or introgression) The incorporation of genes of one species into the gene pool of another species.

invader complex Whereby introduced species preferentially establish interactions with other introduced species.

invasion The spread of a naturalized exotic species into natural or semi-natural habitats, hence 'invasive species'.

island rule The tendency for a graded series of changes in the size of island vertebrate species in relation to mainland congeners, such that small-bodied species tend to get larger, and vice versa.

island sterilization Process occurring when destructive volcanic ash flows are deposited over an entire island, completely eliminating the existing biota. It implies that colonization has to begin again from nothing.

isolating mechanism Any structural, physiological, ecological, or behavioural mechanism that blocks or strongly interferes with gene exchange between two populations.

isostatic sea level changes Changes in sea level due to the relative adjustment of the elevation of the land surface (e.g. (1) due to the removal of mass from the land causing uplift, as when an icecap melts, or (2) by tectonic uplift).

keystone species Species supplying a vital resource or holding a crucial function in an ecosystem, such that alteration in its numbers has a radical impact on key functional properties of the ecosystem.

kipuka Hawaiian term for a remnant of an old forest ecosystem that has been isolated by a new lava flow.

lag time The time taken for the relaxation or decrease of species richness following fragmentation (i.e. time taken to pay the **extinction debt)**.

landscape Spatial concept, in ecology referring to the interacting complex of systems on and close to the Earth's surface, including parts of the atmosphere, biosphere, hydrosphere, lithosphere, and pedosphere.

landslides Gravitational instabilities that in the most spectacular cases can lead to the collapse of the slopes of an island, producing debris avalanches of hundreds of km^3 into the sea. Such collapses may occur unexpectedly and suddenly and may lead to the disappearance of a significant part of an island in minutes.

late successional species Species that occur primarily in, or are dominant in, the late stages of succession.

laurisilva Relict subtropical evergreen montane cloud forest typical of the **Macaronesian islands**, dominated by Laureaceae tree species: formerly much more widely distributed.

lignification The acquirement of a woodiness in insular plant species derived from herbaceous continental ancestors.

Macaronesia Biogeographical region of the Palearctic comprising the Atlantic Ocean volcanic archipelagoes of the Azores, Madeira, the Salvage Islands, the Canaries, and the Cape Verde Islands.

Makatea islands Islands formed by young volcanoes, with raised coral reef uplifted to a few metres above sea level.

marine transgression The flooding of the land by the sea due to a rise in the sea level, either by an absolute sea-level change (interglacial periods) or by tectonic subsidence.

mesopredator release The increasing number of smaller omnivores and predators due to the absence of larger predators: a phenomenon that can result from habitat fragmentation.

metapopulation A population of geographically separated subpopulations interconnected by patterns of gene flow, extinction and recolonization.

minimum viable area (MVA) The area needed for maintaining the minimum viable population of a species.

minimum viable metapopulation (MVM) The minimum number of interacting local populations needed for the long-term persistence of a species in an area.

minimum viable population (MVP) The minimum size of a population that will ensure its long-term survival, for instance, providing a 95% of persistence probability for the next 100 or 1000 years.

molecular clock Based on the idea that proteins and DNA evolve at a more or less constant rate, the extent of molecular divergence (attributable to random mutations) is used as a metric for the timing of events within lineage development: greater confidence can be placed in such molecular clocks if they can be independently calibrated.

monoecy Refers to plant species that have separate male and female flowers on the same individual.

monophyletic group A group of taxa that share a common ancestor and which includes all descendants of that ancestor, also referred to as a **clade**.

natural selection The process of eliminating from a population through differential survival and reproduction those individuals with inferior fitness.

naturalized species A non-native species that has established a breeding population.

neoendemic A species exclusive to an area (island or island group) that has evolved its distinctive characteristics *in situ*, having originally colonized from a continental ancestor (contrast with **palaeoendemic**).

nestedness Where a set of islands is ranked according to species richness, the tendency for each successively smaller assemblage to be a subset of richer sets.

niche The total requirements of a population or a species for resources and physical conditions.

niche expansion An increase in the range of habitats or resources used by a population, which may occur when a potential competitor is absent.

non-adaptive radiation A radiation process driven not by natural selection, but rather by genetic drift or sexual selection.

non-equilibrium A characteristic of communities that change constantly in structure and composition due to the vagaries of the environment and where stable end-points are not achieved.

nuées ardentes See **pyroclastic flow**

oceanic islands Islands originated from submarine volcanic activity, mostly with basaltic foundations, that have never been connected to the continents. Initially populated by species that have dispersed to the islands from elsewhere are subsequently enriched by speciation.

palaeoendemic A species exclusive to an area (island or island group) that has apparently changed little since colonization, and which has disappeared from its continental source area (contrast with **neoendemic**).

Pangaea The supercontinent, comprising all present continents, which coalesced in the late Carboniferous or early Permian and which broke up some 70 million years later in the late Triassic.

panmictic population Population where mating occurs randomly with respect to the distribution of the genotypes in the population.

parallel evolution The evolution of similar or identical features independently in unrelated lineages.

parapatric speciation A mode of speciation in which differentiation occurs when two populations have contiguous but non-overlapping ranges, often representing two different habitat types.

paraphyletic group An incomplete evolutionary unit in which one or more descendants of a particular ancestor have been excluded from the group: also referred to as a **grade**.

parthenogenesis The development of eggs without fertilization by a male gamete.

peninsular effect The hypothesized tendency for species richness to decrease along a gradient from the axis to the most distal point of a peninsula.

phylogeny The evolutionary relationships between an ancestor and all its known descendants.

phylogeographical analysis The analysis of geographically structured genetic signal within a single taxon (e.g. species); the reconstructed genealogical history is used to infer past episodes of population expansion, bottlenecks, vicariance, and colonization, e.g. in an island context, the sequence of interisland dispersal events.

pioneer species Early colonizing species within a succession.

plate boundary The edge of one of the Earth's lithospheric plates: boundaries can be either constructive (divergent), destructive (convergent), or conservative.

Pleistocene The first epoch of the Quaternary period, formally taken as beginning about 1.8 Ma (although the climate cooling that marked the period began perhaps as early as 2.5 Ma), which ended about 11 500 years BP, with the transition to the present **Holocene** epoch.

Pliocene The final epoch of the Tertiary period, spanning approximately 5.3 to 1.8 Ma.

pollination The transfer of pollen grains to a receptive stigma, e.g. by wind or by pollinating animals.

polyphyletic group A taxonomic group comprising several monophyletic groups.

population viability analyses (PVA) Population analyses based on data for key demographic and genetic parameters, which set out to assess the probability that a population of a given size will persist for a particular time period.

productivity rate The rate at which a community of organisms manufactures new biomass.

propagule The minimal number of individuals of a species capable of successfully colonizing a habitable island.

pseudoturnover In island ecological analyses, species appearing to turn over (undergo extinction and re-immigration) due to incomplete census data, when they have actually been residents throughout or, alternatively, never properly colonized.

pyroclastic flow High-speed avalanche of hot ash mobilized by expanding gases and travelling at speeds in excess of 100 km/h.

Quaternary The most recent period of the geological timescale, following the end of the Pliocene and continuing to the present day; and comprising the **Pleistocene** and the **Holocene** epochs.

radiation In evolutionary biology, a term for the expansion of a group, implying that many new species have been produced.

radiation zone The zone within an ocean, near the effective dispersal limits of a higher taxon (e.g. terrestrial mammals), where few lineages colonize and where radiation in those lineages is correspondingly greater as a consequence of the 'empty niche space' encountered.

red (data) book A catalogue listing species that are rare or in danger of becoming extinct locally, nationally or globally (e.g. the IUCN red data books).

refugia Areas of survival of species during adverse conditions, most notably during glaciation episodes.

relictualism The possession by some islands of **palaeoendemic**, i.e. relict taxa, which formerly had a continental distribution, and which have gone extinct from the source region since colonizing the island, e.g. due to climatic or geological changes.

reproductive isolation Inability of individuals from different populations to produce viable offspring.

rescue effect The prevention of extinction of a small insular population by the occasional influx of individuals from another (e.g. mainland) area.

Sahul Pleistocene continent constituted by Australia, New Guinea, and Tasmania.

seamount A mountain beneath the sea (see also **guyot**).

sexual selection Depends on the success of certain individuals over others of the same sex, in relation to the propagation of the species, and can result in the evolution of character traits that carry a fitness cost, and would otherwise not seem advantageous.

shield volcanoes Wide volcanoes, of gradually sloping sides, build up in layers by basaltic magma flows erupting from lengthy fissures or vents.

sink population A term used in the **metapopulation** literature for a population occupying an unfavourable habitat, which maintains its presence by supplementary immigration of offspring from **source populations**.

SLOSS debate The debate as to which strategy is the best for protected area systems: Single Large Or Several Small reserves.

source population A population occupying a favourable habitat and thus producing an excess of offspring that will emigrate to less favourable habitats where **sink populations**, with negative demography, can be maintained.

specialist species A species that uses a relatively small proportion of the available resource types.

speciation The process in which two or more contemporaneous species evolve from a single ancestral population.

species swarm A large number of closely related species occurring together in an area (e.g. an archipelago) and derived by multiple splitting of an ancestral stock.

species–area relationship (SAR) The relationship between the number of species and the area of the sample or island; within this simple idea, a variety of important distinctions exist, which are often not properly recognized.

species turnover The product of opposing rates of species immigration to and species extinction from an island. Where these rates precisely balance this forms a dynamic equilibrium, with species richness remaining constant through time but species composition continually altering.

stepping stones Island or island groups that facilitate the dispersal of species across oceans by their intermediate geographic position between a continental source pool and an island of interest.

stratovolcano The composite cone built by a volcano that emits both molten and solid materials, and builds up a steep-sided cone.

subduction The process by which lithosphere-bearing oceanic crust is destroyed in plate tectonics by the movement of one plate beneath another to form a marine trench. The process leads to the heating and subsequent re-melting of the lower plate, in turn generating volcanic activity, typically forming an arc of volcanic islands above the area of subduction.

subfossil A dead organism that is not truly fossilized.

subsidence The downward movement of an object relative to its surroundings. Subsidence of the lithosphere can be due to increased mass (e.g. increased ice, water, or rock loading) or due to the movement of the island away from mid-ocean ridges and other areas that can support anomalous mass.

succession Refers to directional change in (mostly) plant communities following significant ecosystem disturbance, or following the creation of an entirely new land surface area.

successional turnover In an island ecological context, refers to **species turnover** attributable to successional processes.

supertramp A species that has excellent colonizing abilities but is a poor competitor in diverse communities.

sustainable development Development that meets the needs of the present generation without compromising the ability of future generations to meet their own needs.

sweepstake dispersal route The least favourable of potential migration pathways in that it can only be traversed very infrequently and with great difficulty, or only under circumstances that occur very rarely.

sympatric speciation The differentiation of two reproductively isolated species from one initial population within the same local area, in which much gene flow potentially could or actually does occur.

taxon cycle A proposed series of ecological and evolutionary changes in a newly arrived colonist on an isolated island, from the state of being indistinguishable from its mainland relatives to that of a highly differentiated endemic, to extinction and replacement by new colonists.

tephra A collective term for all the unconsolidated, primary pyroclastic products of a volcanic eruption, independent of grain size.

territoriality Behaviour related to the defence of a specified area (the territory) against intruders.

Tertiary A period of geological time lasting from *c.*65 Ma to the beginning of the **Quaternary**.

translocation programmes Attempts at species conservation involving the movement of a number of individuals into an area from which the species is currently lacking (typically having once been present).

trophic cascade The chain of knock-on extinctions following the loss of one or a few species that play a critical role (e.g. as a pollinator) in ecosystem functioning.

trophic level Position in a food chain determined by the number of energy transfer steps to that level.

turnover See **species turnover**.

vagility The ability to move actively from one place to another.

vicariance The separation of a population or a species into disjunct groups resulting from the formation of a physical barrier.

Wallace's line A boundary, described by Alfred Russel Wallace, across the Indonesian archipelago, separating faunas characteristic of the islands of the Sunda continental shelf from those of the Sahul shelf. Many variants of Wallace's line have been suggested in this region, which has been termed Wallacea.

widow species A species that has lost its unique pollinator (or disperser, prey, host, etc.) and is thus heading for extinction.

zoochory Dispersal of seeds by animals, either through their gut (**endozoochous**) or attached to their skin, hair or feathers (**exozoochorous**).

zygomorphic Having flowers with bilateral symmetry.

References

Abbott, I. (1983). The meaning of *z* in species/area regressions and the study of species turnover in island biogeography. *Oikos*, **41**, 385–390.

Abbott, I. and Black, R. (1980). Changes in species composition of floras on islets near Perth, Western Australia. *Journal of Biogeography*, **7**, 399–410.

Abbott, I. and Grant, P. R. (1976). Nonequilibrial bird faunas on islands. *American Naturalist*, **110**, 507–528.

Abdelkrim, J., Pascal, M., and Samadi, S. (2005). Island colonization and founder effects: the invasion of the Guadeloupe islands by ship rats (*Rattus rattus*). *Molecular Ecology*, **14**, 2923–2931.

Adler, G. H. (1992). Endemism in birds of tropical Pacific islands. *Evolutionary Ecology*, **6**, 296–306.

Adler, G. H. (1994). Avifaunal diversity and endemism on tropical Indian Ocean islands. *Journal of Biogeography*, **21**, 85–95.

Adler, G. H., Austin, C. A., and Dudley, R. (1995). Dispersal and speciation of skinks among archipelagos in the tropical Pacific Ocean. *Evolutionary Ecology*, **9**, 529–541.

Adler, G. H. and Dudley, R. (1994). Butterfly biogeography and endemism on tropical Pacific Islands. *Biological Journal of the Linnaean Society*, **52**, 151–162.

Adler, G. H. and Levins, R. (1994). The island syndrome in rodent populations. *Quarterly Review of Biology*, **69**, 473–490.

Adler, G. H. and Seamon, J. O. (1991). Distribution and abundance of a tropical rodent, the spiny rat, on Islands in Panama. *Journal of Tropical Ecology*, **7**, 349–360.

Adler, G. H. and Wilson, M. L. (1985). Small mammals on Massachusetts islands: the use of probability functions in clarifying biogeographic relationships. *Oecologia*, **66**, 178–186.

Adler, G. H. and Wilson, M. L. (1989). Insular distributions of voles and shrews: the rescue effect and implications for species co-occurence patterns. *Coenoses*, **4**, 69–72.

Adler, G. H. Wilson, M. L. and DeRosa, M. J. (1986). Influence of island area and isolation on population characteristics of *Peromyscus leucopus*. *Journal of Mammalogy*, **67**, 406–409.

Adsersen, H. (1995). Research on islands: classic, recent, and prospective approaches. In *Islands: biological diversity and ecosystem function* (ed. P. M. Vitousek, L. L. Loope, and H. Adsersen), Ecological Studies **115**, pp. 7–21. Springer-Verlag, Berlin.

Aguilera, F., Brito, A., Castilla, C., Díaz, A., Fernández-Palacios, J.M., Rodríguez, A. Sabaté, F., and Sánchez, J. (1994). *Canarias: economía ecología y medio ambiente*. Francisco Lemus Editor, La Laguna.

Alonso, J. C., Alonso, J. A., and Muñoz-Pulido, R. (1994). Mitigation of bird collisions with transmission lines through groundwire marking. *Biological Conservation*, **67**, 129–134.

Anciães, M. and Marini, M. Â. (2000). The effects of fragmentation on fluctuating asymmetry in passerine birds of Brazilian tropical forests. *Journal of Applied Ecology*, **37**, 1013–1028.

Ancochea, E., Hernán, F., Cendrero, A., Cantagrel, J. M., Fúster, J. M., Ibarrola, E., and Coello, J. (1994). Constructive and destructive episodes in the building of a young oceanic island, La Palma, Canary Islands, and the genesis of the Caldera de Taburiente. *Journal of Volcanology and Geothermal Research*, **60**, 243–262.

Anguita, F. and Hernán, F. (1975). A propagating fracture model versus a hot-spot origin for the Canary Islands. *Earth and Planetary Science Letters*, **24**, 363–368.

Anguita, F. and Hernán, F. (2000).The Canary Islands origin: a unifying model. *Journal of Volcanology and Geothermal Research*, **103**, 1–26.

Anguita, F., Márquez, A., Castiñeiras, P., and Hernán, F. (2002). *Los volcanes de Canarias. Guía geológica e itinerarios*. Ed. Rueda, Madrid.

Araújo, M. B., Thuiller, W., Williams, P. H., and Reginster, I. (2005a). Downscaling European species atlas distributions to a finer resolution: implications for conservation planning. *Global Ecology and Biogeography*, **14**, 17–30.

Araújo, M. B., Whittaker, R. J., Ladle, R. J., and Erhard, M. (2005b). Reducing uncertainty in projections of extinction risk from climate change. *Global Ecology and Biogeography*, **14**, 529–538.

Arechavaleta, M., Zurita, N, Marrero, M. C., and Martín, J. L. (2005). *Lista preliminar de especies silvestres de Cabo Verde (hongos, plantas y animales terrestres)*. Consejería de Medio ambiente y Ordenación Territorial, Gobierno de Canarias, Santa Cruz de Tenerife.

Arendt, W. J., Gibbons, D. W., and Gray, G. (1999). Status of the volcanically threatened Montserrat Oriole *Icterus oberi* and other forest birds in Montserrat, West Indies. *Bird Conservation International*, **9**, 351–372.

Arévalo, J. R., Delgado, J. D., Otto, R., Naranjo, A., Salas, M., and Fernández-Palacios, J. M. (2005). Distribution of alien vs. native plant species in roadside communities along an altitudinal gradient in Tenerife and Gran Canaria (Canary Islands). *Perspectives in Plant Ecology, Evolution and Systematics*, **7**, 185–202.

Armstrong, P. (1982). Rabbits *(Oryctolagus cuniculus)* on islands: a case study of successful colonization. *Journal of Biogeography*, **9**, 353–362.

Arnold, E. N. (1979). Indian Ocean giant tortoises: their systematics and island adaptations. *Philosophical Transactions of the Royal Society of London*, B **286**, 127–145.

Arrhenius, O. (1921). Species and area. *Journal of Ecology*, **9**, 95–99.

Ashmole, N. P. (1963). The regulation of numbers of tropical oceanic birds. *Ibis*, **103**, 458–478.

Ashmole, N. P. and Ashmole, M. J. (1988). Insect dispersal on Tenerife, Canary Islands: high altitude fallout and seaward drift. *Arctic and Alpine Research*, **20**, 1–12.

Asquith, A. (1995). Evolution of *Sarona* (Heroptera, Miridae): Speciation on geographic and ecological islands. In *Hawaiian Biogeography: evolution on a hot spot archipelago* (ed. W. L. Wagner and V. A. Funk), pp. 90–120. Smithsonian Institution Press, Washington, DC.

Axelrod, D. I. (1975). Evolution and biogeography of Madrean–Tethyan sclerophyll vegetation. *Annals of the Missouri Botanical Garden*, **62**, 280–334.

Ayers, J. M., Bodmer, R. E., and Mittermeier, R. A. (1991). Financial considerations of reserve design in countries with high primate diversity. *Conservation Biology*, **5**, 109–114.

Backer, C. A. (1929). *The problem of Krakatoa as seen by a botanist*. Published by the author, Surabaya.

Badano, E. I., Regidor, H. A., Núñez, H. A., Acosta, R., and Gianoli, E. (2005). Species richness and structure of ant communities in a dynamic archipelago: effects of island age and area. *Journal of Biogeography*, **32**, 221–227.

Báez, M., Martín, J. L. and Oromí, P (2001). Diversidad taxonómica terrestre. In *Naturaleza de las Islas Canarias: Ecología y conservación*. (ed. J. M. Fernández-Palacios and J. L. Martín Esquivel), pp. 119–125. Turquesa Ediciones, Santa Cruz de Tenerife.

Bahn, P. and Flenley, J. R. (1992). *Easter Island, Earth island*. Thames & Hudson, London.

Baker, H. G. (1955). Self-compatibility and establishment after long-distance dispersal. *Evolution*, **9**, 347–349.

Barrett, S. C. H. (1989). Mating system evolution and speciation in heterostylous plants. In *Speciation and its consequences* (ed. D. Otte and J. A. Endler), pp. 257–283. Sinauer, Sunderland, MA.

Barrett, S. C. H. (1996). The reproductive biology and genetics of island plants. *Philosophical Transactions of the Royal Society of London*, B **351**, 725–733.

Barton, N. H. (1989). Founder effect speciation. In *Speciation and its consequences* (ed. D. Otte and J. A. Endler), pp. 229–256. Sinauer, Sunderland, MA.

Barton, N. H. and Charlesworth, B. (1984). Genetic revolutions, founder effects, and speciation. *Annual Review of Ecology and Systematics*, **15**, 133–164.

Bawa, K. S. (1980). Evolution of dioecy in flowering plants. *Annual Review of Ecology and Systematics*, **11**, 15–19.

Bawa, K. S. (1982). Outcrossing and incidence of dioecism in island floras. *American Naturalist*, **119**, 866–871.

Begon, M., Harper, J. L., and Townsend, C. R. (1986). *Ecology*. Blackwell, Oxford.

Begon, M., Harper, J. L., and Townsend, C. R. (1990). *Ecology* (2nd edn). Blackwell, Oxford.

Bell, M. and Walker, M. J. C. (1992). *Late Quaternary environmental change: physical and human perspectives*. Longman, Harlow.

Beller, W., d'Ayala, P., and Hein, P. (ed.) (1990). *Sustainable development and environmental management of small islands*, Vol. **5**, Man and the Biosphere Series. UNESCO/Parthenon Publishing, Paris.

Bennett, A. F. and Gorman, G. C. (1979). Population density and energetics of lizards on a tropical island. *Oecologia*, **42**, 339–358.

Benton, T. and Spencer, T. (ed.) (1995). *The Pitcairn Islands: biogeography, ecology and prehistory*. Academic Press, London.

Berry, R. J. (1986). Genetics of insular populations of mammals with particular reference to differentiation and founder effects in British small mammals. *Biological Journal of the Linnean Society*, **28**, 205–230.

Berry, R. J. (1992). The significance of island biotas. *Biological Journal of the Linnean Society*, **46**, 3–12.

Berry, R. J., Berry, A. J., Anderson, T. J. C., and Scriven, P. (1992). The house mice of Faray, Orkney. *Journal of Zoology (London)*, **228**, 233–246.

Bevanger, K. (1996). Estimates and population consequences of teraonid mortality caused by collisions with high tension power lines in Norway. *Journal of Applied Ecology*, **32**, 745–753.

Bibby, C. J. (1995). Recent past and future extinctions in birds. In *Extinction rates* (ed. J. H. Lawton and R. M. May), pp. 98–110. Oxford University Press, Oxford.

Bierdermann, R. (2003). Body size and area–incidence relationships: is there a general pattern? *Global Ecology and Biogeography*, **12**, 381–387.

Blackburn, T. M., Cassey, P., Duncan, R. P., Evans, K. L., and Gaston, K. J. (2004). Avian extinctions and mammalian introductions on oceanic islands. *Science*, **305**, 1955–1958.

Blackburn, T. M. and Duncan, R. P. (2001). Establishment patterns of exotic birds are constrained by non-random patterns in introduction. *Journal of Biogeography*, **28**, 927–939.

Blake, J. G. (1991). Nested subsets and the distribution of birds on isolated woodlots. *Conservation Biology*, **5**, 58–66.

Blondel, J. (1985). Breeding strategies of the Blue Tit and Coal Tit (*Parus*) in mainland and island Mediterranean habitats: A comparison. *Journal of Animal Ecology*, **54**, 531–556.

Blondel, J. and Aronson, J. (1999). *Biology and wildlife of the Mediterranean Region*. Oxford University Press, Oxford.

Bodmer, R. E., Eisenberg, J. F., and Redford, K. H. (1997). Hunting and the likelihood of extinction of Amazonian mammals. *Conservation Biology*, **11**, 460–466.

Boecklen, W. J. (1986). Effects of habitat heterogeneity on the species–area relationships of forest birds. *Journal of Biogeography*, **13**, 59–68.

Boecklen, W. J. and Gotelli, N. J. (1984). Island biogeographic theory and conservation practice: species–area or specious–area relationships? *Biological Conservation*, **29**, 63–80.

Boggs, C. L. and Murphy, D. D. (1997). Community composition in mountain ecosystems: climatic determinants of montane butterfly distributions. *Global Ecology and Biogeography Letters*, **6**, 39–48.

Böhle, U.-R., Hilger, H. H., and Martin, W. F. (1996). Island colonization and evolution of the insular woody habit in *Echium* L. (Boraginaceae). *Proceedings of the National Academy of Sciences of the USA*, **93**, 11740–11745.

Borgen, L. (1979). Karyology of the Canarian flora. In *Plants and islands* (ed. D. Bramwell), pp. 329–346. Academic Press, London.

Borges, P. A. V., Cunha, R., Gabriel, R., Frias Martins, A, Silva L., and Vieira, V. (eds.) (2005). *A list of the terrestrial fauna (Mollusca and Arthropoda) and flora (Byrophyta, Pteridophyta and Spermatophyta) from the Azores*. Direcçao Regional de Ambiente and Universidade dos Açores, Horta.

Bramwell, D. (1972). Endemism in the flora of the Canary Islands. In *Taxonomy, phytogeography and evolution* (ed. D. H. Valentine) pp. 141–159. Academic Press, London.

Bramwell, D. (ed.) (1979). *Plants and islands*. Academic Press, London.

Bramwell, D. and Bramwell, Z. I. (1974). *Wild flowers of the Canary Islands*. Stanley Thornes, Cheltenham.

Bravo, T. (1962). El Circo de las Cañadas y sus dependencias. *Boletín de la Real Sociedad Española de Historia Natural*, **60**, 93–108.

Brochmann, C., Rustan, O. H., Lobin, W., and Kilian, N. (1997). The endemic plants of the Cape Verde Islands, W. Africa. *Sommerfeltia*, **24**, 1–356.

Bromham, L. and Woolfit, M. (2004). Explosive radiations and the reliability of molecular clocks: island endemic radiations as a test case. *Systematic Biology*, **53**, 758–766.

Brooks, T. and Balmford, A. (1996). Atlantic forest extinctions. *Nature*, **380**, 115.

Brooks, T. M., Mittermeier, R. A., Mittermeier, C. G., da Fonseca, G. A. B., Rylands, A.B., Konstant, W. R. Flick, P., Pilgrim, J., Oldfield, S., Magin, G., and Hilton-Taylor, C. (2002) Habitat loss and extinction in the hotspots of biodiversity. *Conservation Biology*, **16**, 909–923.

Brooks, T. M., Pimm, S. L., and Collar, N. J. (1997). Deforestation predicts the number of threatened birds in insular Southeast Asia. *Conservation Biology*, **11**, 382–394.

Brooks, T. M., Pimm, S. L., and Oyugi, J. O. (1999). Time lag between deforestation and bird extinction in tropical forest fragments. *Conservation Biology*, **13**, 1140–1150.

Brothers, T. S. and Spingarn, A. (1992). Forest fragmentation and alien plant invasion of Central Indiana old-growth forests. *Conservation Biology*, **6**, 91–100.

Brown, J. H. (1971). Mammals on mountaintops: nonequilibrium insular biogeography. *American Naturalist*, **105**, 467–478.

Brown, J. H. (1981). Two decades of homage to Santa Rosalia: toward a general theory of diversity. *American Zoologist*, **21**, 877–888.

Brown, J. H. (1986). Two decades of interaction between the MacArthur–Wilson model and the complexities of mammalian distribution. *Biological Journal of the Linnean Society*, **28**, 231–251.

Brown, J. H. (1995). *Macroecology*. Chicago University Press, Chicago.

Brown, J. H. (1999). Macroecology: progress and prospect. *Oikos*, **87**, 3–13.

Brown, J. H. and Gibson, A. C. (1983). *Biogeography*. Mosby, St Louis.

Brown, J. H. and Kodric-Brown, A. (1977). Turnover rates in insular biogeography: effect of immigration on extinction. *Ecology*, **58**, 445–449.

Brown, J. H. and Lomolino, M. V. (1989). Independent discovery of the equilibrium theory of island biogeography. *Ecology*, **70**, 1954–1957.

Brown, J. H. and Lomolino, M. V. (1998). *Biogeography*, 2nd edn. Sinauer, Sunderland, MA.

Brown, J. W. (1987). The peninsular effect in Baja California: an entomological assessment. *Journal of Biogeography*, **14**, 359–366.

Brown, J. W. and Opler, P. A. (1990). Patterns of butterfly species density in peninsular Florida. *Journal of Biogeography*, **17**, 615–622.

Brown, M. and Dinsmore, J. J. (1988). Habitat islands and the equilibrium theory of island biogeography: testing some predictions. *Oecologia*, **75**, 426–429.

Brown, P., Sutikna, T., Morwood, M. J., Soejono, R. P., Jatmiko, Saptomo, E. W., and Due, R. A. (2004). A new small-bodied hominin from the late Pleistocene of Flores, Indonesia. *Nature*, **431**, 1055–1061.

Brown, W. L., Jr and Wilson, E. O. (1956). Character displacement. *Systematic Zoology*, **7**, 49–64.

Browne, J. and Neve, M. (1989). Introduction to the abridged version of Charles Darwin's 1839 *Voyage of the Beagle*. Penguin Classics, London.

Brualdi, R. A. and Sanderson, J. G. (1999). Nested species subsets, gaps, and discrepancy. *Oecologia*, **119**, 256–264.

Bruijnzeel, L. A., Waterloo, M. J., Proctor, J., Kuiters, A. T., and Kotterink, B. (1993). Hydrological observations in montane rain forests on Gunung Silam, Sabah, Malaysia, with special reference to the 'Massenerhebung' effect. *Journal of Ecology*, **81**, 145–167.

Buckley, R. C. (1981). Scale-dependent equilibrium on highly heterogeneous islands: plant geography of northern Great Barrier Reef sand cays and shingle islets. *Australian Journal of Ecology*, **6**, 143–148.

Buckley, R. C. (1982). The Habitat-unit model of island biogeography. *Journal of Biogeography*, **9**, 339–344.

Buckley, R. C. (1985). Distinguishing the effects of area and habitat type on island plant species richness by separating floristic elements and substrate types and controlling for island isolation. *Journal of Biogeography*, **12**, 527–535.

Buckley, R. C. and Knedlhans, S. B. (1986). Beachcomber biogeography: interception of dispersing propagules by islands. *Journal of Biogeography*, **13**, 68–70.

Bunce, M., Worthy, T. H., Ford, T., Hoppitt, W., Willerslev, E., Drummond, A., and Cooper, A. (2003). Extreme reversed sexual size dimorphism in the extinct New Zealand moa *Dinornis*. *Nature*, **425**, 172–175.

Burkey, T. V. (1993). Edge effects in seed and egg predation at two neotropical rainforest sites. *Biological Conservation*, **66**, 139–143.

Bush, M. B. (1994). Amazonian speciation—a necessarily complex model. *Journal of Biogeography*, **21**, 5–17.

Bush, M. B. (1996). Amazonian conservation in a changing world. *Biological Conservation*, **76**, 219–228.

Bush, M. B. (2002). Distributional change and conservation on the Andean flank: a palaeoecological perspective. *Global Ecology and Biogeography*, **11**, 463–473.

Bush, M. B. and Whittaker, R. J. (1991). Krakatau: colonization patterns and hierarchies. *Journal of Biogeography*, **18**, 341–356.

Bush, M. B. and Whittaker, R. J. (1993). Non-equilibration in island theory of Krakatau. *Journal of Biogeography*, **20**, 453–458.

Bush, M. B., Whittaker, R. J., and Partomihardjo, T. (1992). Forest development on Rakata, Panjang and Sertung: contemporary dynamics (1979–1989). *Geojournal*, **28**, 185–199.

Buskirk, R. E. (1985). Zoogeographic patterns and tectonic history of Jamaica and the northern Caribbean. *Journal of Biogeography*, **12**, 445–461.

Cabrera, J. C. (2001). Poblamiento e impacto aborigen. In *Naturaleza de las Islas Canarias: Ecología y conservación*. (ed. J. M. Fernández-Palacios and J. L. Martín Esquivel), pp. 241–245. Turquesa Ediciones, Santa Cruz de Tenerife.

Cadena, C. D., Ricklefs, R. E., Jiménez, I., and Bermingham, E. (2005). Is speciation driven by species diversity? Arising from B. C. Emerson & N. Kolm *Nature* **434**, 1015–1017 (2005). *Nature*, **438**, doi:10.1038 / nature04308.

Cameron, R. A. D., Cook, L. M., and Hallows, J. D. (1996). Land snails on Porto Santo: adaptive and non-adaptive radiation. *Philosophical Transactions of the Royal Society of London*, B **351**, 309–337.

Canals, M., Urgelés, R., Masson, D. G., and Casamor, J. L. (2000). Los deslizamientos submarinos de las Islas Canarias. *Makaronesia*, **2**, 57–69.

Carine, M. A., Russell, S. J., Santos-Guerra, A., and Francisco-Ortega, J. (2004). Relationships of the Macaronesian and Mediterranean floras: molecular evidence for multiple colonizations into Macaronesia and back-colonization of the continent in *Convolvulus* (Convolvulaceae). *American Journal of Botany*, **91**, 1070–1085.

Carlquist, S. (1965). *Island life: a natural history of the islands of the world*. Natural History Press, New York.

Carlquist, S. (1974). *Island biology*. Columbia University Press, New York.

Carlquist, S. (1995). Introduction. In *Hawaiian Biogeography: evolution on a hot spot archipelago* (ed. W. L. Wagner and V. A. Funk), pp. 1–13. Smithsonian Institution Press, Washington, DC.

Campbell, K. and Donlan, C. J. (2005). Feral goat eradications on islands. *Conservation Biology*, **19**, 1362–1374.

Carr, G. D., Powell, E. A., and Kyhos, D. W. (1986). Self-incompatibility in the Hawaiian Madiinae (Compositae): an exception to Baker's rule. *Evolution*, **40**, 430–434.

Carr, G. D., Robichaux, R. H, Witter, M. S., and Kyhos D. W. (1989). Adaptive radiation of the Hawaiian silversword alliance (Compositae—Madiinae): a comparsion with Hawaiian picture-winged *Drosophila*. In *Genetics, speciation and the founder principle* (ed. L. Y. Giddings, K. Y. Kaneshiro, and W. W. Anderson), pp. 79–95. Oxford University Press, New York.

Carracedo, J. C. and Tilling, R. I. (2003). *Geología y volcanología de islas volcánicas oceánicas. Canarias–Hawai.* CajaCanarias–Gobierno de Canarias.

Carracedo, J. C. (2003). El volcanismo de las Islas Canarias. In *Geología y volcanología de islas volcánicas oceánicas. Canarias-Hawai* (ed. J. C. Carracedo and R. I. Tilling), pp. 17–38. CajaCanarias–Gobierno de Canarias.

Carracedo, J. C., Day, S., Guillou, H., Badiola, E. R., Canas, J. A., and Torrado, F. J. P. (1998). Hotspot volcanism close to a passive continental margin: the Canary islands. *Geological Magazine*, **135**, 591–604.

Carracedo, J. C., Day, S. J., Guillou, H., and Torrado, F. J. P. (1999). Giant Quaternary landslides in the evolution of La Palma and El Hierro, Canary Islands. *Journal of Volcanology and Geothermal Research*, **94**, 169–190.

Carranza, S., Harris, D. J., Arnold., E. N., Batista, V., and Gonzalez de la Vega, J. P. (2006). Phylogeography of the lacertid lizard *Psammodromos algirus*, in Iberia and across the Strait of Gibraltar. *Journal of Biogeography*, **33**, 1279–1288.

Carson, H. L. (1983). Chromosomal sequences and interisland colonizations in Hawaiian Drosophilidae. *Genetics*, **103**, 465–482.

Carson, H. L. (1992). Genetic change after colonization. *Geojournal*, **28**, 297–302.

Carson, H. L., Hardy, D. E., Spieth, H. T., and Stone, W. S. (1970). The evolutionary biology of the Hawaiian Drosophilidae. In *Essays in evolution and genetics in honor of Theodosius Dobzhansky* (ed. M. K. Hecht and W. C. Steere), pp. 437–543. Appleton-Century-Crofts, New York.

Carson, H. L., Lockwood, J. P., and Craddock, E. M. (1990). Extinction and recolonization of local populations on a growing shield volcano. *Proceedings of the National Academy of Sciences of the USA*, **87**, 7055–7057.

Carson, H. L. and Templeton, A. R. (1984). Genetic revolutions in relation to speciation phenomena: the founding of new populations. *Annual Review of Ecology and Systematics*, **15**, 97–131.

Carvajal, A. and Adler, G. H. (2005). Biogeography of mammals on tropical Pacific islands. *Journal of Biogeography*, **32**, 1561–1569.

Case, T. J., Bolger, D. T., and Richman, A. D. (1992). Reptilian extinctions: the last ten thousand years. In *Conservation Biology: the theory and practice of nature conservation and management* (ed. P. L. Fielder and S. K. Jain), pp. 91–125. Chapman & Hall, New York.

Case, T. J. and Cody, M. L. (1987). Testing theories of island biogeography. *American Scientist*, **75**, 402–411.

Cassey, P., Blackburn, T. M., Duncan, R. P., and Lockwood, J. L. (2005a). Lessons from the establishment of exotic species: a meta-analytical case study using birds. *Journal of Animal Ecology*, **74**, 250–258.

Cassey, P., Blackburn, T. M., Duncan, R. P., and Gaston, K. J. (2005b). Causes of exotic bird establishment across oceanic islands. *Proceedings of the Royal Society of London*, B, **272**, 2059–2063.

Caswell, H. (1978). Predator-mediated coexistence: a non-equilibrium model. *American Naturalist*, **112**, 127–154.

Caswell, H. (1989). Life-history strategies. In *Ecological concepts: the contribution of ecology to an understanding of the natural world* (ed. J. M. Cherrett), pp. 285–307. Blackwell Scientific Publications, Oxford.

Causton, C. (2005). *Galapagos News*, Winter 2005, p.1.

Chang, M. (2000). Exclusive economic zones. In: <http://geography.about.com/library/misc/uceez.htm> (visited March 2006)

Christensen, U. (1999). Fixed hotspots gone with the wind. *Nature*, **391**, 739–740.

Christian, C. S. (1993). The challenge of parrot conservation in St Vincent and the Grenadines. *Journal of Biogeography*, **20**, 463–469.

Christian, C. S., Lacher, T. E., Jr, Zamore, M. P., Potts, T. D., and Burnett, G. W. (1996). Parrot conservation in the Lesser Antilles with some comparison to the Puerto Rican efforts. *Biological Conservation*, **77**, 159–167.

Christiansen, M. B. and Pitter, E. (1997). Species loss in a forest bird community near Lagoa Santa in Southerastern Brazil. *Biological Conservation*, **80**, 23–32.

CI (2005). Conservation International hotspots. <www.biodiversityhotspots.org/xp/Hotspots/hotspotsScience/hotspots_revisited.xml>. Visited 8 June 2005.

Clague, D. A. (1996). The growth and subsidence of the Hawaiian-Emperor volcanic chain. In *The origin and evolution of Pacific Island Biotas, New Guinea to Eastern Polynesia: patterns and processes* (ed. A. Keast, and S. E. Miller), pp. 35–50. SPB Academic Publishing, Amsterdam.

Clarke, B. and Grant, P. R. (ed.) (1996). Evolution on islands. *Philosophical Transactions of the Royal Society of London*, B, **351**, 723–854.

Clarke, B., Johnson, M. S., and Murray, J. (1996). Clines in the genetic distance between two species of island land snails: how 'molecular leakage' can mislead us about speciation. *Philosophical Transactions of the Royal Society of London*, B, **351**, 773–784.

Clegg, S. M., Degnan, S. M., Kikkawa, J., Moritz, C., Estoup, A., and Owens, I. P. F. (2002). Genetic consequences of sequential founder events by an island-colonizing bird. *Proceedings of the National Academy of Sciences of the USA*, **99**, 8127–8132.

Cody, M. L. (1971). Ecological aspects of reproduction. In *Avian Biology* (ed. D. S. Farner and J. R. King). pp. 461–512. Academic Press, London.

Cody, M. L. and Overton, J. McC. (1996). Short-term evolution of reduced dispersal in island plant populations. *Journal of Ecology*, **84**, 53–61.

Colinvaux, P. A. (1972). Climate and the Galápagos Islands. *Nature*, **219**, 590–594.

Clout, M. N. and Craig, J. L. (1995). The conservation of critically endangered flightless birds in New Zealand. *Ibis*, **137**, S181–190.

Colwell, R. K. and Winkler, D. W. (1984). A null model for null models in biogeography. In *Ecological communities: conceptual issues and the evidence* (ed. D. R. Strong, Jr. D. Simberloff, L. G. Abele, and A. B. Thistle), pp. 344–359. Princeton University Press, Princeton, NJ.

Compton, S. G., Ross, S. J., and Thornton, I. W. B. (1994). Pollinator limitation of fig tree reproduction on the island of Anak Krakatau (Indonesia). *Biotropica*, **26**, 180–186.

Conant, S. (1988). Geographic variation in the Laysan finch (*Telepyza cantans*). *Evolutionary Ecology*, **2**, 270–282.

Connell, J. H. (1980). Diversity and the coevolution of competitors, or the ghost of competition past. *Oikos*, **35**, 131–138.

Connor, E. F. and McCoy, E. D. (1979). The statistics and biology of the species–area relationship. *American Naturalist*, **113**, 791–833.

Connor, E. F., McCoy, E. D., and Cosby, B. J. (1983). Model discrimination and expected slope values in species–area studies. *American Naturalist*, **122**, 789–796.

Connor, E. F. and Simberloff, D. (1979). The assembly of species communities: chance or competition? *Ecology*, **60**, 1132–1140.

Cook, L. G. and Crisp, M. D. (2005). Directional asymmetry of long-distance dispersal and colonization could mislead reconstructions of biogeography. *Journal of Biogeography*, **32**, 741–754.

Cook, R. R. (1995). The relationship between nested subsets, habitat subdivision, and species diversity. *Oecologia*, **101**, 204–210.

Cook, R. R. and Quinn, J. F. (1995). The influence of colonization in nested species subsets. *Oecologia*, **102**, 413–425.

Cook, W. M., Lane, K. T., Foster, B. L., and Holt, R. D. (2002). Island theory, matrix effects and species richness patterns in habitat fragments. *Ecology Letters*, **5**, 619–623.

Cordeiro, N. J. and Howe, H. F. (2001). Low recruitment of trees dispersed by animals in African forest fragments. *Conservation Biology*, **15**, 1733–1741.

Corlett, R. T. (1992). The ecological transformation of Singapore, 1819–1990. *Journal of Biogeography*, **19**, 411–420.

Cox, C. B. (2001). The biogeographic regions reconsidered. *Journal of Biogeography*, **28**, 511–523.

Cox, C. B. and Moore, P. D. (1993). *Biogeography: an ecological and evolutionary approach* (5th edn). Blackwell Scientific Publications, Oxford.

Cox, G. W. (1999). *Alien species in North America and Hawaii: Impacts on natural ecosystems*. Island Press, Washington, DC.

Cox, G. W. and Ricklefs, R. E. (1977). Species diversity, ecological release, and community structuring in Caribbean land bird faunas. *Oikos*, **28**, 113–122.

Cox, P. A. and Elmqvist, T. (2000). Pollinator extinction in the Pacific Islands. *Conservation Biology*, **14**, 1237–1239.

Cox, P. A., Elmqvist, T., Pierson, E. D., and Rainey, W. E. (1992). Flying foxes as pollinators and seed dispersers in Pacific Island ecosystems. In *Pacific island flying foxes: Proceedings of an international conservation conference*, Biological Report 90 (23), July 1992 (ed. D. E. Wilson and G. L. Graham), pp. 18–23. US Department of the Interior, Fish and Wildlife Service, Washington, DC.

Cramp, S. and Perrins, C. M. (eds.) (1994). *Handbook of the birds of Europe, the Middle East and North Africa. The birds of the Western Palearctic. Vol. VIII. Crows to finches*. Oxford University Press, Oxford.

Crawford, D. J., Sang, T., Stuessy, T. F., Kim, S. C., and Silva, M. (1998). *Dendroseris* (Asteraceae: Lactucaceae) and *Robinsonia* (Asteraceae: Senecioneae) on the Juan Fernández Islands: similarities and differences in biology and phylogeny. In *Evolution and speciation of island plants*. (ed. T. F. Stuessy and M. Ono) pp. 97–119. Cambridge University Press, Cambridge.

Crawford, D. J., Stuessy, T. F., Haines, D. W., Cosner, M. B. O., Silva, M., and López, P. (1992). Allozyme diversity within and divergence among four species of *Robinsonia* (Asteraceae: Senecioneae), a genus endemic to the Juan

Fernandez islands, Chile. *American Journal of Botany*, **79**, 962–966.

Cronk, Q. C. B. (1987). History of the endemic flora of St. Helena: a relictual series. *New Phytologist*, **105**, 509–520.

Cronk, Q. C. B. (1989). The past and present vegetation of St. Helena. *Journal of Biogeography*, **16**, 47–64.

Cronk, Q. C. B. (1990). History of the endemic flora of St. Helena: late Miocene 'Trochetiopsis-like' pollen from St. Helena and the origin of Trochetiopsis. *New Phytologist*, **114**, 159–165.

Cronk, Q. C. B. (1992). Relict floras of Atlantic islands: patterns assessed. *Biological Journal of the Linnean Society*, **46**, 91–103.

Cronk, Q. C. B. and Fuller, J. L. (1995). *Plant invaders*. Chapman & Hall, London.

Crooks, K. R. and Soulé, M. E. (1999). Mesopredator release and avifaunal extinctions in a fragmented system. *Nature*, **400**, 563–566.

Crowell, K. L. (1962). Reduced interspecific competition among the birds of Bermuda. *Ecology*, **43**, 75–88.

Crowell, K. L. (1986). A comparison of relict versus equilibrium models for insular mammals of the Gulf of Maine. *Biological Journal of the Linnean Society*, **28**, 37–64.

Cruz, F., Donlan, C. J., Campbell, K., and Carrión, V. (2005). Conservation action in the Galápagos: feral pig (*Sus scrofa*) eradication from Santiago Island. *Biological Conservation*, **121**, 473–478.

Cuddihy, L. W. and Stone, C. P. (1990). *Alteration of native Hawaiian vegetation: effects of humans, their activities and introductions*. University of Hawaii Cooperative National Park Resources Studies Unit, 3190 Maile Way, Honolulu, Hawaii.

Cutler, A. (1991). Nested fauna and extinction in fragmented habitats. *Conservation Biology*, **5**, 496–505.

Daily, G. C., Ceballos, G., Pacheco, J., Suzán, G., and Sánchez-Azofeifa, A. (2003). Countryside biogeography of Neotropical mammals: conservation opportunities in agricultural landscapes of Costa Rica. *Conservation Biology*, **17**, 1814–1826.

Daily, G. C., Ehrlich, P. R., and Sánchez-Azofeifa, A. (2001). Countryside biogeography: use of human-dominated habitats by the avifauna of southern Costa Rica. *Ecological Applications*, **11**, 1–13.

D'Antonio, C. M. and Dudley, T. L. (1993). Alien species: the insidious invasion of ecosystems by plants and animals from around the world has become a major environmental problem. *Pacific Discovery* 1993, Summer, 9–11.

D'Antonio, C. M. and Dudley, T. L. (1995). Biological invasions as agents of change on islands versus mainlands. In *Islands: biological diversity and ecosystem function* (ed. P. M. Vitousek, L. L. Loope, and H. Adsersen), Ecological Studies, **115**, pp. 103–121. Springer-Verlag, Berlin.

Dammerman, K. W. (1948). The fauna of Krakatau (1883–1933). *Verhandelingen der Konninklijke Nederlandse Akademie Van Wetenschappen Afdeling Natuurkunde (II)*, **44**, 1–594. N. V. Noord-Hollandsche Uitgevers Maatschappij, Amsterdam.

Damuth, J. (1993). Cope's rule, the island rule and the scaling of mammalian population density. *Nature*, **365**, 748–750.

Danley, P. D. and Kocher, T. D. (2001). Speciation in rapidly diverging systems: Lessons from Lake Malawi. *Molecular Ecology*, **10**, 1075–1086.

Darlington, P. J. (1957). *Zoogeography: the geographical distribution of animals*. Wiley, New York.

Darwin, C. (1842). *The structure and distribution of coral reefs. Being the first part of the geology of the voyage of the Beagle, under the command of Capt. Fitzroy, R. N., during the years 1832–36.* Smith, Elder, & Company, London.

Darwin, C. (1845). *Voyage of the Beagle.* (First published 1839. Abridged version edited by J. Browne and M. Neve (1989). Penguin Books, London.)

Darwin, C. (1859). *On the origin of species by means of natural selection.* J. Murray, London. (Page numbers cited here are from an edition published by Avenel, New York, in 1979 under the title *The origin of species*.)

Davis, S. D., Heywood, V. H., and Hamilton, A. C. (ed.) (1994). *Centres of plant diversity: a guide and strategy for their conservation. Volume 1: Europe, Africa, South West Asia and the Middle East.* WWF and IUCN, Cambridge.

Davis, S. D., Heywood, V. H., and Hamilton, A. C. (ed.) (1995). *Centres of plant diversity: a guide and strategy for their conservation. Volume 2: Asia, Australasia and the Pacific.* WWF and IUCN, Cambridge.

Dawson, D. (1994). Are habitat corridors conduits for animals and plants in a fragmented landscape? A review of the scientific evidence. *English Nature Research Reports*, No. 94.

Decker, R. W. and Decker, B. B. (1991). *Mountains of fire: The nature of volcanoes.* Cambridge University Press, Cambridge.

DeJoode, D. R. and Wendel, J. F. (1992). Genetic diversity and the origin of the Hawaiian Islands cotton *Gossypium tomentosum*. *American Journal of Botany*, **79**, 1311–1319.

Delcourt, H. R. and Delcourt, P. A. (1988). Quaternary landscape ecology: relevant scales in space and time. *Landscape Ecology*, **2**, 23–44.

Delcourt, H. R. and Delcourt, P. A. (1991). *Quaternary ecology: A paleoecological perspective.* Chapman & Hall, London.

Delgado, J. D. (2000). Patterns of insect flower visitation in *Lavandula buchii* Webb (Lamiaceae), an endemic shrub of Tenerife (Canary Islands). *Journal of Natural History*, **34**, 2145–2155.

Delgado, J. D., Arévalo, J. R., and Fernández-Palacios, J. M. (2001). Road and topography effects on invasion: edge effects in rat foraging patterns in two oceanic island forests (Tenerife, Canary Islands). *Ecography*, **24**, 539–546.

Delgado, J. D., Arévalo, J. R., and Fernández-Palacios, J. M. (2005). Patterns of artificial avian nest predation by introduced rats in a fragmented laurel forest (Tenerife, Canary Islands). *Journal of Natural History*, **38**, 2661–2669.

Delgado, J. D., Arévalo, J. R. and Fernández-Palacios, J. M. (in press) Road edge effect on the abundance of the lizard *Gallotia galloti* (Sauria: Lacertidae) in two Canary Islands forests. *Biodiversity and Conservation*. In press.

Desender, K., Baert, L., Maelfait, J.-P. and Verdyck, P. (1999). Conservation on Volcán Alcedo (Galápagos): terrestrial invertebrates and the impact of introduced feral goats. *Biological Conservation*, **87**, 303–310.

Deshaye, J. and Morisset, P. (1988). Floristic richness, area, and habitat diversity in a hemiarctic archipelago. *Journal of Biogeography*, **15**, 721–728.

Diamond, J. M. (1969). Avifaunal equilibria and species turnover rates on the Channel Islands of California. *Proceedings of the National Academy of Sciences of the USA*, **64**, 57–63.

Diamond, J. M. (1972). Biogeographical kinetics: estimation of relaxation times for avifaunas of southwest Pacific islands. *Proceedings of the National Academy of Sciences of the USA*, **69**, 3199–3203.

Diamond, J. M. (1974). Colonization of exploded volcanic islands by birds: the supertramp strategy. *Science*, **184**, 803–806.

Diamond, J. M. (1975a). Assembly of species communities. In *Ecology and evolution of communities* (ed. M. L. Cody and J. M. Diamond), pp. 342–444. Harvard University Press, Cambridge, MA.

Diamond, J. M. (1975b). The island dilemma: lessons of modern biogeographic studies for the design of natural preserves. *Biological Conservation*, **7**, 129–146.

Diamond, J. M. (1977). Continental and insular speciation in Pacific land birds. *Systematic Zoology*, **26**, 263–268.

Diamond, J. M. (1984). 'Normal' extinctions of isolated populations. In *Extinctions* (ed. M. H. Nitecki), pp. 191–246. University of Chicago Press, Chicago.

Diamond, J. M. (1986). The design of a nature reserve system for Indonesian New Guinea. In *Conservation biology: the science of scarcity and diversity* (ed. M. Soulé), pp. 485–503. Sinauer Associates, Sunderland, MA.

Diamond, J. M. (1991a). A new species of rail from the Solomon islands and convergent evolution of insular flightlessness. *Auk*, **108**, 461–470.

Diamond, J. M. (1991b). *The rise and fall of the third chimpanzee*. Vintage, London.

Diamond, J. M. (1996). Daisy gives an evolutionary answer. *Nature*, **380**, 103–104.

Diamond, J. M. (2005). *Collapse: how societies choose to fail or survive*. Allen Lane/Penguin, London.

Diamond, J. M. and Gilpin, M. E. (1982). Examination of the 'null' model of Connor and Simberloff for species co-occurrences on islands. *Oecologia*, **52**, 64–74.

Diamond, J. M. and Gilpin, M. E. (1983). Biogeographical umbilici and the origin of the Philippine avifauna. *Oikos*, **41**, 307–321.

Diamond, J. M. and May, R. M. (1977). Species turnover rates on islands: dependence on census interval. *Science*, **197**, 266–270.

Diamond, J. M. and May, R. M. (1981). Island biogeography and the design of nature reserves. In *Theoretical Ecology* (2nd edn) (ed. R. M. May), pp. 228–252. Blackwell, Oxford.

Diamond, J. and Pimm, S. (1993). Survival times of bird populations: a reply. *American Naturalist*, **142**, 1030–1035.

Diamond, J. M., Pimm, S. L., Gilpin, M. E., and LeCroy, M. (1989). Rapid evolution of character displacement in Myzomelid Honeyeaters. *American Naturalist*, **134**, 675–708.

Dias, E. (1996). *Vegetaçao natural dos Açores. Ecologia e sintaxonomia das florestas naturais*. Doctoral thesis, Azores University.

Diehl, S. and Bush, G. L. (1989). The role of habitat preference in adaptation and speciation. In *Speciation and its consequences* (ed. D. Otte and J. A. Endler), pp. 345–365. Sinauer, Sunderland, MA.

Doak, D. F. and Mills, L. S. (1994). A useful role for theory in conservation. *Ecology*, **75**, 615–626.

Docters van Leeuwen, W. M. (1936). Krakatau 1883–1933. *Annales du jardin botanique de Buitenzorg*, **46–47**, 1–506.

Drotz, M. K. (2003). Speciation and mitochondrial DNA diversification of the diving beetles *Agabus bipustulatus* and *A. wollastoni* (Coleoptera, Dytiscidae) within Macaronesia. *Biological Journal of the Linnean Society*, **79**, 653–666.

Duncan, R. P. and Blackburn, T. M. (2002). Morphological over-dispersion in game birds (Aves: Galliformes) successfully introduced to New Zealand was not caused by interspecific competition. *Evolutionary Ecology Research*, **4**, 551–561.

Duncan, R. P., Blackburn, T. M., and Sol, D. (2003). The ecology of bird introductions. *Annual Review of Ecology, Evolution and Systematics*, **34**, 71–98.

Dunn, C. P. and Loehle, C. (1988). Species–area parameter estimation: testing the null model of lack of relationship. *Journal of Biogeography*, **15**, 721–728.

Dunstan, C. E. and Fox, B. J. (1996). The effects of fragmentation and disturbance of rainforest on ground-dwelling small mammals on the Robertson Plateau, New South Wales, Australia. *Journal of Biogeography*, **23**, 187–201.

Ecker, A. (1976). Groundwater behaviour in Tenerife, volcanic island (Canary Islands, Spain). *Journal of Hydrology*, **28**, 73–86.

Economist (2001). Nauru: Paradise well and truly lost. *The Economist*, 20 December 2001. <www.economist.com/ displaystory.cfm?story_id = 884045> (visited March 2006)

Ehrendorfer, E. (1979). Reproductive biology in island plants. In *Plants and islands* (ed. D. Bramwell), pp. 293–306. Academic Press, London.

Ehrlich, P. R. and Hanski, I. (eds) (2004). *On the wings of checkerspots: a model system for population biology*. Oxford University Press, Oxford.

Elisens, W. J. (1992). Genetic divergence in *Galvezia* (Scrophulariaceae): evolutionary and biogeographic relationships among South American and Galápagos species. *American Journal of Botany*, **79**, 198–206.

Elmqvist, T., Rainey, W. E., Pierson, E. D., and Cox, P. A. (1994). Effects of tropical cyclones Ofa and Val on the structure of a Samoan lowland rain forest. *Biotropica*, **26**, 384–391.

Elvers, I. (1977). Flower-visiting lizards on Madeira. *Botaniska Notiser*, **130**, 231–234.

Emerson, B. C. (2002). Evolution on oceanic islands: molecular phylogenetic approaches to understanding pattern and process. *Molecular Ecology*, **11**, 951–966.

Emerson, B. C. and Kolm, N. (2005a). Species diversity can drive speciation. *Nature*, **434**, 1015–1017.

Emerson, B. C. and Kolm, N. (2005b). Emerson and Kolm reply. doi: 10.1038/nature04309.

Erickson, J. D. and Gowdy, J. M. (2000). Resource use, institutions, and sustainability: a tale of two Pacific island cultures. *Land Economics*, **76**, 345–354.

Ernst, A. (1908). *The new flora of the volcanic island of Krakatau* (translated into English by A. C. Seward). Cambridge University Press, Cambridge.

Erwin, R. M., Hatfield, J. S., and Winters, T. J. (1995). The value and vulnerability of small estuarine islands for conserving metapopulations of breeding waterbirds. *Biological Conservation*, **71**, 187–191.

Estrada, A., Coates-Estrada, R., and Meritt, D. (1993). Bat species richness and abundance in tropical rain forest fragments and in agricultural habitats at Los Tuxtlas, Mexico. *Ecography*, **16**, 309–318.

Ewers, R. M. and Didham, R. K. (2005). Confounding factors in the detection of species responses to habitat fragmentation. *Biological Reviews*, doi: 10.1017/S1464793105006949.

Fahrig, L. and Merriam, G. (1994). Conservation of fragmented populations. *Conservation Biology*, **8**, 50–59.

Feeley, K. (2003). Analysis of avian communities in Lake Guri, Venezuala, using multiple assembly rule models. *Oecologia*, **137**, 104–113.

Fernández-Palacios, J. M. (1992). Climatic responses of plant species on Tenerife, the Canary Islands. *Journal of Vegetation Science*, **3**, 595–602.

Fernández-Palacios, J. M. (2004). Introducción a las islas. In *Ecología Insular/Island Ecology* (ed. J. M. Fernández-Palacios and C. Morici), pp. 21–55. Asociación Española de Ecología Terrestre-Cabildo Insular de La Palma, Santa Cruz de La Palma.

Fernández-Palacios, J. M. and Andersson, C. (2000). Geographical determinants of the biological richness in the Macaronesian region. *Acta Phytogeographica Suecica*, **85**, 41–49.

Fernández-Palacios, J. M. and Arévalo, J. R. (1998). Tree strategies regeneration of the trees in the laurel forest of Tenerife (The Canary Islands). *Plant Ecology*, **127**, 21–29.

Fernández-Palacios, J. M., Arévalo J. R., Delgado, J. D., and Otto, R. (2004a). *Canarias: Ecología, medio ambiente y desarrollo*. Centro de la Cultura Popular de Canarias, La Laguna.

Fernández-Palacios, J. M., Arévalo, J. R., González-Delgado, G., Delgado, J. D., and Otto, R. (2004b). Estrategias de regeneración en la laurisilva. *Makaronesia*, **6**, 90–101.

Fernández-Palacios, J. M. and Dias, E. (2001). El marco biogeográfico macaronésico. In *Naturaleza de las Islas Canarias: Ecología y conservación*. (ed. J. M. Fernández-Palacios and J. L. Martín Esquivel), pp. 45–52. Turquesa Ediciones, Santa Cruz de Tenerife.

Fernández-Palacios, J. M. and de Nicolás, J. P. (1995). Altitudinal pattern of vegetation variation on Tenerife. *Journal of Vegetation Science*, **6**, 183–190.

Fernandopullé, D. (1976). Climate characteristics of the Canary Islands. In *Biogeography and ecology in the Canary Islands* (ed. G. Kunkel), pp. 185–206. W. Junk, The Hague.

Fiedler, P. L. and Jain, S. K. (ed.) (1992). *Conservation Biology: The theory and practice of nature conservation, preservation and management*. Chapman & Hall, London.

Field, R., O'Brien, E. M., and Whittaker, R. J. (2005). Global models for predicting woody plant richness from climate: development and evaluation. *Ecology*, **86**, 2263–2277.

Fischer, J. and Lindenmayer, D. B. (2005). Nestedness in fragmented landscapes: a case study on birds, arboreal marsupials and lizards. *Journal of Biogeography*, **32**, 1737–1750.

Fisher, R. A., Corbet, A. S., and Williams, C. B. (1943). The relation between the number of species and the number of individuals in a random sample of an animal population. *Journal of Animal Ecology*, **12**, 42–58.

Flenley, J. R. (1993). The palaeoecology of Easter Island, and its ecological disaster. In *Easter Island studies. Contributions to the history of Rapanui in memory of William T. Mulloy* (ed. S. R. Fisher), pp. 27–45. Oxbow, Oxford.

Flenley, J. R., King, S. M., Jackson, J., Chew, C., Teller, J. T., and Prentice, M. E. (1991). The Late Quaternary vegetational and climatic history of Easter Island. *Journal of Quaternary Science*, **6**, 85–115.

Forsman, A. (1991). Adaptive variation in head size in *Vipera berus* L. populations. *Biological Journal of the Linnean Society*, **43**, 281–296.

Fosberg, F. R. (1948). Derivation of the flora of the Hawaiian Islands. In *Insects of Hawaii*, Vol. 1 (ed. E. C. Zimmerman), pp. 107–119. University of Hawaii Press, Honolulu.

Fox, B. J. and Fox, M. D. (2000). Factors determining mammal species richness on habitat islands and isolates: habitat diversity, disturbance, species interactions and guild assembly rules. *Global Ecology and Biogeography*, **9**, 19–37.

Francisco-Ortega, J., Jansen, R. K., and Santos-Guerra, A. (1996). Chloroplast DNA evidence of colonization, adaptive radiation and hybridization in the evolution of the Macaronesian flora. *Proceedings of the National Academy of Sciences of the USA*, **24**, 249–266.

Francisco-Ortega, J. and Santos-Guerra, A. (2001). Genes y conservación de plantas vasculares. In *Naturaleza de las Islas Canarias: Ecología y conservación*. (ed. J. M. Fernández-Palacios and J. L. Martín Esquivel), pp. 357–365. Turquesa Ediciones, Santa Cruz de Tenerife.

Francisco-Ortega, J., Santos-Guerra, A., Kim, S. C., and Crawford, D. J. (2000). Plant genetic diversity in the Canary Islands: a conservation perspective. *American Journal of Botany*, **87**, 909–919.

Frankham, R. (1997). Do island populations have less genetic variation than mainland populations? *Heredity*, **78**, 311–327.

Franklin, J. and Steadman, D. W. (1991). The potential for conservation of Polynesian birds through habitat mapping and species translocation. *Conservation Biology*, **5**, 506–521.

Fritts, T. H. and Leasman-Tanner, D. (2001). The brown treesnake on Guam: How the arrival of one invasive species damaged the ecology, commerce, electrical systems, and human health on Guam: A comprehensive information source. Available online: <www.fort.usgs.gov/resources/education/bts/bts_home.asp> (visited June 2005).

Fu, Y. (2000). Toward the phylogeny of the family Lacertidae—why 4708 base pairs of mtDNA sequences cannot draw the picture. *Biological Journal of the Linnean Society*, **71**, 203–217.

Funk, V. A. and Wagner, W. L. (1995). Biogeographic patterns in the Hawaiian Islands. In *Hawaiian biogeography. Evolution on a hot spot archipelago* (ed. W. L. Wagner and V. L. Funk). pp. 379–419. Smithsonian, Washington, DC.

Fujita, M. S. and Tuttle, M. D. (1991). Flying foxes (Chiroptera: Pteropodidae): threatened animals of key ecological and economic importance. *Conservation Biology*, **5**, 455–463.

Futuyma, D. (1986). *Evolutionary biology*. Sinauer Associates, Sunderland, MA.

Galapagos Newsletter (1996). No. 3, Autumn. Galapagos Conservation Trust, 18 Curzon Street, London.

Galley, C. and Linder, H. P. (2006). Geographical affinities of the Cape flora, South Africa. *Journal of Biogeography*, **33**, 236–250.

García-Falcón, J. M. and Medina-Muñoz, D. (1999). Sustainable tourism development in islands: a case study of Gran Canaria. *Business Strategy and the Environment*, **8**, 336–357.

García-Talavera, F. (1999). La Macaronesia. Consideraciones geológicas, biogeográficas y paleoecológicas. In *Ecología y cultura en Canarias*.(ed. J. M. Fernández-Palacios, J. J. Bacallado, and J. A. Belmonte), pp. 39–63. Ed. Organismo Autónomo de Museos y Centros, Cabildo Insular de Tenerife, Santa Cruz de Tenerife.

Gardner, A. S. (1986). The biogeography of the lizards of the Seychelles Islands. *Journal of Biogeography*, **13**, 237–253.

Gascon, C., Lovejoy, T. E., Bierregaard, Jr. R. O., Malcolm, J. R., Stouffer, P. C., Vasconcelos, H., Laurance, W. F., Zimmerman, B., Tocher, M., and Borges, S. (1999). Matrix habitat and species persistence in tropical forest remnants. *Biological Conservation*, **91**, 223–230.

Gaston, K. J. (1991). Estimates of the near imponderable: a reply to Erwin. *Conservation Biology*, **5**, 564–566.

Gates, S. and Donald, P. F. (2000). Local extinction of British farmland birds and the prediction of further loss. *Journal of Applied Ecology*, **37**, 806–820.

Gentile, G. and Argano, R. (2005). Island biogeography of the Mediterranean sea: the species-area relationship for terrestrial isopods. *Journal of Biogeography*, **32**, 1715–1726.

Gibbs, H. L. and Grant, P. R. (1987). Ecological consequences of an exceptionally strong El Niño event on Darwin's finches. *Ecology*, **68**, 1735–1746.

Gibson, C. W. D. (1986). Management history in relation to changes in the flora of different habitats on an Oxfordshire estate, England. *Biological Conservation*, **38**, 217–232.

Gilbert, F. S. (1980). The equilibrium theory of island biogeography, fact or fiction? *Journal of Biogeography*, **7**, 209–235.

Gilpin, M. E. and Diamond, J. M. (1982). Factors contributing to non-randomness in species co-occurences on islands. *Oecologia*, **52**, 75–84.

Gittenberger, E. (1991). What about non-adaptive radiation? *Biological Journal of the Linnean Society*, **43**, 263–272.

Givnish, T. J. (1998). Adaptive plant evolution on islands. In: *Evolution on islands* (ed. P. Grant) pp. 281–304. Oxford University Press, Oxford.

Godwin, H. (1975). *The history of the British flora: a factual basis for phytogeography* (2nd edn). Cambridge University Press, Cambridge.

Golinski, M. and Boecklen, W. J. (2006). A model-independent test for the presence of regulatory equilibrium and non-random structure in island species trajectories. *Journal of Biogeography*, **33**, 1566–1570.

González, P., Pinto, F., Nogales, M., Jiménez-Asensio, J., Hernández, M., and Cabrera, V.M. (1996). Phylogenetic relationships of the Canary Islands endemic lizard genus *Gallotia* (Sauria: Lacertidae), inferred from mitochondrial DNA sequences. *Molecular Phylogenetics and Evolution*, **6**, 63–71.

González-Pérez, M. A., Caujapé-Castells, J., and Sosa, P. A. (2004). Molecular evidence of hybridisation between the endemic *Phoenix canariensis* and the widespread *Phoenix dactylifera* with Random Amplified Polymorphic DNA (RAPD). *Plant Systematics and Evolution*, **247**, 165–175.

Goodfriend, G. A., Cameron, R. A. D., and Cook, L. M. (1994). Fossil evidence of recent human impact on the land snail fauna of Madeira. *Journal of Biogeography*, **21**, 309–320.

Goodwillie, C. (1999). Wind pollination and reproductive assurance in *Linanthus parviflorus* (Polemoniaceae), a self-compatible annual. *American Journal of Botany*, **86**, 948–954.

Gorchov, D. L., Cornejo, F., Ascorra, C., and Jaramillo, M. (1993). The role of seed dispersal in the natural regeneration of rain forest after strip-cutting in the Peruvian Amazon. *Vegetatio*, **107/108**, 339–349.

Gorman, M. L. (1979). *Island ecology*. Chapman & Hall, London.

Gotelli, N. J. (1991). Metapopulation models: the rescue effect, the propagule rain, and the core-satellite hypothesis. *American Naturalist*, **138**, 768–776.

Gotelli, N. J. and Entsminger, G. L. (2001). Swap and fill algorithms in null model analysis: rethinking the knight's tour. *Oecologia*, **129**, 281–291.

Gotelli, N. J. and Graves, G. R. (1996). *Null models in ecology*. Smithsonian Institution Press, Washington, DC.

Gotelli, N. J. and McCabe, D. J. (2002). Species co-occurrence: a meta-analysis of J. M. Diamond's assembly rules model. *Ecology*, **83**, 2091–2096.

Grainger, A. (1993). Rates of deforestation in the humid tropics: estimates and measurements. *Geographical Journal*, **159**, 33–44.

Grant, B. R. and Grant, P. R. (1996*a*). High survival of Darwin's finch hybrids: effects of beak morphology and diets. *Ecology*, **77**, 500–509.

Grant, P. R. (1981). Speciation and the adaptive radiation of Darwin's finches. *American Scientist*, **69**, 653–663.

Grant, P. R. (1984). The endemic land birds. In *Galapagos* (ed. R. Perry), Key Environments Series, pp. 175–189. IUCN/Pergamon Press, Oxford.

Grant, P. R. (1986). Interspecific competition in fluctuating environments. In *Community Ecology* (ed. J. M. Diamond and T. J. Case), pp. 173–191. Harper & Row, New York.

Grant, P. R. (1994). Population variation and hybridization: comparison of finches from two archipelagos. *Evolutionary Ecology*, **8**, 598–617.

Grant, P. R. (1998). Competition exposed by knight? *Nature*, **396**, 216–217.

Grant, P. R. (2002). Founder effects and silvereyes. *Proceedings of the National Academy of Sciences of the USA*, **99**, 7818–7820.

Grant, P. R. and Grant, B. R. (1989). Sympatric speciation and Darwin's finches. In *Speciation and its consequences* (ed. D. Otte and J. A. Endler), pp. 433–457. Sinauer Associates, Sunderland, MA.

Grant, P. R. and Grant, B. R. (1994). Phenotypic and genetic effects of hybridization in Darwin's finches. *Evolution*, **48**, 297–316.

Grant, P. R. and Grant, B. R. (1996*b*). Speciation and hybridization in island birds. *Philosophical Transactions of the Royal Society of London*, B **351**, 765–772.

Graves, G. R. and Gotelli, N. J. (1983). Neotropical land-bridge avifaunas: new approaches to null hypotheses in biogeography. *Oikos*, **41**, 322–333.

Gray, J. S., Ugland, K. I., and Lambshead, J. (2004). On species accumulation and species-area curves. *Global Ecology and Biogeography*, **13**, 567–568.

Grayson, D. L. and Livingston, S. D. (1993). Missing mammals on Great Basin mountains: Holocene extinctions and inadequate knowledge. *Conservation Biology*, **7**, 527–532.

Greenslade, P. J. M. (1968). Island patterns in the Solomon Islands bird fauna. *Evolution*, **22**, 751–761.

Grelle, C. E. V., Alves, M. A. S., Bergallo, H. G., Geise, L., Rocha, C. F. D., Van Sluys, M., and Caramaschi, U. (2005). Prediction of threatened tetrapods based on the species–area relationship in Atlantic forest, Brazil. *Journal of the Zoological Society of London*, **265**, 359–364.

Griffith, B., Scott, J. M., Carpenter, J. W., and Reed, C. (1989). Translocation as a species conservation tool: status and strategy, *Science*, **245**, 477–480.

Griffiths, O., Cook, A., and Wells, S. M. (1993). The diet of the introduced carnivorous snail *Euglandina rosea* in Mauritius and its implications for threatened island gastropod faunas. *Journal of Zoology (London)*, **229**, 78–89.

Groombridge, B. (ed.) (1992). *Global biodiversity: status of the Earth's living resources*. (A report compiled by the World Conservation Monitoring Centre.) Chapman & Hall, London.

Groombridge, B. and Jenkins, M. D. (2002). *World atlas of biodiversity. Earth's living resources in the 21st century*. United Nations Environmental Programme–World Conservation Monitoring Center, University of California Press, Berkeley.

Guevara, S. and Laborde, J. (1993). Monitoring seed dispersal at isolated standing trees in tropical pastures: consequences for local species availability. *Plant Ecology*, **107/108**, 319–338.

Haas, A. C., Dunski, J. F., and Maxson, L. R. (1995). Divergent lineages within the *Bufo margaritifera* Complex (Amphibia: Anura; Bufonidae) revealed by albumin immunology. *Biotropica*, **27**, 238–249.

Haila, Y. (1990). Towards an ecological definition of an island: a northwest European perspective. *Journal of Biogeography*, **17**, 561–568.

Haila, Y. and Hanski, I. K. (1993). Birds breeding on small British island and extinction risks. *American Naturalist*, **142**, 1025–1029.

Haila, Y. and Järvinen, O. (1983). Land bird communities on a Finnish island: species impoverishment and abundance patterns. *Oikos*, **41**, 255–273.

Hair, J. B. (1966). Biosystematics of the New Zealand flora 1945–64. *New Zealand Journal of Botany*, **4**, 559–595.

Hambler, C. (1994). Giant tortoise *Geochelone gigantea* translocation to Curieuse Island (Seychelles): Success or failure? *Biological Conservation*, **69**, 293–299.

Hamnett, M. P. (1990). Pacific island resource development and environmental management. In *Sustainable development and environmental management of small islands*. Vol. **5**, Man and the Biosphere Series (ed. W. Beller, P. d'Ayala, and P. Hein), pp. 227–257. UNESCO/Parthenon Publishing, Paris.

Hanley, K. A., Bolger, D. T., and Case, T. J. (1994). Comparative ecology of sexual and asexual gecko species (*Lepidodactylus*) in French Polynesia. *Evolutionary Ecology*, **8**, 438–454.

Hanski, I. (1986). Population dynamics of shrews on small islands accord with the equilibrium model. *Biological Journal of the Linnean Society*, **28**, 23–36.

Hanski, I. (1992). Inferences from ecological incidence functions. *American Naturalist*, **138**, 657–662.

Hanski, I. (1996). Metapopulation ecology. In *Population dynamics in ecological space and time* (ed. O. E. Rhodes, Jr, R. K. Chesser, and M. H. Smith), pp. 13–43. Chicago University Press, Chicago.

Hanski, I., Moilanen, A., and Gyllenberg, M. (1996). Minimum viable metapopulation size. *American Naturalist*, **147**, 527–541.

Hansson, L. (1998). Nestedness as a conservation tool: plants and birds of oak-hazel woodland in Sweden. *Ecology Letters*, **1**, 142–145.

Harris, L. D. (1984). *The fragmented forest: island biogeography theory and the preservation of biotic diversity*. University of Chicago Press, Chicago.

Harrison, S. (1994). Metapopulations and conservation. In *Large-scale ecology and conservation biology* (ed. P. J. Edwards, R. M. May, and N. R. Webb), pp. 111–128. Blackwell Scientific Publications, Oxford.

Harrison, S., Murphy, D. D., and Ehrlich, P. R. (1988). Distribution of the Bay Checkerspot Butterfly, *Euphydryas editha bayensis*: evidence for a metapopulation model. *American Naturalist*, **132**, 360–382.

Hart, K., Carey, S., Sigurdsson, H., Sparks, R. S. J., and Robertson, R. E. A. (2004). Discharge of pyroclastic flows into the sea during the 1996–1998 eruptions of the Soufriere Hills volcano, Montserrat. *Bulletin of Volcanology*, **66**, 599–614.

Hawkins, B. A., Field, R., Cornell, H. V., Currie, D. J., Guegan, J-F., Kaufman, D. M. Kerr, J. T., Mittelbach, G. G., Oberdorff, T., O'Brien, E. M., Porter, E. E., and Turner, J. R. G. (2003). Energy, water, and broad-scale geographic patterns of species richness. *Ecology*, **84**, 3105–3117.

Heads, M. J. (1990). Mesozoic tectonics and the deconstruction of biogeography: a new model of Australasian biology. *Journal of Biogeography*, **17**, 223–225.

Heads, M. J. (2004). What is a node? *Journal of Biogeography*, **32**, 1883–1891.

Heaney, L. R. (1978). Island area and body size of insular mammals: evidence from the tri-colored squirrel

(*Callosciurus prevosti*) of Southeast Asia. *Evolution*, **32**, 29–44.

Heaney, L. R. (1986). Biogeography of mammals in SE Asia: estimates of rates of colonization, extinction and speciation. *Biological Journal of the Linnean Society*, **28**, 127–165.

Heaney, L. R. (2000). Dynamic disequilibrium: a long-term, large-scale perspective on the equilibrium model of island biogeography. *Global Ecology and Biogeography*, **9**, 59–74.

Heaney, L. R., Walsh, J. S., Jr., and Peterson, A. T. (2005). The roles of geological history and colonization abilities in genetic differentiation between mammalian populations in the Philippine archipelago. *Journal of Biogeography*, **32**, 229–247.

Hearty, J. P. (1997). Boulder deposits from large waves during the last interglaciation on North Eleuthera Island, Bahamas. *Quaternary Research*, **48**, 326–338.

Henderson, S., Dawson, T. P., and Whittaker, R. J. (2006). Progress in invasive plants research. *Progress in Physical Geography*, **30**, 25–46.

Herben, T., Suda, J., and Munclinger, P. (2005). The ghost of hybridization past: niche pre-emption is not the only explanation of apparent monophyly in island endemics. *Journal of Ecology*, **93**, 572–575.

Hernández, E., Nogales, M., and Martín, A. (2000). Discovery of a new lizard in the Canary Islands, with a multivariate analysis of *Gallotia. Herpetelogica*, **56**, 63–76.

Herrera, C. M. (1989). Pollinator abundance, morphology and floral visitation rate: analysis of the quantity component in a plant-pollinator system. *Oecologia*, **80**, 241–248.

Herrero, S. (1997). Galapagos tortoises threatened. *Conservation Biology*, **11**, 305.

Hess, A. L. (1990). Overview: sustainable development and environmental management of small islands. In *Sustainable development and environmental management of small islands*. Vol. **5**, Man and the Biosphere Series (ed. W. Beller, P. d'Ayala, and P. Hein), pp. 3–14. UNESCO/Parthenon Publishing, Paris.

Hess, G. R. (1996). Linking extinction to connectivity and habitat destruction in metapopulation models. *American Naturalist*, **148**, 226–236.

Hess, J., Kadereit, J. W., and Vargas, P. (2000). The colonization history of *Olea europaea* L. in Macaronesia based on internal transcribed spacer 1 (ITS-1) sequences, randomly amplified polymorphic DNAs (RAPD), and intersimple sequence repeats (ISSR). *Molecular Ecology*, **9**, 857–868.

Higuchi, H. (1976). Comparative study on the breeding of mainland and island subspecies of the varied tit (*Parus varius*). *Tori*, **25**, 11–20.

Hille, S. M., Nesje, M., and Segelbacher, G. (2003). Genetic structure of kestrel populations and colonization of the Cape Verde archipelago. *Molecular Ecology*, **12**, 2145–2151.

Hilton, G. M., Atkinson, P. W., Gray, G. A. L., Arendt, W.J., and Gibbons, D. W. (2003). Rapid decline of the volcanically threatened Montserrat oriole. *Biological Conservation*, **111**, 79–89.

Hinsley, S. A., Bellamy, P. E., Newton, I., and Sparks, T. H. (1994). Factors influencing the presence of individual breeding bird species in woodland fragments. *English Nature Research Reports*, No. 99.

Hnatiuk, S. H. (1978). Plant dispersal by the Aldabran Giant tortoise, *Geochelone gigantea* (Schweigger) *Oecologia*, **36**, 345–350.

Hobbs, R. J. and Huenneke, L. F. (1992). Disturbance, diversity, and invasion: implications for conservation. *Conservation Biology*, **6**, 324–337.

Hohmann, H., La Roche, F., Ortega, G., and Barquín, J. (1993). *Bienen, Wespen und Ameisen der Kanarischen Inseln, I-II.* Übersee-Museum, Bremen.

Holdaway, R. N. (1989). New Zealand's pre-human avifauna and its vulnerability. In Rudge, M. R. (ed.), Moas, man and climate in the ecological history of New Zealand. *New Zealand Journal of Ecology*, **12** (Suppl.), 11–25.

Holdaway, R. N. (1990). Changes in the diversity of New Zealand forest birds. *New Zealand Journal of Zoology*, **17**, 309–321.

Holloway, J. D. (1996). The Lepidoptera of Norfolk Island, actual and potential, their origins and dynamics. In *The origin and evolution of Pacific Island Biotas, New Guinea to Eastern Polynesia: Patterns and processes* (ed. A. Keast, and S. E. Miller), pp. 123–151. SPB Academic Publishing, Amsterdam.

Hommel, P. W. F. (1990). Ujung Kulon: landscape survey and land evaluation as a habitat for the Javan rhinoceros. *ITC Journal*, 1990–1, 1–15.

Honnay, O., Hermy, M., and Coppin, P. (1999). Nested plant communities in deciduous forest fragments: species relaxation or nested habitats? *Oikos*, **84**, 119–129.

Hoogerwerf, A. (1953). Notes on the vertebrate fauna of the Krakatau Islands, with special reference to the birds. *Treubia*, **22**, 319–348.

Hubbell, S. P. (2001). *The unified neutral theory of biodiversity and biogeography*. Princeton University Press, Princeton, NJ.

Huggett, R. J. (1995). *Geoecology: an evolutionary approach*. Routledge, London.

Hughes, J. B., Daily, G. C., and Ehrlich, P. R. (2002). Conservation of tropical forest birds in countryside habitats. *Ecology Letters*, **5**, 121–129.

Hughes, L., Dunlop, M., French, K., Leishman, M. R., Rice, B., Rodgerson, L., and Westoby, M. (1994). Predicting dispersal spectra: a minimal set of hypotheses based on plant attributes. *Journal of Ecology*, **82**, 933–950.

Humphreys, W. F. and Kitchener, D. J. (1982). The effect of habitat utilization on species–area curves: implications for optimal reserve area. *Journal of Biogeography*, **9**, 381–396.

Humphries, C. J. (1979). Endemism and evolution in Macaronesia. In *Plants and islands* (ed. D. Bramwell), pp. 171–199. Academic Press, London.

Hunt, G. L., Jr and Hunt, M. W. (1974). Trophic levels and turnover rates: the avifauna of Santa Barbara island, California. *Condor*, **76**, 363–369.

Huntley, B., Berry, P. M., Cramer, W., and McDonald, A. P. (1995). Modelling present and potential future ranges of some European higher plants using climate response surfaces. *Journal of Biogeography*, **22**, 967–1001.

Hürlimann A., Martí J., and Ledesma, A. (2004). Morphological and geological aspects related to large slope failures on oceanic islands—The huge La Orotava landslides on Tenerife, Canary Islands. *Geomorphology*, **62**, 143–158.

Hutchinson, G. E. (1957). Concluding remarks. *Cold Spring Harbor Symposia on Quantitative Biology*, **22**, 415–427.

Inoue, K. and Kawahara, T. (1996). Evolution of *Campanula* flowers in relation to insect pollinators on islands. In *Floral biology: Studies on floral evolution in animal-pollinated plants* (ed. D. G. Lloyd and S. C. H. Barrett) pp. 377–400. Chapman & Hall, New York.

Itow, S. (1988). Species diversity of mainland- and island forests in the Pacific area. *Vegetatio*, **77**, 193–200.

IUCN/SSC Invasive Species Specialist Group (2004). <www.issg.org> (visited June 2005).

Izquierdo, I, Martín, J. L., Zurita, N., and Arechavaleta, M. (eds) (2001). *Lista de especies silvestres de Canarias (hongos, plantas y animales terrestres)*. Consejería de Política Territorial y Medio Ambiente, Gobierno de Canarias, La Laguna, Tenerife.

Izquierdo, I, Martín, J. L., Zurita, N., and Arechavaleta, M. (eds.) (2004). *Lista de especies silvestres de Canarias (hongos, plantas y animales terrestres) 2004*. Consejería de Medio Ambiente y Ordenación Territorial, Gobierno de Canarias, Santa Cruz de Tenerife.

Jackson, E. D., Silver, E. A., and Dalrymple, G. B. (1972). Hawaiian-Emperor chain and its relation to Cenozoic circumpacific tectonics. *Geological Society of America, Bulletin*, **83**, 601–618.

Jackson, M. H. (1995). *Galápagos: a natural history*. University of Calgary Press, Calgary.

James, H. F. (1995). Prehistoric extinctions and ecological changes on Oceanic islands. In *Islands: biological diversity and ecosystem function* (ed. P. M. Vitousek, L. L. Loope, and H. Adersen), Ecological Studies **115**, pp. 87–102. Springer-Verlag, Berlin.

James, H. F. and Olson, S. L. (1991). Descriptions of thirty-two new species of birds from the Hawaiian Islands: Part II. Passeriformes. *Ornithological Monographs*, **46**, 1–88.

Janzen, D. H. (1979). How to be a fig. *Annual Review of Ecology and Systematics*, **10**, 13–51.

Janzen, D. H. (1983). No park is an island: increase in interference from outside as park size increases. *Oikos*, **41**, 402–410.

Janzen, D. H. (1985). On ecological fitting. *Oikos*, **45**, 308–310.

Järvinen, O. and Haila, Y. (1984). Assembly of land bird communities on Northern Islands: a quantitative analysis of insular impoverishment. In *Ecological communities: conceptual issues and the evidence* (ed. D. R. Strong, Jr, D. Simberloff, L. G. Abele, and A. B. Thistle), pp. 138–147. Princeton University Press, Princeton, NJ.

Jenkyns, H. C. and Wilson, P. A. (1999). Stratigraphy, paleoceanography, and evolution of Cretaceous Pacific guyots: relics from a greenhouse earth. *American Journal of Science*, **299**, 341–392.

Johnson, D. L. (1980). Problems in the land vertebrate zoogeography of certain islands and the swimming power of elephants. *Journal of Biogeography*, **7**, 383–398.

Johnson, T. C., Scholz, C. A., and Talbot, M. R. (1996). Late Pleistocene desiccation of Lake Victoria and rapid evolution of cichlid fishes. *Science*, **273**, 1091–1093.

Johnson, T. H. and Stattersfield, A. J. (1990). A global review of island endemic birds. *Ibis*, **132**, 167–180.

Juan, C., Emerson, B. C., Oromí, P., and Hewitt, G. M. (2000). Colonization and diversification: towards a phylogenetic synthesis for the Canary Islands. *Trends in Ecology and Evolution*, **15**, 104–109.

Juvig, J. O. and Austring, A. O. (1979). The Hawaiian avifauna: biogeographical theory in evolutionary time. *Journal of Biogeography*, **6**, 205–224.

Kadmon, R. (1995). Nested species subsets and geographic isolation: a case study. *Ecology*, **76**, 458–465.

Kalmar, A. and Currie, D. J. (2006). A global model of island biogeography. *Global Ecology and Biogeography*, **15**, 72–81.

Kaneshiro, K. Y. (1989). The dynamics of sexual selection and founder effects in species formation. In *Genetics, speciation and the founder principle* (ed. L. V. Giddings, K. Y. Kaneshiro, and W. W. Anderson), pp. 279–296. Oxford University Press, New York.

Kaneshiro, K. Y. (1995). Evolution, speciation, and the genetic structure of island populations. In *Islands: biological diversity and ecosystem function* (ed. P. M. Vitousek, L. L. Loope, and H. Adsersen), Ecological Studies **115**, pp. 22–23. Springer-Verlag, Berlin.

Kaneshiro, K. Y., Gillespie, R. G., and Carson, H. L. (1995). Chromosomes and male genitalia of Hawaiian *Drosophila*: tools for interpreting phylogeny and geography. In *Hawaiian biogeography: evolution on a hot spot archipelago* (ed. W. L. Wagner and V. A. Funk), pp. 57–71. Smithsonian Institution Press, Washington, DC.

Kapos, V. (1989). Effects of isolation on the water status of forest patches in the Brazilian Amazon. *Journal of Tropical Ecology*, **5**, 173–185.

Kayanne, H., Ishi, T., Matsumoto, E., and Yonekura, N. (1993). Late Holocene sea-level change on Rota and Guam, Mariana Islands, and its contraint on geophysical predictions. *Quaternary Research*, **40**, 189–200.

Keast, A. and Miller, S. E. (ed.) (1996). *The origin and evolution of Pacific Island biotas, New Guinea to Eastern Polynesia: patterns and processes*. SPB Academic Publishing, Amsterdam.

Kellman, M. (1996). Redefining roles: plant community reorganization and species preservation in fragmented systems. *Global Ecology and Biogeography Letters*, **5**, 111–116.

Kelly, B. J., Wilson, J. B., and Mark, A. F. (1989). Causes of the species–area relation: a study of islands in Lake Manapouri, New Zealand. *Journal of Ecology*, **77**, 1021–1028.

Kim, S.-C., Crawford, D. J., Francisco-Ortega, J., and Santos-Guerra, A. (1996). A common origin for woody *Sonchus* and five related genera in the Macaronesian islands: Molecular evidence for extensive radiation. *Proceedings of the National Academy of Sciences of the USA*, **93**, 7743–7748.

Kindvall, O. and Ahlén, I. (1992). Geometrical factors and metapopulation dynamics of the Bush Cricket, *Metrioptera bicolor* Philippi (Orthoptera: Tettigoniidae). *Conservation Biology*, **6**, 520–529.

King, C. (1984). *Immigrant killers*. Oxford University Press, Auckland.

King, K. J., Young, K. D., Waters, J. M., and Wallis, G. P. (2003). Preliminary genetic analysis of koaro (*Galaxias brevipinnis*) in New Zealand lakes: Evidence for allopatric differentiation among lakes but little population subdivision within lakes. *Journal of the Royal Society of New Zealand*, **33**, 591–600.

Klein, N. K. and Brown, W. M. (1994). Intraspecific molecular phylogeny in the Yellow Warbler (*Dendroica petechia*), and implications for avian biogeography in the West Indies. *Evolution*, **48**, 1914–1932.

Kohn, D. D. and Walsh, D. M. (1994). Plant species richness—the effect of island size and habitat diversity. *Journal of Ecology*, **82**, 367–377.

Koopowitz, H. and Kaye, H. (1990). *Plant extinction: a global crisis* (2nd edn). Christopher Helm, London.

Korn, H. (1994). Genetic, demographic, spatial, environmental and catastrophic effects on the survival probability of small populations of mammals. In *Minimum animal populations* (ed. H. Remmert), pp. 39–49. Ecological Studies **106**. Springer-Verlag, Berlin.

Krebs, J. R. (1970). Regulation of numbers in the great tit. *Journal of the Zoological Society of London*, **162**, 317–333.

Kunkel, G. (ed.) (1976). *Biogeography and ecology of the Canary Islands*. W. Junk, The Hague.

Kunkel, G. (1993). *Die Kanarischen Inseln und ihre Pflanzenwelt*. Gustav Fisher, Stuttgart.

Kvist, L., Broggi, J., Illera, J. C., and Koivula, K. (2005). Colonisation and diversification of the blue tits (*Parus caeruleus teneriffae*-group) in the Canary Islands. *Molecular Phylogenetics and Evolution*, **34**, 501–511.

Lack, D. (1947*a*). *Darwin's finches: an essay on the general biological theory of evolution*. Cambridge University Press, Cambridge

Lack, D. (1947*b*). The significance of clutch size. *Ibis*, **89**, 302–352.

Lack, D. (1969). The numbers of bird species on islands. *Bird Study*, **16**, 193–209.

Lack, D. (1970). The endemic ducks of remote islands. *Wildfowl*, **21**, 5–10.

Lack, D. (1976). *Island biology illustrated by the land birds of Jamaica*. Blackwell Scientific Publications, Oxford.

Lacy, R. C. (1992). The effects of inbreeding on isolated populations: are minimum viable population sizes predictable? In *Conservation Biology: the theory and practice of nature conservation and management* (ed. P. L. Fiedler and S. K. Jain), pp. 276–320. Chapman & Hall, New York.

Lambshead, P. J. D. and Boucher, G. (2003). Marine nematode deep-sea biodiversity—hyperdiverse or hype? *Journal of Biogeography*, **30**, 475–485.

Laurance, W. F. (2002). Hyperdynamism in fragmented habitats. *Journal of Vegetation Science*, **13**, 595–602.

Law, R. and Watkinson, A. R. (1989). Competition. In *Ecological concepts: the contribution of ecology to an understanding of the natural world* (ed. J. M. Cherrett), pp. 243–284. Blackwell Scientific Publications, Oxford.

Lawesson, J. E., Adsersen, H., and Bentley, P. (1987). An updated and annotated check list of the vascular plants of the Galápagos Islands. *Reports from The Botanical Institute, University of Aarhus*, **16**.

Lawlor, T. E. (1986). Comparative biogeography of mammals on islands. *Biological Journal of the Linnean Society*, **28**, 99–125.

Lawton, J. H. and May, R. M. (ed.) (1995). *Extinction rates*. Oxford University Press, Oxford.

Le Friant, A., Harford, C. L., Deplus, C., Boudon, G., Sparks, R. S. J., Herd, R. A. and Komorowski, J. C. (2004). Geomorphological evolution of Montserrat (West Indies): importance of flank collapse and erosional processes. *Journal of the Geological Society*, **161**, 147–160.

Leader-Williams, N. and Walton, D. (1989). The isle and the pussycat. *New Scientist*, **121** (1651), 11 February, 48–51.

Leberg, P. L. (1991). Influence of fragmentation and bottlenecks on genetic divergence of wild turkey populations. *Conservation Biology*, **5**, 522–530.

Leuschner, C. (1996). Timberline and alpine vegetation on the tropical and warm-temperate oceanic islands of the world: elevation, structure, and floristics. *Vegetatio*, **123**, 193–206.

Lipman, P. W., Normark, W. R., Moore, J. G., Wilson, J. B. and Gutmacher. C. E. (1988). The giant Alika debris slide, Mauna Loa, Hawaii. *Journal of Geophysical Research*, **93**, 4279–4299.

Lister, A. M. (1993). Mammoths in miniature. *Nature*, **362**, 288–289.

Livezey, B. C. (1993). An ecomorphological review of the dodo (*Raphus cucullatus*) and solitaire (*Pezophaps solitaria*), flightless Columbiformes of the Mascarene Islands. *Journal of Zoology (London)*, **230**, 247–292.

Lobin, W. (1982). Untersuchung über flora, vegetation und biogeographische beziehungen der Kapverdischen Inseln. *Courier Forschung Institut Senckenberg*, **53**, 1–112.

Lockwood, J. L. and Moulton, M. P. (1994). Eco-morphological pattern in Bermuda birds: the influence of competition and implications for nature preserves. *Evolutionary Ecology*, **8**, 53–60.

Loehle, C. and Li, B. L. (1996). Habitat destruction and the extinction debt revisited. *Ecological Applications*, **6**, 784–789.

Lomolino, M. V. (1984a). Immigrant selection, predation, and the distribution of *Microtus pennsylvanicus* and *Blarina brevicauda* on islands. *American Naturalist*, **123**, 468–483.

Lomolino, M. V. (1984b). Mammalian island biogeography: effects of area, isolation, and vagility. *Oecologia*, **61**, 376–382.

Lomolino, M. V. (1985). Body size of mammals on islands: the island rule reexamined. *American Naturalist*, **125**, 310–316.

Lomolino, M. V. (1986). Mammalian community structure on islands: the importance of immigration, extinction and interactive effects. *Biological Journal of the Linnean Society*, **28**, 1–21.

Lomolino, M. V. (1990). The target area hypothesis: the influence of island area on immigration rates of non-volant mammals. *Oikos*, **57**, 297–300.

Lomolino, M. V. (1996). Investigating causality of nested-ness of insular communities: selective immigrations or extinctions? *Journal of Biogeography*, **23**, 699–703.

Lomolino, M. V. (2000a). A call for a new paradigm of island biogeography. *Global Ecology and Biogeography*, **9**, 1–6.

Lomolino, M. V. (2000b). A species-based theory of insular zoogeography. *Global Ecology and Biogeography*, **9**, 39–58.

Lomolino, M. V. (2000c). Ecology's most general, yet protean pattern: the species-area relationship. *Journal of Biogeography*, **27**, 17–26.

Lomolino, M. V. (2002). ' . . . there are areas too small, and areas too large, to show clear diversity patterns . . . ' R. H. MacArthur (1972: 191). *Journal of Biogeography*, **29**, 555–557.

Lomolino, M. V. (2005). Body size evolution in insular vertebrates: generality of the island rule. *Journal of Biogeography*, **32**, 1683–1699.

Lomolino, M. V., Brown, J. H., and Davis, R. (1989). Island biogeography of montane forest mammals in the American southwest. *Ecology*, **70**, 180–194.

Lomolino, M. V. and Davis, R. (1997). Biogeographic scale and biodiversity of mountain forest mammals of western North America. *Global Ecology and Biogeography Letters*, **6**, 57–76.

Lomolino, M. V., Riddle, B. R. and Brown, J. H. (2005). *Biogeography* (3rd edn), Sinauer Associates, Sunderland, MA.

Lomolino, M. V., and Weiser, M. D. (2001). Towards a more general species-area relationship: diversity on all islands, great and small. *Journal of Biogeography*, **28**, 431–445.

Long, A. J., Crosby, M. J., Stattersfield, A. J., and Wege, D. C. (1996). Towards a global map of biodiversity: patterns in the distribution of range-restricted birds. *Global Ecology and Biogeography Letters*, **5**, 281–304.

Losos, J. B. (1990). A phylogenetic analysis of character displacement in Caribbean *Anolis* lizards. *Evolution*, **44**, 558–569.

Losos, J. B. (1994). Integrative approaches to evolutionary ecology: *Anolis* lizards as model systems. *Annual Review of Ecology and Systematics*, **25**, 467–493.

Losos, J. B. (1996). Phylogenetic perspectives on community ecology. *Ecology*, **77**, 1344–1354.

Losos, J. B., Jackman, T. R., Larson, A., de Queiroz, K., and Rodríguez-Schettino, L. (1998). Contingency and determinism in replicated adaptive radiations of island lizards. *Science*, **279**, 2115–2118.

Losos, J. B., Warheit, K. I., and Schoener, T. W (1997). Adaptive differentiation following experimental island colonization in *Anolis* lizards. *Nature*, **387**, 70–73.

Lovegrove, T. G. (1996). Island releases of Saddlebacks *Phiesturnus carunculatus* in New Zealand. *Biological Conservation*, **77**, 151–157.

Lovejoy, T. E., Bierregaard, R. O., Rylands, A. B., Malcolm, J. R., Quintela, C. E., Harper, L. H., Brown, K. S., Jr., Powell, A. H., Powell, G. V. N., Schubart, H. O. R., and Hays, M. B. (1986). Edge and other effects of isolation on Amazon forest fragments. In *Conservation biology: the science of scarcity and diversity* (ed. M. Soulé), pp. 257–285. Sinauer Associates, Sunderland, MA.

Ludwig, D. (1996). The distribution of population survival times. *American Naturalist*, **147**, 506–26.

Lugo, A. E. (1988). Ecological aspects of catastrophes in Caribbean Islands. *Acta Cientifica*, **2**, 24–31.

Lüpnitz, D. (1995). Kanarischen Inseln. Florenvielfalt auf engen Raum. *Palmengarten Sonderheft*, **23**, 1–117.

Lynch, J. D. and Johnson, N. V. (1974). Turnover and equilibria in insular avifaunas, with special reference to the California Channel Islands. *Condor*, **76**, 370–84.

Mabberley, D. J. (1979). Pachycaul plants and islands. In *Plants and islands* (ed. D. Bramwell), pp. 259–277. Academic Press, London.

MacArthur, R. H., Diamond, J. M., and Karr, J. (1972). Density compensation in island faunas. *Ecology*, **53**, 330–342.

MacArthur, R. H. and Wilson, E. O. (1963). An equilibrium theory of insular zoogeography. *Evolution*, **17**, 373–387.

MacArthur, R. H. and Wilson, E. O. (1967). *The theory of island biogeography*. Princeton University Press, Princeton, NJ.

Mackay, R. (2002). *The atlas of endangered species. Threatened plants and animals of the world*. Earthscan, London.

Madsen, T., Stille, B., and Shine, R. (1996). Inbreeding depression in an isolated population of adders *Vipera berus*. *Biological Conservation*, **75**, 113–118.

Magurran, A. E. (2004). *Measuring biological diversity*. Blackwell, Oxford.

Mallet, J. (1995). A species definition for the modern synthesis. *Trends in Ecology and Evolution*, **10**, 294–299.

Mangel, M. and Tier, C. (1994). Four facts every conservation biologist should know about persistence. *Ecology*, **75**, 607–614.

Manne, L. L., Pimm, S. L., Diamond, J. M., and Reed, T. M. (1998). The form of the curves: a direct evaluation of MacArthur & Wilson's classic theory. *Journal of Animal Ecology*, **67**, 784–794.

Marra, A. C. (2005). Pleistocene mammals of Mediterranean islands. *Quaternary International*, **129**, 5–14.

Marrero, A. and Francisco-Ortega, J. (2001a). Evolución en islas: la metáfora espacio-tiempo-forma. In *Naturaleza de las Islas Canarias. Ecología y Conservación* (ed. J. M. Fernández-Palacios and J. L. Martín Esquivel), pp. 133–140. Turquesa Ediciones, Santa Cruz de Tenerife.

Marrero, A. and Francisco-Ortega, J. (2001b). Evolución en islas. La forma en el tiempo. In *Naturaleza de las Islas Canarias. Ecología y conservación*. (ed. J. M. Fernández-Palacios and J. L. Martín Esquivel) pp. 141–150. Turquesa Ediciones, Santa Cruz de Tenerife.

Marrero-Gómez, M. V., Bañares-Baudet, A., and Carqué-Alamo, E. (2003). Plant resource conservation planning in protected natural areas: an example from the Canary Islands, Spain. *Biological Conservation*, **113**, 399–410.

Marshall, H. D. and Baker, A. J. (1999). Colonization history of Atlantic Island common chaffinches (*Fringilla coelebs*) revealed by mitochondrial DNA. *Molecular Phylogenetics and Evolution*, **11**, 201–212.

Martin, J.-L., Gaston, A. J., and Hitier, S. (1995). The effect of island size and isolation on old growth forest habitat and bird diversity in Gwaii Haanas (Queen Charlotte Islands, Canada). *Oikos*, **72**, 115–131.

Martín, A. and Lorenzo, J. A. (2001). *Aves del Archipiélago Canario*. Francisco Lemus Ed., La Laguna.

Martin, T. E. (1981). Species-area slopes and coefficients: a caution on their interpretation. *American Naturalist*, **188**, 823–837.

Martín-Esquivel, J.L., Fajardo, S., Cabrera, M.A., Arechavaleta, M., Aguiar, A., Martín, S., and Naranjo, M. (2005). *Evaluación 2004 de especies amenazadas de Canarias. Especies en peligro de extinción, sensibles a la alteración de su hábitat y vulnerables*. Consejería de Medio ambiente y ordenación territorial, Gobierno de Canarias. Santa Cruz de Tenerife.

Martín-Esquivel, J. L., García, H., Redondo, C., García, I., and Carralero, I. (1995). *La red Canaria de espacios naturales protegidos*. Viceconsejería de Medio Ambiente, Gobierno de Canarias. Santa Cruz de Tenerife.

Martínez-Garza, C. and Howe, H. F. (2003). Restoring tropical diversity: beating the time tax on species loss. *Journal of Applied Ecology*, **40**, 423–429.

Masson, D. G., Watts, A. B., Gee, M. J. R., Urgelés, R., Mitchell, N. C., Le Bas, T. P., and Canals, M. (2002). Slope failures in the flanks of the western Canary Islands. *Earth-Science Reviews*, **57**, 1–35.

Matthews, J. A., Bridges, E. M., Caseldine, C. J., Luckman, A. J., Owen, G., Perry, A. H., Shakesby, R. A., Walsh, R. P. D., Whittaker, R. J., and Willis, K. J. (eds) (2001). *The encyclopaedic dictionary of environmental change*. Arnold, London.

Mawdsley, N. A., Compton, S. G., and Whittaker, R. J. (1998). Population persistence, pollination mutualisms, and figs in fragmented tropical landscapes. *Conservation Biology*, **12**, 1416–1420.

Mayr, E. (1942). *Systematics and the origin of species.* Columbia University Press, New York.

Mayr, E. (1954). Change of genetic environment and evolution. In *Evolution as a process* (ed. J. S. Huxley, A. C. Hardy, and E. B. Ford), pp. 156–180. Allen & Unwin, London.

Mayr, E. (1963). *Animal species and evolution.* Harvard University Press, Cambridge, MA.

McCall, R. A. (1997). Implications of recent geological investigations of the Mozambique Channel for the mammalian colonization of Madagascar. *Proceedings of the Royal Society of London,* B **264**, 663–665.

McCall, R. A. (1998). The role of wing length in the evolution of avian flightlessness. *Evolutionary Ecology*, **12**, 569–580.

McDowall, R. M. (2005). Falklands: fact, fiction or fiddlesticks? *Journal of Biogeography*, **32**, 2187.

McGlone, M. S., Duncan, R. P. and Heenan, P. B. (2001). Endemism, species selection and the origin and distribution of the vascular plant flora of New Zealand. *Journal of Biogeography*, **28**, 199–216.

McGlone, M. S. (2005). Goodbye Gondwana. *Journal of Biogeography*, **32**, 739–740.

McGuinness, K. A. (1984). Equations and explanations in the study of species–area curves. *Biological Reviews*, **59**, 423–440.

McMullen, C. K. (1987). Breeding systems of selected Galápagos islands angiosperms. *American Journal of Botany*, **74**, 1694–1705.

Means, D. B. and Simberloff, D. (1987). The peninsula effect: habitat-correlated species decline in Florida's herpetofauna. *Journal of Biogeography*, **14**, 551–568.

Meiri, S., Dayan, T. and Simberloff, D. (2006). The generality of the island rule re-examined. *Journal of Biogeography*, **33**, 1571–1577.

Menard, H. W. (1986). *Islands.* Scientific American Library, New York.

Menges, E. S. (1992). Stochastic modeling of extinction in plant populations. In *Conservation Biology: the theory and practice of nature conservation, preservation and management* (ed. P. L. Fiedler and S. K. Jain), pp. 253–275. Chapman & Hall, London.

Merlen, G. (1995). Use and misuse of the seas around the Galápagos Archipelago. *Oryx*, **29**, 99–106.

Meyer, J.-Y. and Florence, J. (1996). Tahiti's native flora endangered by the invasion of *Miconia calvescens* DC. (Melastomataceae). *Journal of Biogeography*, **23**, 775–781.

Micol, T. and Jouventin, P. (1995). Restoration of Amsterdam Island, South Indian Ocean, following control of feral cattle. *Biological Conservation*, **73**, 199–206.

Mielke, H. W. (1989). *Patterns of life: biogeography of a changing world.* Unwin Hyman, Boston, MA.

Milberg, P. and Lamont, B. B. (1995). Fire enhances weed invasion of roadside vegetation in Southwestern Australia. *Biological Conservation*, **73**, 45–49.

Milberg, P. and Tyrberg, T. (1993). Naive birds and noble savages—a review of man-caused prehistoric extinctions of island birds. *Ecography*, **16**, 229–250.

Millien-Parra, V. and Jaeger, J.-J. (1999). Island biogeography of the Japanese terrestrial mammal assemblages: an example of a relict fauna. *Journal of Biogeography*, **26**, 959–972.

Miskelly, C. M. (1990). Effects of the 1982–83 El Niño event on two endemic landbirds on the Snares Islands, New Zealand. *Emu*, **90**, 24–27.

Moore, D. M. (1979). The origins of temperate island floras. In *Plants and islands* (ed. D. Bramwell), pp. 69–86. Academic Press, London.

Moore, G. J. and Clague, D. A. (1992). Volcano growth and evolution of the island of Hawaii. *Bulletin of the Geological Society of America*, **104**, 1471–1484.

Moore, G. J, Clague, D. A., Holcomb, R. T., Lipman, P. W., Normark, W. R., and Torressan, M. E. (1989). Prodigious submarine landslides on the Hawaiian ridge. *Journal of Geophysical Research*, **94**, 14465–14484.

Moore, G. J., Normark, W. R., Holcomb, R. T. (1994). Giant Hawaiian underwater landslides. *Science*, **264**, 46–47.

Morand, S. (2000). Geographic distance and the role of island area and habitat diversity in the species–area relationships of four Lesser Antillean faunal groups: a complementary note to Ricklefs and Lovette. *Journal of Animal Ecology*, **69**, 1117–1119.

Morgan, G. S. and Woods, C. A. (1986). Extinction and zoogeography of West Indian land mammals. *Biological Journal of the Linnean Society*, **28**, 167–203.

Morici, C. (2004). Palmeras e islas: la insularidad en una de las familias más diversas del reino vegetal. In *Ecología Insular/Island Ecology* (ed. J. M. Fernández-Palacios and C. Morici), pp. 81–122. Asociación Española de Ecología Terrestre-Cabildo Insular de La Palma, Santa Cruz de La Palma.

Morin, M. P. (1992). The breeding biology of an endangered Hawaiian Honeycreeper, the Laysan Finch. *Condor*, **94**, 646–667.

Morrison, L. W. (1997). The insular biogeography of small Bahamian cays. *Journal of Ecology*, **85**, 441–454.

Morrison, L. W. (2002*a*). Island biogeography and metapopulation dynamics of Bahamian ants. *Journal of Biogeography*, **29**, 387–394.

Morrison, L. W. (2002*b*). Determinants of plant species richness on small Bahamian islands. *Journal of Biogeography*, **29**, 931–941.

Morrone, J. J. (2005). Falklands: facts and fiction. *Journal of Biogeography*, **32**, 2183–2187.

Morwood, M. J., Soejono, R. P., Roberts, R. G., Sutikna, T., Turney, C. S. M., Westaway, K. E., Rink, W. J., Zhao, J.-X., van Den Bergh, G. D., Rokus Awe Due, Hobbs, D. R., Moore, M. W., Bird, M. I., and Fifield, L. K. (2004). Archeology and age of a new hominin from Flores in Eastern Indonesia. *Nature*, **431**, 1087–1091.

Moulton, M. P., Sanderson, J. G., and Labisky, R. F. (2001). Patterns of success in game bird (Aves: Galliformes) introductions to the Hawaiian islands and New Zealand. *Evolutionary Ecology Research*, **3**, 507–519.

Moya, O., Contreras-Díaz, H. G., Oromí, P., and Juan, C. (2004). Genetic structure, phylogeography and demography of two ground-beetle species endemic to the Tenerife laurel forest (Canary Islands). *Molecular Ecology*, **13**, 3153–3167.

Mueller-Dombois, D. (1975). Some aspects of island ecosystem analysis. In *Tropical ecological systems: trends in terrestrial and aquatic research* (ed. F. B. Golley and E. Medina), pp. 353–366. Springer-Verlag, New York.

Mueller-Dombois, D. and Fosberg, F. R. (1998). *Vegetation of the tropical Pacific islands*. Springer-Verlag, New York.

Munroe, E. G. (1948). The geographical distribution of butterflies in the West Indies. PhD dissertation, Cornell University, Ithaca, NY.

Munroe, E. G. (1953). The size of island faunas. In *Proceedings of the Seventh Pacific Science Congress of the Pacific Sciences Association*, vol IV, Zoology, pp. 52–53. Whitcome and Tombs, Auckland, New Zealand.

Myers, A. A. (1991). How did Hawaii accumulate its biota? A test from the Amphipoda. *Global Ecology and Biogeography Letters*, **1**, 24–29.

Myers, A. A. and Giller, P. S. (ed.) (1988). *Analytical biogeography*. Chapman & Hall, London.

Myers, N., Mittermeier, R. A., Mittermeier, C. G., da Fonseca, G. A. B., and Kent, J. (2000). Biodiversity hotspots for conservation priorities. *Nature*, **403**, 853–859.

Newmark, W. D. (1991). Tropical forest fragmentation and the local extinction of understorey birds in the Eastern Usambara Mountains, Tanzania. *Conservation Biology*, **5**, 67–78.

Nicholson, K. E., Glor, R. E., Kolbe, J. J., Larson, A., Hedges, S. B., and Losos, J. B. (2005). Mainland colonization by island lizards. *Journal of Biogeography*, **32**, 929–938.

Nicolás, J. P. de, Fernández-Palacios, J. M., Ferrer, F. J., and Nieto, E. (1989). Inter-island floristic similarities in the Macaronesian Region. *Vegetatio*, **84**, 117–125.

Nilsson, I. N. and Nilsson, S. G. (1985). Experimental estimates of census efficiency and pseudo-turnover on islands: error trend and between-observer variation when recording vascular plants. *Journal of Ecology*, **73**, 65–70.

Nogales, M., Hernández, E. C., and Valdés, F. (1999). Seed dispersal by common ravens *Corvus corax* among island habitats (Canarian Archipelago). *Ecoscience*, **5**, 56–61.

Nogales, M., Nieves, C., Illera, J. C., Padilla, D. P., and Traveset, A. (2005). Effect of native and alien vertebrate frugivores on seed viability and germination patterns of *Rubia fruticosa* (Rubiaceae) in the eastern Canary Islands. *Functional Ecology*, **19**, 429–436.

Nores, M. (1995). Insular biogeography of birds on mountain-tops in north western Argentina. *Journal of Biogeography*, **22**, 61–70.

Nores, M. (2004). The implications of Tertiary and Quaternary sea level rise events for avian distribution patterns in the lowlands of northern South America. *Global Ecology and Biogeography*, **13**, 149–161.

Nunn, P. D. (1990). Recent environmental changes on Pacific Islands. *Geographical Journal*, **156**, 125–140.

Nunn, P. D. (1994). *Oceanic islands*. Blackwell, Oxford.

Nunn, P. D. (1997). Late Quaternary environmental changes on Pacific islands: controversy, certainty and conjecture. *Journal of Quaternary Science*, **12**, 443–450.

Nunn, P. D. (2000). Illuminating sea-level fall around AD 1220–1510 (730–440 cal yr BP) in the Pacific Islands: implications for environmental change and cultural transformation. *New Zealand Geographer*, **56**, 46–54.

Nunn, P. D. (2004). Through a mist on the ocean: human understanding of island environments. *Tijdschrift voor Economische en Sociale Geografie*, **95**, 311–325.

Nyhagen, D. R., Kragelund, C., Olesen, J. M., and Jones, C. G. (2001). Insular interactions between lizards and flowers: flower visitation by an endemic Mauritian gecko. *Journal of Tropical Ecology*, **17**, 755–761.

O'Brien, E. M. (1993). Climatic gradients in woody plant species richness: towards an explanation based on an analysis of southern Africa's woody flora. *Journal of Biogeography*, **20**, 181–198.

O'Brien, E. M. (1998). Water–energy dynamics, climate, and prediction of woody plant species richness: an interim general model. *Journal of Biogeography*, **25**, 379–398.

O'Dea, N., Araújo, M. B., and Whittaker, R. J. (2006). How well do Important Bird Areas represent species and minimise conservation conflict in the tropical Andes? *Diversity and Distributions*, **12**, 205–214.

O'Dowd, D. J., Green, P. T. and Lake, P. S. (2003). Invasional 'meltdown' on an oceanic island. *Ecology Letters*, **6**, 812–817.

Olesen, J. M., Eskildsen, L. I., and Venkatasamy, S. (2002). Invasion of pollination networks on oceanic islands: importance of invader complexes and endemic supergeneralists. *Diversity and Distributions*, **8**, 181–192.

Olesen, J. M. and Jordano, P. (2002). Geographical patterns in plant-pollinator mutualistic networks. *Ecology*, **83**, 2416–2424.

Olesen, J. M. and Valido, A. (2003). Lizards as pollinators and seed dispersers: an island phenomenon. *Trends in Ecology and Evolution*, **18**, 177–181.

Olesen, J. M. and Valido, A. (2004). Lizards and birds as generalized pollinators and seed dispersers of island plants. In *Ecología Insular/Island Ecology* (ed. J. M. Fernández-Palacios and C. Morici), pp. 229–249. Asociación Española de Ecología Terrestre–Cabildo Insular de La Palma, Santa Cruz de La Palma.

Oliver, D. L. (2003). *Las islas del Pacífico* (3rd Ed). Melusina, Barcelona.

Ollier, C. D. (1988). *Volcanoes*. Basil Blackwell, Oxford.

Olson, S. L. and James, H. F. (1982). Fossil birds from the Hawaiian Islands: evidence for wholesale extinction by man before western contact. *Science*, **217**, 633–635.

Olson, S. L. and James, H. F. (1991). Descriptions of thirty-two new species of birds from the Hawaiian Islands: Part I. Non-passeriformes. *Ornithological Monographs*, **45**, 1–88.

Olson, S. L. and Jouventin, P. (1996). A new species of small flightless duck from Amsterdam Island, Southern Indian Ocean (Anatidae: *Anas*). *Condor*, **98**, 1–9.

Oromí, P. and Báez, M. (2001). Fauna invertebrada nativa terrestre. In *Naturaleza de las Islas Canarias. Ecología y Conservación* (ed. J. M. Fernández-Palacios and J. L. Martín Esquivel). pp. 205–211. Editorial Turquesa, Santa Cruz de Tenerife.

Otte, D. (1989). Speciation in Hawaiian crickets. In *Speciation and its consequences* (ed. D. Otte and J. A. Endler), pp. 482–526. Sinauer Associates, Sunderland, MA.

Otte, D. and Endler, J. A. (ed.) (1989). *Speciation and its consequences*. Sinauer Associates, Sunderland, MA.

Otto, R., Fernández-Palacios, J. M., and Krüsi, B. O. (2001). Variation in species composition and vegetation structure of succulent scrub on Tenerife in relation to environmental variation. *Journal of Vegetation Science*, **12**, 237–248.

Otto, R., Krüsi, B. O., Burga, C. A., and Fernández-Palacios, J. M. (2006). Old-field succession along a precipitation gradient in the semi-arid coastal region of Tenerife. *Journal of Arid Environments*, **65**, 156–178.

Paine, R. T. (1985). Re-establishment of an insular winter wren population following a severe freeze. *Condor*, **87**, 558–559.

Panero, J. L., Francisco-Ortega, J., Jansen, R. K., and Santos-Guerra, A. (1999). Molecular evidence for multiple origins of woodiness and a New World biogeographic connection of the Macaronesian Islands endemic *Pericallis* (Asteraceae: Senecioneae). *Proceedings of the National Academy of Sciences of the USA*, **96**, 13886–13891.

Pannell, C. (1989). The role of animals in natural regeneration and the management of equatorial rain forests for conservation and timber production. *Commonwealth Forestry Review*, **68**, 309–313.

Parrish, T. (2002). *Krakatau: genetic consequences of island colonization*. University of Utrecht and Netherlands Institute of Ecology, Heteren.

Parrish, T. L., Koelewijn, H. P., and van Dijk, P. J. (2003). Genetic evidence for natural hybridization between species of dioecious *Ficus* on island populations. *Biotropica*, **35**, 333–343.

Partomihardjo, T., Mirmanto, E., and Whittaker, R. J. (1992). Anak Krakatau's vegetation and flora circa 1991, with observations on a decade of development and change. *Geojournal*, **28**, 233–248.

Paton, P. W. C. (1994). The effect of edge on avian nest success: how strong is the evidence? *Conservation Biology*, **8**, 17–26.

Patrick, L. (2001). Introduced Species Summary Project: Brown tree snake (*Boiga irregularis*). <www.columbia.edu/itc/cerc/danoff-burg/invasion_bio/inv_spp_summ/boiga_irregularis.html>. (Last visited March 2006).

Patterson, B. D. (1990). On the temporal development of nested subset patterns of species composition. *Oikos*, **59**, 330–342.

Patterson, B. D. and Atmar, W. (1986). Nested subsets and the structure of insular mammalian faunas and archipelagos. *Biological Journal of the Linnean Society*, **28**, 65–82.

Paulay, G. (1994). Biodiversity on oceanic islands: its origin and extinction. *American Zoologist*, **34**, 134–144.

Pearson, R. G. and Dawson, T. P. (2003). Predicting the impacts of climate change on the distribution of species: are bioclimatic envelope models useful? *Global Ecology and Biogeography*, **12**, 361–371.

Peres, C. A. (2001). Synergistic effects of subsistence hunting and habitat fragmentation on Amazonian forest vertebrates. *Conservation Biology*, **15**, 1490–1505.

Peltonen, A. and Hanski, A. (1991). Patterns of island occupancy explained by colonization and extinction rates in shrews. *Ecology*, **72**, 1698–1708.

Pérez-Mellado, V. and Traveset, A. (1999). Relationships between plants and Mediterranean lizards. *Natura Croatica*, **8**, 275–285.

Perry, R. (ed.) (1984). *Galapagos*, Key Environments Series. IUCN/Pergamon Press, Oxford.

Petren, K., Grant, B. R., and Grant, P. R. (1999). A phylogeny of Darwin's finches based on microsatellite DNA variation. *Proceedings of the Royal Society of London*, B **266**, 321–329.

Pickett, S. T. A., Parker, V. T., and Fiedler, P. L. (1992). The new paradigm in ecology: implications for conservation biology above the species level. In *Conservation biology: the theory and practice of nature conservation and management* (ed. P. L. Fiedler and S. K. Jain), pp. 91–125. Chapman & Hall, New York.

Pickett, S. T. A. and Thompson, J. N. (1978). Patch dynamics and the design of nature reserves. *Biological Conservation*, **13**, 27–37.

Pickett, S. T. A. and White, P. S. (ed.) (1985). *The ecology of natural disturbance and patch dynamics*. Academic Press, Orlando, FL.

Pierson, E. D., Elmqvist, T., Rainey, W. E., and Cox, P. A. (1996). Effects of tropical cyclonic storms on flying fox populations on the South Pacific islands of Samoa. *Conservation Biology*, **10**, 438–451.

Pimm, S. L. (1991). *The balance of nature? Ecological issues in the conservation of species and communities*. University of Chicago Press, Chicago.

Pimm. S. L. and Askins, R. A. (1995). Forest losses predict bird extinctions in eastern North America. *Proceedings of the National Academy of Sciences of the USA*, **92**, 9343–9347.

Pimm, S. L., Jones, H. L., and Diamond, J. (1988). On the risk of extinction. *American Naturalist*, **132**, 757–785.

Pimm, S. L., Moulton, M. P., and Justice, L. J. (1995). Bird extinctions in the central Pacific. In *Extinction rates* (ed. J. H. Lawton and R. M. May), pp. 75–87. Oxford University Press, Oxford.

Pole, M. (1994). The New Zealand flora—entirely long-distance dispersal? *Journal of Biogeography*, **21**, 625–635.

Porter, D. M. (1979). Endemism and evolution in Galapagos islands vascular plants. In *Plants and islands* (ed. D. Bramwell), pp. 225–256. Academic Press, London.

Porter, D. M. (1984). Endemism and evolution in terrestrial plants. In *Galapagos*, Key Environments Series (ed. R. Perry), pp. 85–99. IUCN/Pergamon Press, Oxford.

Power, D. M. (1972). Numbers of bird species on the California islands. *Evolution*, **26**, 451–463.

Pratt, H. D. (2005). *The Hawaiian honeycreepers*. Oxford University Press, Oxford.

Pratt, H. D., Bruner, P. L., and Berrett, D. G. (1987). *A field guide to the birds of Hawaii and the tropical Pacific*. Princeton University Press, Princeton, NJ.

Pregill, G. K. (1986). Body size of insular lizards: a pattern of Holocene dwarfism. *Evolution* **40**, 997–1008.

Pregill, G. K. and Olson, S. L. (1981). Zoogeography of West Indian vertebrates in relation to Pleistocene climate cycles. *Annual Review of Ecology and Systematics*, **12**, 75–98.

Preston, F. W. (1948). The commonness, and rarity, of species. *Ecology*, **48**, 254–283.

Preston, F. W. (1962). The canonical distribution of commonness and rarity. *Ecology*, **43**, part I, pp. 185–215; part II, pp. 410–432.

Price, J. P. and Clague, D. A. (2002). How old is the Hawaiian biota? Geology and phylogeny suggest recent divergence. *Proceedings of the Royal Society of London*, B **269**, 2429–2435.

Price, J. P. and Elliott-Fisk, D. (2004). Topographic history of the Maui Nui complex, Hawai'i, and its implications for biogeography. *Pacific Science*, **58**, 27–45.

Price, J. P. and Wagner, W. L. (2004). Speciation in Hawaiian angiosperm lineages: cause, consequence, and mode. *Evolution*, **58**, 2185–2200.

Price, O., Woinarski, J. C. Z., Liddle, D. L., and Russell-Smith, J. (1995). Patterns of species composition and reserve design for a fragmented estate: monsoon rainforests in the Northern Territory, Australia. *Biological Conservation*, **74**, 9–19.

Primack, R. B. (1993). *Essentials of conservation biology*. Sinauer Associates, Sunderland, MA.

Primack, R. B. and Miao, S. L. (1992). Dispersal can limit local plant distribution. *Conservation Biology*, **6**, 513–519.

Primack, R. B. and Ros, J. (2002). *Introducción a la biología de la conservación*. Ariel Ciencia, Barcelona.

Pukrop, M. E. (1997). Phosphate mining in Nauru. <www.american.edu/projects/mandala/TED/nauru.htm> (Visited June 2006.)

Pulliam, H. R. (1996). Sources and sinks: empirical evidence and population consequences. In *Population dynamics in ecological space and time* (ed. O. E. Rhodes, Jr. R. K. Chesser, and M. H. Smith), pp. 45–69. Chicago University Press, Chicago.

Racine, C. H. and Downhower, J. F. (1974). Vegetative and reproductive strategies of *Opuntia* (Cactaceae) in the Galapagos Islands. *Biotropica*, **6**, 175–186.

Rafe, R. W., Usher, M. B., and Jefferson, R. G. (1985). Birds on reserves: the influence of area and habitat on species richness. *Journal of Applied Ecology*, **22**, 327–335.

Rainey, W. E., Pierson, E. D., Elmqvist, T., and Cox, P. A. (1995). The role of flying foxes (Pteropodidae) in oceanic island ecosystems of the Pacific. *Symposium of the Zoological Society of London*, **67**, 47–62.

Ralph, C. J. and Fancy, S. G. (1994). Demography and movements of the Omao (*Myadestes obscurus*). *Condor*, **96**, 503–511.

Rassmann, K., Trillmich, F., and Tautz, D. (1997). Hybridization between the Galápagos land and marine iguana (*Conlophus subcristatus* and *Amblyrhynchus cristatus*) on Plaza Sur. *Journal of Zoology (London)*, **242**, 729–739.

Ratcliffe, D. A. (ed.) (1977). *A nature conservation review*. Cambridge University Press, Cambridge.

Rawlinson, P. A., Zann, R. A., van Balen, S., and Thornton, I. W. B. (1992). Colonization of the Krakatau islands by vertebrates. *Geojournal*, **28**, 225–231.

Reed, D. H., O'Grady, J. J., Brook, B. W., Ballou, J. D., and Frankham, R. (2003). Estimates of minimum viable population sizes for vertebrates and factors influencing those estimates. *Biological Conservation*, **113**, 23–34.

Rees, D. J., Emerson, B. C., Oromí, P., and Hewitt, G. M. (2001). The diversification of the genus *Nesotes* (Coleoptera: Tenebrionidae) in the Canary Islands: evidence from mtDNA. *Molecular Phylogenetics and Evolution*, **21**, 321–326.

Reijnen, R., Foppen, R., and Meeuwsen, H. (1996). The effects of traffic on the density of breeding birds in Dutch agricultural grasslands. *Biological Conservation*, **75**, 255–260.

Renvoize, S. A. (1979). The origins of Indian Ocean island floras. In *Plants and islands* (ed. D. Bramwell), pp. 107–129. Academic Press, London.

Rey, J. R. (1984). Experimental tests of island biogeographic theory. In *Ecological communities: conceptual issues and the evidence* (ed. D. R. Strong, Jr, D. Simberloff, L. G. Abele, and A. B. Thistle), pp. 101–112. Princeton University Press, Princeton, NJ.

Rey, J. R. (1985). Insular ecology of salt marsh arthropods: species level patterns. *Journal of Biogeography*, **12**, 97–107.

Reyment, R. A. (1983). Palaeontological aspects of island biogeography: colonization and evolution of mammals on Mediterranean islands. *Oikos*, **41**, 299–306.

Ribon, R., Simon, J. E., and de Mattos, G. T. (2003). Bird extinctions in Atlantic forest fragments of the Viçosa region, southeastern Brazil. *Conservation Biology*, **17**, 1827–1839.

Ricketts, T. H. (2001). The matrix matters: effective isolation in fragmented landscapes. *American Naturalist*, **158**, 87–99.

Ricketts, T. H. (2004). Tropical forest fragments enhance pollinator activity in nearby coffee crops. *Conservation Biology*, **18**, 1262–1271.

Ricklefs, R. E. (1980). Geographical variation in clutch size among passerine birds: Ashmole's hypothesis. *Auk*, **97**, 38–49.

Ricklefs, R. E. (1989). Speciation and diversity: the integration of local and regional processes. In *Speciation and its consequences* (ed. D. Otte and J. A. Endler), pp. 599–622. Sinauer Associates, Sunderland, MA.

Ricklefs, R. E. and Bermingham, E. (2001). Nonequilibrium diversity dynamics of the Lesser Antillean avifauna. *Science*, **294**, 1522–1524.

Ricklefs, R. E. and Bermingham, E. (2002). The concept of the taxon cycle in biogeography. *Global Ecology and Biogeography*, **11**, 353–362.

Ricklefs, R. E. and Cox, G. W. (1972). Taxon cycles in the West Indian avifauna. *American Naturalist*, **106**, 195–219.

Ricklefs, R. E. and Cox, G. W. (1978). Stage of taxon cycle, habitat distribution, and population density in the avifauna of the West Indies. *American Naturalist*, **112**, 875–895.

Ricklefs, R. E. and Lovette, I. J. (1999). The roles of island area *per se* and habitat diversity in the species-area relationships of four Lesser Antillean faunal groups. *Journal of Animal Ecology*, **68**, 1142–1160.

Ridley, H. N. (1930). *The dispersal of plants throughout the world*. Reeve, Ashford, England.

Ridley, M. (ed.) (1994). *A Darwin selection*. Fontana Press, London.

Ridley, M. (1996). *Evolution* (2nd edn). Blackwell Science, Cambridge, MA.

Roberts, A. and Stone, L. (1990). Island-sharing by archipelago species. *Oecologia*, **83**, 560–567.

Rocha, S., Carretero, M. A., Vences, M., Glaw, F., and Harris, D. J. (2006). Deciphering patterns of trans-oceanic dispersal: the evolutionary origin and biogeography of coastal lizards (*Cryptoblepharus*) in the western Indian Ocean region. *Journal of Biogeography*, **33**, 13–22.

Rodda, G. H., Fritts, T. H., McCoid, M. J., and Campbell, E. W. III (1999). An overview of the biology of the brown treesnake (*Boiga irregularis*), a costly introduced pest on Pacific Islands. In *Problem snake management: the habu and the brown treesnake*, (eds Rodda, G. H., Sawai, Y., Chiszar, D., and Tanaka, H.), pp 44–80. Cornell University Press, Ithaca, NY.

Rodríguez de la Fuente, F. (1980). *Enciclopedia Salvat de la Fauna*. Salvat Ediciones, Pamplona.

Rodríguez-Estrella, R., Leon de la Luz, J. L., Breceda, A., Castellanos, A., Cancino, J., and Llinas, J. (1996). Status, density and habitat relationships of the endemic terrestrial birds of Socorro island, Revillagigeo Islands, Mexico. *Biological Conservation*, **76**, 195–202.

Rodríguez-Gironés, M. A. and Santamaría, L. (2006). A new algorithm to calculate the nestedness temperature of presence-absence matrices. *Journal of Biogeography*, **33**, 924–935.

Roff, D. A. (1991). The evolution of flightlessness in insects. *Ecological Monographs*, **60**, 389–421.

Roff, D. A. (1994). The evolution of flightlessness: is history important? *Evolutionary Ecology*, **8**, 639–657.

Rogers, R. R., Hartman, J. H., and Krause, D. W. (2000). Stratigraphic analysis of Upper Cretaceous rocks in the Mahajanga Basin, northwestern Madagascar: implications for ancient and modern faunas. *Journal of Geology*, **108**, 275–301.

Rosenzweig, M. L. (1978). Competitive speciation. *Biological Journal of the Linnean Society*, **10**, 275–289.

Rosenzweig, M. L. (1995). *Species diversity in space and time.* Cambridge University Press, Cambridge.

Rosenzweig, M. L. (2003). *Win-win ecology: How the Earth's species can survive in the midst of human enterprise.* Oxford University Press, New York.

Rosenzweig, M. L. (2004). Applying species-area relationships to the conservation of diversity. In *Frontiers of biogeography: new directions in the geography of Nature.* (eds M. V. Lomolino and L. R. Heaney), pp. 325–343. Sinauer Associates, Sunderland, MA.

Rosenzweig, M. L. and Clark, C. W. (1994). Island extinction rates from regular censuses. *Conservation Biology*, **8**, 491–494.

Ross, K. A., Fox, B. J., and Fox, M. D. (2002). Changes to plant species richness in forest fragments: fragment age, disturbance and fire history may be as important as area. *Journal of Biogeography*, **29**, 749–765.

Roughgarden, J. (1989). The structure and assembly of communities. In *Perspectives in Ecological Theory* (ed. J. Roughgarden, R. M. May, and S. A. Levin), pp. 203–226. Princeton University Press, Princeton, NJ.

Roughgarden, J. and Pacala, S. (1989). Taxon cycle among *Anolis* lizard populations: review of evidence. In *Speciation and its consequences* (ed. D. Otte and J. A. Endler), pp. 403–432. Sinauer Associates, Sunderland, MA.

Rudge, M. R. (ed.) (1989). Moas, man and climate in the ecological history of New Zealand. *New Zealand Journal of Ecology*, **12** (Suppl.), 11–25.

Rummel, J. D. and Roughgarden, J. D. (1985). A theory of faunal buildup for competition communities. *Evolution*, **39**, 1009–1033.

Runhaar, J., van Gool, C. R., and Groen, C. L. G. (1996). Impact of hydrological changes on nature conservation areas in the Netherlands. *Biological Conservation*, **76**, 269–276.

Russell, G. J., Diamond, J. M., Pimm, S. L., and Reed, T. M. (1995). A century of turnover: community dynamics at three timescales. *Journal of Animal Ecology*, **64**, 628–641.

Russell, R. W., Carpenter, F. L., Hixon, M. A., and Paton, D. C. (1994). The impact of variation in stopover habitat quality on migrant rufous hummingbirds. *Conservation Biology*, **8**, 483–490.

Safford, R. J. (1997). Distribution studies on the forest-living native passerines of Mauritius. *Biological Conservation*, **80**, 189–198.

Sakai, A. K., Wagner, W. L., Ferguson, D. M., and Herbst, D. R. (1995*a*). Origins of dioecy in the Hawaiian flora. *Ecology*, **76**, 2517–2529.

Sakai, A. K., Wagner, W. L., Ferguson, D. M., and Herbst, D. R. (1995*b*). Biogeographical and ecological correlates of dioecy in the Hawaiian flora. *Ecology*, **76**, 2530–2543.

Sanderson, J. G., Moulton, M. P., and Selfridge, R. G. (1998). Null matrices and the analysis of species co-occurrences. *Oecologia*, **116**, 275–283.

Sanmartín, I. and Ronquist, F. (2004). Southern hemisphere biogeography inferred by event-based models: plant versus animal patterns. *Systematic Biology*, **53**, 216–243.

Santos-Guerra, A. (1990). *Bosques de laurisilva en la región Macaronésica.* Consejo de Europa, Strasbourg.

Santos-Guerra, A. (1999). Origen y evolución de la flora canaria. In *Ecología y cultura en Canarias* (ed. J. M. Fernández-Palacios, J. J. Bacallado, and J. A. Belmonte), pp. 107–129. Museo de las Ciencias y el Cosmos, Cabildo Insular de Tenerife, Santa Cruz de Tenerife.

Santos-Guerra, A. (2001). Flora vascular nativa. In *Naturaleza de las Islas Canarias. Ecología y Conservación.* (ed. J. M. Fernández-Palacios and J. L. Martín Esquivel), pp 185–192. Turquesa Ediciones, Santa Cruz de Tenerife.

Sarre, S. and Dearn, J. M. (1991). Morphological variation and fluctuating asymmetry among insular populations of the Sleepy Lizard, *Trachydosurus rugosus* Gray (Squamata: Scincidea). *Australian Journal of Zoology*, **39**, 91–104.

Sarre, S., Schwaner, T. D., and Georges, A. (1990). Genetic variation among insular populations of the Sleepy Lizard, *Trachydosaurus rugosus* Gray (Squamata: Scincidea). *Australian Journal of Zoology*, **38**, 603–616.

Sato, A., O'hUigin, C., Figueroa, F., Grant, P. R., Grant, B. R., Tichy, H., and Klein, J. (1999). Phylogeny of Darwin's finches as revealed by mtDNA sequences. *Proceedings of the National Academy of Sciences of the USA*, **96**, 51010–5106.

Sauer, J. D. (1969). Oceanic island and biogeographic theory: a review. *Geographical Review*, **59**, 582–593.

Sauer, J. D. (1990). Allopatric speciation: deduced but not detected. *Journal of Biogeography*, **17**, 1–3.

Saunders, D. A. and Hobbs, R. J. (1989). Corridors for conservation. *New Scientist*, **121** (1648), 28 January, 63–68.

Saunders, D. A., Hobbs, R. J., and Margules, C. R. (1991). Biological consequences of ecosystem fragmentation: a review. *Conservation Biology*, **5**, 18–32.

Saunders, N. E. and Gibson, D. J. (2005). Breeding system, branching processes, hybrid swarm theory, and the humped-back diversity relationship as additional explanations for apparent monophyly in the Macaronesian island flora. *Journal of Ecology*, **93**, 649–652.

Sax, D. F., Brown, J. H., and Gaines, S. D. (2002). Species invasions exceed extinctions on islands world-wide: a comparative study of plants and birds. *American Naturalist*, **160**, 776–783.

Scatena, F. N. and Larsen, M. C. (1991). Physical aspects of Hurricane Hugo in Puerto Rico. *Biotropica*, **23**, 317–323.

Scheiner, S. M. (2003). Six types of species-area curves. *Global Ecology and Biogeography*, **12**, 441–447.

Scheiner, S. M. (2004). A mélange of curves—further dialogue about species–area relationships. *Global Ecology and Biogeography*, **13**, 479–484.

Schluter, D. (1988). Character displacement and the adaptive divergence of finches on islands and continents. *American Naturalist*, **131**, 799–824.

Schmidt, N. M. and Jensen, P. M. (2003). Changes in mammalian body length over 175 years—adaptations to a fragmented landscape? *Conservation Ecology*, **7**, 6 (online journal).

Schmitt, S. and Whittaker, R. J. (1998). Disturbance and succession on the Krakatau Islands, Indonesia. In *Dynamics of tropical communities*. British Ecological Society Symposium, Vol. 37 (ed. D. M. Newbery, H. N. T. Prins, and N. D. Brown), pp. 515–548. Blackwell Science, Oxford.

Schoener, T. W. (1975). Presence and absence of habitat shift in some widespread lizard species. *Ecological Monographs*, **45**, 233–258.

Schoener, T. W. (1983). Rate of species turnover decreases from lower to higher organisms: a review of the data. *Oikos*, **41**, 372–377.

Schoener, T. W. (1986). Patterns in terrestrial vertebrate versus arthropod communities: do systematic differences in regularity exist? In *Community ecology* (ed. J. M. Diamond and T. J. Case), pp. 556–586. Harper & Row, New York.

Schoener, T. W. (1988). Testing for non-randomness in sizes and habitats of West Indian lizards: choice of species pool affects conclusions from null models. *Evolutionary Ecology*, **2**, 1–26.

Schoener, T. W. (1989). The ecological niche In *Ecological concepts: the contribution of ecology to an understanding of the natural world* (ed. J. M. Cherrett), pp. 79–113. Blackwell Scientific Publications, Oxford.

Schoener, T. W. and Gorman, G. C. (1968). Some niche differences in three Lesser Antillean lizards of the genus *Anolis. Ecology*, **49**, 819–830.

Schoener, T. W. and Schoener, A. (1983). Distribution of vertebrates on some very small islands. I. Occurrence sequences of individual species. *Journal of Animal Ecology*, **52**, 209–235.

Schoener, T. W. and Spiller, D. A. (1987). High population persistence in a system with high turnover. *Nature*, **330**, 474–477.

Schumm, S. A. (1991). *To interpret the Earth: ten ways to be wrong*. Cambridge University Press, Cambridge.

Schüle, W. (1993). Mammals, vegetation and the initial human settlement of the Mediterranean islands: a palaeoecological approach. *Journal of Biogeography*, **20**, 399–411.

Schwaner, T. D. and Sarre, S. D. (1988). Body size of Tiger Snakes in Southern Australia, with particular reference to *Notechis ater serventyi* (Elapidae) on Chappell Island. *Journal of Herpetology*, **22**, 24–33.

Scott, T. A. (1994). Irruptive dispersal of black-shouldered kites to a coastal island. *Condor*, **96**, 197–200.

Sequeira, A. S., Lanteri, A. A., Scataglini, M. A., Confalonieri, V. A., and Farrell, B. D. (2000). Are flightless *Galapaganus* weevils older than the Galápagos Islands they inhabit? *Heredity*, **85**, 20–29.

Sfenthourakis, S. (1996). The species–area relationship of terrestrial isopods (Isopoda; Oniscidea) from the Aegean archipelago (Greece): a comparative study. *Global Ecology and Biogeography Letters*, **5**, 149–157.

Shafer, C. L. (1990). *Nature reserves: island theory and conservation practice*. Smithsonian Institution Press, Washington, DC.

Shennan, I. (1983). Flandrian and late Devensian sea-level changes and crustal movements in England and Wales. In *Shorelines and isostacy* (ed. D. E. Smith and A. G. Dawson), pp. 255–284. Academic Press, London.

Shepherd, U. L. and Brantley, S. L. (2005). Expanding on Watson's framework for classifying patches: when is an island not an island? *Journal of Biogeography*, **32**, 951–960.

Shine, T., Böhme, W. Nickel, H., Thies, D. F., and Wilms, T. R. (2001). Rediscovery of relict populations of the Nile crocodile *Crocodylus niloticus* in south-eastern Mauritania, with observations on their natural history. *Oryx*, **35**, 260–262.

Shilton, L. A. (1999). Seed dispersal by fruit bats on the Krakatau Islands, Indonesia. Ph.D. thesis, School of Biology, University of Leeds.

Shilton, L. A., Altringham, J. D., Compton, S. G., and Whittaker, R. J. (1999). Old World fruit bats can be long-distance seed dispersers through extended retention of viable seeds in the gut. *Proceedings of the Royal Society of London*, B **266**, 219–223.

Short, J. and Turner, B. (1994). A test of the vegetation mosaic hypothesis: a hypothesis to explain the decline and extinction of Australian mammals. *Conservation Biology*, **8**, 439–449.

Shrader-Frechette, K. S. and McCoy, E. D. (1993). *Method in ecology: strategies for conservation*. Cambridge University Press, Cambridge.

Silvertown, J. (2004). The ghost of competition past in the phylogeny of island endemic plants. *Journal of Ecology*, **92**, 168–173.

Silvertown, J., Francisco-Ortega, J., and Carine, M. (2005). The monophyly of island radiations: an evaluation of niche pre-emption and some alternative explanations. *Journal of Ecology*, **93**, 653–657.

Simberloff, D. (1976). Species turnover and equilibrium island biogeography. *Science*, **194**, 572–578.

Simberloff, D. (1978). Using island biogeographic distributions to determine if colonization is stochastic. *American Naturalist*, **112**, 713–726.

Simberloff, D. (1983). When is an island community in equilibrium? *Science*, **220**, 1275–1277.

Simberloff, D. (1992). Do species–area curves predict extinction in fragmented forests? In *Tropical deforestation and species extinction* (ed. T. C. Whitmore and J. A. Sayer), pp. 119–142. Chapman & Hall, London.

Simberloff, D., Farr, J. A., Cox, J., and Mehlman, D. W. (1992). Movement corridors: conservation bargains or poor investments? *Conservation Biology*, **6**, 493–504.

Simberloff, D. and Levin, B. (1985). Predictable sequences of species loss with decreasing island area—land birds in two archipelagoes. *New Zealand Journal of Ecology*, **8**, 11–20.

Simberloff, D. and Martin, J.-L. (1991). Nestedness of insular avifaunas: simple summary statistics masking complex species patterns. *Ornis Fennica*, **68**, 187–192.

Simberloff, D. and Wilson, E. O. (1969). Experimental zoogeography of islands. The colonisation of empty islands. *Ecology*, **50**, 278–296.

Simberloff, D. and Wilson, E. O. (1970). Experimental zoogeography of islands. A two year record of colonization. *Ecology*, **51**, 934–937.

Simkin, T. (1984). Geology of Galápagos Islands. In *Galapagos*, Key Environments Series (ed. R. Perry), pp. 15–41. IUCN/Pergamon Press, Oxford.

Simpson, B. B. (1974). Glacial migration of plants: island biogeographical evidence. *Science*, **185**, 698–700.

Sohmer, S. H. and Gustafson, R. (1993). *Plants and flowers of Hawaii* (3rd printing). University of Hawaii Press, Honolulu.

Sorensen, M. F. L. (1977). Niche shifts of Coal Tits *Parus ater* in Denmark. *Journal of Avian Biology*, **28**, 68–72.

Sosa, P. (2001). Genes, poblaciones y especies. In: *Naturaleza de las Islas Canarias. Ecología y conservación* (ed. J. M. Fernández-Palacios and J. L. Martín-Esquivel) pp. 151–155. Turquesa Ediciones, Santa Cruz de Tenerife.

Soulé, M. (ed.) (1986). *Conservation biology: the science of scarcity and diversity*. Sinauer Associates, Sunderland, MA.

Soulé, M. E., Bolger, D. T., Alberts, A. C., Wright, J., Sorice, M., and Hill, S. (1988). Reconstructed dynamics of rapid extinctions of Chaparral-requiring birds in urban habitat islands. *Conservation Biology*, **2**, 75–92.

Spellerberg, I. F. and Gaywood, M. J. (1993). Linear features: linear habitats and wildlife corridors. *English Nature Research Reports*, No. 60.

Spencer-Smith, D., Ramos, S. J., McKenzie, F., Munroe, E., and Miller, L. D. (1988). Biogeographical affinities of the butterflies of a 'forgotten' island: Mona (Puerto Rico). *Bulletin of the Allyn Museum*, No. **121**, pp. 1–35.

Spiller, D. A. and Schoener, T. W. (1995). Longterm variation in the effect of lizards on spider density is linked to rainfall. *Oecologia*, **103**, 133–139.

Stace, C. A. (1989). Dispersal versus vicariance—no contest! *Journal of Biogeography*, **16**, 201–202.

Stamps, J. A. and Buechner, M. (1985). The territorial defence hypothesis and the ecology of insular vertebrates. *Quarterly Review of Biology*, **60**, 155–181.

Stanley, S. M. (1999). *Earth system history*. W. H. Freeman, New York.

Steadman, D. W. (1997a). Human-caused extinctions of birds. In *Biodiversity II: understanding and protecting our biological resources* (ed. M. L. Reaka-Kudla, W. E. Wilson, and W. O. Wilson), pp. 139–161. Joseph Henry Press, Washington, DC.

Steadman, D. W. (1997b). The historic biogeography and community ecology of Polynesian pigeons and doves. *Journal of Biogeography*, **24**, 737–753.

Steadman, D. W., Pregill, G. K., and Burley, D. V. (2002). Rapid prehistoric extinction of iguanas and birds in Polynesia. *Proceedings of the National Academy of Sciences of the USA*, **99**, 3673–3677.

Steadman, D. W., White, J. P. and Allen, J. (1999). Prehistoric birds from New Ireland, Papua New Guinea: extinctions on a large Melanesian island.

Proceedings of the National Academy of Sciences of the USA, **96**, 2563–2568.

Stearn, W. T. (1973). Philip Barker Webb and Canarian Botany. *Monographiae Biologicae Canarienses,* **4**, 15–29.

Steers, J. A. and Stoddart, D. R. (1977). The origin of fringing reefs, barrier reefs and atolls. In *Biology and geology of coral reefs* (ed. O. A. Jones and R. Endean), pp. 21–57. Academic Press, New York.

Stoddart, D. R. and Walsh, R. P. D. (1992). Environmental variability and environmental extremes as factors in the island ecosystem. *Atoll Research Bulletin,* No. 356.

Stone, L. and Roberts, A. (1990). The checkerboard score and species distributions. *Oecologia,* **85**, 74–79.

Stone, L. and Roberts, A. (1992). Competitive exclusion, or species aggregation? *Oecologia,* **91**, 419–424.

Storey, B. C. (1995). The role of mantle plumes in continental breakup: case histories from Gondwanaland. *Nature,* **377**, 310–318.

Stouffer, P. C. and Bierregaard, R. O., Jr (1995). Use of Amazonian forest fragments by understory insectivorous birds. *Ecology,* **76**, 2429–2445.

Stouffer, P. C. and Bierregaard, R. O., Jr (1996). Effects of forest fragmentation on understorey hummingbirds in Amazonian Brazil. *Conservation Biology,* **9**, 1085–1094.

Streelman, J. T. and Danley, P. D. (2003). The stages of vertebrate evolutionary radiation. *Trends in Ecology and Evolution,* **18**, 126–131.

Strong, D. R., Jr, Simberloff, D., Abele, L. G., and Thistle, A. B. (ed.) (1984). *Ecological communities: conceptual issues and the evidence.* Princeton University Press, Princeton, NJ.

Stuessy, T. F., Crawford, D. J., and Marticorena, C. (1990). Patterns of phylogeny in the endemic vascular flora of the Juan Fernandez Islands, Chile. *Systematic Botany,* **15**, 338–346.

Stuessy, T. F., Crawford, D. J., and Marticorena, C. and Rodríguez, R. (1998). Island biogeography of angiosperms of the Juan Fernandez archipelago. In *Evolution and speciation of island plants.* (ed. T. F. Stuessy and M. Ono), pp. 121–138. Cambridge University Press, Cambridge.

Stuessy, T. F., Jakubowsky,G. Salguero Gómez, R., Pfosser, M., Schlüter, P. M., Fer, T., Sun, B.-Y., and Kato, H. (2006). An alternative model for plant evolution in islands. *Journal of Biogeography,* **33**, 1259–1265.

Sugihara, G. (1981). $S = CA^z$, $z = 1/4$: a reply to Connor and McCoy. *American Naturalist,* **117**, 790–793.

Sunding, P. (1979). Origins of the Macaronesian Flora. In *Plants and islands* (ed. D. Bramwell), pp. 13–40. Academic Press, London.

Sziemer, P. (2000). *Madeira's natural history in a nutshell.* Ribeiro & Filhos, Funchal.

Tarbuck, E. J. and Lutgens, F. K. (1999). *Earth: introduction to physical geology,* 6th edn. Prentice-Hall, Englewood Cliffs, NJ.

Tarr, C. L. and Fleischer, R. C. (1995). Evolutionary relationships of the Hawaiian Honeycreepers (Aves, Drepanidae). In *Hawaiian biogeography: evolution on a hot spot archipelago* (ed. W. L. Wagner and V. A. Funk), pp. 147–159. Smithsonian Institution Press, Washington, DC.

Tauber, C. A. and Tauber, M. J. (1989). Sympatric speciation in insects: perception and perspective. In *Speciation and its consequences* (ed. D. Otte and J. A. Endler), pp. 307–344. Sinauer Associates, Sunderland, MA.

Taylor, B. W. (1957). Plant succession on recent volcanoes in Papua. *Journal of Ecology,* **45**, 233–243.

Tellería, J. L. and Santos, T. (1995). Effects of forest fragmentation on a guild of wintering passerines: the role of habitat selection. *Biological Conservation,* **71**, 61–67.

Templeton, A. R. (1981). Mechanisms of speciation—a population genetic approach. *Annual Review of Ecology and Systematics,* **12**, 23–48.

Terborgh, J. (1992). Maintenance of diversity in tropical forests. *Biotropica,* **24**, 283–292.

Terrell, J. (1986). *Prehistory in the Pacific Islands: A study of variation in language, customs, and human biology.* Cambridge University Press, Cambridge.

Thomas, C. D. (1994). Extinction, colonization, and metapopulations: environmental tracking by rare species. *Conservation Biology,* **8**, 373–378.

Thomas, C. D., Cameron, A., Green, R E., Bakkenes, M., Beaumont, L. J., Collingham, Y. C., Erasmus, B. F. N., de Siquiera, M. F., Grainger, A., Hannah, L., Hughes, L., Huntley, B., van Jaarsveld, A. S., Midgley, G. F., Miles, L., Ortega-Huerta, M. A., Peterson, A. T., Phillips, O., and Williams, S. E. (2004). Extinction risk from climate change. *Nature,* **427**, 145–148.

Thomas, C. D. and Harrison, S. (1992). Spatial dynamics of a patchily distributed butterfly species. *Journal of Animal Ecology,* **61**, 437–446.

Thomas, C. D., Singer, M. C., and Boughton, D. A. (1996). Catastrophic extinction of population sources in a butterfly metapopulation. *American Naturalist,* **148**, 957–975.

Thornton, I. W. B. (1992). K. W. Dammerman—forerunner of island equilibrium theory. *Global Ecology and Biogeography Letters,* **2**, 145–148.

Thornton, I. W. B. (1996). *Krakatau—the destruction and reassembly of an island ecosystem.* Harvard University Press, Cambridge, MA.

Thornton, I. W.B., Cook, S., Edwards, J. S., Harrison, R. D., Schipper, C., Shanahan, M., Singadan, R., and Yamuna, R. (2001). Colonization of an island volcano, Long Island, Papua New Guinea, and an emergent island,

Motmot, in its caldera lake. VII. Overview and discussion. *Journal of Biogeography*, **28**, 1389–1408.

Thornton, I. W. B., Partomihardjo, T., and Yukawa, J. (1994). Observations on the effects, up to July 1993, of the current eruptive episode of Anak Krakatau. *Global Ecology and Biogeography Letters*, **4**, 88–94.

Thornton, I. W. B. and Walsh, D. (1992). Photographic evidence of rate of development of plant cover on the emergent island Anak Krakatau from 1971 to 1991 and implications for the effect of volcanism. *Geojournal*, **28**, 249–259.

Thornton, I. W. B., Ward, S. A., Zann, R. A., and New, T. R. (1992). Anak Krakatau—a colonization model within a colonization model? *Geojournal*, **28**, 271–286.

Thornton, I. W. B., Zann, R. A., and van Balen, S. (1993). Colonization of Rakata (Krakatau Is.) by non-migrant land birds from 1883 to 1992 and implications for the value of island equilibrium theory. *Journal of Biogeography*, **20**, 441–452.

Thornton, I. W. B., Zann, R. A., and Stephenson, D. G. (1990). Colonization of the Krakatau islands by land birds, and the approach to an equilibrium number of species. *Philosophical Transactions of the Royal Society of London*, B **328**, 55–93.

Thorpe, R. S. and Malhotra, A. (1996). Molecular and morphological evolution within small islands. *Philosophical Transactions of the Royal Society of London*, **351**, 815–822.

Thorpe, R. S., McGregor, D. P., Cumming, A. M., and Jordan, W. C. (1994). DNA evolution and colonization sequence of island lizards in relation to geological history: mtDNA RFLP, cytochrome B, cytochrome oxidase, 12s RRNA sequence and nuclear RAPD analysis. *Evolution*, **48**, 230–240.

Tjørve, E. (2003). Shapes and functions of species–area curves: a review of possible models. *Journal of Biogeography*, **30**, 827–835.

Tidemann, C. D., Kitchener, D. J., Zann, R. A., and Thornton, I. W. B. (1990). Recolonization of the Krakatau islands and adjacent areas of West Java, Indonesia, by bats (Chiroptera) 1883–1986. *Philosophical Transactions of the Royal Society of London*, B **328**, 121–130.

Toft, C. A. and Schoener, T. W. (1983). Abundance and diversity of orb spiders on 106 Bahamian islands: biogeography at an intermediate trophic level. *Oikos*, **41**, 411–426.

Townsend, C.R., Begon, M., and Harper J.L. (2003). *Essentials of ecology*, 2nd edn. Blackwell Publishing, Oxford.

Tracy, C. R. and George. L. (1992). On the determinants of extinction. *American Naturalist*, **139**, 102–122.

Traveset, A. (2001). Ecología reproductiva de plantas en condiciones de insularidad: consecuencias ecológicas y evolutivas del aislamiento geográfico. In *Ecosistemas mediterráneos: análisis funcional* (ed. R. Zamora and F. Puignaire). pp. 269–289. Consejo Superior de Investigaciones Científicas–Asociación Española de Ecología Terrestre, Granada.

Traveset, A. and Sáez, E. (1997). Pollination of *Euphorbia dendroides* by lizards and insects: spatio-temporal variation in patterns in flower visitation. *Oecologia*, **111**, 241–248.

Traveset, A. and Santamaría, L. (2004). Alteración de mutualismos planta-animal debido a la introducción de especies exóticas en ecosistemas insulares. In *Ecología Insula /Island Ecology* (ed. J. M. Fernández-Palacios and C. Morici), pp. 251–276. Asociación Española de Ecología Terrestre–Cabildo Insular de La Palma, Santa Cruz de La Palma.

Trewick, S. A. (1997). Flightlessness and phylogeny amongst endemic rails (Aves: Railidae) of the New Zealand region. *Philosophical Transactions of the Royal Society of London*, B **352**, 429–456.

Triantis, K. A., Mylonas, M., Lika, K., and Vardinoyannis, K. (2003). A model for the species–area–habitat relationship. *Journal of Biogeography*, **30**, 19–27.

Triantis, K. A., Mylonas, M., Weiser, M. D., Lika, K., and Vardinoyannis, K. (2005). Species richness, environmental heterogeneity and area: A case study based on land snails in Skyros archipelago (Aegean Sea, Greece). *Journal of Biogeography*, **32**, 1727–1735.

Triantis, K. A., Vardinoyannis, K., Tsolaki, E. P., Botsaris, I. Lika, K., and Mylonas, M. (2006). Re-approaching the small island effect. *Journal of Biogeography*, **33**, 914–923.

Trillmich, F. (1992). Conservation problems on Galápagos: the showcase of evolution in danger. *Naturwissenschaften*, **79**, 1–6.

Trusty, J. L., Olmstead, R. G., Santos-Guerra, A., Sá-Fontinha, S., and Francisco-Ortega, J. (2005). Molecular phylogenetics of the Macaronesian-endemic genus *Bystropogon* (Lamiaceae): palaeo-islands, ecological shifts and interisland colonizations. *Molecular Ecology*, **14**, 1177–1189.

Tudge, C. (1991). Time to save rhinoceroses. *New Scientist*, **131** (1788), 30–35.

Turner, I. M. (1996). Species loss in fragments of tropical rain forest: a review of the evidence. *Journal of Applied Ecology*, **33**, 200–209.

Turner, W. R. and Tjørve, E. (2005). Scale-dependence in species-area relationships. *Ecography*, **28**, 721–730.

Tye, A. (2006). Can we infer island introduction and naturalization rates from inventory data? Evidence from introduced plants in Galápagos. *Biological Invasions*, **8**, 201–215.

Valido, A., Dupont, Y. L., and Hansen, D. M. (2002). Native birds and insects, and introduced honey bees visiting *Echium wildpretii* (Boraginaceae) in the Canary Islands. *Acta Oecologica*, **23**, 413–419.

Valido, A. and Nogales, M. (1994). Frugivory and seed dispersal by the lizard *Gallotia galloti* (Lacertidae) in a xeric habitat of the Canary Islands. *Oikos*, **70**, 403–411.

Valido, A., Nogales, M., and Medina, F. M. (2003). Fleshy fruits in the diet of Canarian lizards *Gallotia galloti* (Lacertidae) in a xeric habitat of the island of Tenerife. *Journal of Herpetology*, **37**, 741–747.

Valido, A., Rando, J. C., Nogales, M., and Martín, A. (2000). 'Fossil' lizard found alive in the Canary Islands. *Oryx*, **34**, 71–72.

van Balgooy, M. M. J., Hovenkamp, P. H., and van Welzen, P. C. (1996). Phytogeography of the Pacific—floristic and historical distribution patterns in plants. In *The origin and evolution of Pacific island biotas, New Guinea to Eastern Polynesia: patterns and processes* (ed. A. Keast and S. E. Miller), pp. 191–213. SPB Academic Publishing, Amsterdam.

van Riper, C. III, van Riper, S. G., Goff, M. L., and Laird, M. (1986). The epizootiology and ecological significance of Malaria in Hawaiian land birds. *Ecological Monographs*, **56**, 327–344.

van Steenis, C. G. G. J. (1972). *The mountain flora of Java*. E. J. Brill, Leiden.

van Valen, L. (1973). A new evolutionary law. *Evolutionary Theory*, **1**, 1–30.

Vartanyan, S. L., Garutt, V. E., and Sher, A. V. (1993). Holocene dwarf mammoths from Wrangel Island in the Siberian Arctic. *Nature*, **362**, 337–340.

Villa, F., Rossi, O., and Sartore, F. (1992). Understanding the role of chronic environmental disturbance in the context of island biogeographic theory. *Environmental Management*, **16**, 653–666.

Vincek, V., O'Huigin, C., Satta, Y., Takahata, Y., Boag, P. T., Grant, P. R., Grant, B. R., and Klein, J. (1997). How large was the founding population of Darwin's finches? *Proceedings of the Royal Society of London*, B **264**, 111–118.

Vitousek, P. M. (1990). Biological invasion and ecosystem processes: towards an integration of population biology and ecosystem studies. *Oikos*, **57**, 7–13.

Vitousek, P. M., Loope, L. L., and Adsersen, H. (ed.) (1995). *Islands: biological diversity and ecosystem function*, Ecological Studies 115. Springer-Verlag, Berlin.

Vitousek, P, Loope, L. L., and Stone, C. P. (1987a). Introduced species in Hawaii: biological effects and opportunities for ecological research. *Trends in Ecology and Evolution*, **2**, 224–227.

Vitousek, P.M. and Walker, L.R. (1989). Biological invasion by *Myrica faya* in Hawai'i: plant demography, nitrogen fixation, ecosystem effects. *Ecological Monographs*, **59**, 247–265.

Vitousek, P. M., Walker, L. R., Whiteaker, L. D., Mueller-Dombois, D., and Matson, P. A. (1987b). Biological invasion by *Myrica faya* alters ecosystem development in Hawaii. *Science*, **238**, 802–804.

Wagner, W. L. and Funk, V. A. (ed.) (1995). *Hawaiian biogeography: evolution on a hot spot archipelago*. Smithsonian Institution Press, Washington, DC.

Waide, R. B. (1991). Summary of the responses of animal populations to hurricanes in the Caribbean. *Biotropica*, **23**, 508–512.

Walker, L. R., Brokaw, N. V. L., Lodge, D. J., and Waide, R. B. (ed.) (1991b). Special issue: ecosystem, plant, and animal responses to hurricanes in the Caribbean. *Biotropica*, **23**, 313–521.

Walker, L. R., Lodge, D. J., Brokaw, N. V. L., and Waide, R. B. (1991a). An introduction to hurricanes in the Caribbean. *Biotropica*, **23**, 313–316.

Walker, L. R., Silver, W. L., Willig, M. R., and Zimmerman, J. K. (ed.) (1996). Special issue: long term responses of Caribbean ecosystems to disturbance. *Biotropica*, **28**, 414–613.

Wallace, A. R. (1878). *Tropical nature and other essays*. Macmillan, London.

Wallace, A. R. (1902). *Island life* (3rd edn). Macmillan, London.

Walter, H. (1998). Driving forces of island biodiversity: an appraisal of two theories. *Physical Geography*, **19**, 351–377.

Walter, K. S. and Gillet. H. J. (eds.) (1998). *1997 IUCN Red List of Threatened Plants*. Compiled by the WCMC–IUCN–The World Conservation Union, Gland. (See also <www.redlist.org>.)

Ward, S. N., and Day, S. (2001). Cumbre Vieja Volcano—Potential collapse and tsunami at La Palma, Canary Islands. *Geophysical Research Letters*, **28**, 3397–3400.

Wardle, D. A., Zackrisson, O., Hörnberg, G., and Gallet, C. (1997). The influence of island area on ecosystem properties. *Science*, **277**, 1296–1299.

Warner, R. E. (1968). The role of introduced diseases in the extinction of the endemic Hawaiian avifauna. *Condor*, **70**, 101–120.

Waterloo, M. J., Schelleken, J., Bruijnzeel, L. A., Vugts, H. F., Assenberg, P. N., and Rawaqa, T. T. (1997). Chemistry of bulk precipitation in southwestern Viti Levu, Fiji. *Journal of Tropical Ecology*, **13**, 427–447.

Watson, J. E. M. (2004). Bird responses to habitat fragmentation: illustrations from Madagascan and Australian case studies. D.Phil. thesis, University of Oxford.

Watson, J. E. M., Whittaker, R. J., and Dawson, T. P. (2004a). Habitat structure and proximity to forest edge affect the abundance and distribution of forest-dependent birds in tropical coastal forests of southeastern Madagascar. *Biological Conservation*, **120**, 311–327.

Watson, J. E. M., Whittaker, R. J., and Dawson, T. P. (2004*b*). Avifaunal responses to habitat fragmentation in the threatened littoral forests of south-eastern Madagascar. *Journal of Biogeography*, **31**, 1791–1807.

Watson, J. E. M., Whittaker, R. J., and Freudenberger, D. (2005). Bird community responses to habitat fragmentation: how consistent are they across landscapes? *Journal of Biogeography*, **32**, 1353–1370.

Watts, D. (1970). Persistence and change in the vegetation of oceanic islands: an example from Barbados, West Indies. *Canadian Geographer*, **14**, 91–109.

Watts, D. (1987). *The West Indies: patterns of development, culture and environmental change since 1492*. Cambridge University Press, Cambridge.

Weaver, M. and Kellman, M. (1981). The effects of forest fragmentation on woodlot tree biotas in Southern Ontario. *Journal of Biogeography*, **8**, 199–210.

Webb, C. J. and Kelly, D. (1993). The reproductive biology of the New Zealand flora. *Trends in Ecology and Evolution*, **8**, 442–447.

Weiher, E. and Keddy, P. A. (1995). Assembly rules, null models, and trait dispersion: new questions from old patterns. *Oikos*, **74**, 159–164.

Weiher, E. and Keddy, P. A. (eds) (1999). *Ecological assembly rules: perspectives, advances, retreats*. Cambridge University Press, Cambridge.

Weins, J. (1984). On understanding a non-equilibrium world: myth and reality in community patterns and processes. In *Ecological communities: conceptual issues and the evidence* (ed. D. R. Strong, Jr, D. Simberloff, L. G. Abele, and A. B. Thistle), pp. 439–457. Princeton University Press, Princeton, NJ.

Weisler, M. I. (1995). Henderson Island prehistory: colonization and extinction on a remote Polynesian island. *Biological Journal of the Linnean Society*, **56**, 377–404.

Werner, T. K. and Sherry, T. W. (1987). Behavioural feeding specialization in *Pinaroloxias inornata*, the 'Darwin's Finch' of Cocos Island, Costa Rica. *Proceedings of the National Academy of Sciences of the USA*, **84**, 5506–5510.

Westerbergh, A. and Saura, A. (1994). Genetic differentiation in endemic *Silene* (Caryophyllaceae) on the Hawaiian islands. *American Journal of Botany*, **81**, 1487–1493.

Whelan, F. and Kelletat, D. (2003). Submarine slides on volcanic islands—a source for mega-tsunamis in the Quaternary. *Progress in Physical Geography*, **27**, 198–216.

Whitaker, A. H. (1987). The roles of lizards in New Zealand plant reproductive strategies. *New Zealand Journal of Botany*, **25**, 315–328.

White, G. B. (1981). Semispecies, sibling species and superspecies. In *The evolving biosphere (chance, chance and challenge)* (ed. P. L. Forey), pp. 21–28. British Museum (Natural History) and Cambridge University Press, Cambridge.

Whitehead, D. R. and Jones, C. E. (1969). Small islands and the equilibrium theory of insular biogeography. *Evolution*, **23**, 171–179.

Whitmore, T. C. (1984). *Tropical rain forests of the Far East* (2nd edn). Clarendon Press, Oxford.

Whitmore, T. C. (ed.) (1987). *Biogeographical evolution of the Malay Archipelago*, Oxford Monographs on Biogeography No. 4. Oxford University Press, Oxford.

Whitmore, T. C. and Sayer, J. A. (ed.) (1992). *Tropical deforestation and species extinction*. Chapman & Hall, London.

Whittaker, R. H. (1977). Evolution of species diversity in land communities. In *Evolutionary Biology*, vol. **10** (ed. M. K. Hecht, W. C. Steere, and B. Wallace), pp. 250–268. Plenum Press, New York.

Whittaker, R. J. (1992). Stochasticism and determinism in island ecology. *Journal of Biogeography*, **19**, 587–591.

Whittaker, R. J. (1995). Disturbed island ecology. *Trends in Ecology and Evolution*, **10**, 421–425.

Whittaker, R. J. (2004). Dynamic hypotheses of richness on islands and continents. In *Frontiers of Biogeography: new directions in the geography of Nature* (eds M. V. Lomolino and L. R. Heaney), pp. 211–231. Sinauer Associates, Sunderland, MA.

Whittaker, R. J., Araújo, M. B., Jepson, P., Ladle, R. J., Watson, J. E. M., and Willis, K. J. (2005). Conservation biogeography: assessment and prospect. *Diversity and Distributions*, **11**, 3–23.

Whittaker, R. J. and Bush, M. B. (1993). Anak Krakatau and old Krakatau: a reply. *Geojournal*, **29**, 417–420.

Whittaker, R. J., Bush, M. B., Partomihardjo, T., Asquith, N. M., and Richards, K. (1992*a*). Ecological aspects of plant colonisation of the Krakatau Islands. *Geojournal*, **28**, 201–211.

Whittaker, R. J., Bush, M. B., and Richards, K. (1989). Plant recolonization and vegetation succession on the Krakatau Islands, Indonesia. *Ecological Monographs*, **59**, 59–123.

Whittaker, R. J., Field, R., and Partomihardjo, T. (2000). How to go extinct: lessons from the lost plants of Krakatau. *Journal of Biogeography*, **27**, 1049–1064.

Whittaker, R. J. and Jones, S. H. (1994*a*). Structure in re-building insular ecosystems: an empirically derived model. *Oikos*, **69**, 524–529.

Whittaker, R. J. and Jones, S. H. (1994*b*). The role of frugivorous bats and birds in the rebuilding of a tropical forest ecosystem, Krakatau, Indonesia. *Journal of Biogeography*, **21**, 689–702.

Whittaker, R. J., Jones, S. H., and Partomihardjo, T. (1997). The re-building of an isolated rain forest assemblage: how disharmonic is the flora of Krakatau? *Biodiversity and Conservation*, **6**, 1671–1696.

Whittaker, R. J., Partomihardjo, T., and Riswan, S. (1995). Surface and buried seed banks from Krakatau, Indonesia: implications for the sterilization hypothesis. *Biotropica*, **27**, 346–354.

Whittaker, R. J., Schmitt, S. F., Jones, S. H., Partomihardjo, T., and Bush, M. B. (1998). Stand biomass and tree mortality from permanent forest plots on Krakatau, 1989–1995. *Biotropica*, **30**, 519–529.

Whittaker, R. J., Walden, J., and Hill, J. (1992*b*). Post-1883 ash fall on Panjang and Sertung and its ecological impact. *Geojournal*, **28**, 153–171.

Whittaker, R. J., Willis, K. J., and Field, R. (2001). Scale and species richness: towards a general, hierarchical theory of species diversity. *Journal of Biogeography*, **28**, 453–470.

Wiggins, D. A., Moller, A. P., Sorensen, M. F. L., and Brand, L. (1998). Island biogeography and the reproductive ecology of great tits *Parus major*. *Oecologia*, **115**, 478–482.

Wilcove, D. S., McLellan, C. H., and Dobson, A. P. (1986). Habitat fragmentation in the temperate zone. In *Conservation biology: the science of scarcity and diversity* (ed. M. Soulé), pp. 237–256. Sinauer Associates., Sunderland, MA.

Wiles, G. J., Schreiner, I. H., Nafus, D., Jurgensen, L. K., and Manglona, J. C. (1996). The status, biology, and conservation of *Serianthes nelsonii* (Fabaceae), an endangered Micronesian tree. *Biological Conservation*, **76**, 229–239.

Williams, E. E. (1972). The origin of faunas. Evolution of lizard congeners in a complex island fauna: A trial analysis. *Evolutionary Biology*, **6**, 47–88.

Williamson, M. H. (1981). *Island populations*. Oxford University Press, Oxford.

Williamson, M. H. (1983). The land-bird community of Skokholm: ordination and turnover, *Oikos*, **41**, 378–384.

Williamson, M. H. (1984). Sir Joseph Hooker's lecture on insular floras. *Biological Journal of the Linnean Society*, **22**, 55–77.

Williamson, M. H. (1988). Relationship of species number to area, distance and other variables. In *Analytical Biogeography, an integrated approach to the study of animal and plant distributions* (ed. A. A. Myers and P. S. Giller), pp. 91–115. Chapman & Hall, London.

Williamson, M. H. (1989*a*). The MacArthur and Wilson theory today: true but trivial. *Journal of Biogeography*, **16**, 3–4.

Williamson, M. H. (1989*b*). Natural extinction on islands. *Philosophical Transactions of the Royal Society of London*, B **325**, 457–468.

Williamson, M. H. (1996). *Biological invasions*. Population and Community Biology Series **15**, Chapman & Hall, London.

Williamson, M., Gaston, K. J., and Lonsdale, W. M. (2001). The species–area relationship does not have an asymptote! *Journal of Biogeography*, **28**, 827–830.

Williamson, M., Gaston, K. J., and Lonsdale, W. M. (2002). An asymptote is an asymptote and not found in species–area relationships. *Journal of Biogeography*, **29**, 1713.

Willis, K. J., Gillson, L., and Brncic, T. M. (2004). How 'virgin' is virgin rainforest? *Science*, **304**, 402–403.

Willis, K. J. and Whittaker, R. J. (2002). Species diversity-scale matters. *Science*, **295**, 1245–1248.

Wilson, D. E. and Graham, G. L. (ed.) (1992). *Pacific island flying foxes: proceedings of an international conservation conference*. US Fish and Wildlife Service, Biological Report **90** (23), Washington, DC.

Wilson, E. O. (1959). Adaptive shift and dispersal in a tropical ant fauna. *Evolution*, **13**, 122–144.

Wilson, E. O. (1961). The nature of the taxon cycle in the Melanesian ant fauna. *American Naturalist*, **95**, 169–193.

Wilson, E. O. (1969). The species equilibrium. *Brookhaven Symposia in Biology*, **22**, 38–47.

Wilson, E. O. (1995). *Naturalist*. Allen Lane, The Penguin Press, London.

Wilson, E. O. and Bossert, W. H. (1971). *A primer of population biology*. Sinauer Associates, Stamford, CT.

Wilson, E. O. and Willis, E. O. (1975). Applied biogeography. In *Ecology and evolution of communities* (ed. M. L. Cody and J. M. Diamond), pp. 522–534. Harvard University Press, Cambridge, MA.

Wilson, J. B. and Roxburgh, S. H. (1994). A demonstration of guild-based assembly rules for a plant community, and the determination of intrinsic guilds. *Oikos*, **69**, 267–276.

Wilson, J. B. and Whittaker, R. J. (1995). Assembly rules demonstrated in a saltmarsh community. *Journal of Ecology*, **83**, 801–807.

Wilson, J. T. (1963). A possible origin of the Hawaiian islands. *Canadian Journal of Physics*, **41**, 863–870.

Wilson, M. H., Kepler C. B., Snyder, N. F. R., Derrickson, S. R., Dein, F. J., Wiley, J. W., Wunderle, J. M., Lugo, A. E., Graham, D. L., and Toone, W. D. (1994). Puerto Rican parrots and potential limitations of the metapopulation approach to species conservation. *Conservation Biology*, **8**, 114–123.

Winkworth, R. C., Wagstaff, S. J., Glenny, D., and Lockhart, P. J. (2002). Plant dispersal NEWS from New Zealand. *Trends in Ecology and Evolution*, **17**, 514–520.

Woinarski, J. C. Z., Whitehead, P. J., Bowman, D. M. J. S., and Russell-Smith, J. (1992). Conservation of mobile species in a variable environment: the problem of reserve design in the Northern Territory, Australia. *Global Ecology and Biogeography Letters*, **2**, 1–10.

Worthen, W. B. (1996). Community composition and nested-subset analyses: basic descriptors for community ecology. *Oikos*, **76**, 417–426.

Wragg, G. M. (1995). The fossil birds of Henderson Island, Pitcairn Group: natural turnover and human impact, a synopsis. *Biological Journal of the Linnean Society*, **56**, 405–414.

Wright, D. H. (1983). Species–energy theory: an extension of species–area theory. *Oikos*, **41**, 496–506.

Wright, D. H., Currie, D. J., and Maurer, B. A. (1993). Energy supply and patterns of species richness on local and regional scales. In *Species diversity in ecological communities: historical and geographical perspectives* (ed. R. Ricklefs and D. Schluter), pp. 66–74. University of Chicago Press, Chicago.

Wright, D. H., Patterson, B. D., Mikkelson, G. M., Cutler, A., and Atmar, W. (1998). A comparative analysis of nested subset patterns of species composition. *Oecologia*, **113**, 1–20.

Wright, D. H. and Reeves, J. J. (1992). On the meaning and measurement of nestedness of species assemblages. *Oecologia*, **92**, 416–428.

Wright, S. (1932). The roles of mutation, inbreeding, crossbreeding, and selection in evolution. *Proceedings of the XI International Congress of Genetics* **1**, 356–366.

Wright, S. D., Yong, C. G., Dawson, J. W., Whittaker, D. J., and Gardner, R. C. (2000). Riding the ice age El Niño? Pacific biogeography and evolution of *Metrosideros* subg. *Metrosideros* (Myrtaceae) inferred from nuclear ribosomal DNA. *Proceedings of the National Academy of Sciences of the USA*, **97**, 4118–4123.

Wright, S. J. (1980). Density compensation in island avifaunas. *Oecologia*, **45**, 385–389.

Wyatt-Smith, J. (1953). The vegetation of Jarak Island, Straits of Malacca. *Journal of Ecology*, **41**, 207–225.

Wylie, J. L. and Currie, D. J. (1993*a*). Species–energy theory and patterns of species richness: I. Patterns of bird, angiosperm, and mammal richness on islands. *Biological Conservation*, **63**, 137–144.

Young, T. P. (1994). Natural die-offs of large mammals: implications for conservation. *Conservation Biology*, **8**, 410–418.

Zann, R. A. and Darjono (1992). The birds of Anak Krakatau: the assembly of an avian community. *Geojournal*, **28**, 261–270.

Zavaleta, E. S., Hobbs, R. J., and Money, H. A. (2001). Viewing invasive species removal in a whole-ecosystem context. *Trends in Ecology and Evolution*, **16**, 454–459.

Zavodna, M., Arens, P., van Dijk, P. J., Partomihardjo, T., Vosman, B., and van Damme, J. M. M. (2005). Pollinating fig wasps: genetic consequences of island recolonization. *Journal of Evolutionary Biology*, **18**, 1234–1243.

Zimmerman, B. L. and Bierregaard, R. O., Jr (1986). Relevance of the equilibrium theory of island biogeography and species–area relations to conservation with a case from Amazonia. *Journal of Biogeography*, **13**, 133–143.

Zink, R. M. (2002). A new perspective on the evolutionary history of Darwin's finches. *Auk*, **119**, 864–871.

Index